Applied Statistics for Business and Management using Microsoft Excel

Linda Herkenhoff • John Fogli

Applied Statistics for Business and Management using Microsoft Excel

Second Edition

Springer

Linda Herkenhoff
Saint Mary's College
Moraga, CA, USA

John Fogli
Diablo Valley College
Pleasant Hill, CA, USA

ISBN 978-3-031-46370-9 ISBN 978-3-031-46371-6 (eBook)
https://doi.org/10.1007/978-3-031-46371-6

This Springer imprint is published by the registered company Springer Nature Switzerland AG
The registered company address is: Gewerbestrasse 11, 6330 Cham, Switzerland

To "my boys" $\sum_{i=1}^{4} b_i = \text{Fun} =$ *where* b_1 = *Frederic,* b_2 = *Brett,* b_3 = *Kyle,* b_4 = *Eric, a family set that truly enjoys the spirit of numbers*

Linda Herkenhoff

To my parents, wife, kids, and friends who put up with my use of sensical and nonsensical numbers in life.

John Fogli

Preface

Mathematical formulas and Greek letters seem to intimidate many people, so this book focuses on providing clear word descriptions and step-by-step Excel instructions, rather than including lots of x's and y's everywhere. We have specifically avoided including a lot of technical information. This book was written for those who want a basic understanding of how to use statistics in the business world without all the details of statistical theory. It is designed to enhance a practical learning of statistics by non-statisticians and to make the experience a little easier, maybe even fun.

This book has been written with managers in mind who may need a quick refresher on how to complete Excel analysis without having to wade through pages of technical explanations. Those more detailed explanations can be found in traditional statistics textbooks. The key concepts in this book are more complex than the presentations in this book might suggest. We have aimed at concisely defining the concepts and providing simple explanations on how to use them.

For instructors, this publication is designed as a companion guide to your core text. Students can practice their skills that have been introduced to them in the classroom. However, at the beginning of each chapter, the key concepts are summarized to highlight their importance.

We believe that problems and case studies should involve actual data from the business community. Therefore, there is a continuing analysis throughout several of the chapters using a real business case from Infinity Auto Insurance. The data sets for the other practice problems in each chapter have been kept intentionally small so that the user can easily maneuver within the data. The rigorous sample size rules have been put aside to accommodate this goal. The same Excel procedures apply to all sizes of data sets, small or large.

We have only included work place relevant Excel tools. This approach should work well for those having to run their own analysis and for those needing to better understand statistical reports.

Although it may prove helpful, no prior Excel experience is necessary. Simple instructions are provided under the assumption that you are a first-time user.

Some of the traditional statistical texts include Excel instructions. However, these are typically embedded deep in the chapter as one of several different software instruction packets. This book focuses only on Excel and provides easy to find instructions in each chapter.

In addition to offering this book in traditional textbook form, it is also available as a environmentally friendly e-book.

The authors bring a unique perspective as guides through the World of Statistics. Their combined skills and expertise in the workplace/classroom equip them to understand the challenges students/managers face as they navigate the World of Statistics. Their active consulting and teaching lives keep them up to date on the common real-world issues in understanding/applying statistical concepts.

We have worked diligently to make sure that this book is accurate and error free. But if you happen to find any errors or have suggestions for improvements, please contact us at jfogli@hotmail.com or lherkenh@stmarys-ca.edu.

Many of you have survived statistics courses through pure memorization, but hopefully this book will encourage both logical reasoning and intuitive understanding. We believe that logical reasoning is fundamental to retaining analytical skills that can be practiced in the real world. Enjoy....

Orinda, CA, USA Linda Herkenhoff
Pleasant Hill, CA, USA John Fogli

Acknowledgments

We would like to thank several of our colleagues who tirelessly reviewed every word and number in our second edition. Your fastidious reviews let us sleep at night. But, we had even more overseas colleagues who generously reviewed specific sections of the book and to them we owe our thanks: Dr. Peter Hosie, Dr. Madeline Wu, and Dr. Joann Heydenfeldt.

We express appreciation to all of our past Statistics students who allowed us to better understand the challenges in bringing Statistics alive as a powerful business tool. This book reflects our many years of student input and observation, both in the United States and abroad.

Special recognition is paid to Infinity Life Insurance Company for allowing us to use their data for our real-world case study.

Contents

Chapter 1
Data and Statistics

Linda Herkenhoff and John Fogli

Case Study: Exploring New Marketing Strategies

INFINITY®
Auto Insurance

Infinity Property and Casualty Corporation (Infinity Auto Insurance) is headquartered in Birmingham, Alabama. It provides personal automobile insurance with a concentration on nonstandard auto insurance. Nonstandard insurance serves individuals unable to obtain coverage through standard insurance companies, which can be due to a driving record with accidents and/or tickets, prior DUI, the driver's age, vehicle type, etc. Infinity Auto Insurance's products include personal automobile insurance for individuals, commercial vehicle insurance for businesses, and classic collector insurance for individuals with classic and antique automobiles. Infinity Auto Insurance distributes its products primarily through independent agencies and brokers. Infinity Auto Insurance, a top-performing Infinity brand, provides nonstandard car insurance through more than 12,500 independent agents. Infinity Auto Insurance utilizes Internet-based software applications to provide many of its agents with real-time underwriting, claims, and policy information. The

Supplementary Information The online version contains supplementary material available at https://doi.org/10.1007/978-3-031-46371-6_1.

L. Herkenhoff (✉)
Saint Mary's College, Moraga, CA, USA

J. Fogli
Diablo Valley College, Pleasant Hill, CA, USA

L. Herkenhoff, J. Fogli (eds.), *Applied Statistics for Business and Management using Microsoft Excel*, https://doi.org/10.1007/978-3-031-46371-6_1

company is licensed to write insurance in all 50 states and the District of Columbia. The company traces its roots back to 1955 doing business as Dixie Insurance company, later known as The Infinity Group. In 1991, Pennsylvania Company (American Premier Underwriters) purchased The Infinity Group. American Premier Underwriters, Inc. and the American Financial Corporation merged and became the American Financial Group (AFG) in 1995. The company was officially founded in 2002 when AFG transferred all common stock of Infinity and its sister companies to IPCC. In 2003, IPCC was listed on the NASDAQ as a registered public holding.

Food for Thought

- Who are some of Infinity Auto Insurance's direct auto insurance competitors?

How large (annual revenue) is the auto insurance industry?
Discuss what type of customers Infinity Auto Insurance might target.
If you were a marketing consultant to Infinity Auto Insurance, what market areas would you suggest to target to access the customers identified above?

Possible Answers

- Who are some of Infinity Auto Insurance's direct auto insurance competitors?
- GEICO, Progressive, Mercury, AIG

How large (annual revenue) is the auto insurance industry?
 Annual revenue for the auto insurance industry ranges in billions of dollars every year.
Discuss what type of customers Infinity Auto Insurance might target.
 High-risk auto insurance individuals
If you were a marketing consultant to Infinity Auto Insurance, what market areas would you suggest to target to access the customers identified above?
 Large metropolitan areas or lower-income areas

Key Concepts

Business statistics, Constant, Continuous data, Cross-sectional data, Data, Descriptive statistics, Discrete data, Distribution, Frequency, Inferential statistics, Longitudinal data, Population, Probability, Qualitative data, Quantitative data, Random sample, Sample, Scales, Statistics, and Variables.

Discussion

This chapter will include some of the basic concepts and definitions necessary for understanding and applying statistical methods in business. One of the challenges in statistical analysis and subsequent report writing is the confusion over what words and statistical terms really mean. In statistics, the mathematical meaning may not match up with the everyday usage of the same term, which can lead to confusion. For example, in our daily communications the term "significant" means important. To illustrate this point, a CEO's attendance at an award ceremony would be very significant to the press. In statistics, when a statistic is "significant" it simply means that you are very sure that the statistic is reliable or real and not just due to chance. It doesn't mean the finding is important nor that it has any decision-making utility.

Statistics is often said to be more open to misuse, both deliberately and naively, than any other area in business. Part of that misuse may stem from inconsistency in statistical definitions. The next few pages will attempt to firmly establish the lexicon for moving into the next chapters, but each chapter will also provide content-specific definitions. It is in your best interest to take the time to read through these terms and not assume you have a complete/accurate understanding of these key concepts.

Business statistics: Business statistics include the areas of descriptive statistics, probability statistics, and inferential statistics applied to business.

Constant: A value that remains unchanged. For example, let **C** be a constant, **C** = number of years of university education required for all entry-level accountants in your company, and in this case **C** = 4 years.

Cross-sectional data: Cross-sectional data is data that is collected from participants at one point in time (this data should generally be collected within 6 or less months). Time is not considered one of the study variables in a cross-sectional research design. In a cross-sectional study, time is assumed to be a random effect that produces only variance, not bias. For example, the number of IPOs in the San Francisco area.

Data: Data is specific values of the variables of interest that are collected, analyzed, and summarized for presentation/interpretation.

Descriptive statistics: The statistics associated with organizing, summarizing, and presenting data in a convenient/informative way. Both graphical and numerical techniques are employed.

Distribution: An arrangement of values of a variable showing their observed or theoretical frequency of occurrence.

Frequency: How often a value occurs.

Inferential statistics: The statistics associated with making an estimate, prediction, or decision from a small group to draw conclusions about a larger group. For example, in a random sample, we note that 40% of the voters do not want an increase in business taxes. If the sample has been selected without bias, we can **infer** the population would also vote about 40% against an increase in business taxes.

Longitudinal data or time-series data: Data collected over several time periods. For example, the number of IPOs in the San Jose area over the past 5 years. In this example, the time period would be defined as 1 year and we are collecting data over 5 of these time periods, or years. Note the term longitudinal has nothing to do with geospatial data on the globe.

Population: All the existing individuals, items, or data about which you want to draw a conclusion. For example, all of the MBA graduates in the United States, or all the copy machines owned by your company in the past 5 years.

Probability: The analysis associated with the chance of occurrence of one or more possible results of an unpredictable event. When potential students apply to an MBA school, they consider the chance or probability of being accepted.

Qualitative data: Categorical, non-numeric data such as gender and industry type. There is no mathematical relationship between the data values. The data labels can be numeric or non-numeric, and are descriptors of the data. For example, the labels may be True/False or could be shown as 1/2. But there is no mathematical relationship between 1 and 2 in this case, they are just labels. In other words, 2 is not twice as big as 1.

Quantitative data: Numerical data such as salaries or sales. There is a mathematical relationship between the data values. The data labels must always be numeric. However, quantitative data can be either discrete or continuous. **Continuous** quantitative data arises from a measurement process. Continuous data is information that can be measured on a continuum or scale. Continuous data can have almost any numeric value and can be meaningfully subdivided into finer and finer increments, depending on the precision of the measurement system. For example, **time** may be considered continuous quantitative data. **Discrete** quantitative data arises from a counting process. An example includes how many text messages you sent this past week.

Random sample: Data selected from a population in a way that ensures each data value has an equal opportunity of being selected.

Sample: A portion of a population selected for analysis.

Scales (measurement): A scale may be defined as any set of items that are progressively arranged according to value or magnitude into which an item can be placed according to its quantification. In other words, a scale is a continuous spectrum or series of categories. The purpose of scaling is to represent, usually quantitatively, an item's, a person's, or an event's place in the spectrum. The four types of scales are nominal, ordinal, interval, and ratio.

Scale Type	Numerical Operations	Descriptive Statistics
Nominal	Counting	Frequency in each category Percentage in each category Mode
Ordinal	Rank ordering	Median Range Percentile ranking
Interval	Arithmetic operations on intervals between numbers	Mean Standard deviation Variance
Ratio	Arithmetic operations on actual quantities	Geometric mean Coefficient of variation

Fig. 1.1 Descriptive statistics for types of scales

- **Nominal scale**
- A scale in which the numbers or letters assigned to objects serve as labels for identification or classification. An example of this scale includes MBA schools.
- **Ordinal scale**
- A scale that arranges objects or alternatives according to their ranking. An example includes the Business Week ratings of MBA schools.
- **Interval scale**
- A scale that not only arranges objects according to their magnitudes, but also distinguishes this ordered arrangement in units or equal intervals. The interval between measurements is a meaningful value. It does not involve a true zero point. Standardized test scores in MBA entrance exams is an example of this scale.
- **Ratio scale**
- A scale having absolute rather than relative quantities and possessing an absolute zero, where there is an absence of a given attribute. The number of years an MBA program offers is measured on a ratio scale.

These scales and their associated descriptive statistics are summarized in Fig. 1.1. All statistics that are appropriate for lower-order scales (nominal is the lowest) are appropriate for higher-order scales (ratio is the highest) but not vice versa.

Statistics: Values derived from the data collected for a random sample.

Variables: Characteristics of interest that may change, unlike a constant, within the scope of a given problem. For example, the variable $\mathbf{s}$ = executive salary and the associated values include $\mathbf{s}_1$ = \$200,000, $\mathbf{s}_2$ = \$580,000, and $\mathbf{s}_3$ = \$610,000.

Common Pitfalls

- Be careful of longitudinal data. Make sure the conditions for the data collection process have remained constant. Often environmental factors may change and thus bias the data.
- You cannot convert qualitative data to quantitative data by changing the labels to numeric format.
- Don't show too many decimal places. You cannot make your answer more precise than the data that was used to calculate it.
- Make sure the scales have the correct units of measurement and variable name.

Final Thoughts and Activities

Practice Problems

1. Longitudinal data: Take a look at what has been happening with world record times in the men's mile run since 1900. Extrapolate forward 5 years from today by simply continuing the same trend of the existing line. Are you surprised?
2. What could be some of the issues with the following statements from a statistical standpoint?
   a) There has been no global warming since 1998.
   b) The number of hurricanes has doubled in the last 50 years!
   c) Carbon emissions are down 20% in country X!
   d) Climate change is causing 500% increase in wildfires!

Questions

1. Why is comparing data from countries with different populations/demographics and comparing regional data with single-country data misleading?
2. What other data do you need to address the issue that 40% of midwives in Zambia were found to be HIV positive?
3. The average life expectancy in the worst affected sub-Saharan countries has fallen by 20 years. The biggest increase in deaths among 20 to 49-year-old adults accounts for 60% of all deaths. To better understand this data and its impact, what other data do you need?
4. Should more quantitative data in the area of economics be provided? Justify your answer.

Discussion Boards

1. Often political goals can get in the way of accurate data reporting. Discuss this and any other issues that might be obstacles in getting complete and accurate data on the unrest in Afghanistan.

2. This year's top mutual funds will likely slide in rank next year. Do you agree/disagree? Use statistical arguments to support your position.
3. Statistics can change your life … disagree or agree? Provide detailed examples to support your position.
4. Why is *Moneyball* still such a popular book and movie?

Group Activity

Productivity Growth, defined as getting more output per hour worked by an employee, has been identified as a key goal in companies within the United States. Often companies provide managers and workers with rewards/incentives for resulting gains in productivity.

Search the web to find supporting data on one of the positions (pro or con) in addressing the question: Are productivity gains always in the best interest of society?

Parting Thought

It has been proven that the celebration of birthdays is healthy. Statistics show that those people who celebrate the most birthdays are the oldest …

Problem Solutions

Problem #1: Longitudinal Data

Answer: (Answers may vary) The world record times have decreased in time since the 1900s. Runners have gotten faster.

Time	Athlete	Nationality	Date	Venue
4:14.4	John Paul Jones	The United States	31 May 1913[a]	Allston, Mass.
4:12.6	Norman Taber	The United States	16 July 1915[a]	Allston, Mass.
4:10.4	Paavo Nurmi	Finland	23 August 1923[a]	Stockholm
4:09.2	Jules Ladoumègue	France	4 October 1931[a]	Paris
4:07.6	Jack Lovelock	New Zealand	15 July 1933[a]	Princeton, N.J.
4:06.8	Glenn Cunningham	The United States	16 June 1934[a]	Princeton, N.J.
4:06.4	Sydney Wooderson	The United Kingdom	28 August 1937[a]	Motspur Park
4:06.2	Gunder Hägg	Sweden	1 July 1942[a]	Göteborg
4:06.2	Arne Andersson	Sweden	10 July 1942[a]	Stockholm
4:04.6	Gunder Hägg	Sweden	4 September 1942[a]	Stockholm
4:02.6	Arne Andersson	Sweden	1 July 1943[a]	Göteborg

Time	Athlete	Nationality	Date	Venue
4:01.6	Arne Andersson	Sweden	18 July 1944[a]	Malmö
4:01.4	Gunder Hägg	Sweden	17 July 1945[a]	Malmö
3:59.4	Roger Bannister	The United Kingdom	6 May 1954[a]	Oxford
3:58.0	John Landy	Australia	21 June 1954[a]	Turku
3:57.2	Derek Ibbotson	The United Kingdom	19 July 1957[a]	London
3:54.5	Herb Elliott	Australia	6 August 1958[a]	Santry, Dublin
3:54.4	Peter Snell	New Zealand	27 January 1962[a]	Wanganui
3:54.1	Peter Snell	New Zealand	17 November 1964[a]	Auckland
3:53.6	Michel Jazy	France	9 June 1965[a]	Rennes
3:51.3	Jim Ryun	The United States	17 July 1966[a]	Berkeley, Cal.
3:51.1	Jim Ryun	The United States	23 June 1967[a]	Bakersfield, Cal.
3:51.0	Filbert Bayi	Tanzania	17 May 1975[a]	Kingston
3:49.4	John Walker	New Zealand	12 August 1975[a]	Göteborg
3:49.0	Sebastian Coe	The United Kingdom	17 July 1979[a]	Oslo
3:48.8	Steve Ovett	The United Kingdom	1 July 1980[a]	Oslo
3:48.53	Sebastian Coe	The United Kingdom	19 August 1981[a]	Zürich
3:48.40	Steve Ovett	The United Kingdom	26 August 1981[a]	Koblenz
3:47.33	Sebastian Coe	The United Kingdom	28 August 1981[a]	Bruxelles
3:46.32	Steve Cram	The United Kingdom	27 July 1985[a]	Oslo
3:44.39	Noureddine Morceli	Algeria	5 September 1993[a]	Rieti
3:43.13	Hicham El Guerrouj	Morocco	7 July 1999[a]	Rome

[a]Referenced from "12th IAAF World Championships in Athletics: IAAF Statistics Handbook. Berlin 2009." (PDF). Monte Carlo: IAAF Media & Public Relations Department. 2009. pp. 546, 549–50. Retrieved August 4, 2009

Excel Output

Let's look at how often the record actually changes and by how much it changes. It seems that the changes are very small. But even so, how fast can humans actually run … there must be some limit. The time between records seems to almost follow a cycle, where several records get set close together and then it takes a while for a new breakthrough time. The last time the record was set was in 1999; based on the pattern, it seems that we are long overdue for a new record and should expect one in the next 5 years, but not by much.

Time	Difference in new record	Date	#Years between records
04:14.4		31 May 1913	
04:12.6	00:01.8	16 July 1915	2
04:10.4	00:02.2	23 August 1923	8
04:09.2	00:01.2	4 October 1931	8
04:07.6	00:01.6	15 July 1933	2
04:06.8	00:00.8	16 June 1933	2
04:06.4	00:00.4	28 August 1937	3

Time	Difference in new record	Date	#Years between records
04:06.2	00:00.2	1 July 1942	5
04:06.2	00:00.0	10 July 1942	0
04:04.6	00:01.6	4 September 1942	0
04:02.6	00:02.0	1 July 1943	1
04:01.6	00:01.0	18 July 1944	1
04:01.4	00:00.2	17 July 1945	1
03:59.4	00:02.0	6 May 1954	9
03:58.0	00:01.4	21 June 1954	10
03:57.2	00:00.8	19 July 1957	3
03:54.5	00:02.7	6 August 1958	1
03:54.4	00:00.1	27 January 1962	4
03:54.1	00:00.3	17 November 1964	2
03:53.6	00:00.5	9 June 1965	1
03:51.3	00:02.3	17 July 1966	1
03:51.1	00:00.2	23 June 1967	1
03:51.0	00:00.1	17 May 1975	8
03:49.4	00:01.6	12 August 1975	0
03:49.0	00:00.4	17 July 1979	4
03:48.8	00:00.2	1 July 1980	1
03:48.5	00:00.3	19 August 1981	1
03:48.4	00:00.1	26 August 1981	0
03:47.3	00:01.1	28 August 1981	0
03:46.3	00:01.0	27 July 1985	4
03:44.4	00:01.9	5 September 1993	8
03:43.1	00:01.3	7 July 1999	6

Problem #2: What could be some of the issues with the following statements from a statistical standpoint?

Answers

1. a) **"There has been no global warming since 1998." Possible Issues:** 1998 was an extreme **El Niño year** (a temporary warming event). Also selecting **only one year** as a starting point ignore the long-term trend. **Better Approach:** Look at the **30+ year trend** rather than cherry-picking a single year. Also use temperature **anomalies** over time instead of raw values.
2. b) **"The number of hurricanes has doubled in the last 50 years!" Possible Issues: Better technology** (satellites) means we **detect more hurricanes now** than in the past. Also the increase could be due to **reporting changes**, not actual storm frequency. **Better Approach:** Compare **major hurricanes (Category 3+)** instead of counting weaker storms. Also adjust for **improvements in detection technology** over time.
3. c) **"Carbon emissions are down 20% in Country X!" Possible Issues:** What's the **baseline year?** If it was an economic crisis year (e.g., COVID-19 lockdowns), the drop isn't meaningful. Also are they counting **total emissions** or just

certain sectors? **Better Approach:** Show **multi-year trends**, not just one year's decline. Also specify **absolute vs. relative** reductions (Did total emissions drop, or just emissions per capita?)

4. d) "**Climate change is causing a 500% increase in wildfires!**" **Possible Issues:** Are wildfires increasing because of **climate** or **land management issues** (like more homes built near forests)? Also is the percentage increase from a **small base number**? (e.g., from 2 to 10 fires). **Better Approach:** Compare **long-term wildfire trends**, adjusting for population and reporting bias. Also look at **burned area, not just the number of fires**.

Chapter 2
Introduction to Excel and Basic Charts

Linda Herkenhoff and John Fogli

Key Concepts

Area charts, Bar and Column charts, Bubble charts, Filtering, Formatting, Line charts, Pie charts, Pivot Tables, Sorting, Radar charts.

Discussion

In this chapter, the key concepts are all about the commands in Excel. The first step is to load the **Data Analysis ToolPak**. The standard data structure used throughout this book consists of rows of observations and columns of variables. Some of the basic functionality will be introduced to familiarize you in navigating through Excel spreadsheets.

In this chapter, the emphasis is on the basic Excel procedures to allow you to successfully analyze and chart your data. These basic procedures will form the framework for future chapters that have more complicated Excel functionalities. The instructions are provided in a detailed step-by-step process, so make sure you do not skip steps. One of the most common frustrations is Excel refusing to compute

Supplementary Information The online version contains supplementary material available at https://doi.org/10.1007/978-3-031-46371-6_2.

L. Herkenhoff (✉)
Saint Mary's College, Moraga, CA, USA

J. Fogli
Diablo Valley College, Pleasant Hill, CA, USA

L. Herkenhoff, J. Fogli (eds.), *Applied Statistics for Business and Management using Microsoft Excel*, https://doi.org/10.1007/978-3-031-46371-6_2

or complete your graph, and usually it is because a vital step has been ignored by the user.

There are many versions of Excel for different types of computers under various operating systems. This book has been compiled using Microsoft Office 2021 Professional and Microsoft 365 with Windows 11 operating system. However, the instructions and flow remain very similar regardless of the Excel version you may be using.

These Excel functions and charts will provide a useful companion to your business applications in the workplace or to your business statistics textbook applications in the classroom.

Basic Concepts

Start Up

- Click on the **Start button** in the lower-left corner of your Windows desktop; the Windows Start menu will open
- Select **Excel 2024** (or whatever version you are using)

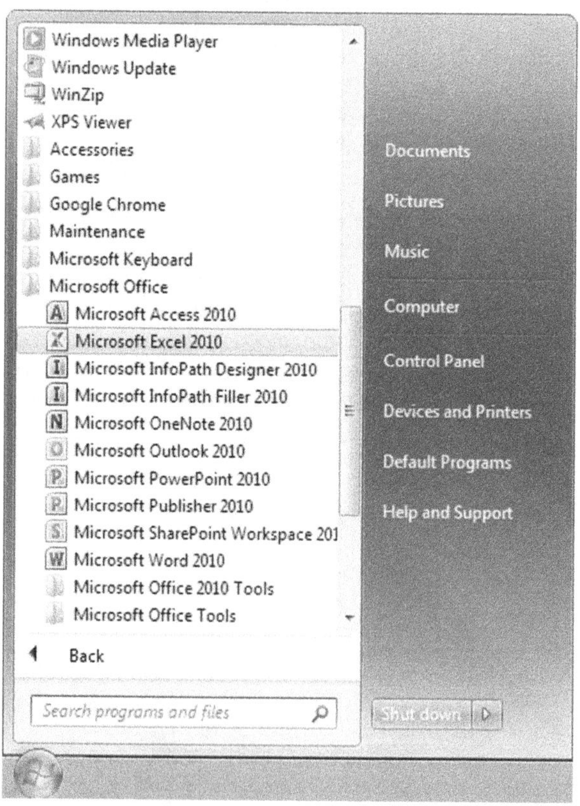

Adding Data Analysis ToolPak

- Click the **File** tab on the upper left hand of the Excel window
- Select **Options** near the bottom of the command menu

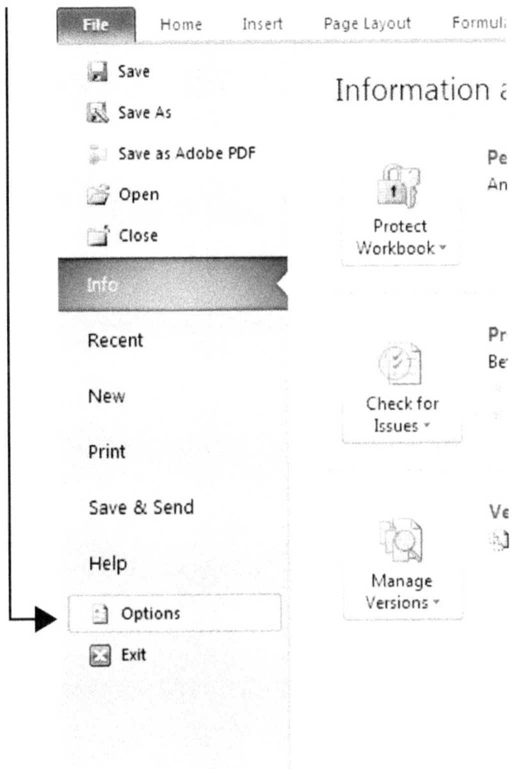

- Select **Add-Ins** on the left-hand side

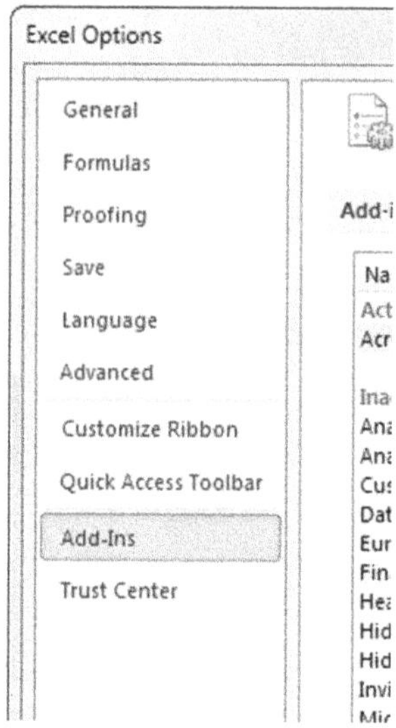

- Select **Excel Add-ins**, under "Manage:"

- Click **Go…**
- Check **Analysis ToolPak**

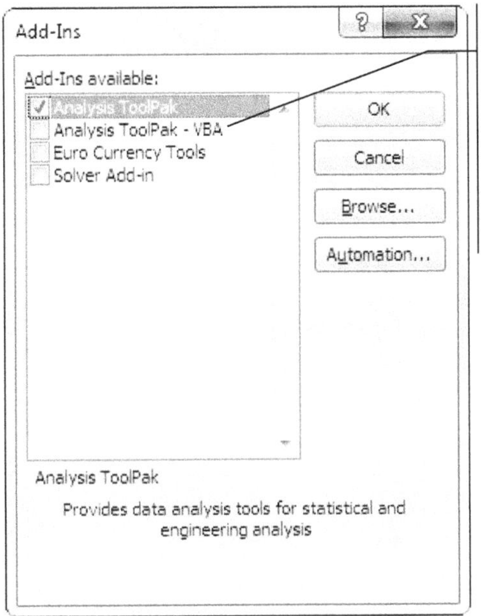

Note: Even though there are two (2) **Analysis ToolPaks**, you only need to select the first one. You will not need to install the **Analysis ToolPak - VBA** because that ToolPak uses a programing language called Visual Basic, which will not be covered in this book.

- Click **OK**

Once the Analysis ToolPak is installed, it will launch automatically upon opening Excel.

- Sometimes the Add-in may seem to vanish. This tool is available to you to reload as necessary each time you launch Excel.

Excel Elements

Element	Purpose
Title bar	Displays the name of the application and the current Excel document
Menu bar	Contains Excel menus with commands for Excel tasks
Toolbar	Contains buttons that allow one click access to Excel commands and features
Formula bar	Displays the formula or value entered into the currently selected cell
Worksheet	Displays the contents of a chart or Excel spreadsheet results
Cells	Stores individual text or numeric entries
Column headings	Organizes cells into lettered columns
Row headings	Organizes cells into numeric rows
Scroll bar	Used to view cell entries that lie outside the main Excel window
Name box	Displays the names or references of the currently selected object or cell

(continued)

Element	Purpose
Task pane	Displays commonly used commands
Sheet tabs	Click to display individual worksheets

Entering Formulas

- Click inside the formula bar
- Type in an "=" sign and type in your formula or click on the ***fx*** symbol to choose from the drop-down menu
  - To make the formula easier to use, make sure you input the ***cell location*** where your variable is located, ***not the actual value*** of the variable. This is done by left clicking on the relevant cell. If multiple cells need to be referenced hold control and left click on each additional cell. You will know that you are performing this correctly if each subsequent cell is in a different color.

  When you input an Excel function, the following symbols can be used:

Operator	Description
+	Addition
−	Subtraction
/	Division
*	Multiplication
^	Exponentiation

Cell References

- For a list of data values, use a colon. If you copy and paste a formula in a row or column, the cell reference in the new formulas will shift along with the cell. In this example, we are asking Excel to sum up all of the data from cell A2 to cell A11.

  Example: SUM(A2:A11)

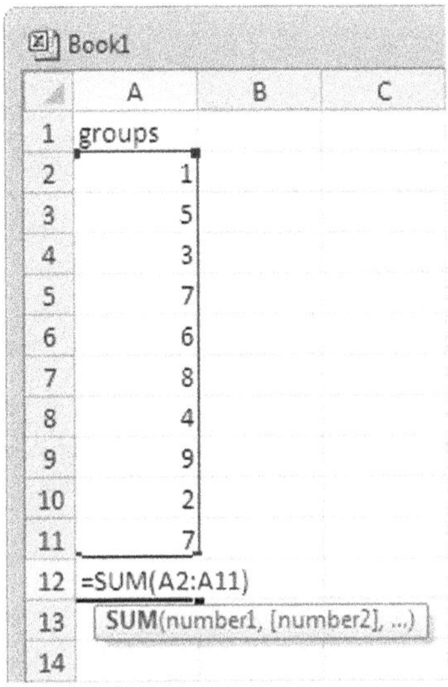

- For a fixed reference, use dollar signs in front of the column letter and row number. If you copied this formula into other cells, it would still point to cells A2:A11 and would not be shifted to reflect a new cell location. Note that if you only place a dollar sign in front of the first letter, only the column will be fixed. Thus, if you move your formula up or down a cell, the referenced row will change. Similarly if you only place a dollar sign in front of the first number, only the row will be fixed. Thus, if you move your formular left or right a cell, the referenced column will be changed.

  Example SUM($A$2:$A$11)

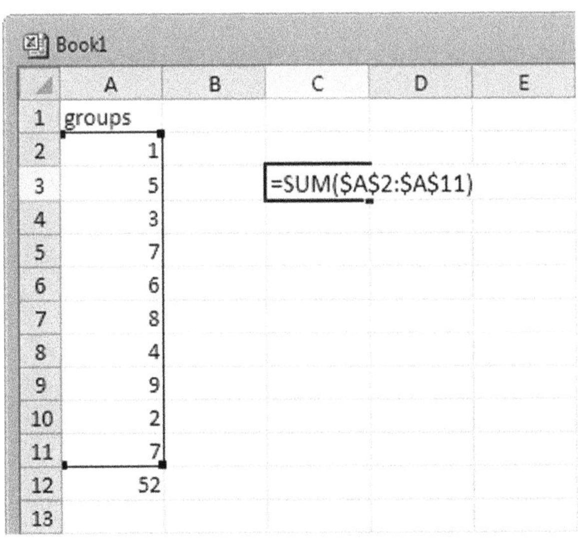

Sorting Data

To sort a column of data:

- Click on the **Data** tab

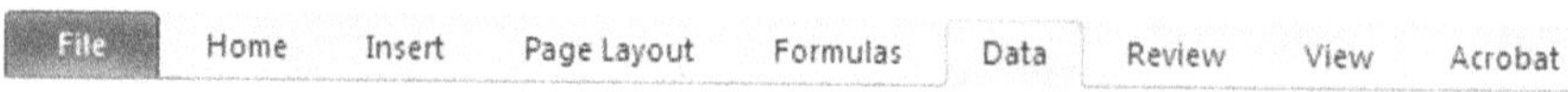

- Highlight the previous data set above or create your own data set
  - If you want to preserve the original data set/order, you need to copy and paste it in another column; otherwise, Excel will replace the unordered column with the ordered data.
- Choose the **Sort** function

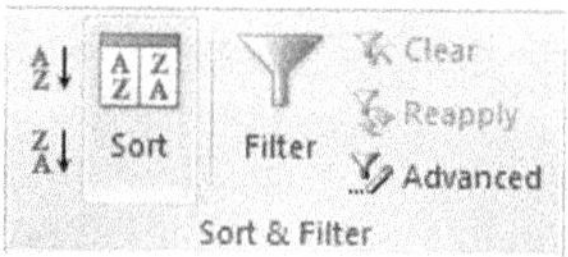

- Select the type of sorting
- You can sort by more than one level. Excel will list the values from top to bottom as the following 1, 2, 3, 4, 5, 6, 7, 7, 8, 9.

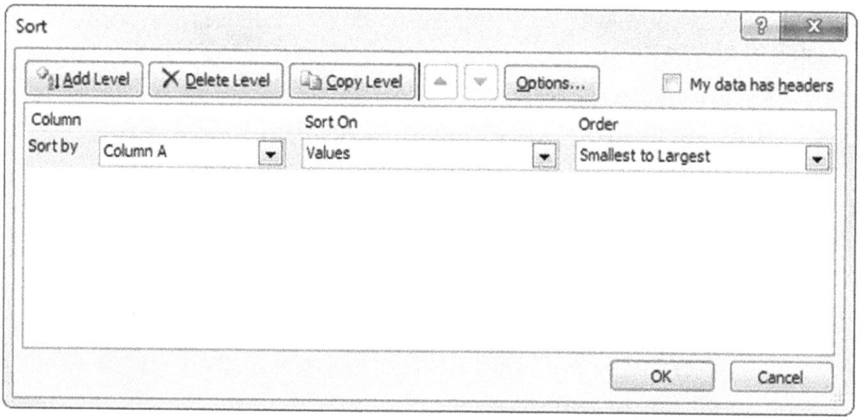

Filtering Data

- Highlight the data set above or create your own data set that you want to filter.
  - If you want to preserve the original data set, you need to copy and paste it in another column. Excel will replace the unordered column with the ordered data.

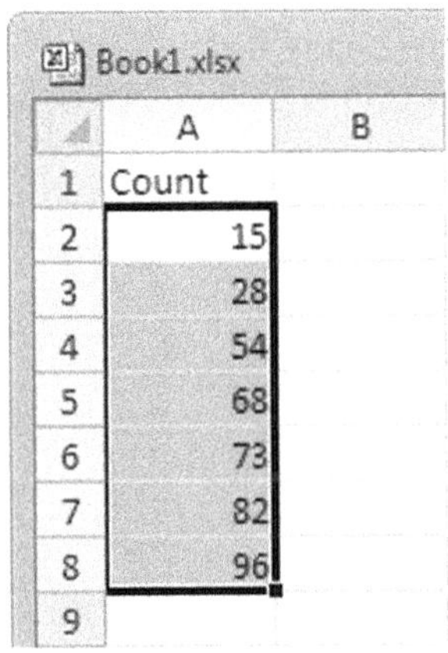

- Select the first row with the data label
- Click on the **Data** tab

- Click on **Filter** function

- Click on the drop-down arrow on the first cell

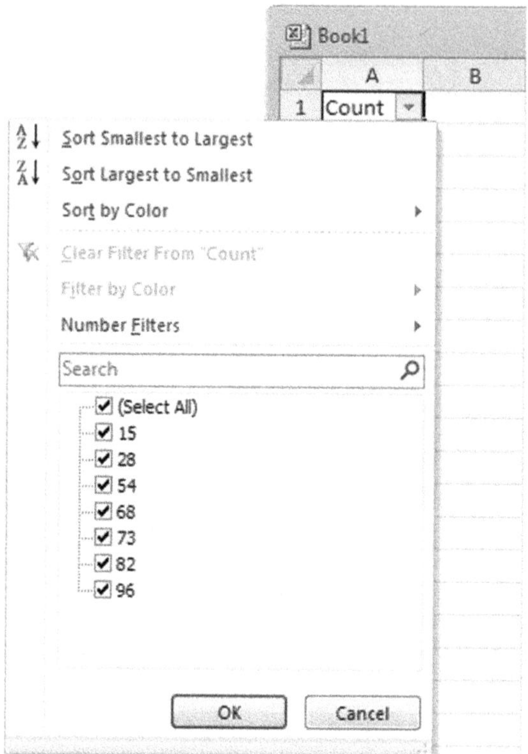

You can apply a sorting function or a filter(s).

Examples
Choose all values greater than 30. Excel will list the values 54, 68, 73, 82, and 96.
Choose all values ≤75 and ≥25. Excel will list the values 28, 54, 68, and 73.

Getting Excel Help

- Click on the input box with a bulb under the worksheet name

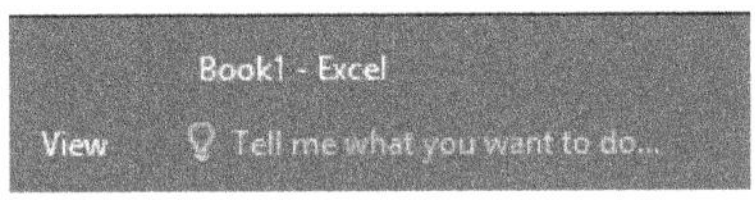

- Type in "help" and select the question mark

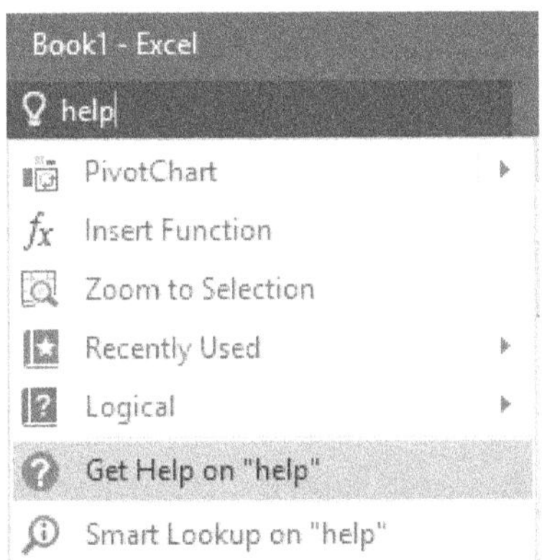

- You can type in the concept or the actual question

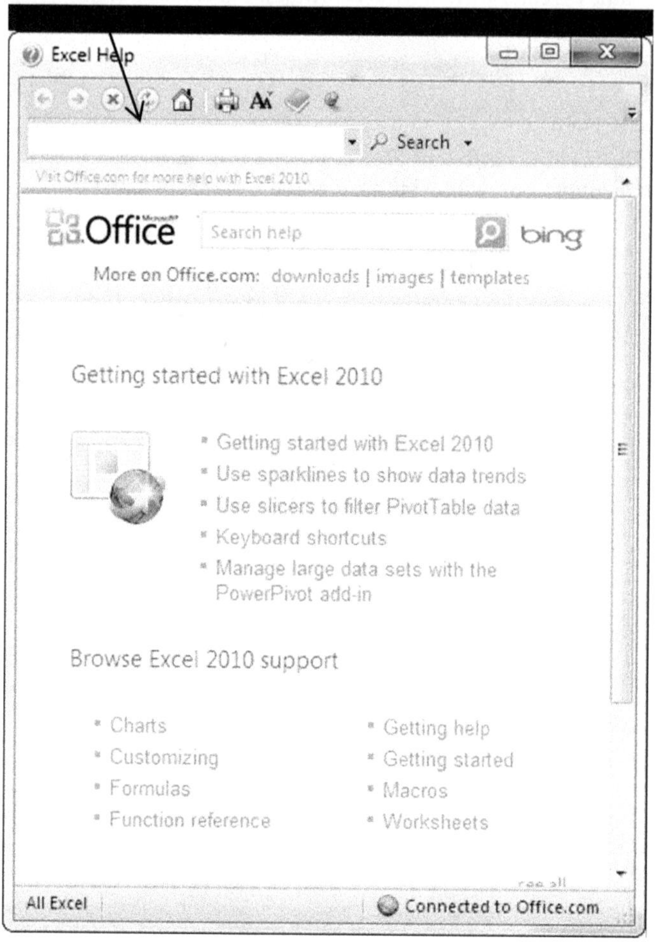

Statistical Tools

- Click on the **Data** tab

- Click in the **Data Analysis** function

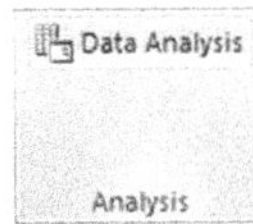

- Highlight the statistical function you need. If the function is not displayed in the list box, click on the vertical scroll bar on the right of the box or scroll to it.

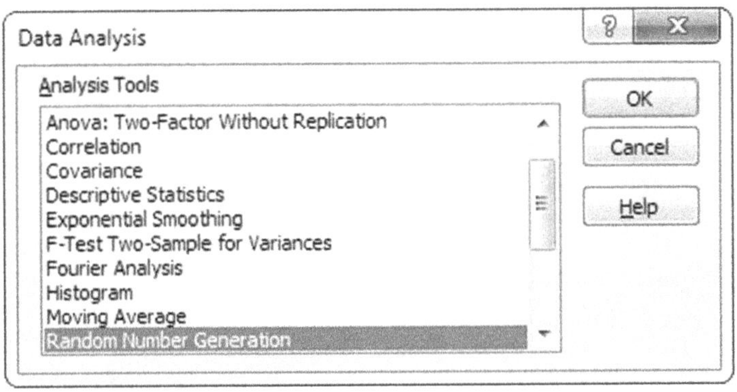

Random Number Generation

- **Number of Variables** is the number of columns
- **Number of Random Numbers** is N
- **Distribution** has seven types: Uniform, Normal, Bernoulli, Binomial, Poisson, Patterned, Discrete
- **Parameters** is changed according to different distributions
- **Random Seed** is optional
- **Output Range** is the top-left cell of the outputs

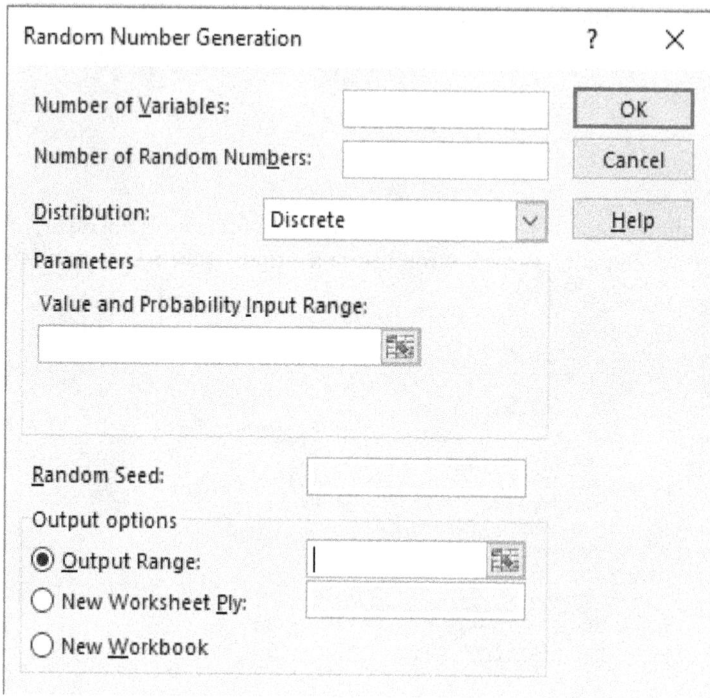

Predefined or Built-In Formulas

- Click on the **Formulas** tab

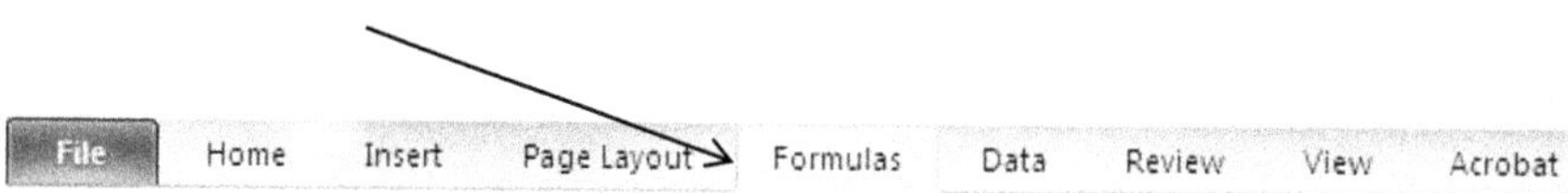

- Click on the **fx** or **Insert Function** button

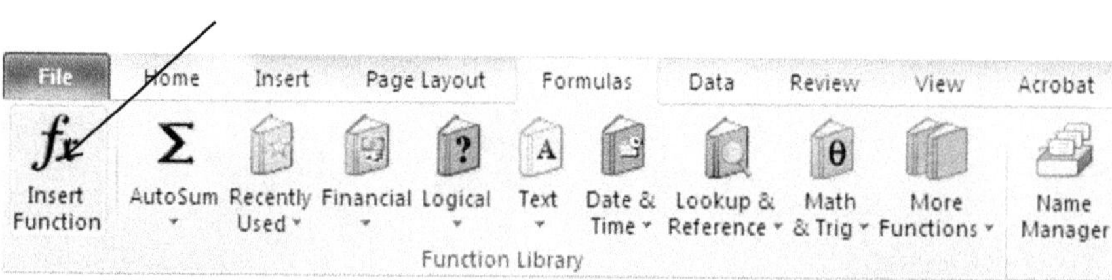

- Click on **Statistical** from the drop-down menu to indicate the category
- Highlight the function you want or type in a description of what you want the function to do

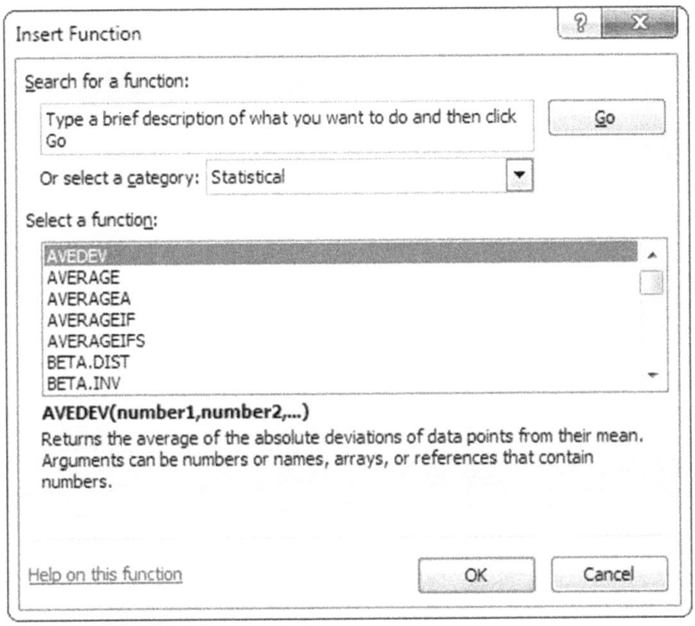

Formatting Data

- Right click on the cells you wish to format
- Select the **Format Cells** option

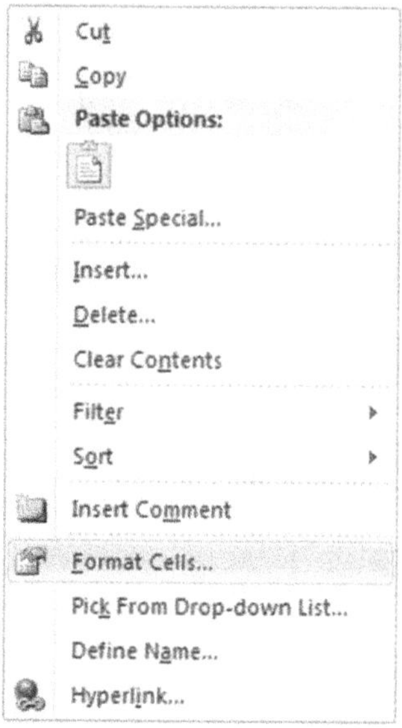

- Select **Number** under Category

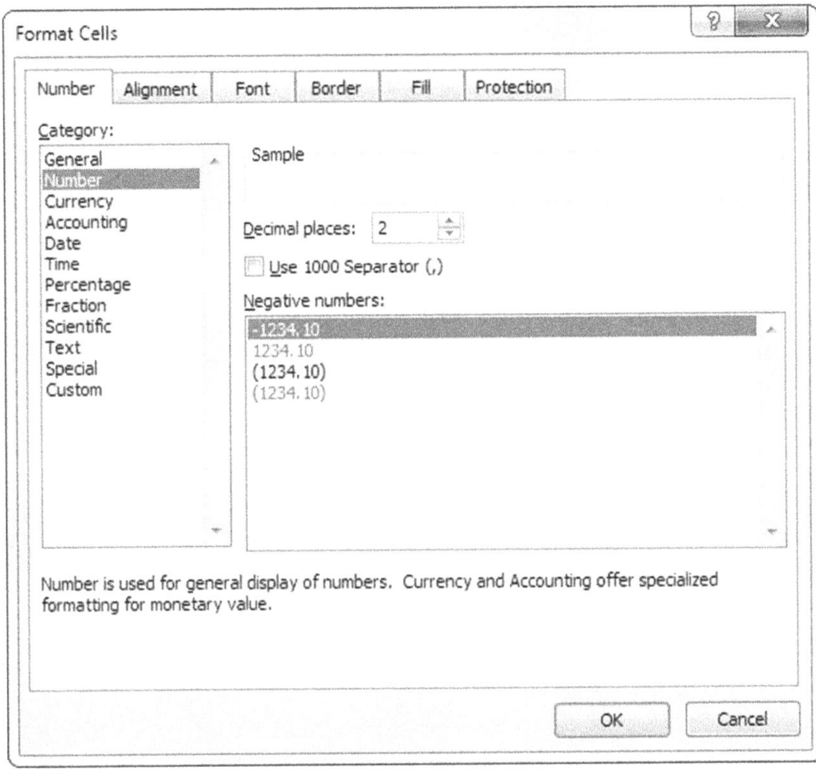

- Choose the correct number of decimal places, usually 2 is sufficient

Chart Wizard

To access the chart wizard:

- Click on the **Insert** tab

- Select the type of chart you want from the "Charts" section
- You can select any of the following chart types listed (see Fig. 2.1)

Formatting of Graphs

To change graph titles or scales:

Name	Icon	Description
Column		Shows how data changes over time or between categories. Values are displayed vertically, categories horizontally.
Line		Shows trends in data spaced at equal intervals. Can also be used to compare values between groups.
Pie	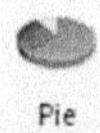	Shows the proportional size of items that make up the whole. The chart is limited to one data series.
Bar		Shows how data changes over time or between categories. Values are displayed horizontally, categories vertically.
Area		Displays the magnitude of change over time or between categories. You can also display the sum of group values, showing the relationship of each part to the whole.
Scatter	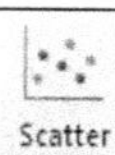	Displays the relationship between numeric values in several data series. This chart is commonly referred to as a scatter plot.
Other Charts		Displays a sub-set of charts that are less commonly used.
Stock	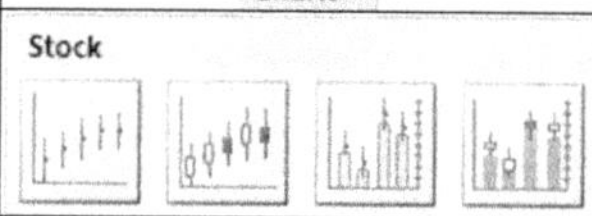	Displays stock market data, including opening, closing, low and high daily values.
Surface	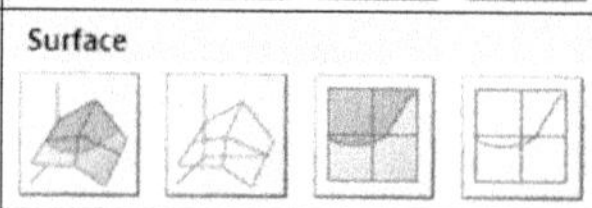	Shows the value of a data series in relation to the combination of the values of two other data series.
Doughnut		Shows the proportional size of items relative to the whole. It can display more than one data series at a time.
Bubble	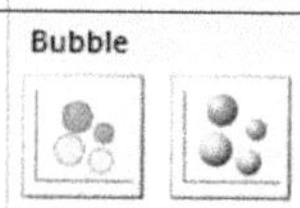	This is a type of scatterplot but the size of the bubbles is proportional to the value of the third data series.
Radar	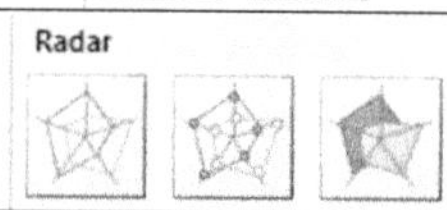	Shows values from different categories radiating from a center point. Lines connect the values within each data series.

Fig. 2.1 Chart types

- Click on the graph
- Click on **Design** and then on the tab **Add Chart Element**

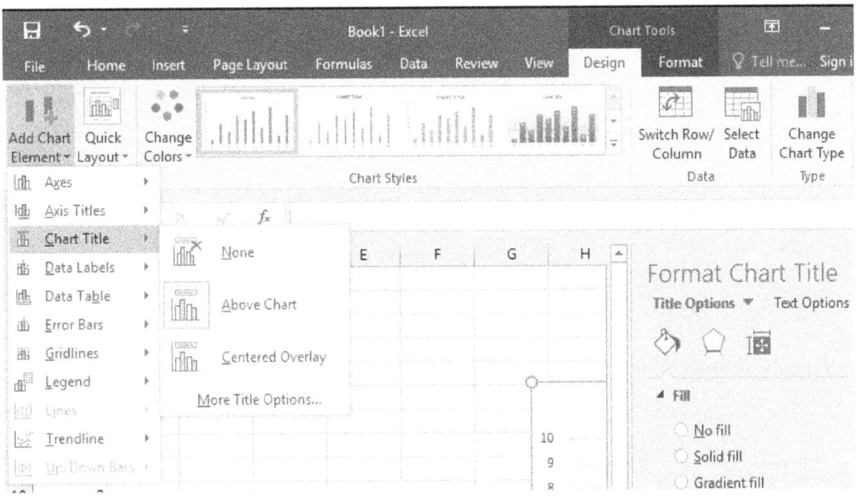

- OR just Click on actual **Chart Title** or **Data Labels** or other aspects of the graph you wish to format and edit

Bar and Column Charts

A bar chart or bar graph is a chart with rectangular bars with lengths proportional to the values that they represent. The bars can be plotted vertically or horizontally. A vertical bar chart is sometimes called a column chart.

Example Problem
You own a small used car lot and want to do some analysis with car sales and the days of the week that the cars were sold.

- Enter the text and numbers in columns A and B

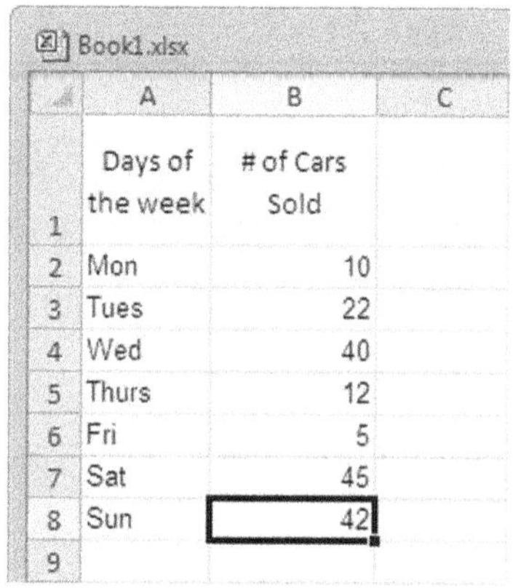

- Highlight the two columns of data

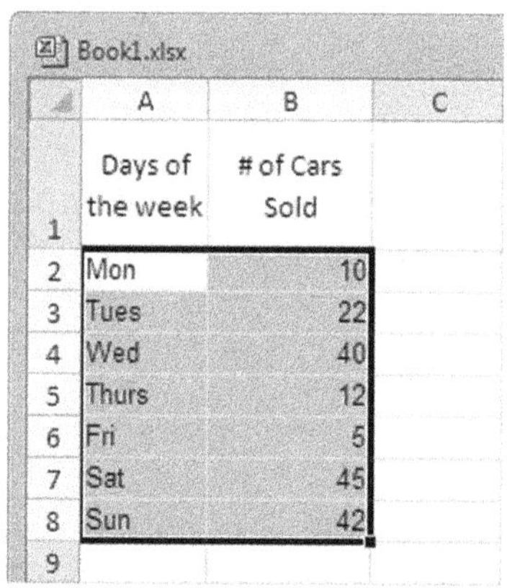

- Select the **Insert** tab

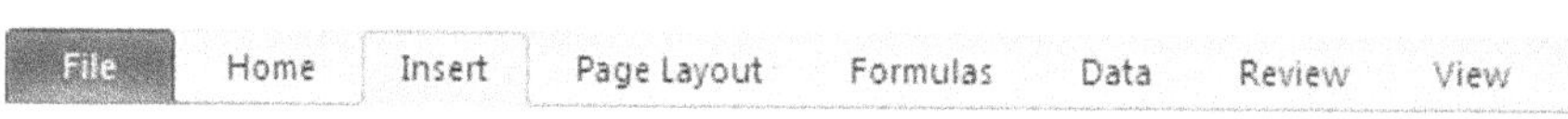

- Select the **Column** chart function, then select the type of Column chart desired

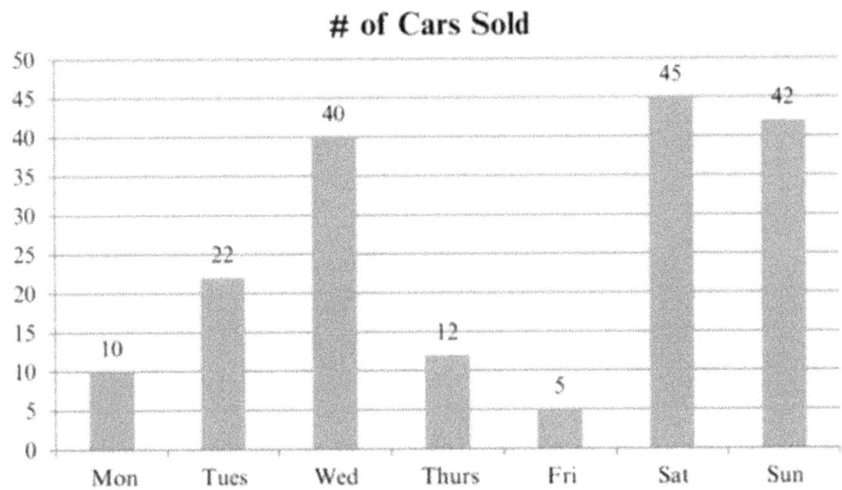

- Click anywhere on the chart
- Click the **Design** tab, then **Add Chart Element > Data Labels > Outside End** to remove all data labels.

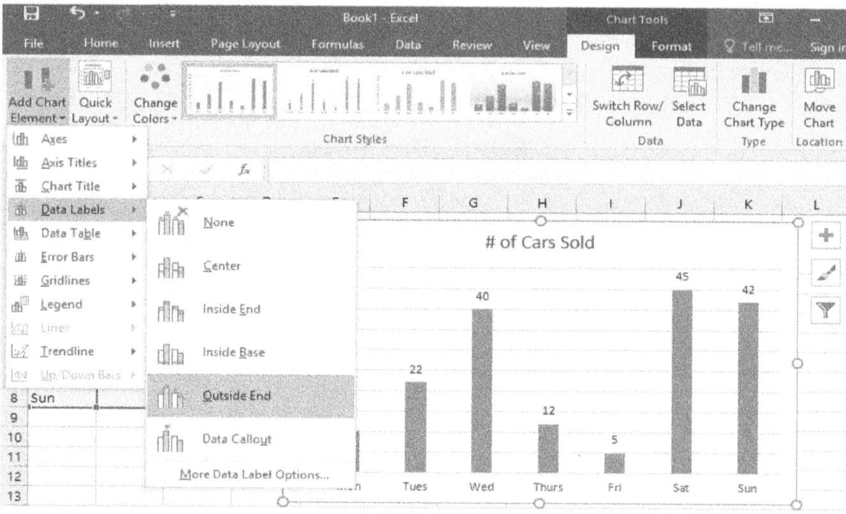

- Click the **Plus** icon in the right head of the chart, then the **Axis Titles** function to add the appropriate vertical and horizontal labels. Double click the appropriate axis title and enter the axis label.

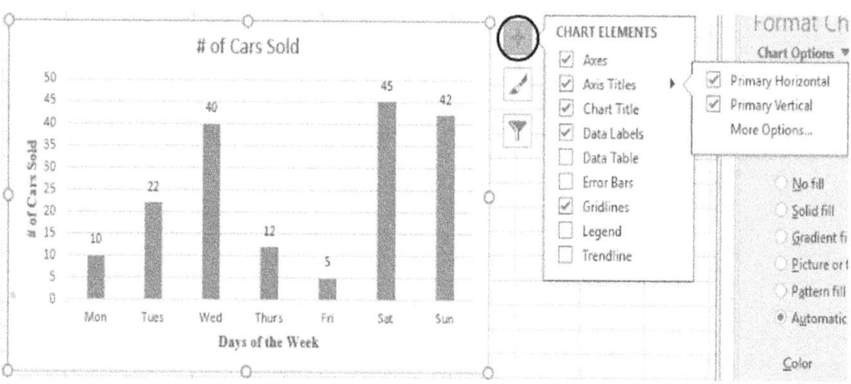

Pie Charts

In the pie chart, the sum of the percentages on the chart should add to 100% if you have used all of the input data correctly. In this example, the input data is given as percent of total sales. In the next chapter, you will learn how to calculate the percentages from the raw data.

Example Problem

As the store manager you wish to complete some sales analysis on each of the departments in your store. The sales data is "% of total sales" by "department."

- Enter the text and numbers in columns A and B
  - Excel will assume the first column is labels, so be careful how you enter the data. Make sure you have changed the number type to percentage before entering in the values. This is done by selecting the home tab, going to the number section, then selecting percentage from the drop down menu. Now enter in each departments "% of total sales".

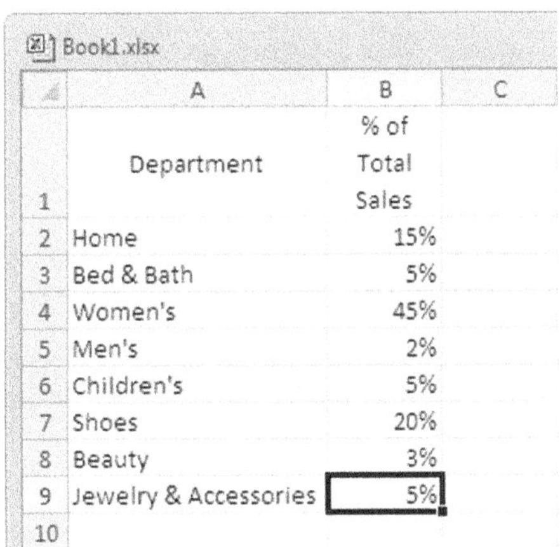

Book1.xlsx

	A	B	C
1	Department	% of Total Sales	
2	Home	15%	
3	Bed & Bath	5%	
4	Women's	45%	
5	Men's	2%	
6	Children's	5%	
7	Shoes	20%	
8	Beauty	3%	
9	Jewelry & Accessories	5%	
10			

- Highlight the two columns of data

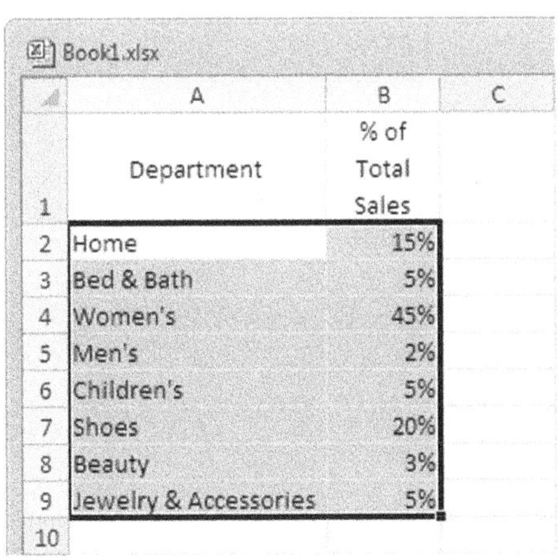

Book1.xlsx

	A	B	C
1	Department	% of Total Sales	
2	Home	15%	
3	Bed & Bath	5%	
4	Women's	45%	
5	Men's	2%	
6	Children's	5%	
7	Shoes	20%	
8	Beauty	3%	
9	Jewelry & Accessories	5%	
10			

- Select the **Insert** tab

- Select the **Pie** chart function, then select the type of Pie chart desired

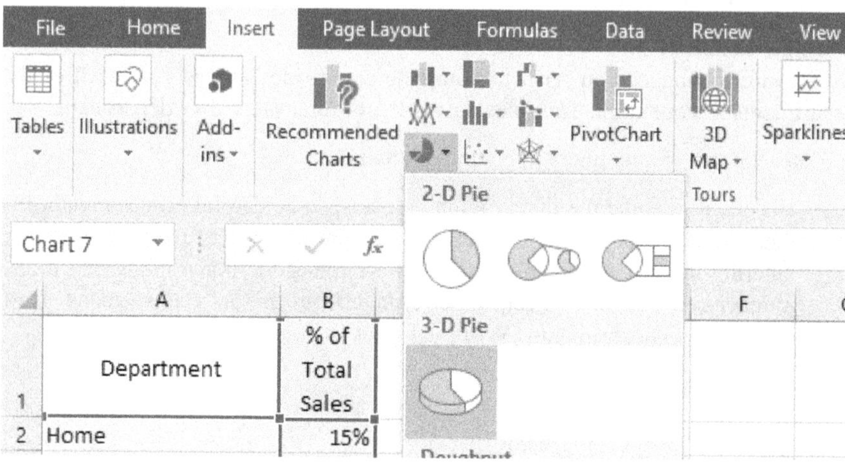

- Click anywhere on the chart

Click the Design tab, then Add Chart Element > Data Labels > Best Fit to add data labels at the proper places

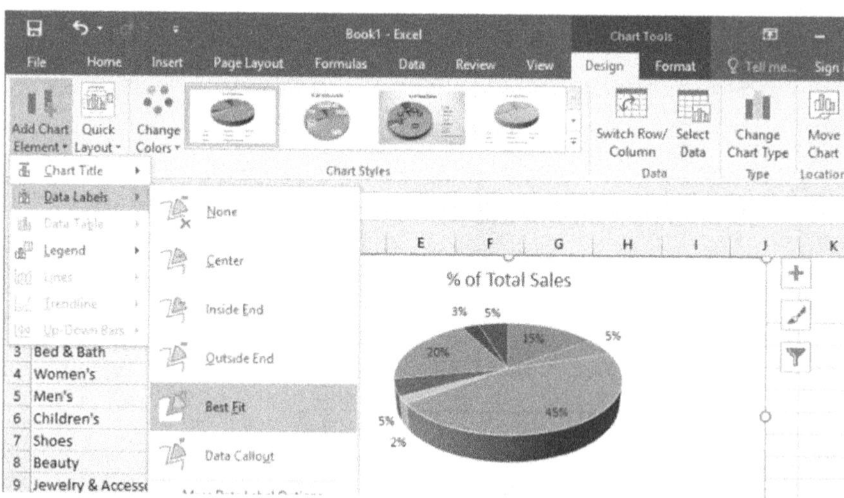

Line Charts and Area Charts

A line chart is a type of chart that displays information as a series of data points connected by straight-line segments. An area chart is based on the line chart. The area between the axis and the data line is commonly filled in with colors, textures, and hatchings.

Example Problem

You own a small used car lot and want to do some analysis with car sales and the days of the week that the cars were sold.

- Enter the text and numbers in columns A and B

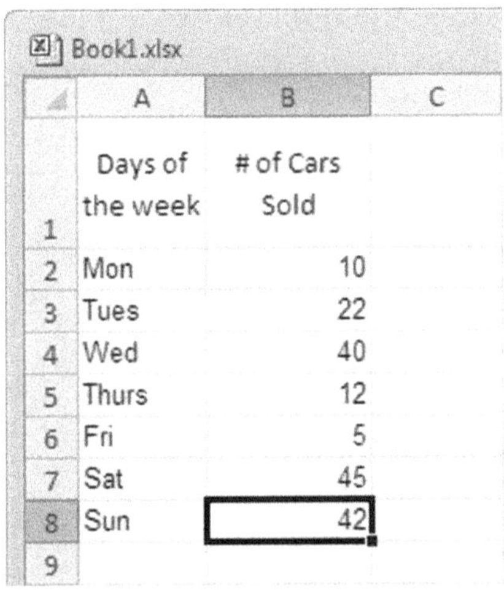

Book1.xlsx

	A	B	C
1	Days of the week	# of Cars Sold	
2	Mon	10	
3	Tues	22	
4	Wed	40	
5	Thurs	12	
6	Fri	5	
7	Sat	45	
8	Sun	42	
9			

- Highlight the two columns of data

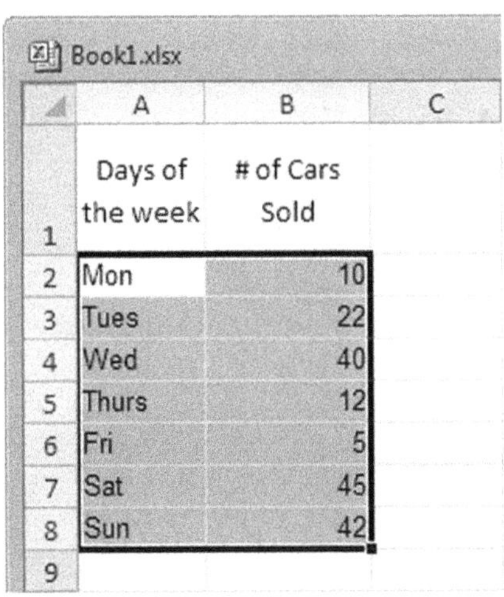

Book1.xlsx

	A	B	C
1	Days of the week	# of Cars Sold	
2	Mon	10	
3	Tues	22	
4	Wed	40	
5	Thurs	12	
6	Fri	5	
7	Sat	45	
8	Sun	42	
9			

- Select the **Insert** tab
- Click the **Arrow** icon at the bottom-right corner of the **Charts** tab

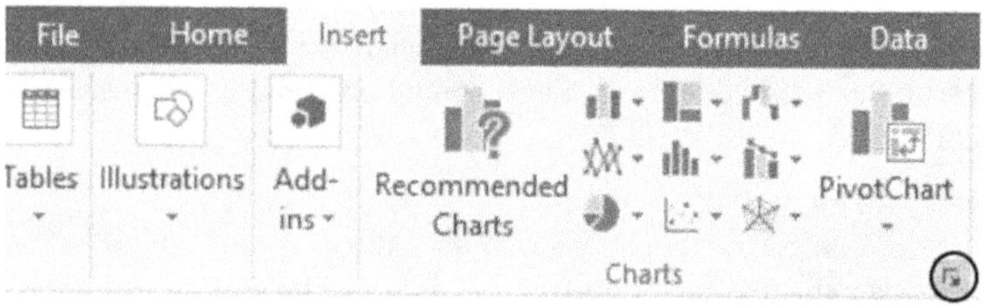

- Select the **Line or Area** function from the **All Charts** tab
- Select the type of Line or Area chart desired

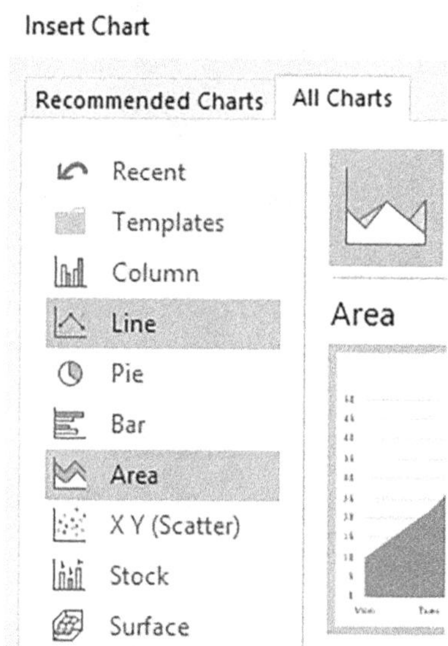

Line Graph Example

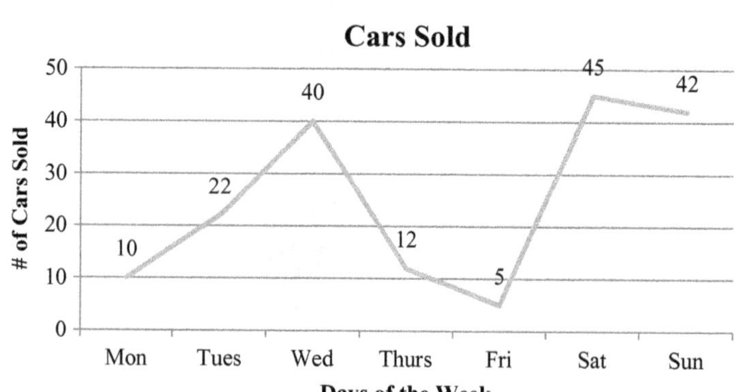

Area Chart Example

The area chart just fills in the area below the line.

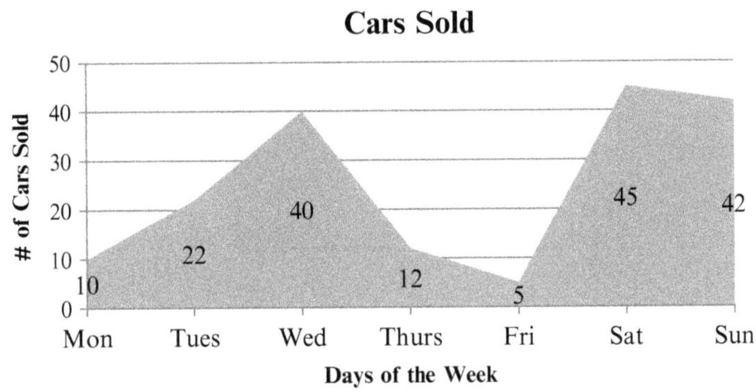

Other Charts

Bubble Chart

A bubble chart plots three-dimensional data as a two-dimensional graph using the size of the bubble to depict the third dimension.

- You need to have three arrays of data: x, y, size of bubble.

Example Problem

The analysis you wish to complete involves reviewing the product sales and market share by location. The data that is available to you includes the number of different products sold in each of your four geographic areas of operation, the sales generated by those products, and the percent of your total market each location represents.

- Enter the text and numbers in columns A, B, and C

Book1.xlsx

	A	B	C	D
1	# of Products	Sales	Market Share %	
2	14	$21,416	15%	
3	20	$60,000	23%	
4	18	$24,400	10%	
5	22	$32,000	42%	
6				

- Highlight the three columns of data

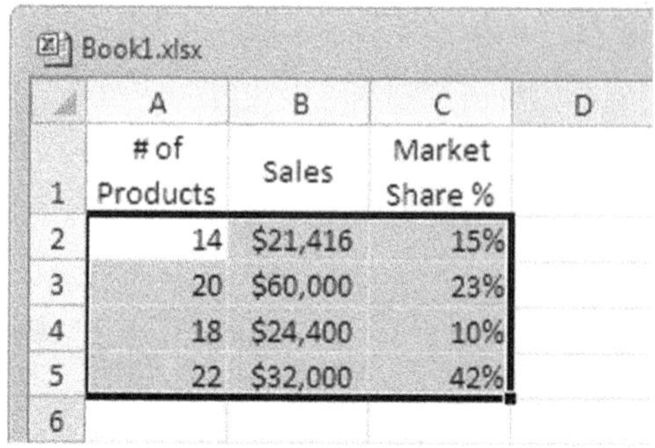

- Select the **Insert** tab
- Select the **Bubble** chart function, then choose the 3-D Bubble type of Bubble chart

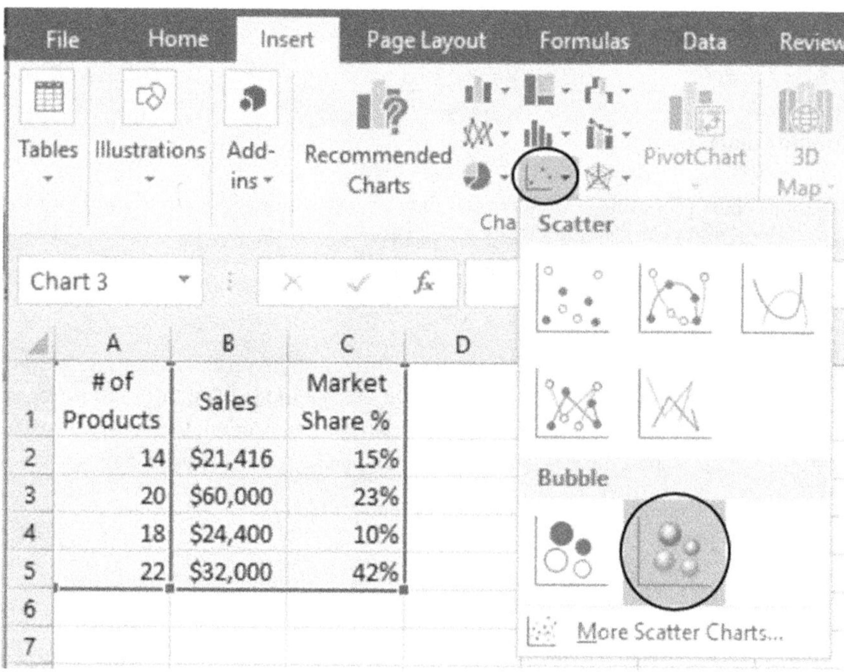

- Click anywhere on the chart
- Click the **Design** tab, then **Add Chart Element > Data Labels** to add data labels at the proper places

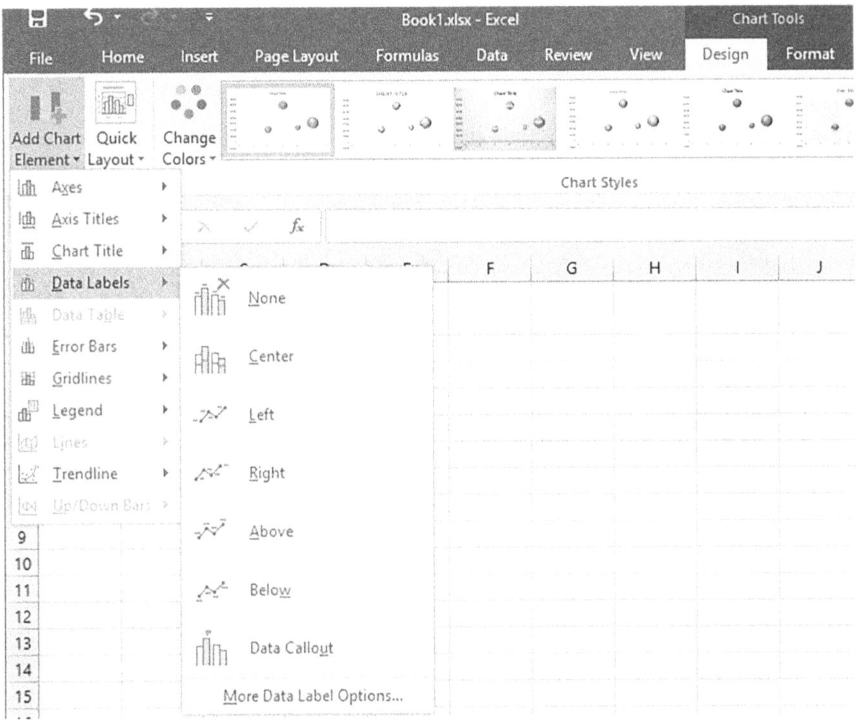

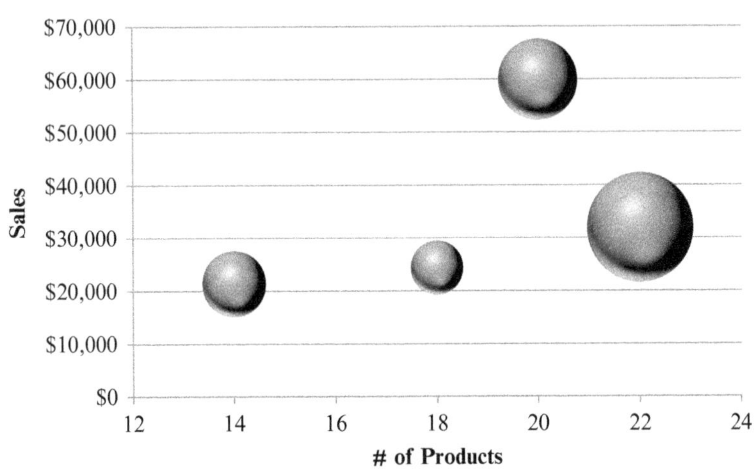

The size of the bubbles is in proportion to the percent of the market share that each store holds. The bubble chart is included in this book as an example of the different charts that can be created using Excel; however, this specific chart is difficult to use and may not have many applications.

Radar Chart

A radar chart plots multivariate data in the form of a two-dimensional chart of three or more quantitative variables represented on axes starting from the same point. It is also known as a spider chart, web chart, or star chart. It consists of a sequence of equi-angular spokes, called radii, with each spoke representing one of the variables. The length of a spoke is proportional to the size of the variable for the data point relative to the maximum size of the variable across all data points. A line is drawn connecting the data values for each spoke. This gives the plot a star-like appearance.

Example Problem

You own a small used car lot and want to do some analysis with car sales and the day of the week that the car was sold.

- Enter the text and numbers in columns A and B

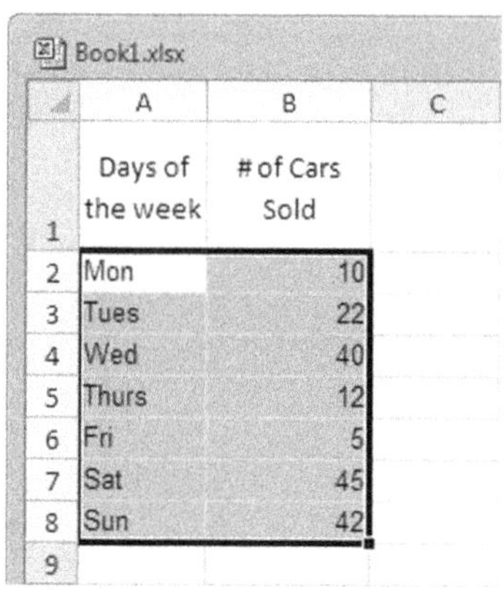
Book1.xlsx

	A	B	C
1	Days of the week	# of Cars Sold	
2	Mon	10	
3	Tues	22	
4	Wed	40	
5	Thurs	12	
6	Fri	5	
7	Sat	45	
8	Sun	42	
9			

- Select the **Insert** tab
- Choose a **Radar** chart

- Click anywhere on the chart
- Click the **Design** tab, then **Add Chart Element** > **Data Labels** > **None** to remove all data labels.

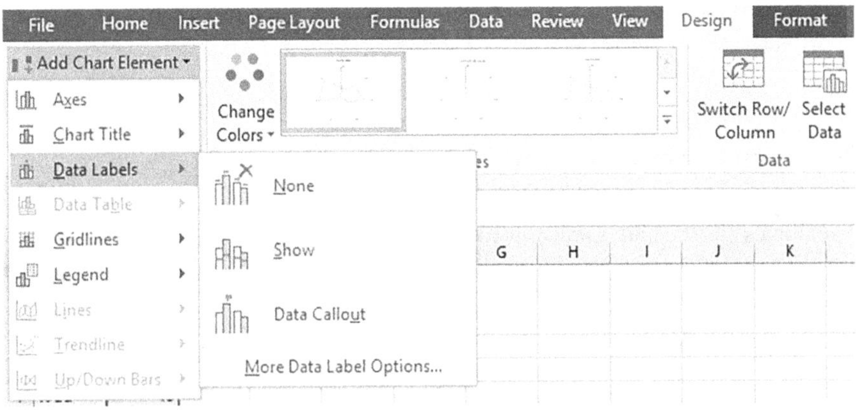

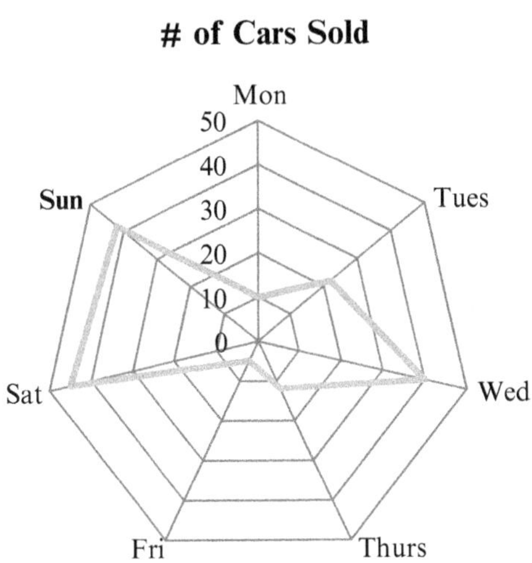

The intercepts (the point where the blue line intersects the radial black lines) in the radar chart indicate how many cars were sold on that specific day of the week.

So if we want to check how many cars were sold on Monday, look along the Monday line that goes into the center of the graph. The radial lines represent the scale from 0 to 50 (# of cars). The blue line intersects the Monday line at 10 units. On Monday 10 cars were sold.

PivotTables (Aka Crosstabs)

PivotTables (as they are known in Excel) or crosstabs allow you to show how two different variables are related to each other in a table format. They are also used in frequency analysis.

Example Problem

Your survey company has been asked to collect data on the importance of name brands to consumers with special attention to the age of the respondents. The consumers in your sample indicate their importance rating using a scale from 1 to 5, with 5 being the highest possible score and 1 being the lowest possible score. Input the following sample data you collected:

Age	Importance of name brands
18–21	5
18–21	5
18–21	5
18–21	5
18–21	4
18–21	4
18–21	4
18–21	4
18–21	3
18–21	3
18–21	2
18–21	1
22–25	1
22–25	2
22–25	3
22–25	3
22–25	4
22–25	4
22–25	2

To insert a pivot table:

- Click on the **Insert** tab

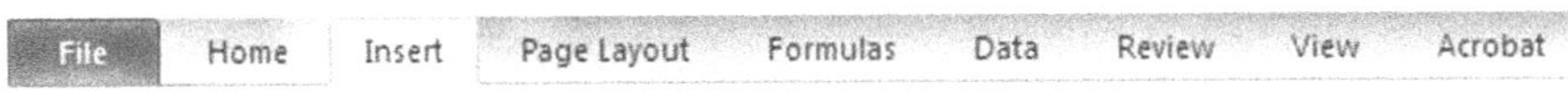

- Select the **PivotTable** function

- Select the appropriate data for **Table/Range**

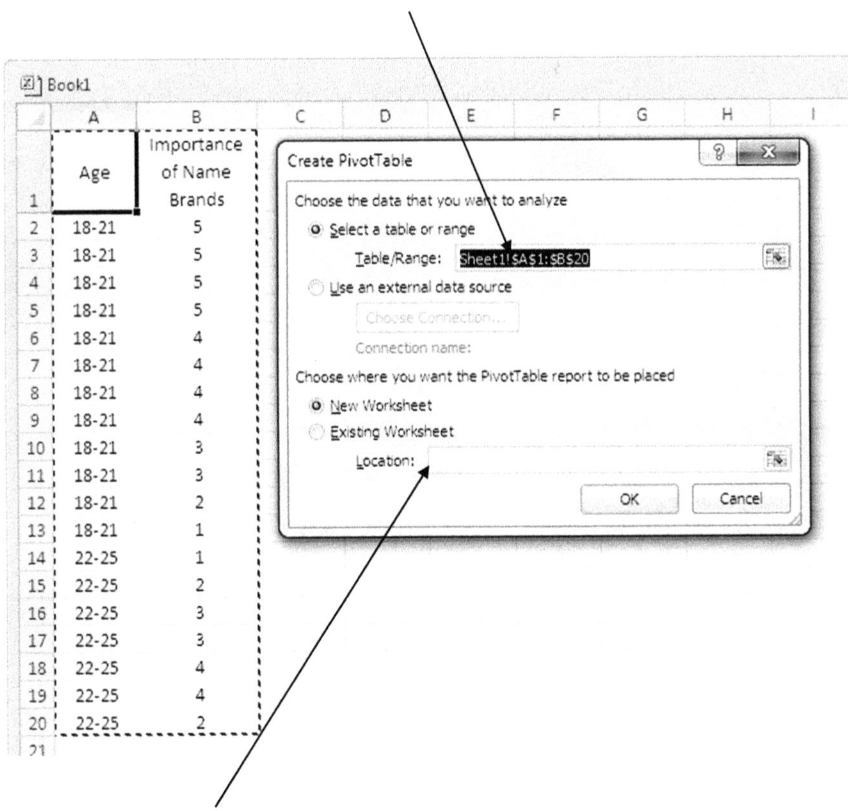

- Choose where you want the PivotTable report to be placed
- Click **OK**
- The PivotTable Field List will appear

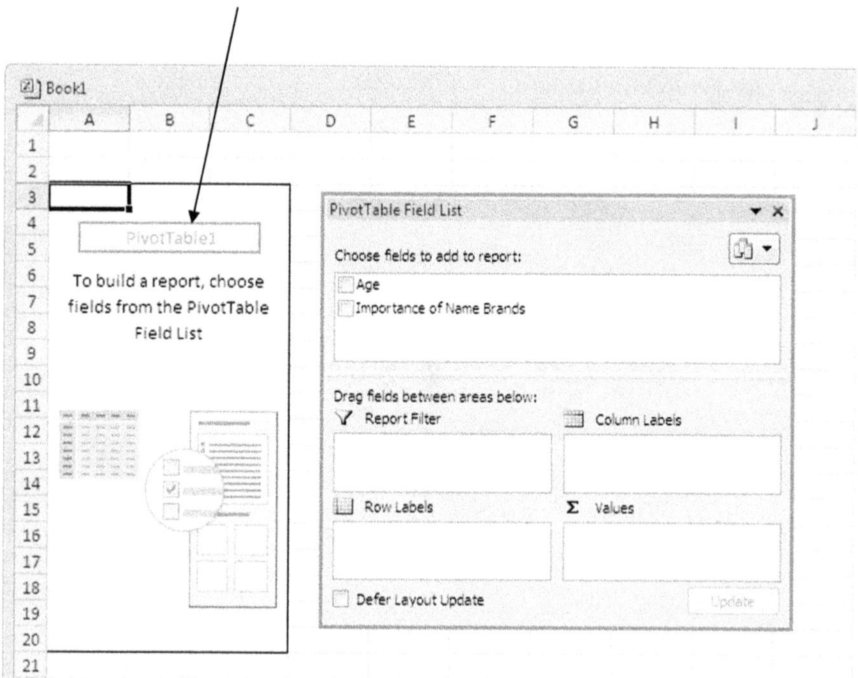

- Set up your table

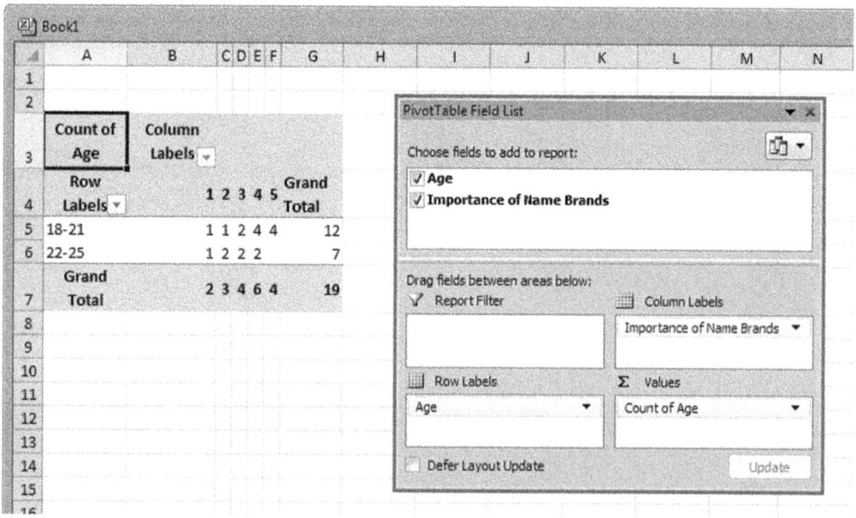

- Click the **Value Field Settings** to edit the table with the appropriate data from the drop-down menu on the PivotTable Field list

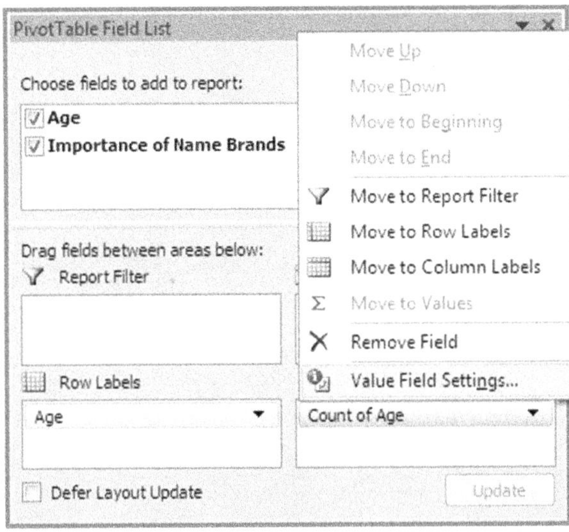

- In the **Summarize value field by tab** select Count. Then Click the Show values as tab

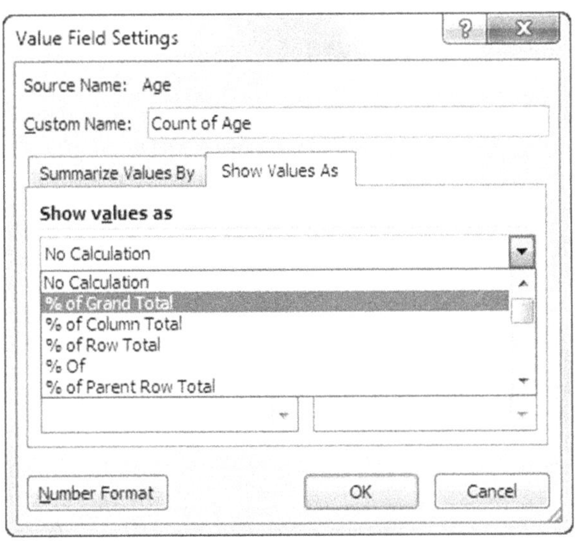

- Select the appropriate calculation

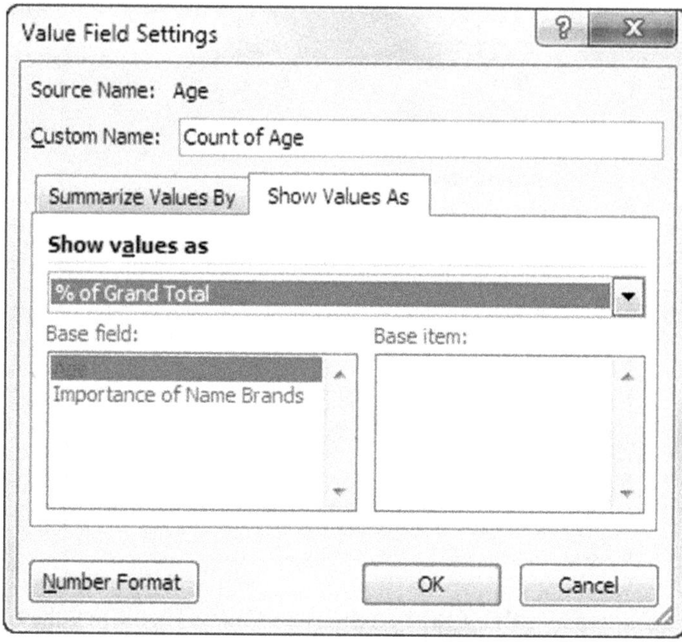

- Click **OK**

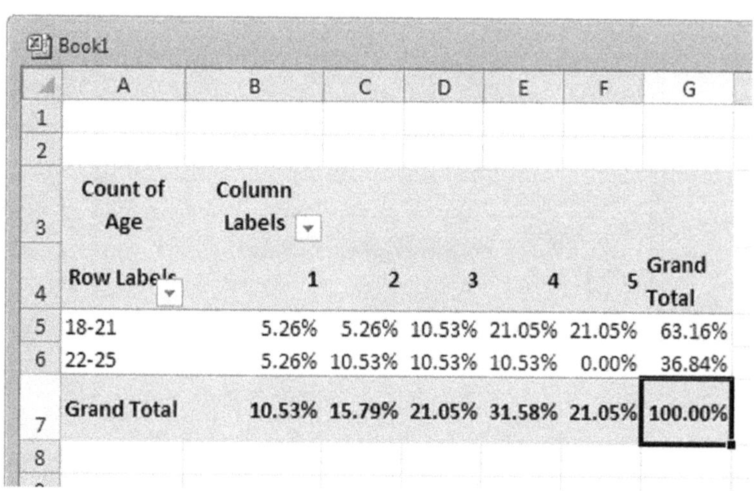

The frequency data for each importance rating value is given within each of the two defined age categories. It seems that the younger age group contains a larger sample, 63% versus 37%. The younger age group indicated higher ratings in the four and five categories of importance. This means younger consumer behaviors may be influenced more by brand names, but we would need much more data to be certain.

Excel

Common Pitfalls

- When doing a pivot table, place the variables in the appropriate boxes for Report Filter, Column Labels, Row Labels, and Values. If you just check the variables, they will end up in Row Labels and Values by default.
- Bubble charts are difficult to use unless you have three interconnected variables. Use caution when deciding to use a bubble chart.
- Some charts will not display the complete variable labels if the labels are long and exceed a certain character limit. If you want to create a chart that will fit a longer variable label, consider using a Column Chart and sizing the graph so that you can expose more text, or changing the text size.
- Don't forget to include the "=" sign when you input a formula. Also make sure to fix cell references in formulas if necessary (with $) when copying and pasting.
- Include labels and units for all axes on your charts. By default, Excel will not label your axes.

Final Thoughts and Activities

Practice Problems

1. Prepare a pie chart showing annual sales as a proportion of total sales. There are several options for developing different pie charts from the data.

Region	Annual sales ($)
North	$18,000
South	$52,000
East	$26,000
West	$30,000

2. Prepare a bar chart to show the number of mistakes made in a year by employees with different levels of expertise. There are several options for developing different bar charts from the data.

Expertise level	# of mistakes
Novice	36
Apprentice	20
Tradesmen	9
Master tradesman	3

3. Prepare a scatter plot with straight lines and markers to show dollars spent on Valentine's Day flowers over time.

Year	Sale of Valentine's Day flowers
2011	$15,730
2010	$14,160
2009	$14,750
2008	$17,200
2007	$16,890

4. Consider the following chart of sales data from a time period. What data is missing for you to fully understand the data?

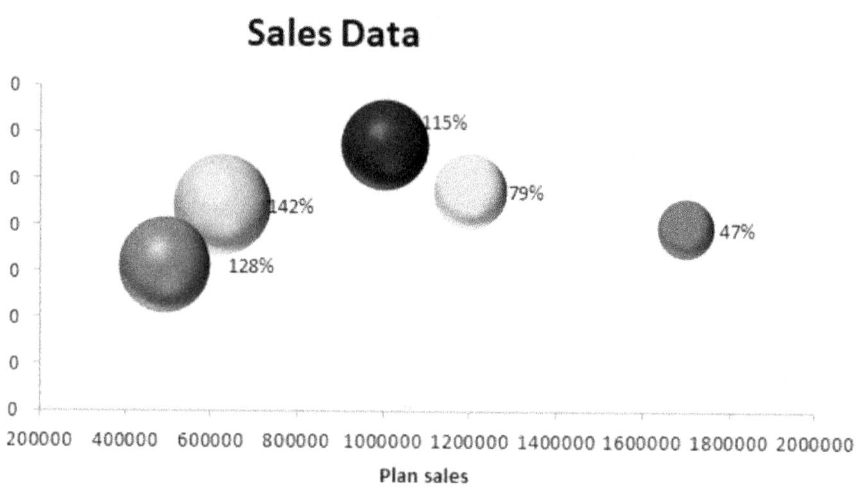

5. Interpret this chart.
China, orange; India, green; Australia, yellow; Korea, blue; Ireland, red

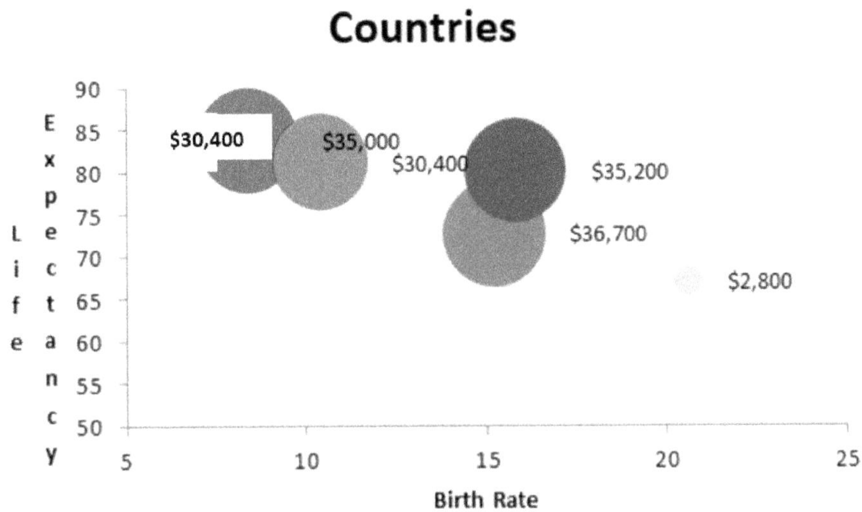

6. What type of charts might be applicable to easily display this data?

	A	B	C	D
1	Telephony System Review Scores			
2				
3		**Vendor 1**	**Vendor 2**	**Vendor 3**
4	**Reliability**	65	73	49
5	**Customer Service**	68	66	72
6	**Customer Satisfaction**	80	54	92
7	**Expandability**	70	68	62
8	**Cost**	66	70	52
9	**Quality**	64	73	54
10	**Features**	67	66	72
11	**Maintenance Costs**	72	59	80

7. Create a radar chart with the following data

	A	B	C	D	E
1	Year	Q - 1 Sales	Q - 2 Sales	Q - 3 Sales	Q - 4 Sales
2	2013	15,489,511	10,930,960	14,120,052	232,599
3	2014	15,060,947	14,147,585	12,829,331	6,435,349
4	2015	19,899,937	12,006,025	10,942,732	4,893,709
5	2016	15,426,148	2,413,549	8,588,270	2,926,127
6	2017	17,376,464	6,954,401	5,281,969	3,052,677
7	2018	9,265,713	8,189,278	7,083,311	6,551,136

8. Create a doughnut chart for the following data. (To merge two cells together to produce the "Project Status" label, first click on A1 and enter "Project Status". Then highlight both A1 and B1. Then under the home tab go to the alignment section. Then click the Merge & Center button.)

	A	B
1	Project Status	
2	Completed	40%
3	Work inprogress	10%
4	Incomplete	50%

Discussion Boards

1. Bubble charts and Radar charts are helpful in displaying data. Why are they not more common?
2. Pie charts can be a powerful tool for managers to use on a daily basis but can be easily misused. Explain.

Group Activity

Applied statistical research is an important support activity for the marketing department. There are consulting companies that provide this type of data. What types of statistical analysis do these companies advertise on their websites? What are the challenges in getting managers to accept this data?

Parting Thought

Facts are stubborn things, but statistics are more pliable.

Problem Solutions

1. Prepare a pie chart showing annual sales as a proportion of total sales.

Region	Annual sales ($)
North	$18,000
South	$52,000
East	$26,000
West	$30,000

Answer:

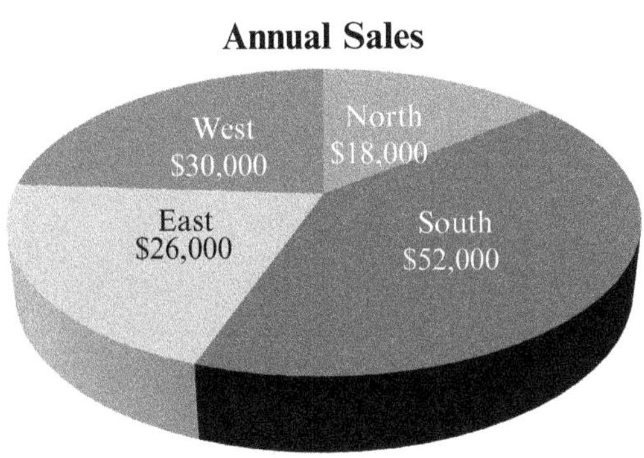

2. Prepare a bar chart to show the number of mistakes made in a year by employees with different levels of expertise.

Expertise level	# of mistakes
Novice	36
Apprentice	20
Tradesmen	9
Master tradesman	3

Answer:

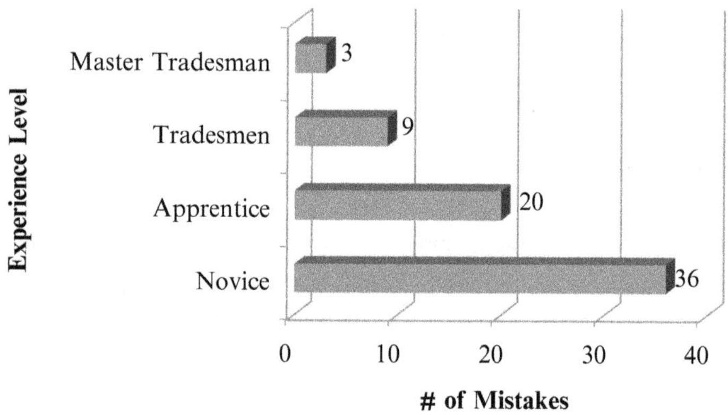

3. Prepare a scatter plot with straight lines and markers to show dollars spent on Valentine's Day flowers over time.

Year	Sale of Valentine's Day flowers
2011	$15,730
2010	$14,160
2009	$14,750
2008	$17,200
2007	$16,890

Answer:

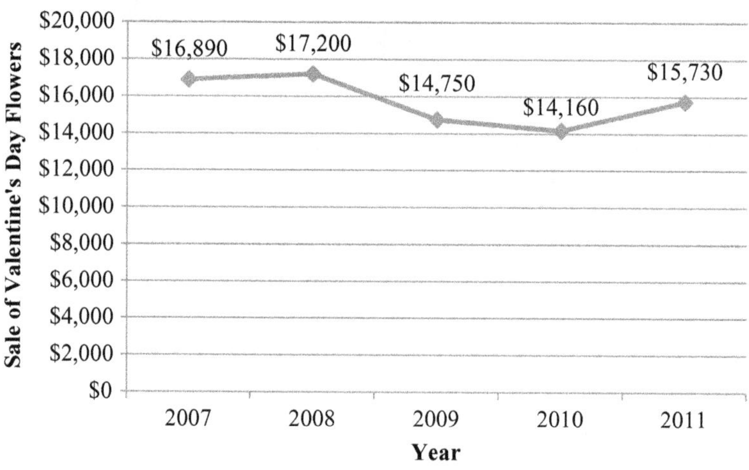

4. Consider the following chart of sales data from a time period. What data is missing for you to fully understand the data?

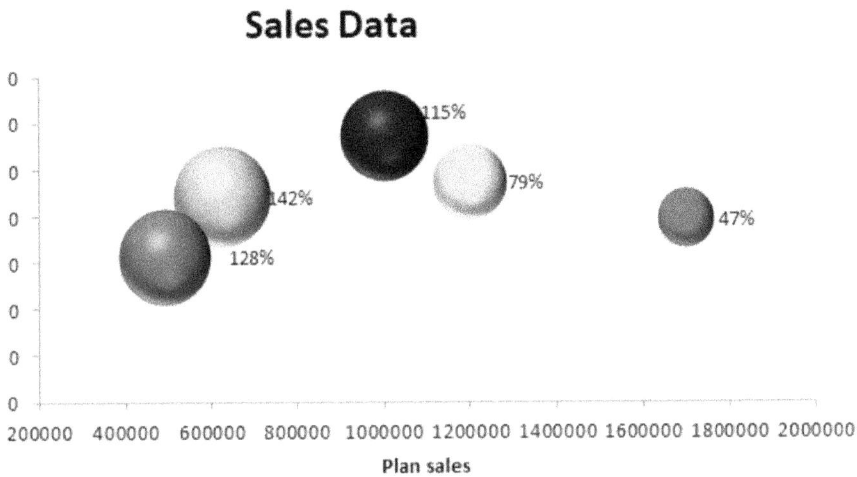

Answer:
The *y* axis units/labels are missing, the currency symbols are missing, and the chart title is vague in terms of what time period the data refers to.

5. Interpret this chart that shows GDP data in $ for each country next to the associated bubble.

China, orange; India, green; Australia, yellow; Korea, blue; Ireland, red

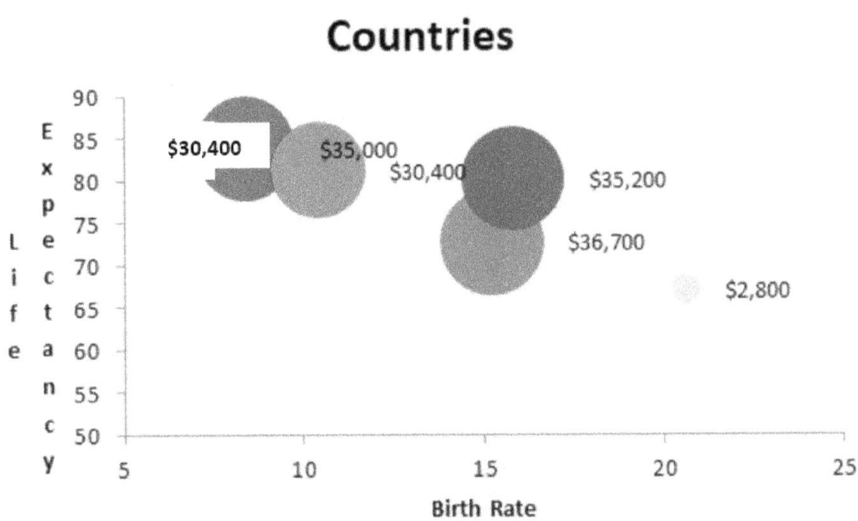

Answer:
China has the highest GDP with a reasonable life expectancy. Australia has the lowest GDP with the lowest life expectancy. However, their birth rate is the highest, which means they have a potential for a growing workforce to counteract the effect of the low life expectancy on the GDP. But the overall takeaways on this graph are quickly understood across all three factors. A bubble chart allows for easy understanding of more than two variables simultaneously.

6. What type of charts might be applicable to easily display this data?

	A	B	C	D
1	Telephony System Review Scores			
2				
3		Vendor 1	Vendor 2	Vendor 3
4	Reliability	65	73	49
5	Customer Service	68	66	72
6	Customer Satisfaction	80	54	92
7	Expandability	70	68	62
8	Cost	66	70	52
9	Quality	64	73	54
10	Features	67	66	72
11	Maintenance Costs	72	59	80

Answer:
Bubble and radar charts

7. Create a radar chart with the following data

	A	B	C	D	E
1	Year	Q - 1 Sales	Q - 2 Sales	Q - 3 Sales	Q - 4 Sales
2	2013	15,489,511	10,930,960	14,120,052	232,599
3	2014	15,060,947	14,147,585	12,829,331	6,435,349
4	2015	19,899,937	12,006,025	10,942,732	4,893,709
5	2016	15,426,148	2,413,549	8,588,270	2,926,127
6	2017	17,376,464	6,954,401	5,281,969	3,052,677
7	2018	9,265,713	8,189,278	7,083,311	6,551,136

Answer:

8. Create a doughnut chart for the following data

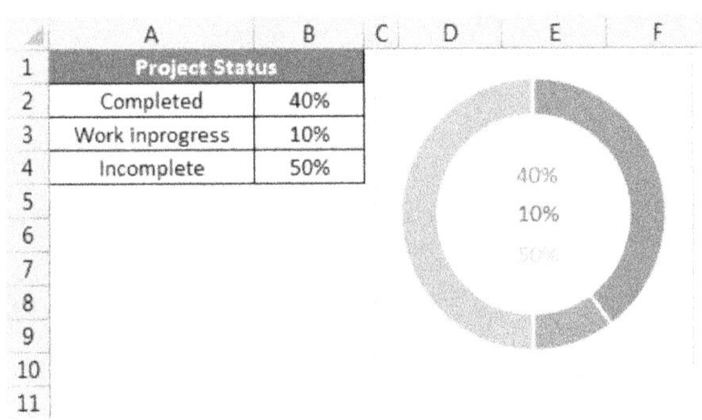

Chapter 3
Summarizing Data: Descriptive Statistics and Histograms

Linda Herkenhoff and John Fogli

Key Concepts

Central tendency, Dispersion, Histogram, Kurtosis, Mean, Median, Mode, Range, Sample variance, Skewness, Standard deviation, and Standard error.

Discussion

This chapter examines the analysis tools used for summarizing single-variable (univariate) data and contains a detailed discussion of **histograms**. The concepts of central tendency and dispersion are explored. Excel output associated with the generation of the histogram provides all of the key descriptive statistics associated with a univariate data set. These are discussed in detail. The access to Excel statistical functions is explained with applied examples.

This chapter concludes with an important section on the pitfalls when working with histograms and the associated basic statistics.

Central Tendency has to do with how the distribution of data clusters around the middle. We typically refer to this position using three important statistics: **mean**, **median**, or **mode**.

Supplementary Information The online version contains supplementary material available at https://doi.org/10.1007/978-3-031-46371-6_3.

L. Herkenhoff (✉)
Saint Mary's College, Moraga, CA, USA

J. Fogli
Diablo Valley College, Pleasant Hill, CA, USA

L. Herkenhoff, J. Fogli (eds.), *Applied Statistics for Business and Management using Microsoft Excel*, https://doi.org/10.1007/978-3-031-46371-6_3

- **Mean**: The mean of a data set refers to the arithmetic average of all the data values. To calculate the mean, sum all of the data and divide the sum by the number of data values.
- **Median**: The median of a data set refers to the value in the middle when the data values are arranged in ascending order. Half of the values are greater than this value and half are less than this value. The median may or may not correspond to a value in your data set; it is just a location.
- **Mode**: The mode of a data set refers to the value that occurs with the greatest frequency. If all of the data values occur with the same frequency, there will not be a mode.

Dispersion refers to how spread out the data is as a distribution. We use **range** and **standard deviation** as a means of describing dispersion.

- **Range**: The range of a data set refers to the difference between the largest and smallest data values. The range is the simplest measure of dispersion. The range can be described by stating the smallest and the largest data values (i.e., $25–$100), or by stating the actual number of units between these end points (i.e., $75).
- **Standard deviation (σ, SD, or sd)**: The standard deviation refers to the most commonly used statistic to measure the variation in any data set. This is a group statistic that summarizes the overall variation of your data. In other words, it indicates how much your data bounces around the average value. Excel provides this statistic in the same units as the original values. Figure 3.1 depicts the concept of standard deviation and provides the formula for a sample.

By combining the information about the **central tendency** and **dispersion**, you get a good idea of how the data is distributed even without the aid of graphs.

The concepts of a constant mean and varying spread (dispersion) are shown in Fig. 3.2. Both distributions have the same center value (the average or mean) but the second distribution is more spread out (greater dispersion). The more spread out a distribution is, the larger the standard deviation and range.

Other less used statistics that are provided in the Excel Data Analysis output include

$$sd = \sqrt{\frac{\sum_{i=1}^{n}(x_i - \bar{x})^2}{n-1}}$$

Fig. 3.1 Standard deviation

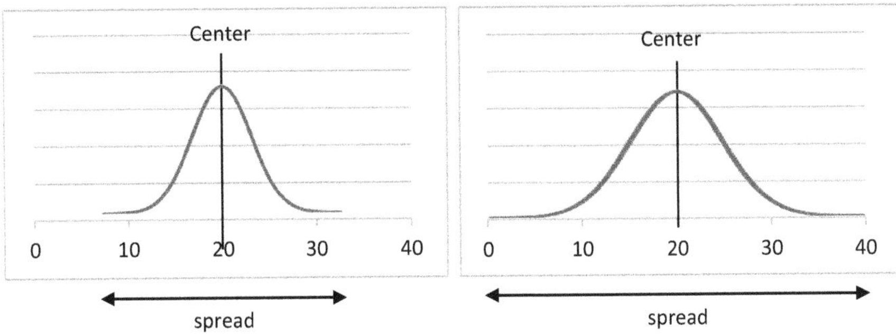

Fig. 3.2 Center and spread

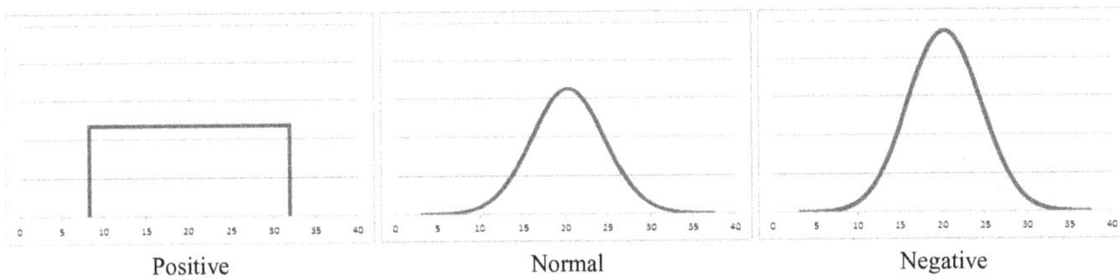

Fig. 3.3 Kurtosis

Kurtosis: Kurtosis refers to how peaked a distribution is or conversely how flat it is. If there are more data values in the tails than what you expect from a normal distribution, the kurtosis is positive. Conversely if there are less data values in the tails than you would expect in a normal distribution, the kurtosis is negative (Fig. 3.3).

- Excel cannot calculate this statistic unless you have at least four data values.

Skewness: Skewness describes the lack of symmetry in the distribution. If the data is pulled to the right, creating a right-hand tail, it is a positive skew. Positively skewed data has a mean greater than the median. Conversely if the data is pulled to the left, creating a left-hand tail, it is a negative skew and the mean is less than the median (Fig. 3.4).

- Excel cannot calculate this if you input less than three data values.

Standard error: This is a measure of the uncertainty about the mean. This becomes an important statistic when working with inference analysis, which is discussed in Chapter 7.

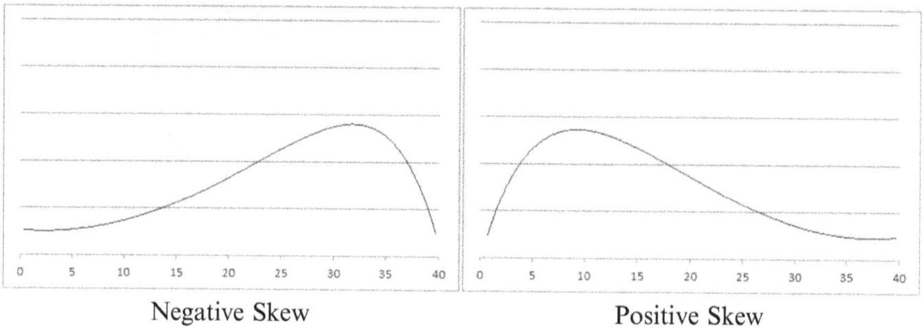

Fig. 3.4 Skewness

Symbols

Σ = Greek capital letter Sigma, which means "take the sum of" or "sum together" everything that follows the sign

$\bar{x}$ or **μ** = average, mean

N or **n** = number of data values in your data set; to be accurate, we use capital **N** for a population (all the possible data) and lower case **n** for a sample (some of the data)

σ or **SD** or **sd** = standard deviation; to be accurate, we use capital **SD** for the standard deviation of a population (all possible data) and lower case **sd** for a sample (some of the data)

Sample variance: This is just the standard deviation squared.

The Histogram

Histograms are used for charting frequency data. These graphs or charts use individual categories, bins, or classes to count how many times a variable occurs. We can also think of the histogram as more of a continuous or connected graph (see Fig. 3.5).

We see from Fig. 3.5 that if we use an infinite or unlimited number of narrower and narrower bins we would eventually approximate a continuous distribution.

Excel

You can run on several columns simultaneously.

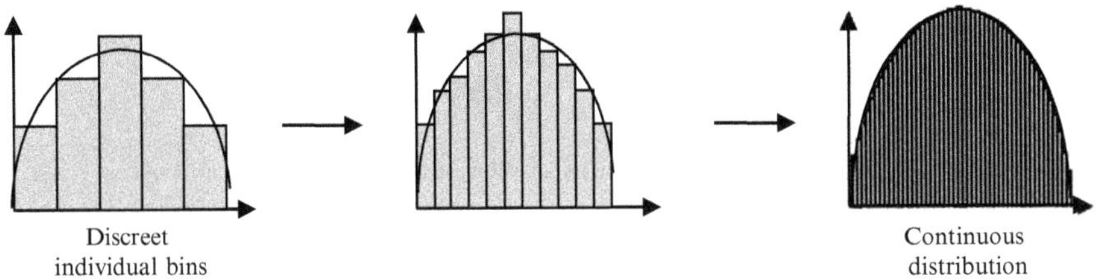

Fig. 3.5 Histogram

Descriptive Statistics

- Input the following data into Column A cells A1 through A12: 99, 150, 300, 780, 350, 500, 510, 360, 200, 175, 196, 450
- From the **Data** tab, choose the **Data Analysis** function
- Select the **Descriptive Statistics** option, under Analysis Tools
- Click **OK**

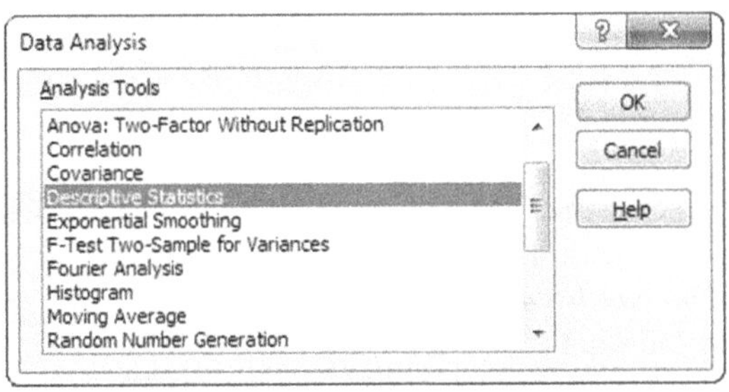

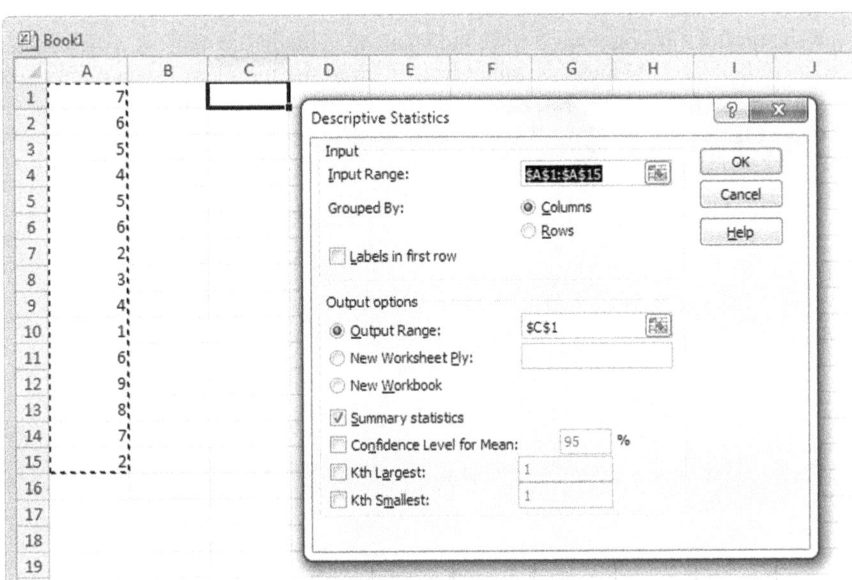

- Enter the range of data in **Input Range**:

- Choose the option that fits your data in **Grouped By**: [columns or rows]

- Select **Output Range**:
- Click inside the text box.
  - Important to do this ***first*** or you will create problems in your **Input Range**. Then select the output range.

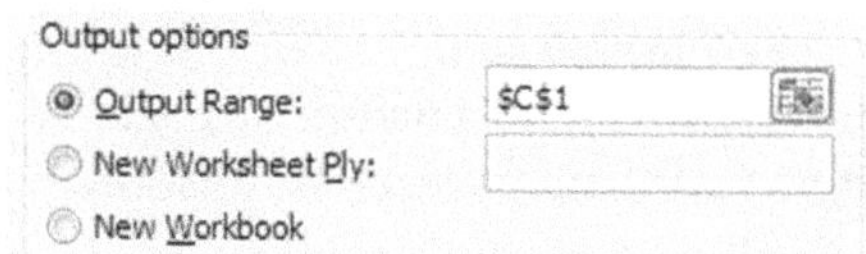

- Click **Summary Statistics**

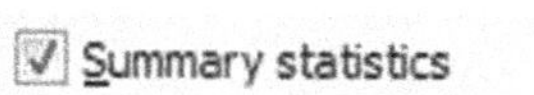

- **Confidence Level for the Mean**: Excel default is 95%, which should be acceptable for all of your work. However, you can input any value greater than 0 and less than 100. ***You can leave this unchecked.***
- **Kth largest**: If you want the third largest value in the data set, you input 3. The default is 1. Values can range from 1 up to the number of data points you have in the distribution. ***You can leave this unchecked.***
- **Kth smallest**: If you want the second smallest value in the data set, you input 2. The default is 1. Values can range from 1 up to the number of data points you have in the distribution. ***You can leave this unchecked.***
- Click **OK**

Descriptive Output Results

Book1

	A	B	C	D	E
1	7		Column1		
2	6				
3	5		Mean	5	
4	4		Standard I	0.601585	
5	5		Median	5	
6	6		Mode	6	
7	2		Standard I	2.329929	
8	3		Sample Vi	5.428571	
9	4		Kurtosis	-0.75623	
10	1		Skewness	-0.11729	
11	6		Range	8	
12	9		Minimum	1	
13	8		Maximum	9	
14	7		Sum	75	
15	2		Count	15	
16					

The most important output value is the **Count**; this is your check that all of the data has actually been input and used in the calculations.

But when it is difficult to read the names of the statistics, we need to adjust the column width.

Changing the Width of the Column Output

Let's make the column width large enough to show all the words.

- Place your cursor on the line at the far right of the top cell in the column you want to make wider. In this case, we want Column C to be wider.
  Before:

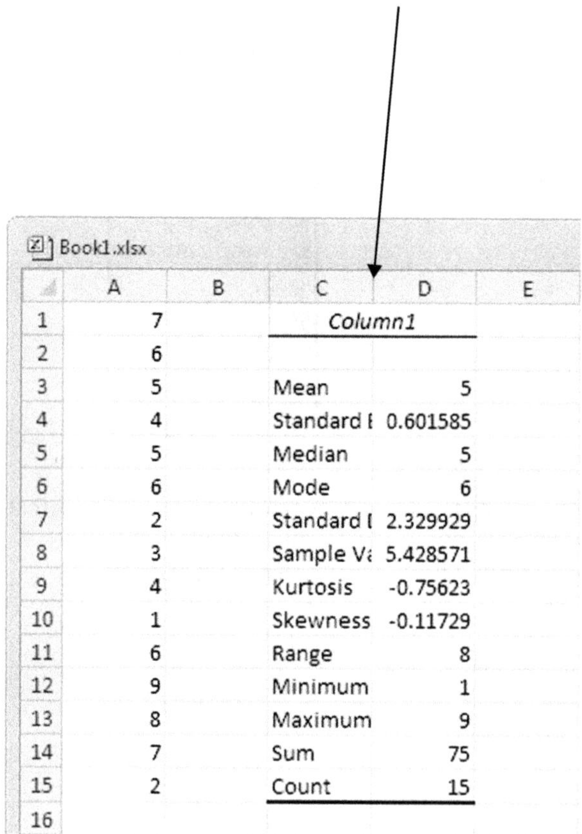

Book1.xlsx

	A	B	C	D	E
1	7		*Column1*		
2	6				
3	5		Mean	5	
4	4		Standard E	0.601585	
5	5		Median	5	
6	6		Mode	6	
7	2		Standard I	2.329929	
8	3		Sample Va	5.428571	
9	4		Kurtosis	-0.75623	
10	1		Skewness	-0.11729	
11	6		Range	8	
12	9		Minimum	1	
13	8		Maximum	9	
14	7		Sum	75	
15	2		Count	15	
16					

- Drag this border line to the width you want for the column.

After:

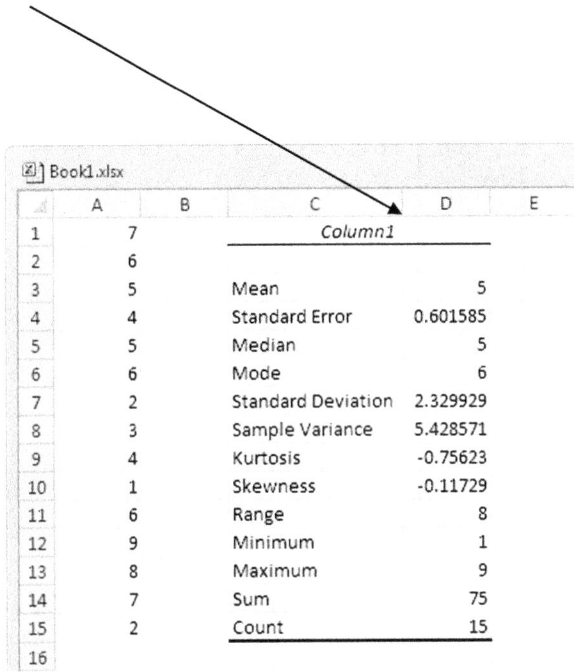

OR:

- Right click the top cell of the column you want to make wider. Select **Column Width.** You can make the column as wide as you need.

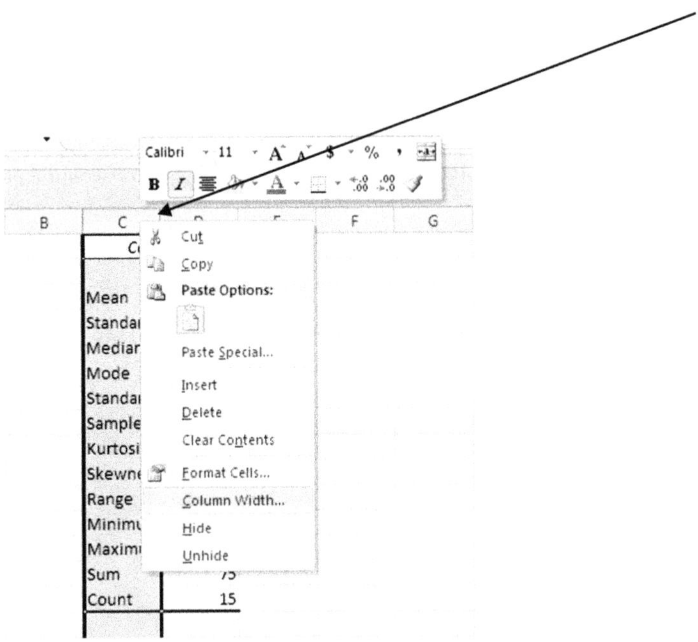

- Type in the desired column width in the popup box.

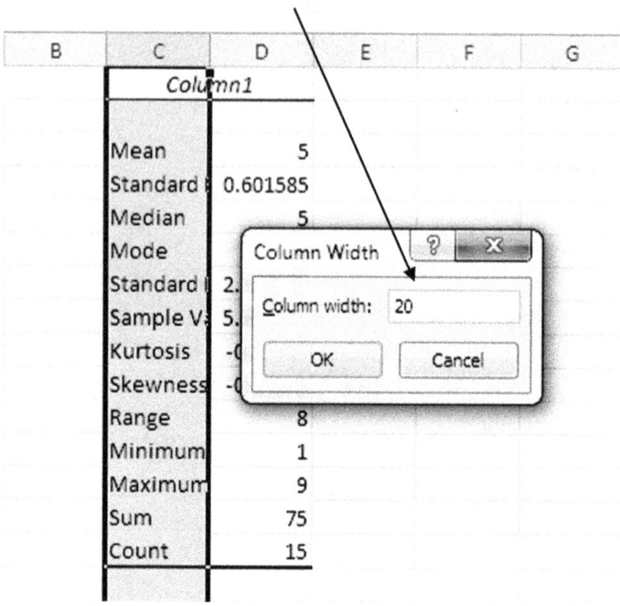

- You can also auto format the column width by double clicking the line on the right side of the column label, which is above the first cell, of the column you want to make wider.

Calculate Rank and Percentile (Rank and Percentile)

Example Problem

You would like to calculate the rank and percentile costs of airplane tickets purchased by employees over the past 3 months. The average cost is $485 and the standard deviation is $260. You know that the ticket prices are normally distributed based on historical data. Input the following ticket prices that are sorted from smallest to largest:

Cost of airplane tickets ($)
65
105
145
185
225
265
305
345

(continued)

Cost of airplane tickets ($)
385
425
465
505
545
585
625
665
705
745
785
825
865
905

- Select the **Data** tab

- Open the **Data Analysis** function

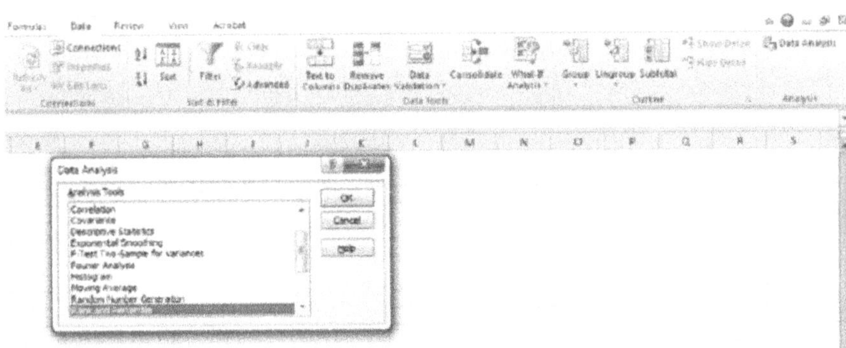

- Select the **Rank and Percentile** function

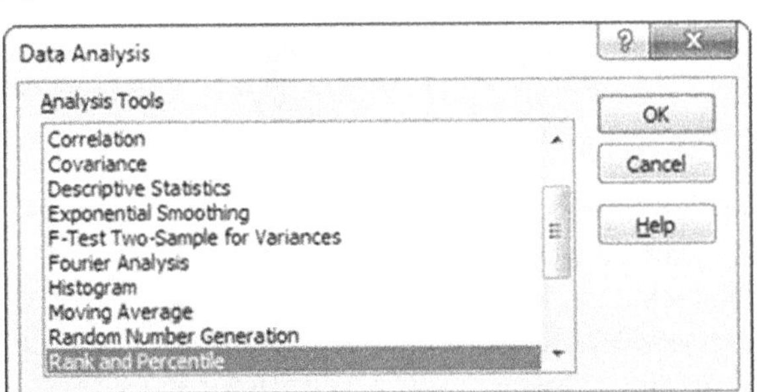

- Type in your column of x data

- Select the how your data is grouped: **rows** or **columns**

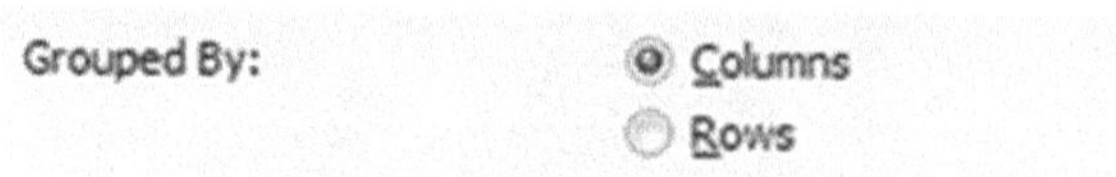

- Select an output location on your Excel spreadsheet

- Click **OK**

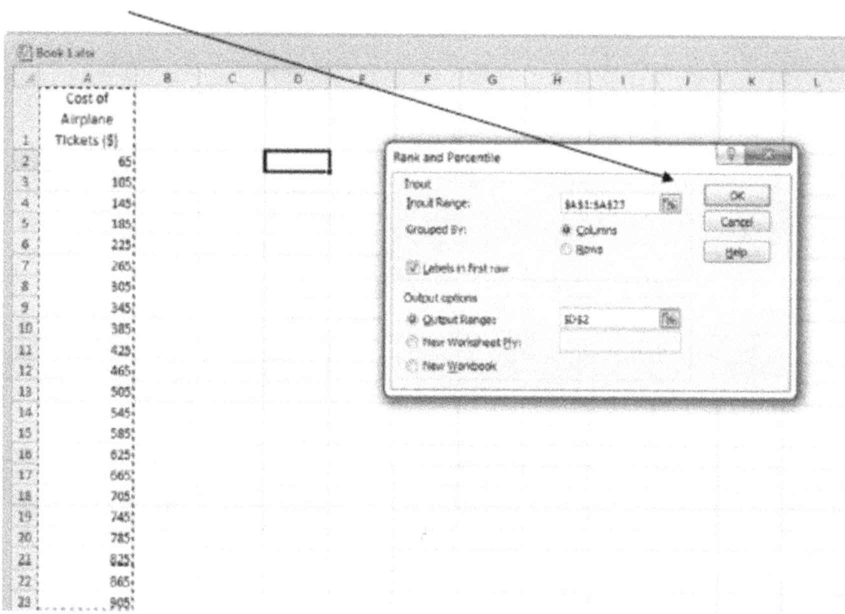

Output

- The first column of data is just the rank order from highest to lowest.
- The second column includes the actual data value.
- The third column has the **Rank** from lowest to highest.
- The final column is the **Percentile** moving from the highest value (right-hand side of the distribution) to the lowest *x* value (left-hand side of the distribution) in the distribution.

Point	Cost of airplane tickets ($)	Rank	Percent
22	905	1	100.00
21	865	2	95.20
20	825	3	90.40
19	785	4	85.70
18	745	5	80.90
17	705	6	76.10
16	665	7	71.40
15	625	8	66.60
14	585	9	61.90
13	545	10	57.10
12	505	11	52.30
11	465	12	47.60
10	425	13	42.80
9	385	14	38.00
8	345	15	33.30
7	305	16	28.50
6	265	17	23.80
5	225	18	19.00
4	185	19	14.20
3	145	20	9.50
2	105	21	4.70
1	65	22	0.00

From the output, we can determine that 47.6% of the prices were at the mean ($485) or less. Similarly we note that about 90% are $825 or less. The 10th percentile is approximately $145. The ticket price of $705 is the sixth most expensive ticket. There are numerous analytical conclusions that can be drawn from this simple data output.

Using Excel Functions

We can also calculate the descriptive statistics one by one using the separate functions in Excel.

- Click the ***fx*** symbol in the white function bar

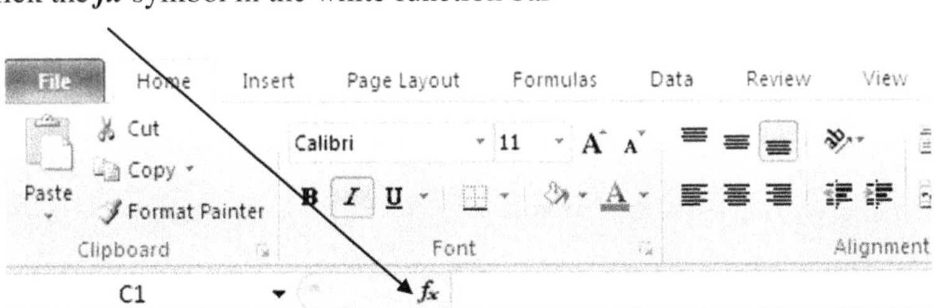

- Input the description of what you want to do if you do not know the name of the statistical function, **OR** click on the next box and select **Statistical**, if you do know the name of the function
- Select the function you need

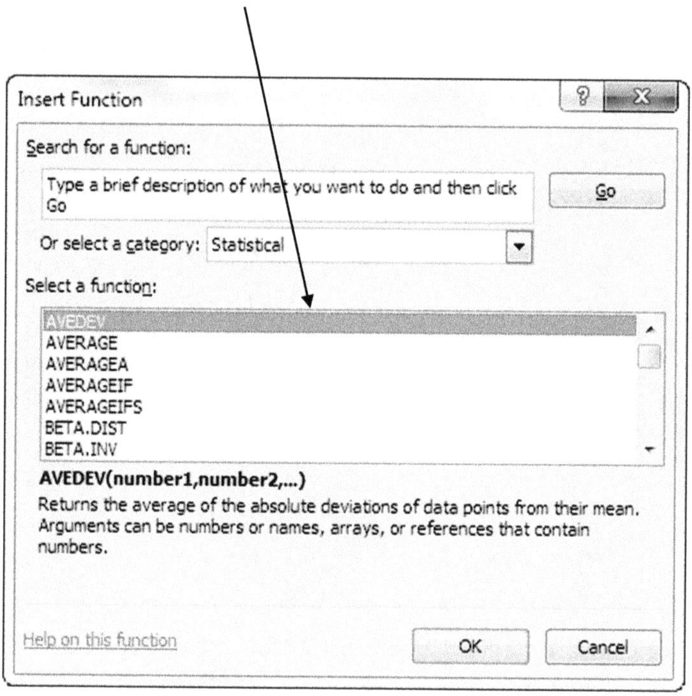

- Click **OK**

Be careful to choose the exact function as listed in Fig. 3.6. Excel has many functions that are only different by one letter and each of those functions will deliver a different result.

Histograms

Excel provides a table with the frequency data as well as the actual histogram graph. Remember a histogram does not have spaces between the bars because both the *x* and *y* axes are scales and show quantitative data.

Statistic	Function name	Calculates
Mean	AVERAGE	Arithmetic average of the data set
Median	QUARTILE QUARTILE.INC QUARTILE.EXC (You need to input 2 for the quartile value)	Median of the data set
Mode	MODE MODE.MULT MODE.SNGL	Mode of the data set
Standard Deviation	STDEV STDEV.P STDEV.S STDEVA STDEVPA STDEVP	Standard deviation of the data set (both population standard deviation and sample standard deviation)
Range	MIN MAX	Range of the variables
Count	COUNT	Total count
Skewness	SKEW	Calculates how skewed the data is
Kurtosis	KURT	Calculates the kurtosis of the data
Kth largest	LARGE	Displays the "K" largest number in your data set
Kth smallest	SMALL	Displays the "K" smallest number in your data set

Fig. 3.6 Functions

Example Problem

You have collected some proficiency scores for the administrative assistants in your department. The proficiency scale ranges from 0 to 1000, with 0 being the lowest possible score and 1000 being the highest possible score. You decide that the best way to get a sense of the data is to graph the data in a histogram. Here are the steps you would need to take.

Input the following data in column B

Data
99
150
300
780
350
500
510
360
200
175
196
450

Setting Up the Bin Ranges

Before you create the histogram, you need to identify your bin/category/class ranges. As a guiding principle, charts have somewhere between 5 and 15 bins. Too few bins or too many bins creates confusion.

Arbitrarily choose a certain number of bins. Let's choose seven bins. To get the width of those bins, determine the range of the data and divide by the number of bins you want. In this example, the range goes from \$99 to \$780 (Fig. 3.7).

$$7\,\text{bins}$$
$$\text{Range } \$99 - \$780$$
$$\text{Bin width} = (\$780 - \$99)/7 = \$97.28$$

Because using bin widths of \$97.28 would be confusing, we can round up or down to choose an easier bin width, such as \$100.

In a new column, type in the largest number in each of the bins to indicate the end point of each bin. In Fig. 3.7, all of the data ≤ 100 will be reported in that first bin. All of the data greater than 100 and ≤ 200 will be reported in that second bin. Remember these are the end points of the bins you input to Excel.

Keep track of where you have entered your bin ranges as you will need this as an input field for the histogram.

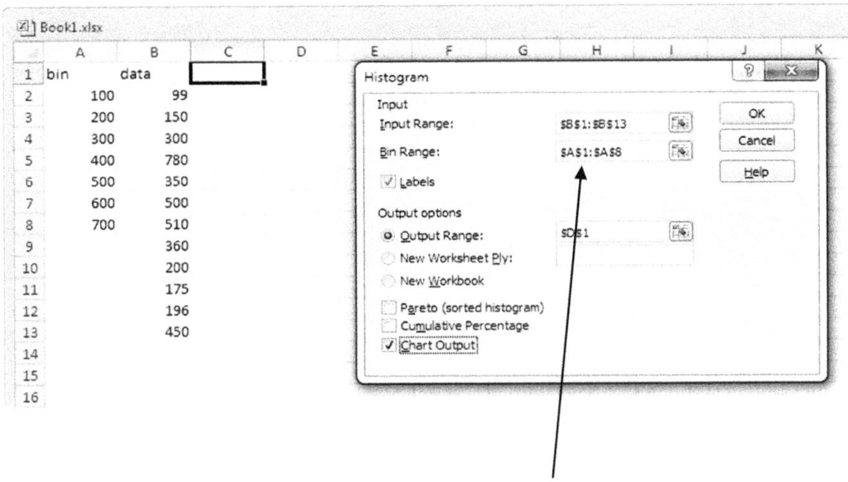

Fig. 3.7 Input bin range

Creating the Histogram Chart

Now you are ready to create a histogram.

- Under the **Data** tab click on the **Data Analysis** function on the far right
- Select **Histogram** from the list of options under Analysis Tools and click **OK**
- Highlight the **Input Range** for the data

- Highlight the range of cells where you have typed the end points of the bins

- Click on the **Output Range**: button and then *click inside the box*. Now highlight a cell where you want the output to start

- Check **Chart Output** or you won't get a graph

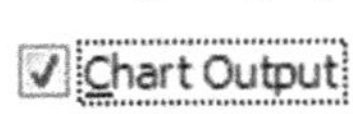

- Click **OK**

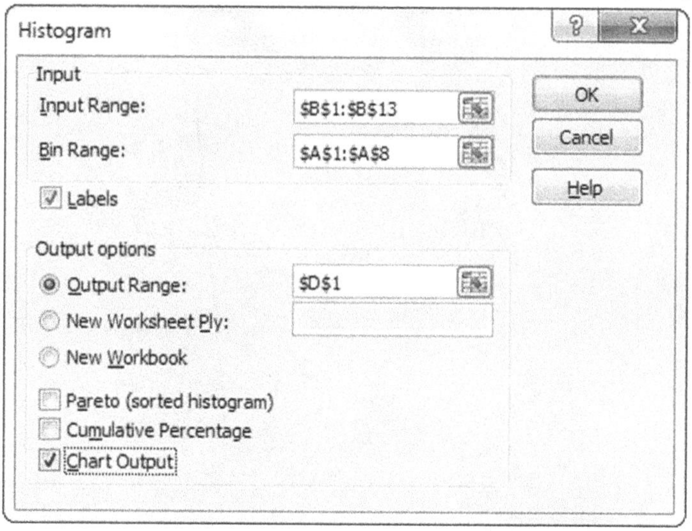

Histogram Clean Up

The actual histogram needs some cleaning up.

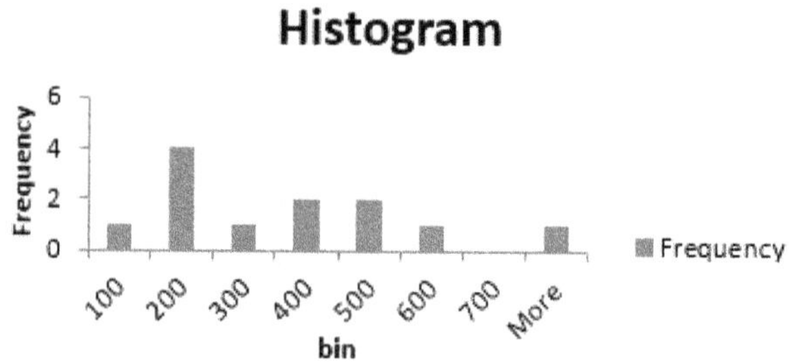

Closing Gaps Between Bars

- Right Click in any column on the graph
- Click the **Format Data Series** option from the menu

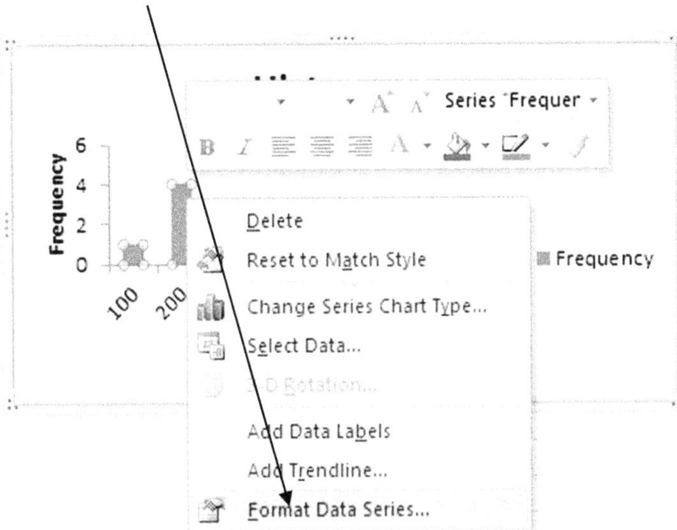

- Go to **Series Options > Gap Width**. *Set to 0%*

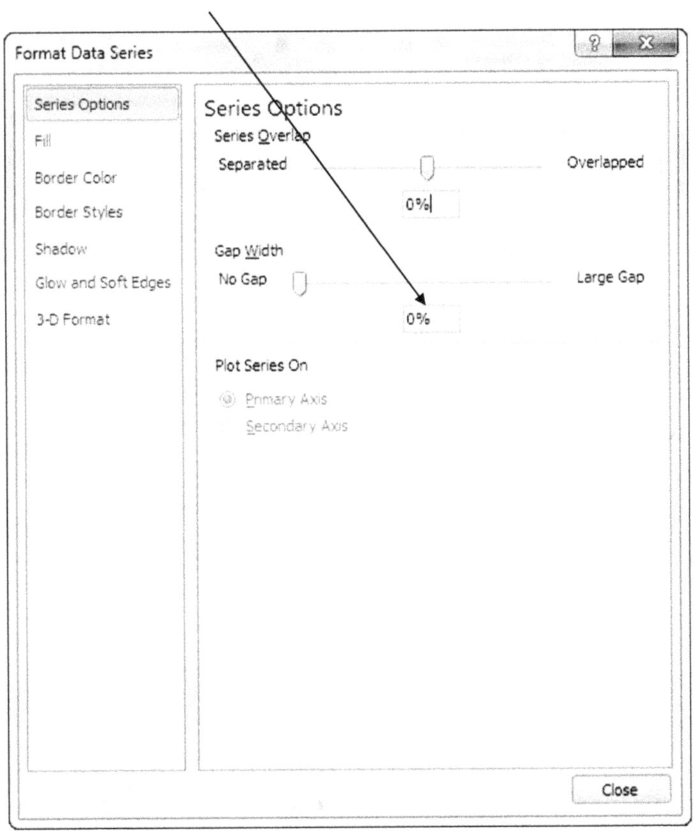

Change Labels on x-*Axis*

- Change the *x*-axis values from ***upper limits*** of the bins to ***midpoints*** of the bins
- Create the column of midpoint labels you want added to your graph
  - ***NOTE***: The data labels on the *x*-axis indicate the upper limits of the bins. In order to change this to the midpoints, first you need to create a column of data with the midpoints of the bins.

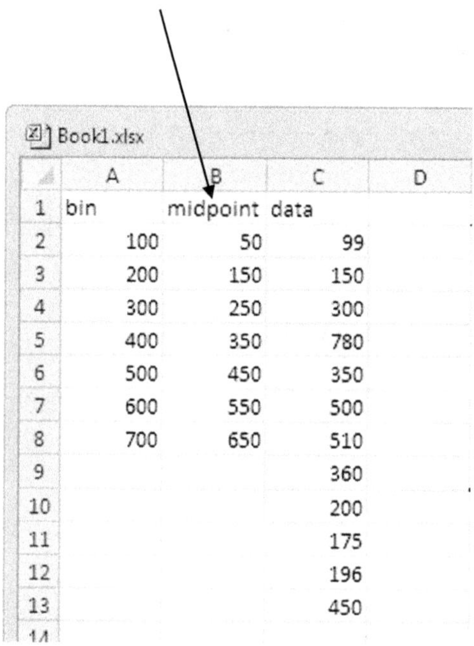

	A	B	C	D
1	bin	midpoint	data	
2	100	50	99	
3	200	150	150	
4	300	250	300	
5	400	350	780	
6	500	450	350	
7	600	550	500	
8	700	650	510	
9			360	
10			200	
11			175	
12			196	
13			450	

- Right click on any column on the graph
- Choose the **Select Data** option

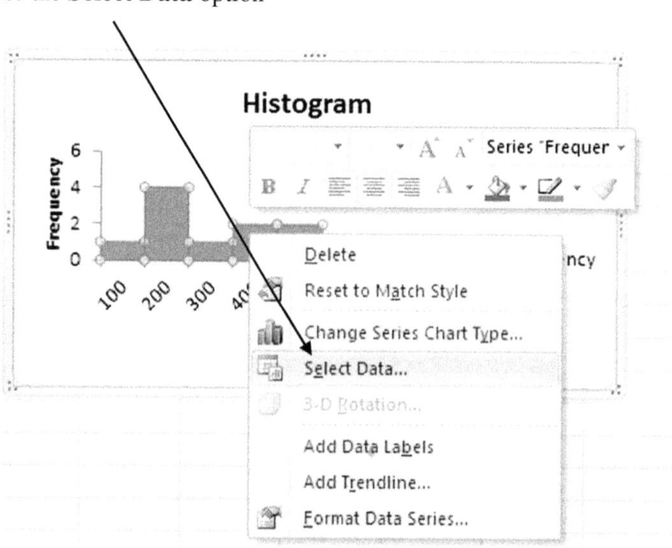

- Under **Horizontal Axis Labels,** click the **Edit** option

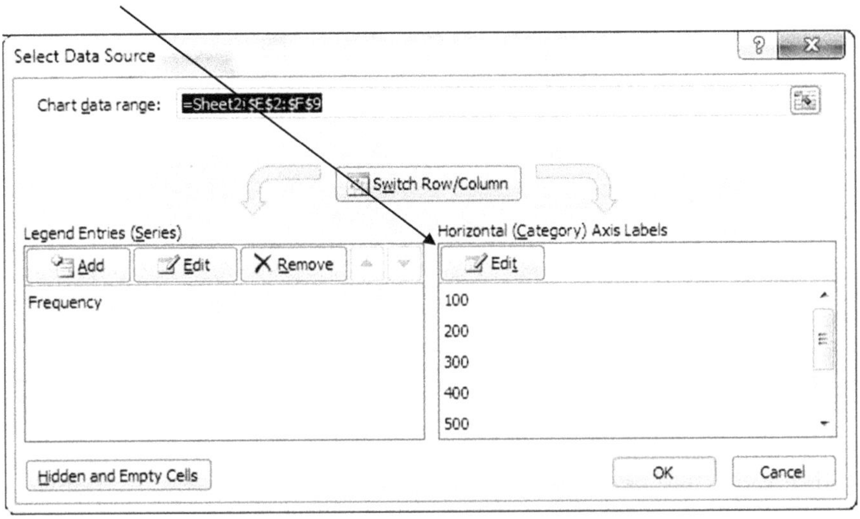

- Click inside the **Axis label range** box. Highlight the new column of midpoint labels

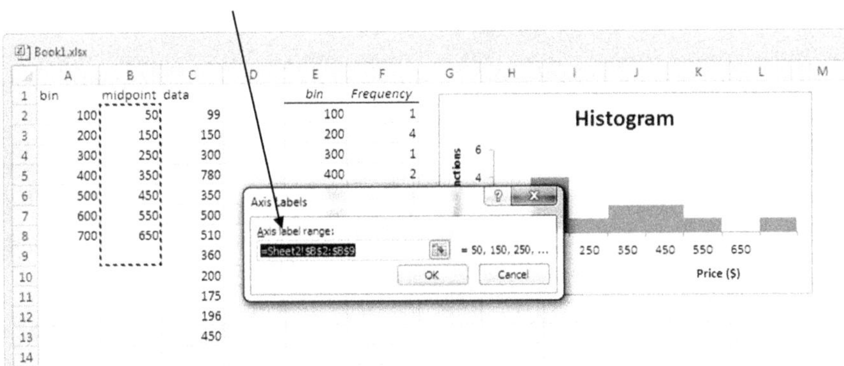

- Click **OK**

Removing More from the Chart and Labeling the Last Column

The word **More** may appear on your graph.

The last column may not be assigned a numeric label. This is because there are data values greater than the last bin value you originally input. In Fig. 3.8, the last bin end value originally selected was 700. In Fig. 3.9, this was changed to a midpoint bin value of 650. But as you can see from the data, there is one value greater than 700. Excel automatically names this last bin **More** and will not assign a numeric label for it on the x-axis.

You can go back and add more bins to ensure all of your data is being included on your graph with the proper labels. In this case, we added **750**.

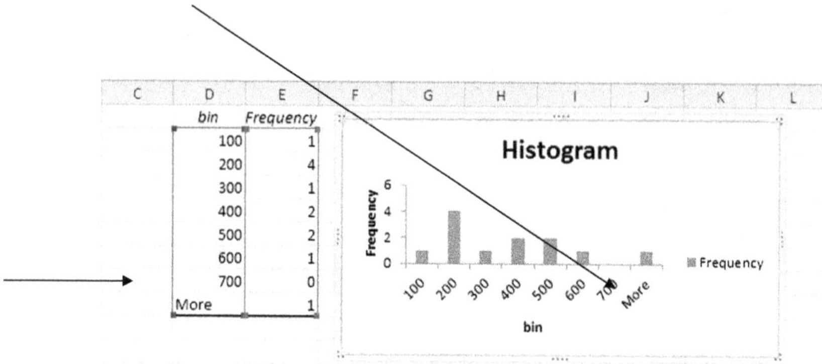

Fig. 3.8 End point bin values

Fig. 3.9 Midpoint bin values

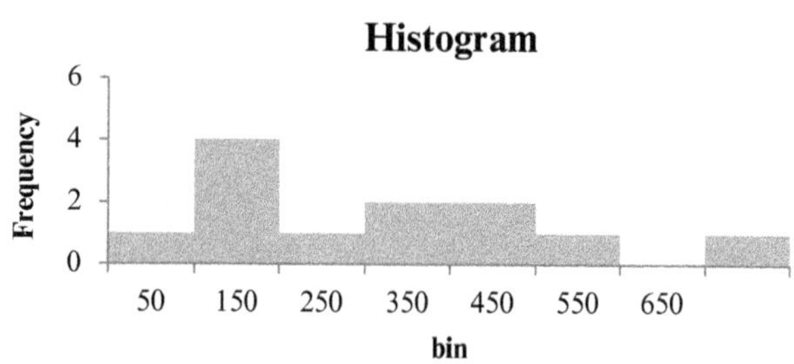

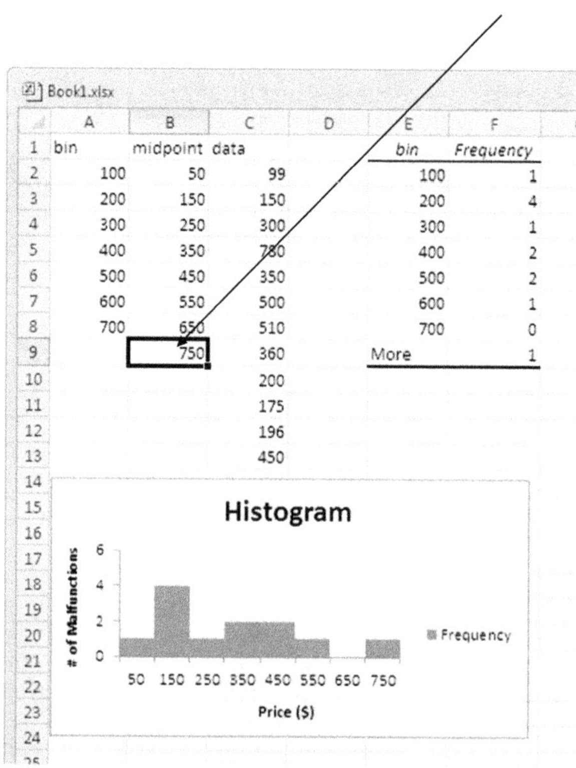

Once you have added the extra bins, edit the bin values that will remove the word "More" from the graph.

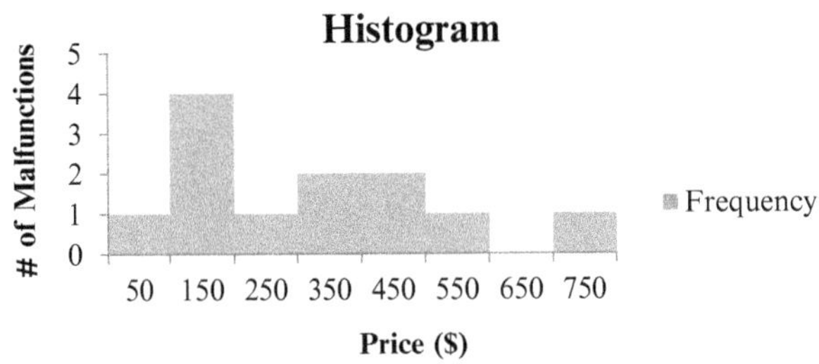

- Changing the bin values will only change the labels on your x-axis. The frequency will still count the data using the initial points you entered, not the midpoints.

Remove Legend

- Right click on the legend and click the **Delete** option

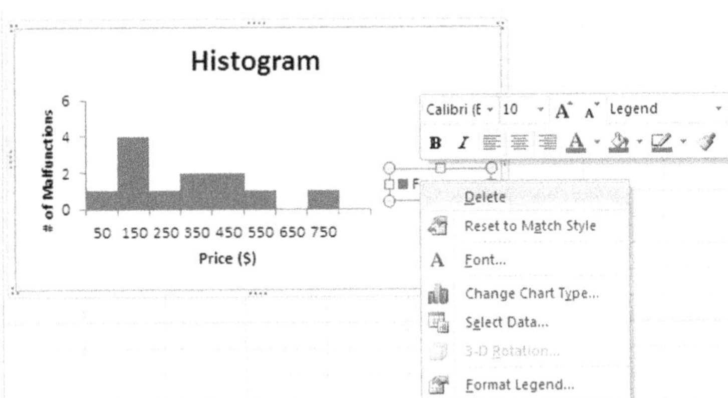

Axes Labels

- To change the axes labels, click the frequency label (y) or bin label (x) and type the appropriate titles

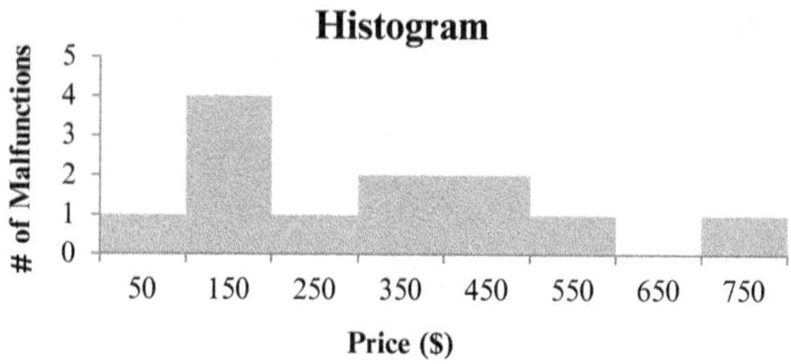

Moving Axes Labels

Left click on the axis you want to move; drag and drop

Before

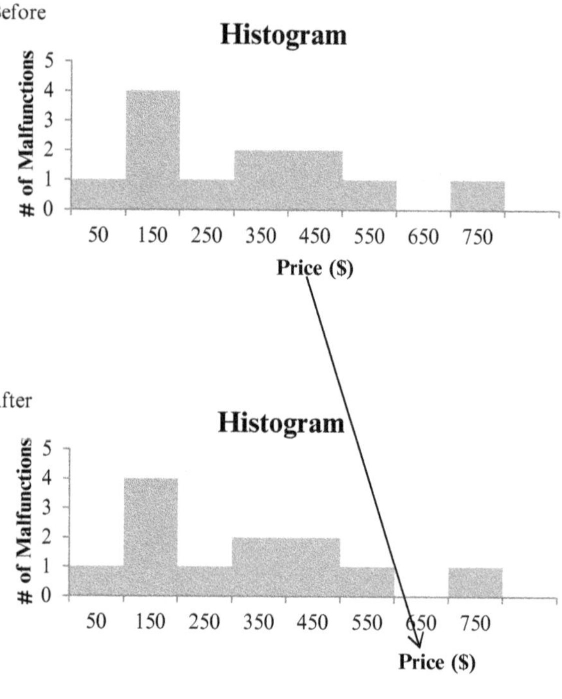

After

Changing the Bar Color

- Right click on any column. Choose the **Format Data Series** option

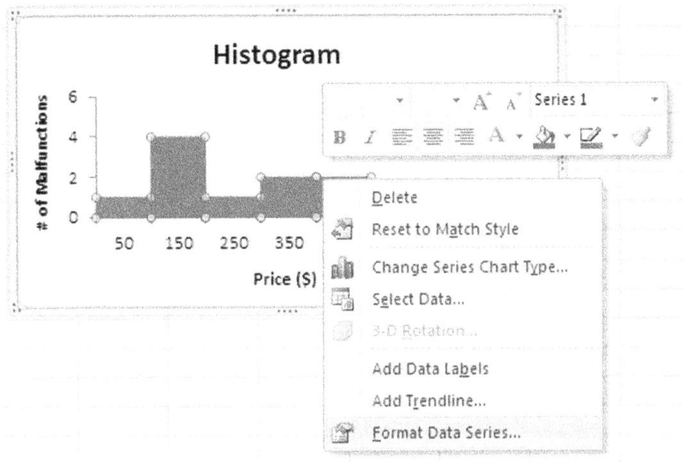

- Click **Paint Bucket** icon, then click **Fill**

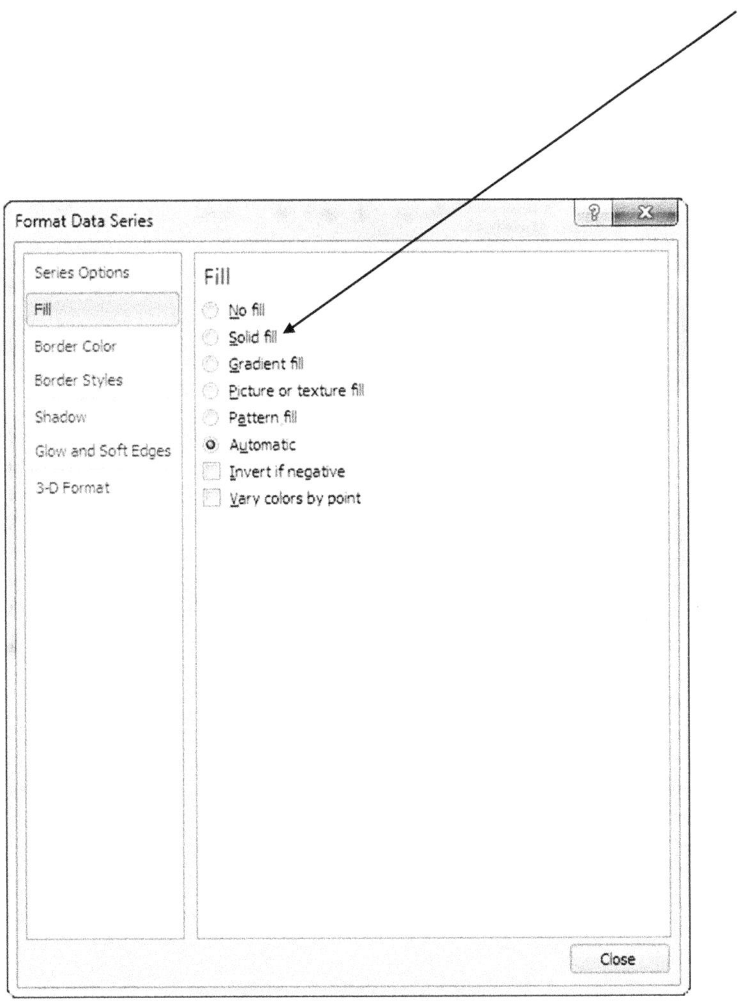

- Change from **Automatic** to the fill you prefer for the bins; if you click on **Solid Fill,** you can choose from a variety of solid colors by clicking on the color

Changing Chart Title

- Right click on the chart title. Select the **Edit Text** option; you can retype the title as you choose but make sure to click outside the text box when done typing

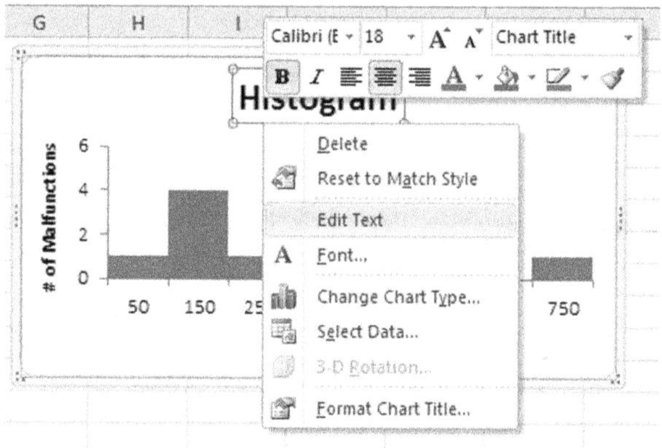

Changing Chart Background Fill

- Right click on the chart background. Click the **Format Chart Area** option

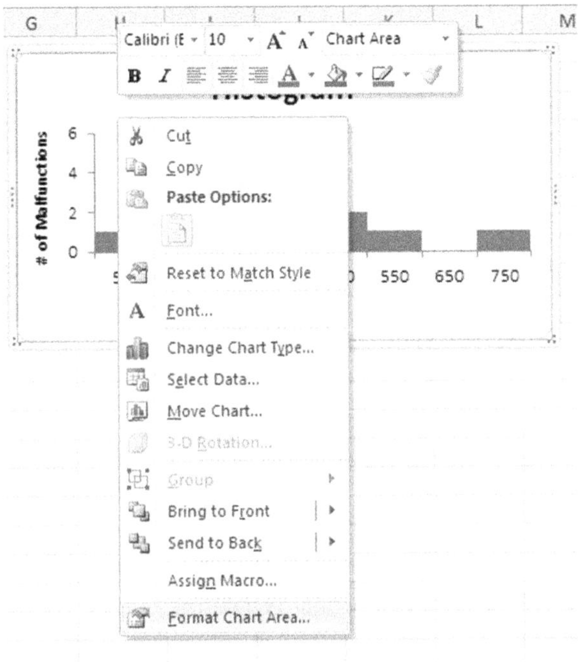

- Click **Paint Bucket** icon, then click **Fill**

- Select the type of fill and color options you desire

Rotating the y*-Axis Label from Vertical to Horizontal*

- Right click on the *y*-axis label. Click on the **Format Axis Title** option

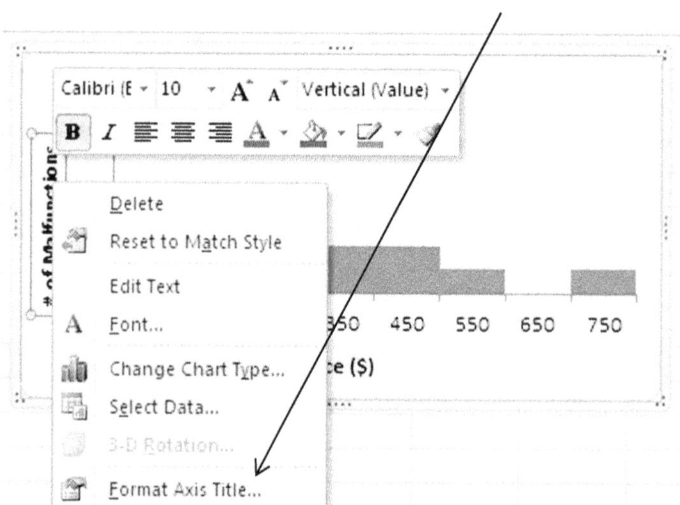

- Click **Size & Properties** icon from the right-hand menu; you can change the alignment

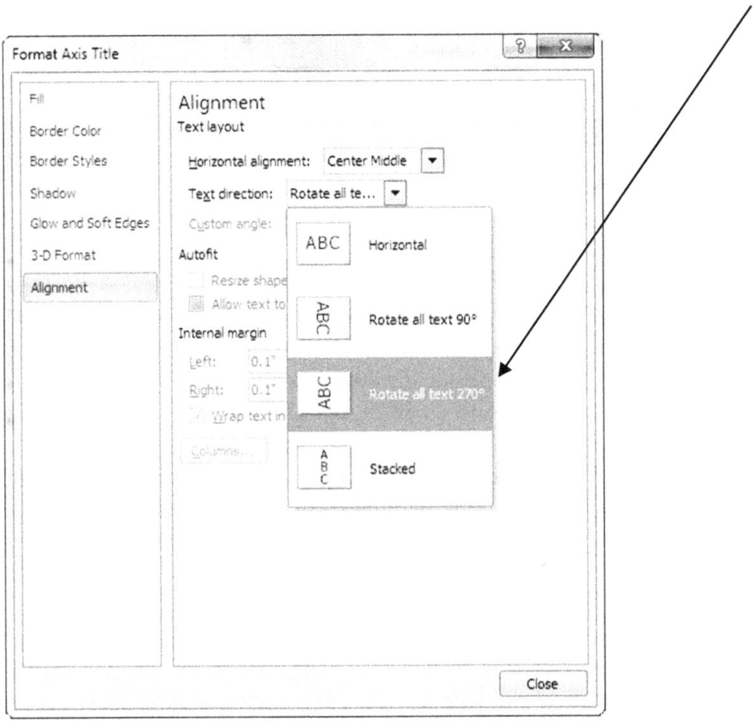

Common Pitfalls

- Be very careful when calculating the mode. When there is more than one value with the same frequency (more than one mode) in a data set, Excel will report the value that appears first in your data set. If each data value appears with the same frequency, Excel will report **#N/A**.
  - {1,1,1,2,2,2,3,3,4} Excel will report that the mode is 1
  - {1,2,3,3,3,4,5,6,7} Excel will report that the mode is 3
  - {1, 2, 3, 4, 5, 6} Excel will report #NA
- The formula used in Excel for STDEV.S to calculate the standard deviation uses a denominator of $(n - 1)$, whereas STDEV.P uses the denominator of (N). The P stands for population and the S stands for sample. When you calculate the standard deviation by hand, you typically use (N) in the denominator of a population and $(n - 1)$ for a sample. These two methods will yield slightly different answers, so be careful when deciding which formula to use.

- Label all axes with variable names and units. Excel will not label the axes by default.
- Excel reports the maximum value for each interval as the bin value not the midpoint.
- Excel will include a bin labeled **More**. Rename the **More** bin as a numerical label to avoid confusion.
- Excel labels all frequency graphs as histograms. Remember bar graphs can have spaces between bins. Histograms have no spaces between bins. You need to close the gaps in the histogram to zero.
- Be careful to tell Excel if you are including labels as the first set of cells in the input. Always check the count (n) to make sure you have all of your data is included in the Excel calculations.
- If you expand your output area, Excel will keep repeating the results to fill up the area. In Fig. 3.10, the data labeled as Column 1 is the same output as Column 2. There is no need to show duplicate data. You can avoid this by keeping the output area large enough for just one set of output data. To expand the columns, you need to right click on the top of the column and change the width. If you simply drag the lowered left corner of the output box, you will generate repeated data.
- Too many decimal places. Excel will calculate as many decimal places as you indicate in the output format. If no decimal limit is set, Excel will fill the cell and show nine decimal places. In Fig. 3.11, the input data does not have any decimal places and yet the Excel output is reported in nine decimal places. This obviously does not make any sense; Excel is reporting a greater number of significant figures than we had recorded in the data. You cannot magically create a higher level of accuracy and precision in your data than exists in the data you are using for the analysis. The rule of thumb is to allow one more decimal place in the output than exists in the input data.

Column1		*Column2*		*Column3*
Mean	5.333333333	Mean	5.333333333	Mean
Standard Error	0.503952631	Standard Erroı	0.503952631	Standard Error
Median	5	Median	5	Median
Mode	5	Mode	5	Mode
Standard Deviation	1.951800146	Standard Devi	1.951800146	Standard Deviation
Sample Variance	3.80952381	Sample Variar	3.80952381	Sample Variance
Kurtosis	-0.464423077	Kurtosis	-0.464423077	Kurtosis
Skewness	0.123160466	Skewness	0.123160466	Skewness
Range	7	Range	7	Range
Minimum	2	Minimum	2	Minimum
Maximum	9	Maximum	9	Maximum
Sum	80	Sum	80	Sum
Count	15	Count	15	Count

Fig. 3.10 Output area. Output from dragging the lower right-hand corner of the output area

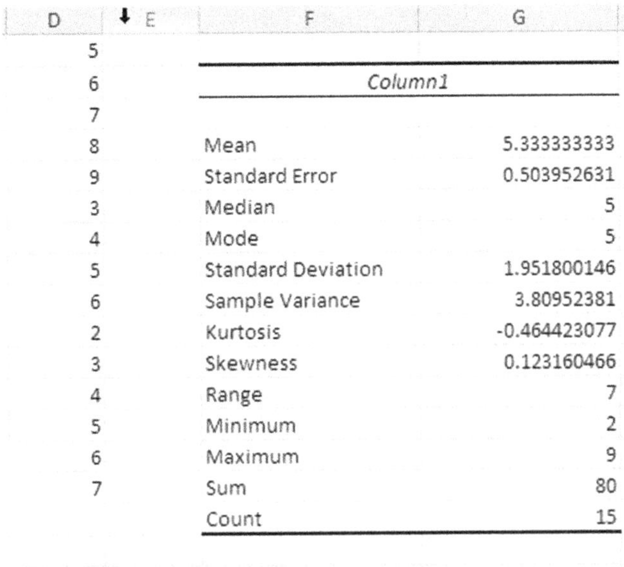

Fig. 3.11 Decimal places

- If you check **Labels** in the histogram input window, you need to include a label in your end point bin data; otherwise, Excel will assume the first bin end point is a label and will not use this value in the calculations.

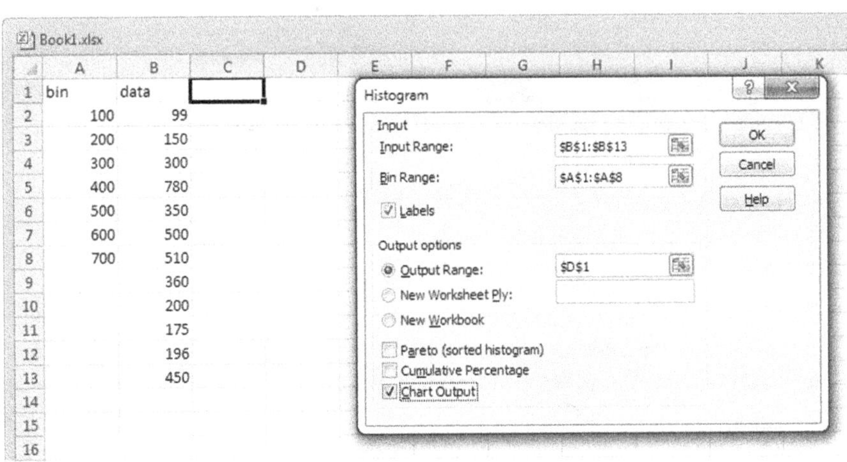

- If you don't check the Labels box, and you include labels in your input range, Excel will create a warning message.

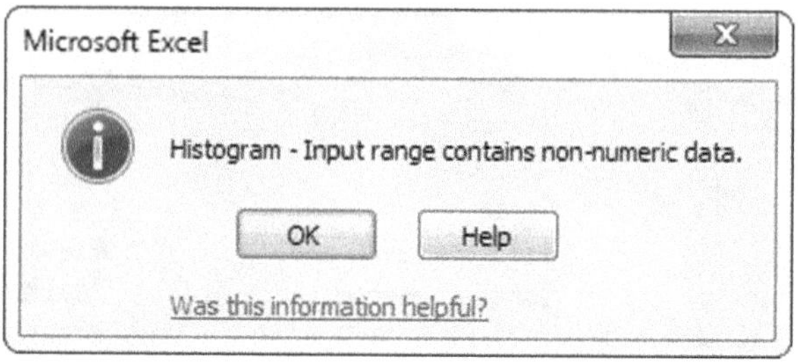

Histogram Checklist

- Do not include a legend unless it is really necessary.
- Label both axes with units and variable names.
- Make the graph easy to read; clean it up.
- If you will be copying the graph in black and white, make sure you can still differentiate data without the color. You can use different patterns rather than different colors.

Final Thoughts and Activities

Practice Problems

1. The data in the **Beverage** file represents the amount of fluid in a sample of 50x 2-L tincture medicine bottles. Using the data from the file, complete the following exercises.
   (a) Construct a histogram.
   (b) From the output, what is the mean?
   (c) From the output, what is the standard deviation?
2. The data in the **Electric** file represents the cost of electricity in June for a random sample of 50x one-bedroom apartments in a large city in the southern portion of the United States. Using the data from the file, complete the following exercises.
   (a) Construct a histogram and run descriptive statistics.
   (b) Around what amount does the monthly electricity cost seem to center?

3. Match the histograms to their associated distribution types: right skew, bimodal, random, left skew, normal

4. What histogram bin sizes would you select for the following data set, which shows student scores on a test?

	A	B
2	Student_Id	Scores
3	1001	41
4	1002	54
5	1003	68
6	1004	92
7	1005	77
8	1006	36
9	1007	54
10	1008	39
11	1009	20
12	1010	59
13	1011	65
14	1012	49
15	1013	23
16	1014	71
17	1015	52
18	1016	84
19	1017	83
20	1018	68
21	1019	63
22	1020	59
23	1021	27
24	1022	50
25	1023	99
26	1024	24
27	1025	61
28	1026	91
29	1027	27
30	1028	53
31	1029	44
32	1030	81

5. If this output was presented for your review, what suggestions might you provide to improve the final chart? Provide your reasons.

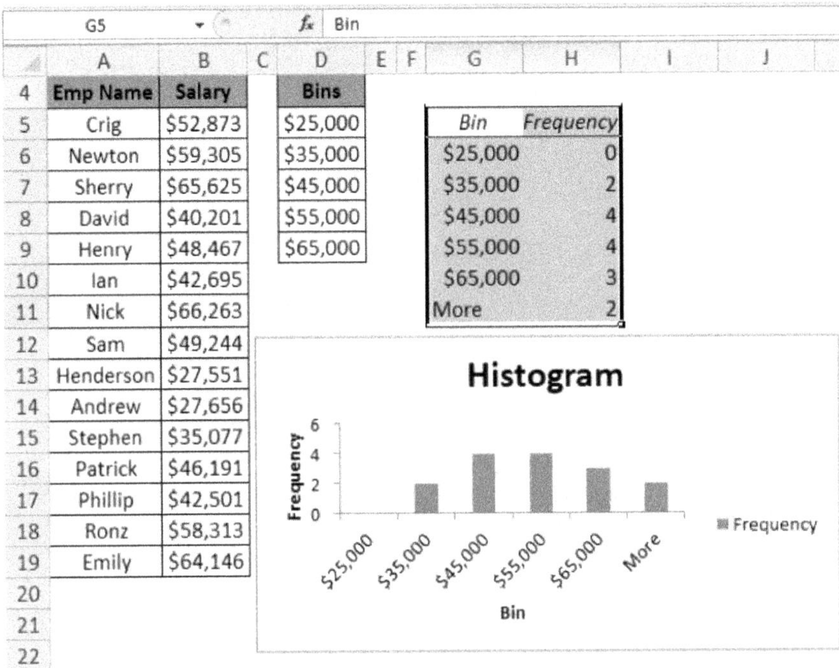

G5 | f_x Bin

	A	B	C	D
4	Emp Name	Salary		Bins
5	Crig	$52,873		$25,000
6	Newton	$59,305		$35,000
7	Sherry	$65,625		$45,000
8	David	$40,201		$55,000
9	Henry	$48,467		$65,000
10	Ian	$42,695		
11	Nick	$66,263		
12	Sam	$49,244		
13	Henderson	$27,551		
14	Andrew	$27,656		
15	Stephen	$35,077		
16	Patrick	$46,191		
17	Phillip	$42,501		
18	Ronz	$58,313		
19	Emily	$64,146		

Bin	Frequency
$25,000	0
$35,000	2
$45,000	4
$55,000	4
$65,000	3
More	2

6. This histogram tells an effective story without numbers; why does this work in this case?

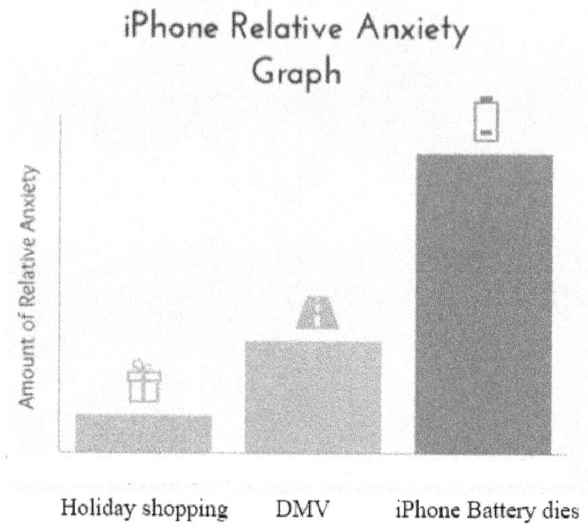

7. What's wrong with this histogram?

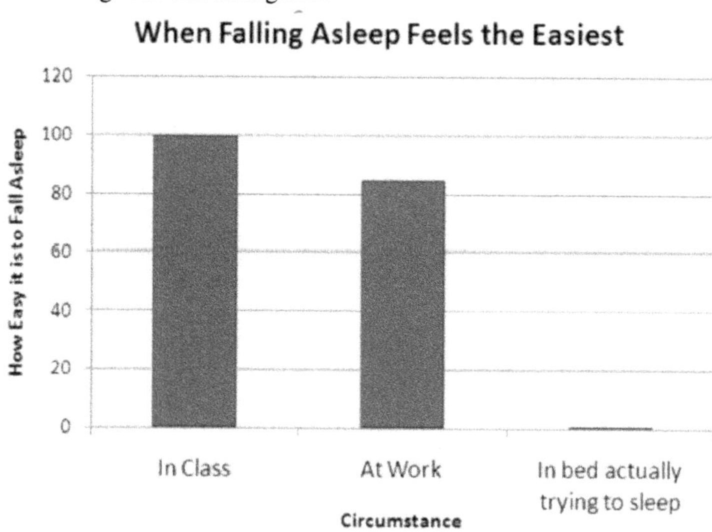

8. Data was charted to show the % sales for four different employees in the same division of the company. Although both graphs use the same data set, employee 2 prefers the first chart be shown. What's going on here? How could this problem be avoided by using a histogram?

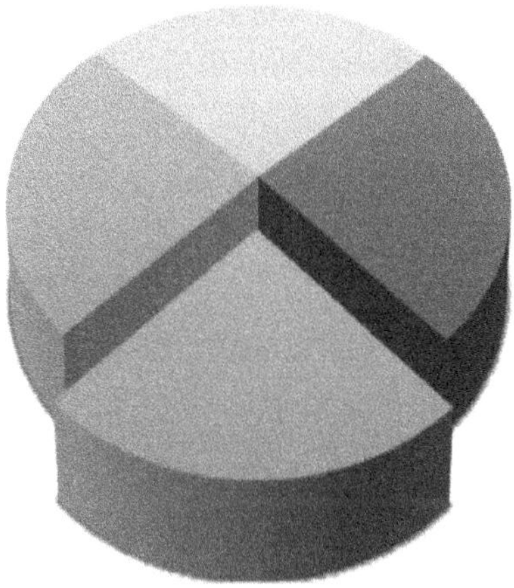

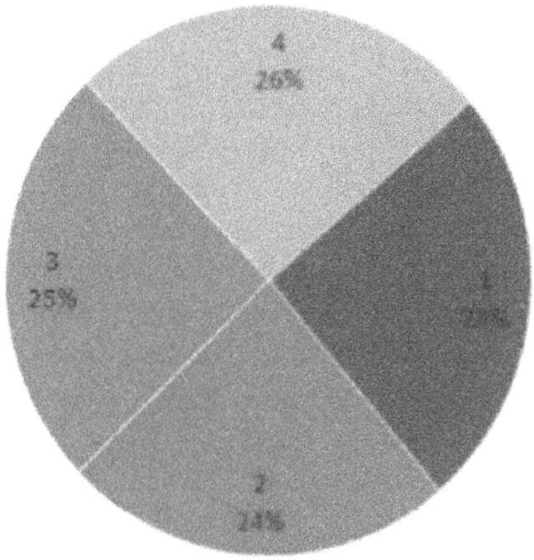

Discussion Boards

1. Today no one wants to be considered average, but in 1835 the average man was a symbol of an egalitarian society. According to Quetelet, a French statistician in the nineteenth century, "If an individual at any given epoch in society possessed all the qualities of the average man, he would represent all that is good, great or beautiful." How would you define the average employee in your current department?
2. Are TV commercials getting longer, or does it just seem that way? In 1990, 6% of commercials were a maximum of 1 min long. Recently, out of a sample of 50 commercials only 16 were 1 min or shorter.
3. Class size paradox. There are two ways to obtain a mean class size and you need to choose one for your training department's report. In one approach, you take the number of employees in all 25 seminars offered this year and average those results. The second approach is on an employee by employee basis, compile a list of class sizes for all seminars he or she attended during the year and average those results. The results for the two approaches will be different. Discuss.

Group Activity

1. The Central Intelligence Agency has specialists who analyze the frequencies of letters of the alphabet in an attempt to decipher intercepted messages. For example, in Standard English text, the letter "r" is used at the rate of 7.7%. Choose another letter and using the Internet determine its usage rate. In an intercepted message sent to Iraq, a page of 2000 characters is found to have the letter "f" occurring 42 times. Discuss if this is unusual.
2. What is meant by "Six degrees of separation" and "The small world problem?"
3. The Cost of Laughing Index (CLI) is developed using the same approach as is used to develop the Consumer Price Index. While standard scores and percentiles allow us to compare different values, they do not take into account time. Index numbers such as the CLI allow you to compare the current value to a value at some base time period. Describe the CLI methodology and what it is this year in the United States.

Parting Thought

Figures don't lie, but liars can figure.

Problem Solutions

1. The data in the **Beverage** file represents the amount of fluid in a sample of 50x 2-L bottles. Using the data from the file, complete the following exercises.

   (a) Construct a histogram. Remember you can use different bin widths. There are multiple solutions to this problem.

Bin	Frequency
2.25	1
2.28	1
2.31	6
2.34	11
2.37	15
2.40	9
2.43	5
More	2

   (b) From the output, what is the mean?
   Answer: The mean is 2.36 tincture medicine bottles but we can't have 0.36 bottles so just 3 bottles

   (c) From the output, what is the standard deviation?
   Answer: sd = 0.04 bottles

Amount	
Mean	2.35672
Standard error	0.006302254
Median	2.36
Mode	2.368
Standard deviation	0.044563662
Sample variance	0.00198592
Kurtosis	0.080400572
Skewness	0.008832396
Range	0.215
Minimum	2.25
Maximum	2.465
Sum	117.836
Count	50

2. The data in the **Electric** file represents the cost of electricity in June for a random sample of 50x one-bedroom apartments in a large city in the southern portion of the United States. Using the data from the file, complete the following exercises.

   (a) Construct a histogram.

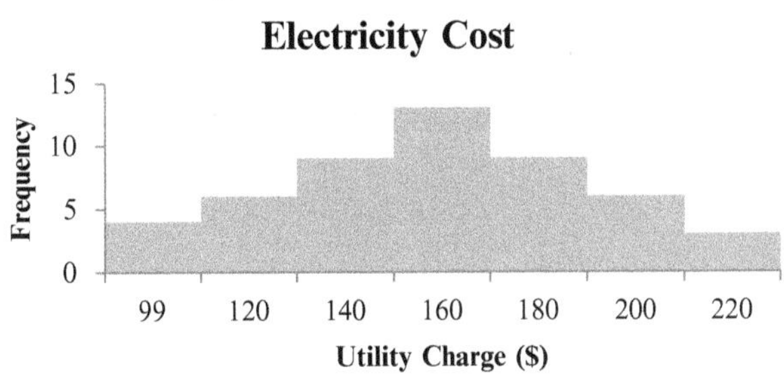

   (b) Around what amount does the monthly electricity cost seem to center around?

   Answer: The majority of utility charges are clustered around $160.

3. Match the histograms to their associated distribution types: right skew, bimodal, random, left skew, normal

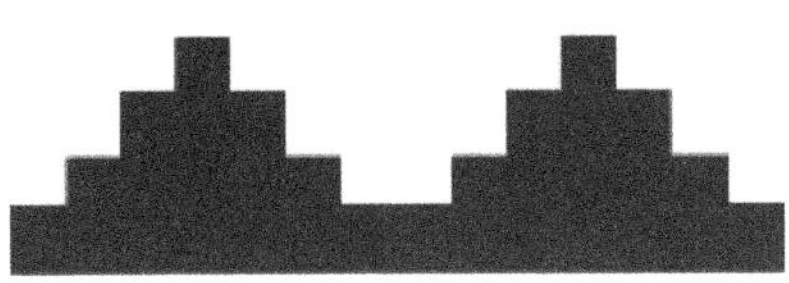

4. What histogram bin sizes would you select for the following data set, which shows student scores on a test?

	A	B
2	Student_Id	Scores
3	1001	41
4	1002	54
5	1003	68
6	1004	92
7	1005	77
8	1006	36
9	1007	54
10	1008	39
11	1009	20
12	1010	59
13	1011	65
14	1012	49
15	1013	23
16	1014	71
17	1015	52
18	1016	84
19	1017	83
20	1018	68
21	1019	63
22	1020	59
23	1021	27
24	1022	50
25	1023	99
26	1024	24
27	1025	61
28	1026	91
29	1027	27
30	1028	53
31	1029	44
32	1030	81

Answer:

	D
2	Bins/Intervals
3	30
4	40
5	50
6	60
7	70
8	80
9	90

5. If this output was presented for your review, what suggestions might you provide to improve the final chart? Provide your reasons.

G5 fx Bin

	A	B	C	D	E	F	G	H
4	Emp Name	Salary		Bins				
5	Crig	$52,873		$25,000			Bin	Frequency
6	Newton	$59,305		$35,000			$25,000	0
7	Sherry	$65,625		$45,000			$35,000	2
8	David	$40,201		$55,000			$45,000	4
9	Henry	$48,467		$65,000			$55,000	4
10	Ian	$42,695					$65,000	3
11	Nick	$66,263					More	2
12	Sam	$49,244						
13	Henderson	$27,551						
14	Andrew	$27,656						
15	Stephen	$35,077						
16	Patrick	$46,191						
17	Phillip	$42,501						
18	Ronz	$58,313						
19	Emily	$64,146						
20								
21								
22								

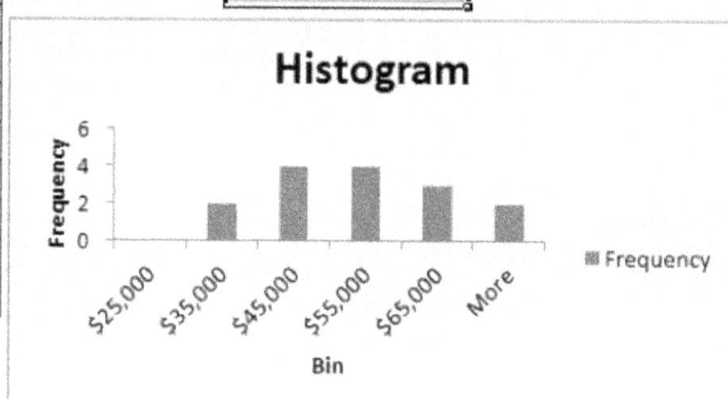

Answer:

- Change the label "bin" on x-axis to "salary (US$)".
- Remove $ signs from x labels (redundant).
- Use a more useful title for the chart.
- Close the gaps between the bins to make the x-axis more accurate.
- Remove the frequency box on the right-hand side of graph.

- Remove the word "more"—when this appears, it makes your audience worried there is more data but you chose not to show it. This bin includes all data over $65,000 so add some more bins to take care of this issue.

6. This histogram tells an effective story without numbers, why does this work in this case?

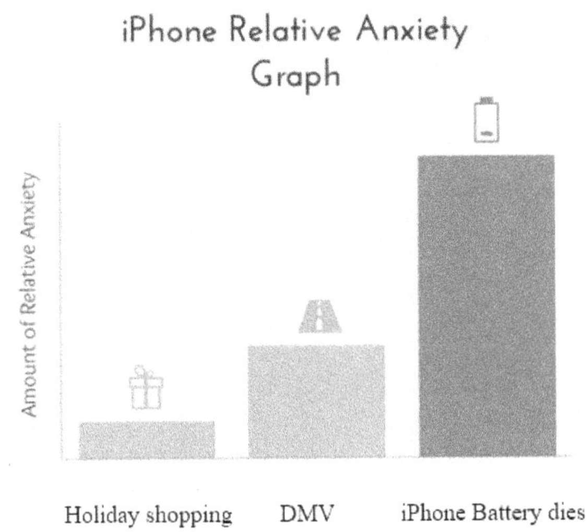

Answer:

We don't need numbers on the horizontal scale as the items are qualitative in nature. The vertical axis explains in the label it is "relative" anxiety as anxiety is a difficult scale to define. So once again the concept is more qualitative than quantitative in this particular chart. Sometimes this type of histogram is referred to as a pictograph.

7. What's wrong with this histogram?

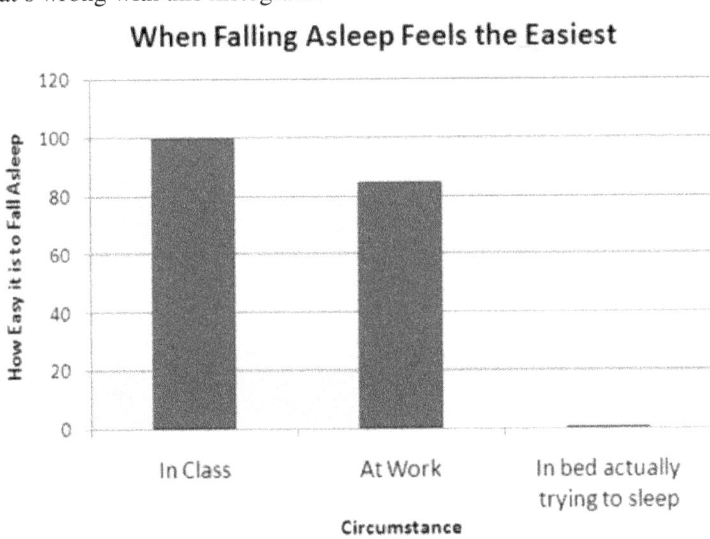

Answer:

- The vertical scale is numeric but lacks units. If it is a percentage then the scale should stop at 100, not 120.

8. Data was charted to show the % sales for four different employees in the same division of the company. Although both graphs use the same data set, employee 2 prefers the first chart be shown. What's going on here? How could this problem be avoided by using a histogram?

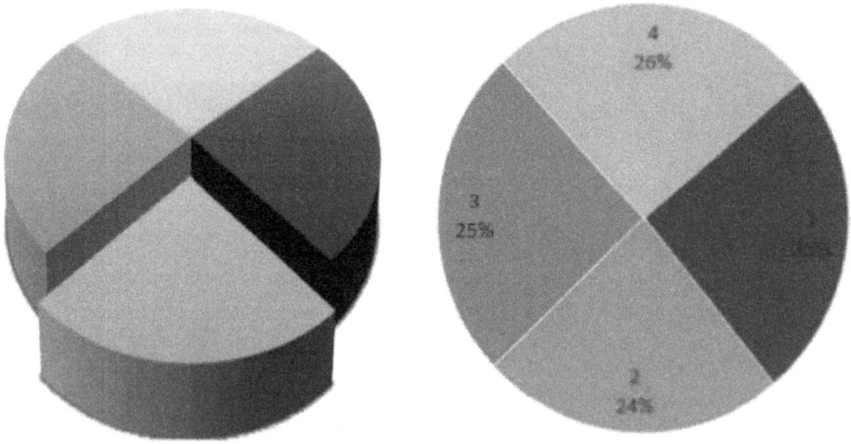

Answer:

The 3-D chart on the left misleads the audience because the orange slice (rep#2) is larger than the other slices. This type of projection confuses any clear visual interpretation of the data. Also the 3-D projection lacks any type of quantitative labels. By using a histogram one can eliminate any visual misleading of the data. Histograms are easier to interpret than 3-D projections.

Chapter 4
Normal Distributions

Linda Herkenhoff and John Fogli

Key Concepts

Chebyshev approximation, Normal distribution, Percentiles, Standard deviation, Standard units, and z-score.

Discussion

This chapter will discuss one of the most common distributions in the business world, the "normal curve." This is another example where the everyday usage of the word means something quite different from the statistical definition. In statistics, the term "normal" refers to a specific mathematically defined curve that looks bell-shaped. It does not mean an everyday, anticipated distribution. An entire chapter is dedicated to this topic because of the normal distribution's usefulness in different applications throughout the business world. The area under the curve is often described as the probability that a specific outcome could occur within a given experiment and will be discussed in more detail in Chapter 8.

As a basis for comparison, we provide a brief description of a non-normal technique for analysis, the Chebyshev approximation.

Supplementary Information The online version contains supplementary material available at https://doi.org/10.1007/978-3-031-46371-6_4.

L. Herkenhoff (✉)
Saint Mary's College, Moraga, CA, USA

J. Fogli
Diablo Valley College, Pleasant Hill, CA, USA

L. Herkenhoff, J. Fogli (eds.), *Applied Statistics for Business and Management using Microsoft Excel*, https://doi.org/10.1007/978-3-031-46371-6_4

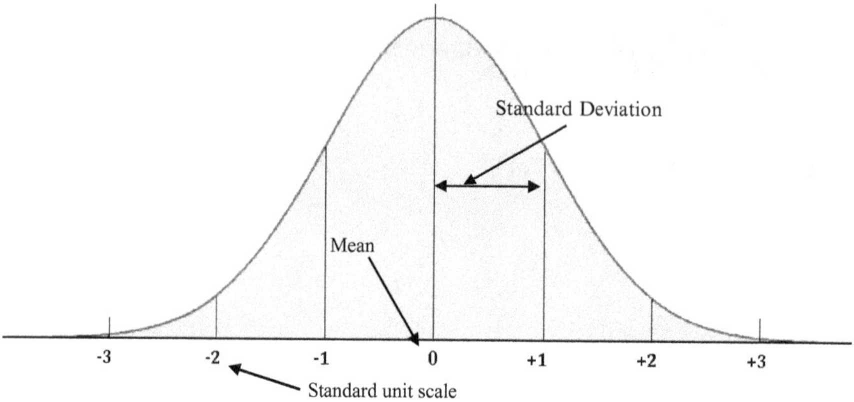

Fig. 4.1 Normal curve

Normal distribution: This is a bell-shaped or Gaussian distribution (Fig. 4.1).

The units of the distribution are standard units(su).

The normal distribution is often described as a central region with a left-hand tail and right-hand tail. The boundary between the central region and the tails is usually defined by the particular problem you are trying to solve.

The **mean** and the **standard deviation** of the distribution describe the bell curve.

Mean ($\bar{x}$): The arithmetic average value of the x variable. This pins down the center of the bell shape. When using the standard unit scale, the mean is always zero.

Standard deviation (σ, sd, SD): The sd defines the width of the bell shape for a sample.

$$\text{sd} = \sqrt{\frac{\sum_{i=1}^{n}\left(x_i - \bar{x}\right)^2}{n-1}}$$

where

Σ is the capital Greek letter sigma, which indicates summing up all variables that follow it. The subscript and superscript indicate you should add the first value through the "*n*th" value. Often the short-hand version just shows the summing sign without the sub/superscripts.

x_i is the variable you are summing up the values for and the subscript is just a place holder. So if the subscript was "1," it means the first value of x.

$\bar{x}$ is the average or mean value of x.

n is the number of values you are summing together, more commonly referred to as the sample size.

Graphically the distributions with smaller standard deviations (SD) will be narrower than those with larger standard deviations.

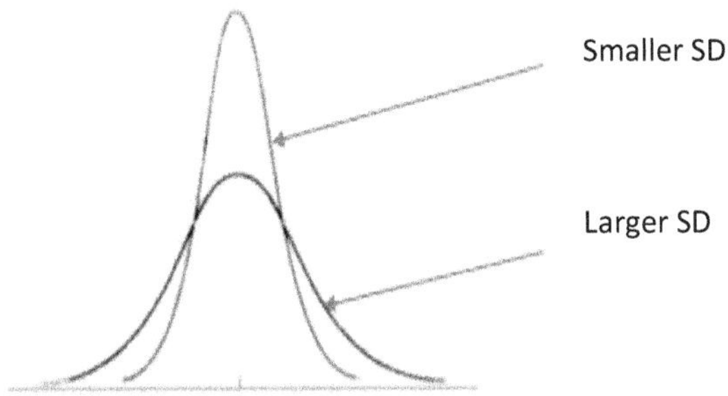

Several features of the normal distribution are very important.

1. The graph is symmetric about 0. The part of the curve on the left side of 0 is the mirror image of the right side of the curve.
2. The total area under the curve is 100%.

Gaussian approximation: This allows us to quickly estimate how much data falls between values on the normal distribution graph, which has a scale of standard units.

- The area under the curve between −1 and +1 std units is about 68%.
- The area under the curve between −2 and +2 std units is about 95%.
- The area under the curve between −3 and +3 std units is about 99%.

An individual value is converted to standard units (*z*-score) by seeing how many standard deviations it is above or below the average. Remember the standard deviation is a group statistic that allows you to convert individual values in the data set to standard units. In addition, *z*-scores provide the key to using a table of standard normal probabilities to perform calculations related to normal distributions.

Percentile: Percentile calculation can be useful in calculating data distributions under the normal curve. The value of a variable, below which a certain percent of observations fall. For example, the 20th percentile is the value (or score) which indicates that 20% of the observations are found below (or less than) that value.

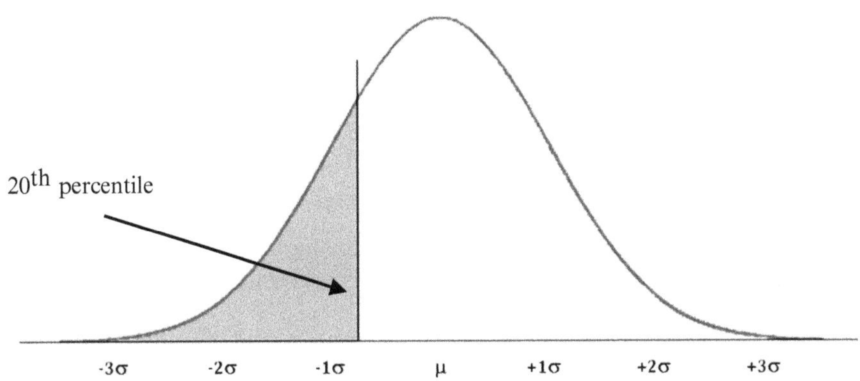

Non-normal Distribution

Chebyshev approximation: This approximation can be used with any distribution. Not all distributions are normally distributed, or you simply may not know what the distribution looks like. To err on the side of conservative estimates, it is better to use the **Chebyshev approximation** in these cases. Because of its general application across all distributions, the answers are not as precise. By this, we mean that the answer you calculate will indicate the MINIMUM amount of data; in other words, it defines the lowest value of how much data you should expect for your range of interest. So, for example, if you want to know how much data falls between ±2 standard deviations in your data set, the solution will indicate ***at least*** 75% of the data will be located between these values, regardless of the distribution. For non-normal problems, we use the symbol ***k*** for the standard deviation. Thanks to Chebyshev, we know that for any non-normal distribution:

At least $(1 - 1/k^2)$ of the items in any data set will be within k standard deviations of the mean, where k is any value greater than 1.

- *At least* 75% of the items must be within $k = \pm 2$ standard deviations of the mean.
- *At least* 89% of the items must be within $k = \pm 3$ standard deviations of the mean.
- *At least* 94% of the items must be within $k = \pm 4$ standard deviations of the mean.

Mechanics of the Normal World

Converting Measured Units to Std Units

A value is converted to standard units (z-scores) by seeing how many SDs it is above or below the average. It is useful to standardize the values (raw scores) of a normal distribution by converting them into z-scores because

(a) It allows researchers to calculate the probability of a score occurring within a standard normal distribution
(b) Enables us to compare two scores that are from different samples (which may have different means and standard deviations)

For converting to std units, the SD and the average are the required statistics:

Step 1: Calculate σ (standard deviation) and average for the *entire* data set.
Step 2: Use this σ and average to convert *each* data point in the data set to std units.

Sample data set: 10, 8.5, 3.25, 4.6, 1

Step 1: Calculate SD by either using SD formula from the previous page, or input data to Excel and run the function **STDEV.S**. Calculate the mean by hand or run the Excel function **AVERAGE**.

$$SD = 3.72 \quad \bar{x} = 5.47$$

Step 2: To convert any one of the values in the data set to standard units, subtract the average from the data value ($x - \bar{x}$) and divide by the SD. Let's convert 10 to standard units.

$$\begin{aligned}&(x - \bar{x}) / \sigma \\ &(10 - 5.47) / 3.72 = 1.22 \text{std units}\end{aligned}$$

Or input the value(s) to Excel and run **STANDARDIZE**.

To summarize, all we are doing is determining how far away the data value is from the mean: ($x - \bar{x}$) = Δx, and how many SDs are in this difference: $\Delta x \div \sigma$.

Example: Mary is 66 inches tall, and the average height of business travelers on her plane is 71 inches. The standard deviation of the business travelers' heights is 2 inches. What is her height in standard units?

Solution: Step 1: (66 − 71) = −5

Step 2: −5/2… or how many 2 inches are there in −5 or how many SDs are there in 5?
Answer: −2.5… Double check the sign. Because her height is less than the average value, we should have a negative sign. Her height of 71 inches is the same as −2.5 std units or a *z*-score of −2.5.

Converting Std Units to Measured Units

We need to already have σ and $\bar{x}$ (calculate by hand or Excel)

1. Multiply the value by σ: ($x \times \sigma$)
2. Add this amount back to the mean: ($x \times \sigma$) + $\bar{x}$

Example 1. Bob's height is associated with a *z*-score of 0.52 standard units (SU). The average height of his fellow business travelers is 61 inches, and their standard deviation is 2 inches. How tall is Bob in inches?

Solution: Step 1: 0.52 × 2 = 1.04 inches…this is a positive value so he must be taller than average

Step 2: 1.04 + 61 = 62.04 inches
Answer: So Bob is 62.04″ tall.

Calculate Area Under the Normal Curve

We can find the area under the normal curve using several methods, which include

1. Use interactive graphic
2. Use a normal/z table
3. Use the Empirical rule
4. Use Excel

1. Interactive Graphic: Real-Time Modeling
Open the website:

http://www.mathsisfun.com/data/standard-normal-distribution-table.html
Move the lines on the graphic and understand this powerful tool….
Z value is the value in std units

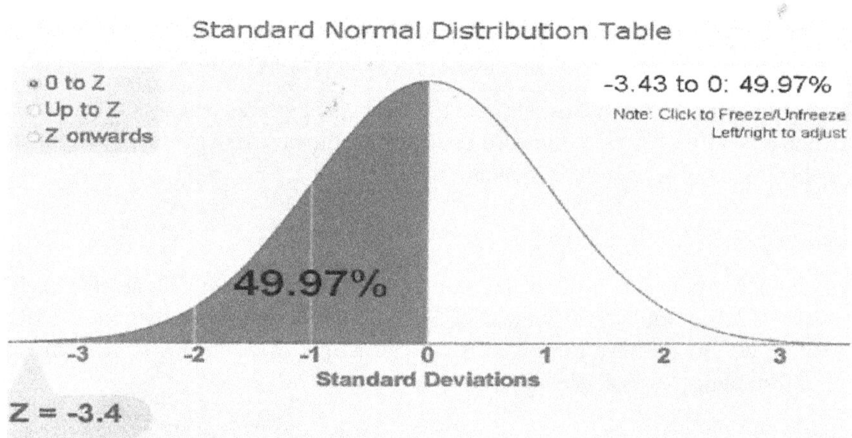

Let's Practice

1. How much data falls between the mean (0) and +2.6 su (z)?
   49.52%
2. How much data falls between the mean (0) and −2.6 su (z)?
   49.52%
3. How much data falls between −.13 su and 2.6 su?
   5.17% + 49.53% = 54.70%
4. How much data is greater than 1.85 su?
   3.22%
5. How much data is less than −1.88 su?
   3.01%

2. z-Tables

The values in the z-table (Appendix 1) represent the area (amount of data) between 0 std units and z std units. It could be $-z$ to zero or $+z$ to zero since we know the normal distribution is symmetric around zero. The bold vertical column in the table is the z value to the first decimal place. The bold horizontal top row is the subsequent decimal places in the z value, allowing you to be more exact. The corresponding values in the table can be multiplied by 100 to get the area in percentage. A value of 0.0398 indicates that about 4% of the data falls between the given z value and zero on the normal distribution for this particular data set. Note z-scores are not usually written with "su" following the numeric.

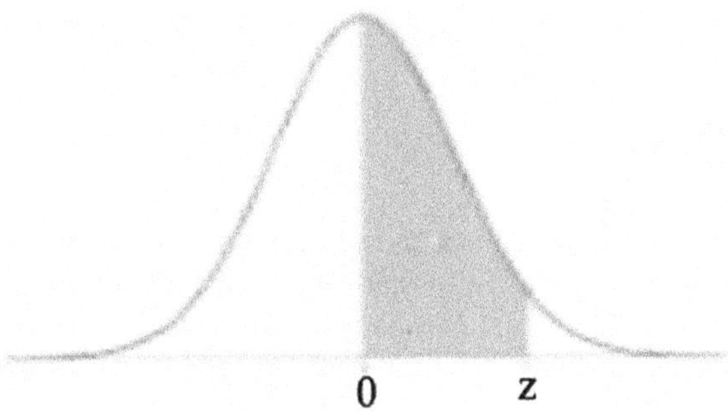

Example: How much data falls between ±2 standard units (su), which is the same as ±z-scores of 2.

- Look in the first column for a value of 2.0.
- Read the corresponding area value of 0.4772.

This means 47.7% of the data in this distribution is estimated to fall between +2 su and zero but we must add the area between −2 su and zero, which is also 47.7%. The total area being 95.4%. This shouldn't surprise us, as we know from Gaussian distribution estimations that 95% of the data falls between ±2 su.

Example: How much data falls between $z = 1.06$ to zero?

- Look in the bold vertical for 1.0.
- Look in the bold horizontal top row for .06.
- Read the value where the 1.0 row intersects the 0.06 column to get .3554.

This means 35.54% of the data falls between 0 and 1.06 standard units on this normal distribution graph.

Example: Find the area between $z = -2$ and $z = 1$ under the normal curve using the standard normal distribution.

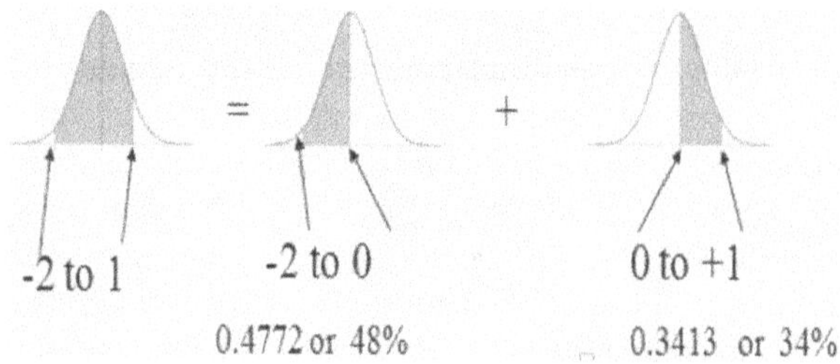

The area between −2 and 1 is approximately 82% (48% + 34%). In other words, 82% of the data falls between −2 and +1 std units.

Percentile Problems

We are often interested in percentiles and percentage distribution. Remember a percentile indicates all data to the left of that value. The 80th percentile indicates that 80% of the data lies to the left of the value. Sometimes we are interested in how data lies to the right of the percentile. The simplest way we can work with these is using standard units.

Example: In the sample of business travelers, what is the % of heights above 72 inches in a data set with sd = 2 inches and avg = 66 inches?

Calculate inches → standards units → %

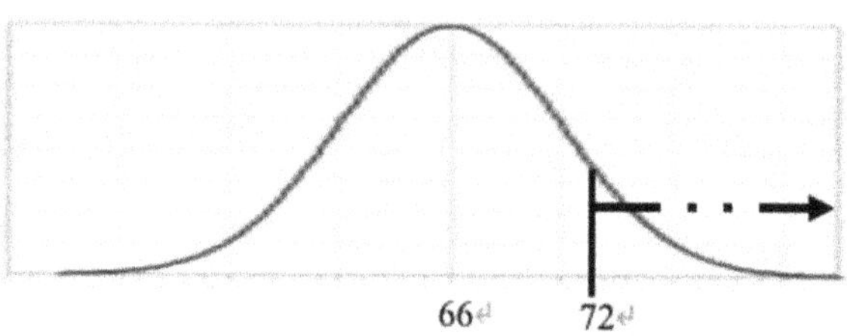

Convert 72 into standard units

$$(x - \bar{x}) / \sigma$$
$$(72 - 66) / 2 = 3 \text{std units}$$
$$z = 3$$

Calculate what % data falls between $z = -3$ and $z = +3$. Using the empirical rule (Gaussian approximations), the interactive graphic or the z-table.

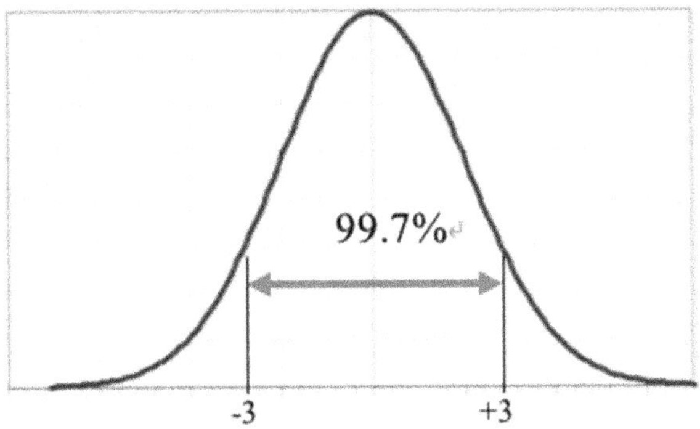

Since we assume that 100% of the data falls under this normal curve, we know that the two tails have 100% − 99.7% = .3%

The right-hand tail has .3%/2 = .15%

Only 0.15% are taller than 72 inches

Example: On one of your competency tests, the average score was 535 pts and the SD was 100. The distribution of scores followed the normal curve. Estimate the 95th percentile of the score distribution.

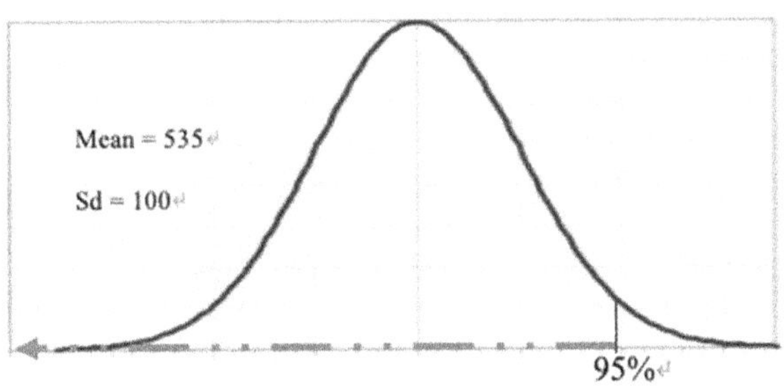

Solution 1: Using the interactive graphic, $z = 1.64$.

Convert std units back to measure units

$$(x \times \sigma) + \bar{x}$$
$$(1.64 \times 100) + 535 = 699\text{pts}$$

Solution 2: Using z-table

If 95% of the data falls to the left of this line because remember it is a percentile, then the amount of data in the right-hand tail is right hand

$$100\% - 95\% = 5\%$$

Calculate the area between zero and the 95% percentile

$$50\% - 5\% = 45\%$$

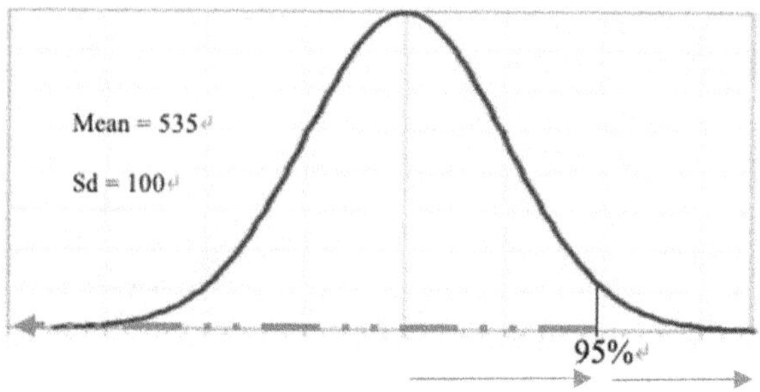

Look up what z value corresponds to 45%

$$1.65\text{su}$$

Convert 1.65 to measured units

$$(1.65 \times 100) + 535 = 700\text{pts}$$

The 95 percentile is about 700 pts.

Using the Empirical Rule to Detect Outliers

An *outlier* is an unusually small or unusually large value in a data set.

Outlier: z-score less than −3 su or greater than +3 su

- It might be an incorrectly recorded data value.
- It might be a data value that was incorrectly included in the data set.
- It might be a correctly recorded data value that belongs in the data set!

Excel

When we have normally distributed data, the normal curve allows us to calculate how much data falls between any two points, and how much data exists in one or both tails. We can also easily calculate specific percentiles. But remember, we need to work in **standard units**. When you have all of the raw data, you can use Excel to calculate the standard deviation and the mean. However, if you don't have the raw data, you need to be given the mean and the standard deviation to use these functions.

The functions we will use in this section apply to both cases where you do not have the raw data and also to those cases where you do have all of the raw data available in Excel:

- **NORM.DIST** calculates the probability or distribution of data to the left of your value and requires the mean and standard deviation statistics. In other words, this calculates the percentile. The data is input using the following format.

  *NORM.DIST (the **x value** in measured units, the **mean** in measured units, **standard deviation** in measured units, **1**)*

The argument 1 (TRUE) tells Excel to compute the normal cumulative distribution. If the last argument is 0 (FALSE), Excel performs a probability density function. The normal cumulative distribution function generates the probability that a variable will be equal to or less than a chosen value. Whereas the probability density function calculates the likelihood that a random variable becomes a certain value around a specific point.

  – The answer is given as a decimal, so you will need to multiply by 100 to get the percentage answer.

- **NORM.INV** will convert the percentile value to measured units. The data is input using the following format.

  *NORM.INV (the **percentile** to be converted to measured units and expressed as a decimal value, the **mean** in original measured units, **standard deviation** in original measured units)*

These Excel functions are summarized by way of examples in Fig. 4.2. In these examples, we assume a data set that has a normal distribution of employee ages.

- Note that other Normal functions exist in Excel (i.e., **NORM.S.DIST**, **NORM.S.INV**) in which all input/output are in standard units. However, normal distribution analysis can be adequately handled using the **NORM.DIST** and **NORM.INV** functions.

  The problem types using these Excel functions include

1. Percentile Calculation Problems (NORM.DIST) including creating area graphs
2. Converting Percentiles to Measured Units (NORM.INV)
3. Converting Measured units to z-scores (STANDARDIZE)
4. Rank and Percentile (RANK, PERCENTILE)
5. Evaluating Normality
6. Non-normal Distributions

1. **Percentile Calculation Problems (NORM.DIST)**
   There are several distribution (area) types that use this function. These include
   (a) Calculating the area to the left of a value
   (b) Calculating the area between two values
   (c) Calculating the area to the right of a value
   (d) Creating a normal distribution or area graph

Example Problem

Your transportation company provides delivery service 7 days a week to stores selling synthetic fire logs. You need to determine how many trucks and drivers to have on hand to deliver the product. The data you have collected from last year indicates the number of stores requiring deliveries on a daily basis during the months of December and January. This data seems to follow a normal distribution. The average number of stores requiring deliveries on any given day during this time is 100. The standard deviation is 15 stores.

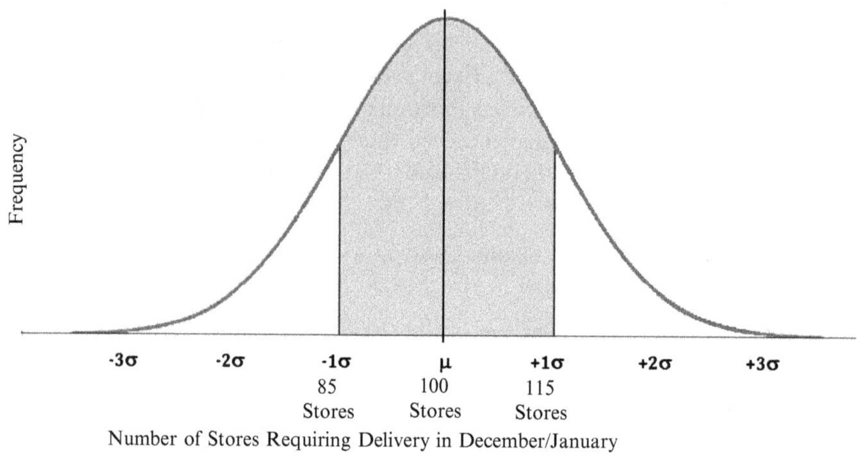

◈ Note to answer these questions you don't actually need a data set as long as you know the data follows a normal distribution. Very powerful and helpful analysis tool.

(a) Calculating the area to the left of a value

What % of the time did you deliver to less than (<) 90 stores during last December and January? In other words, how much data is in the left-hand tail?

Function	**NORM.DIST**		
	Input	**Output**	**Required data**
	Measured units	Percentile units	Mean, std dev for data
Example data	45 yrs old	60th percentile (.60)	
Function	**NORM.INV**		
	Input	**Output**	**Required data**
	percentile	Measured units	Mean, std dev for data
Example data	87th percentile(.87)	56 years old	

Fig. 4.2 Sample input/output

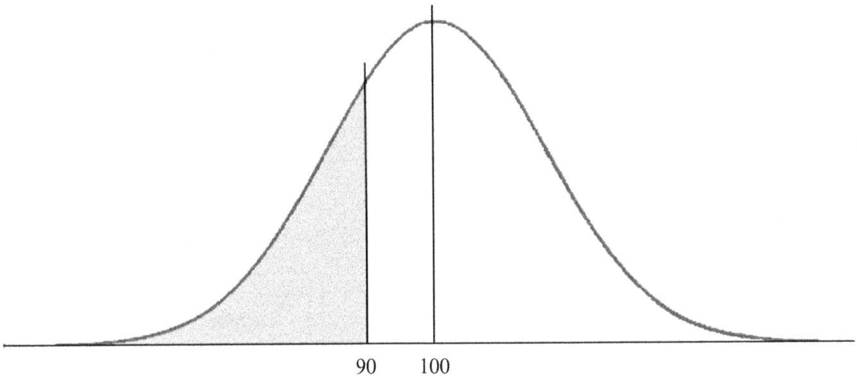

- Click on an empty cell where you want the answer to be output
- Click on the ***fx*** button

- Select the **Statistical** category

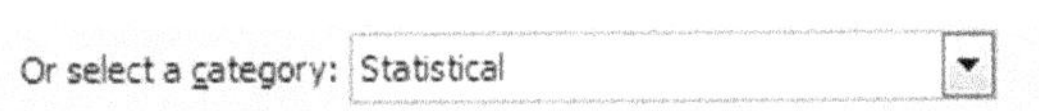

- Highlight the **NORM.DIST** function from the list or just type in the **NORM.DIST** function in the function bar

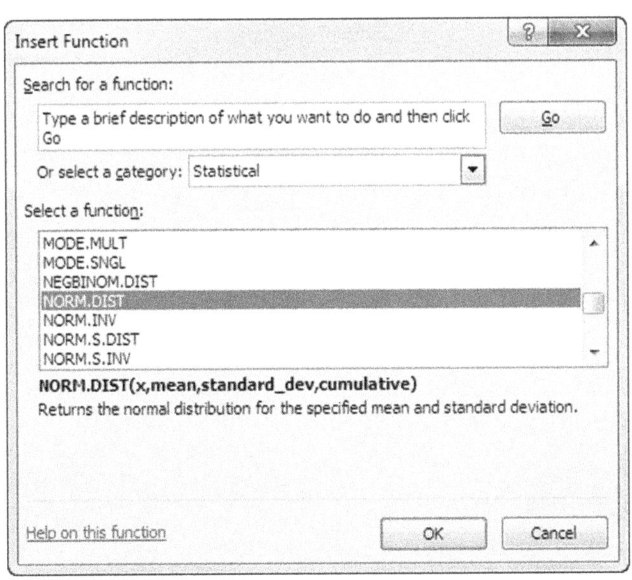

- Click **OK**

- Insert the value of *X* in original measured units. This is the upper limit of the area for which you want to calculate the distribution of data. In other words, this is the value you want to convert to a percentile

- Insert the **Mean** value

Note: You must already have the *Mean* value.

- Insert the **Standard_dev** value

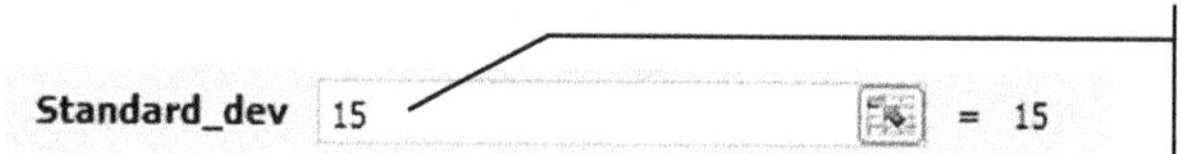

Note: You must already have the *Standard_dev* value.

- Input the value of "**1**" or the word "**TRUE**" for normal cumulative calculations such as this one. Since this is a percentile calculation, we select **1** as the last value in the function input.

- Click **OK**

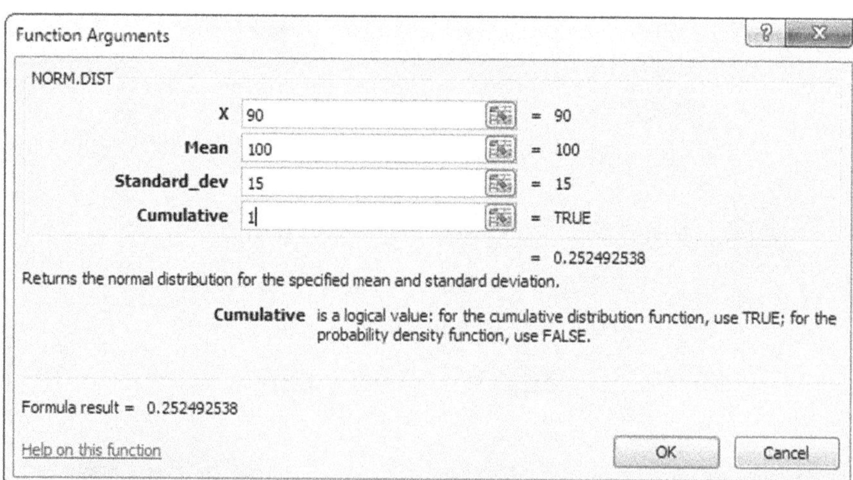

Based on this data, the answer is given as .252493, which translates to about 25% of the time you delivered fire logs to less than 90 stores.

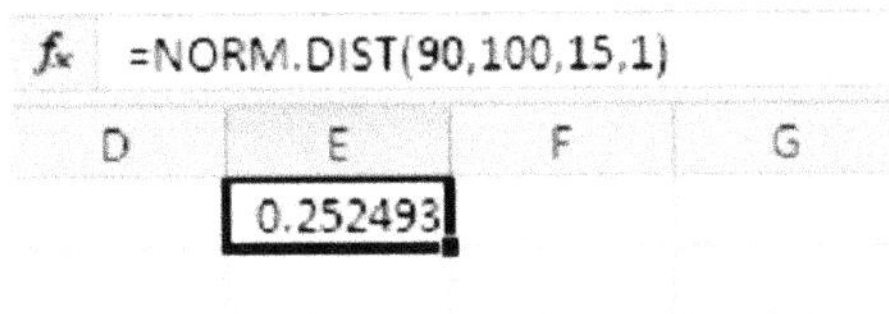

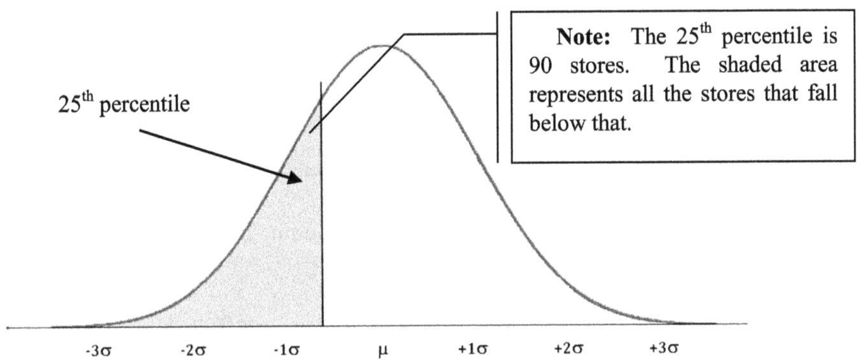

(b) Calculating the area between two values

What % of the time did you deliver fire logs to between 90 and 120 stores during last December and January?

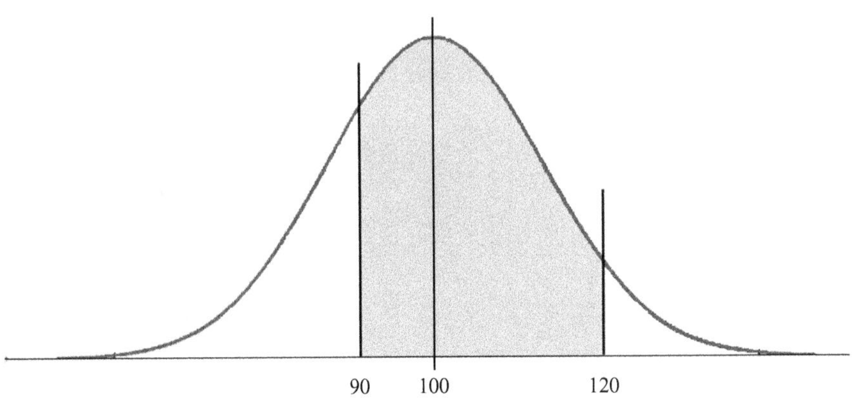

- Click on the empty cell where you want the answer
- Type "=" inside the ***fx*** box (this prepares Excel to accept a function format)

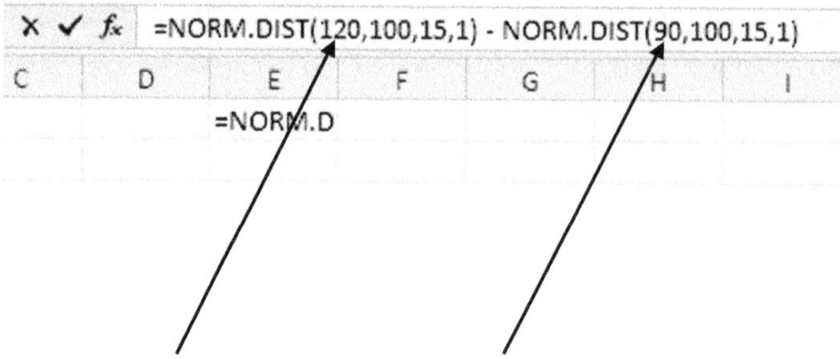

- Type in the formula **NORM.DIST** (limit value, mean, sd, 1)
- Input the *upper limit* value and the *lower limit* value of the area where you want to calculate the data distribution
- Input the *mean* and *standard deviation* from your data set inside both sets of brackets as the next two inputs
- Input the value of **1** or the word **TRUE** for normal cumulative calculations inside both sets of brackets

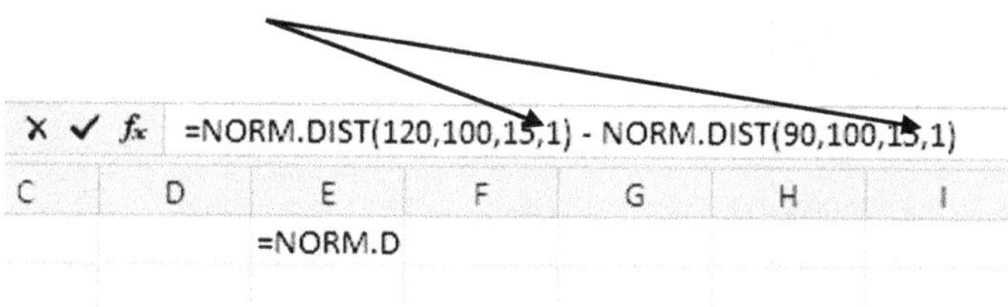

- Click the **checkmark** to the left of ***fx***

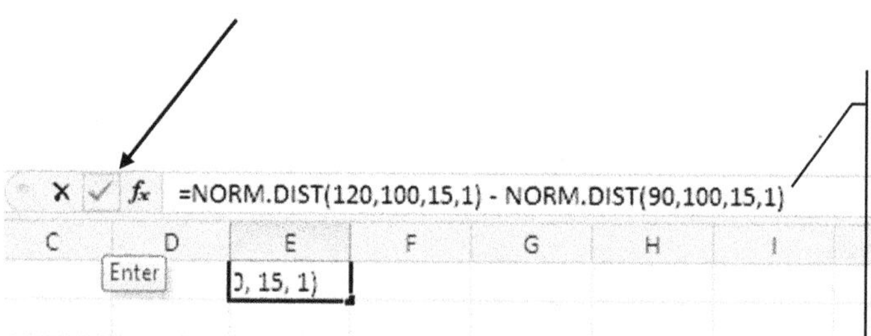

Note: The formula is entered **"NORM.DIST (Limit, Mean, SD, 0/1)"**. Enter either **1** for <u>true</u> or **0** for <u>false</u>.

The answer is given as 0.656296, which means that during last December and January you delivered fire logs to between 90 and 120 stores about 66% of the time.

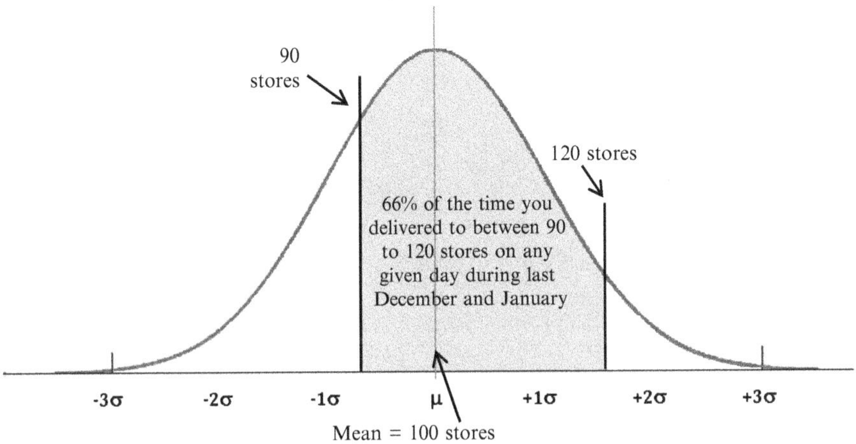

(c) Calculating the area to the right of a value

What % of the time did you deliver to 130 or more stores (≥) during last December and January? In other words, how much data is in the right-hand tail?

- Note in these types of problems the Excel answer will include when you are delivering to exactly 130 stores and when you are delivering to more than 130 stores.

- Click on an empty cell where you want the answer
- Click on the ***fx*** function

- Type the "=" in the white input box
- Type the number "**1**", which equates to 100%
- Type the function or highlight the function from the drop-down menu and subtract from **1**
- Input the upper limit of the right-hand tail

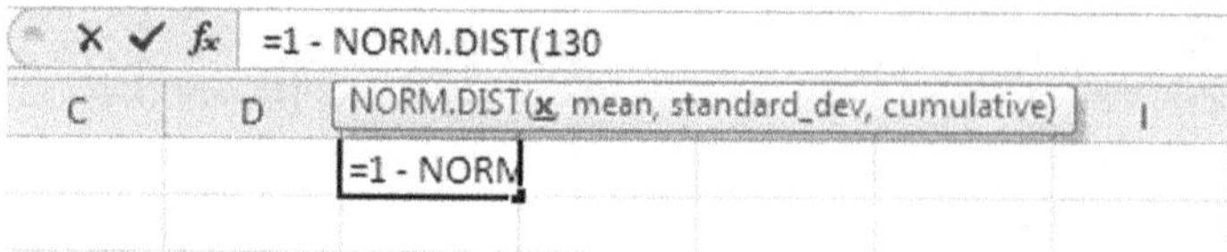

- Input the mean from your data set

- Input the standard deviation from your data set

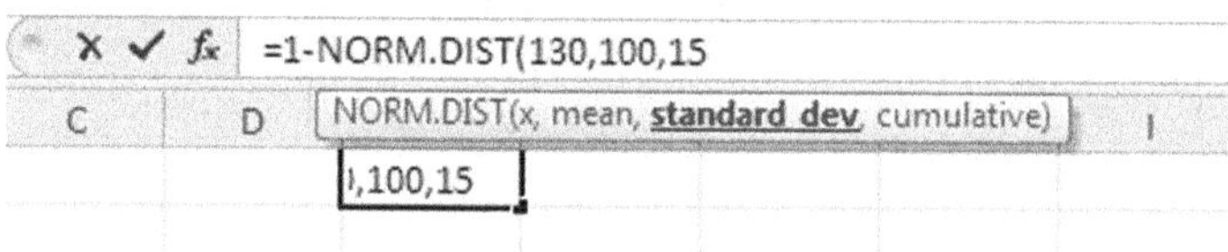

- Input the value of "**1**" or the word "**TRUE**" for the normal cumulative calculation
- Click the **checkmark** to the left of ***fx***

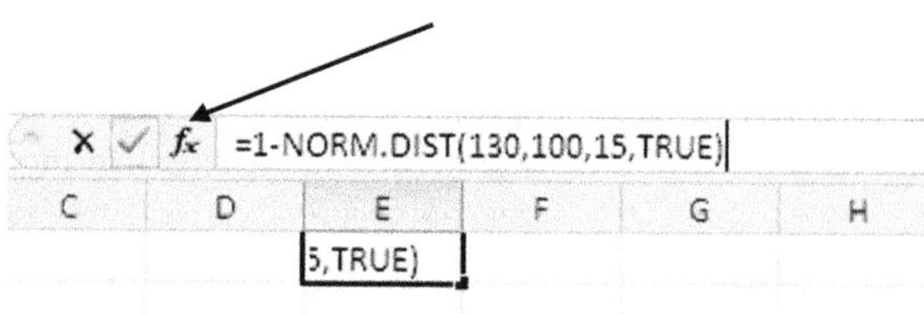

The answer is given as 0.02275, which translates to about 2.3% of the time you delivered fire logs to 130 or more stores on any given day. If we expect the same results this year, we could state that we will be delivering to *at least* 130 stores on any given day about 2.3% of the time this coming December and January.

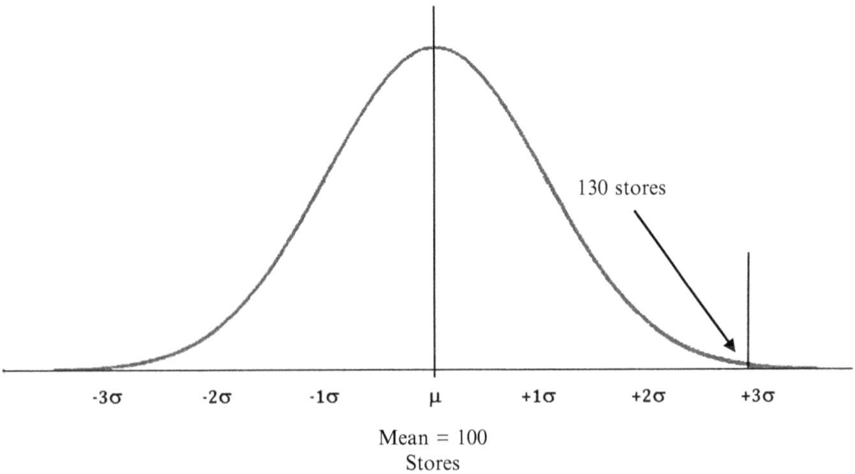

(d) Graphing a normal distribution (Area Graph)

There are three steps in this process:

Step 1: Sort the data into a sequential order (SORT)

Step 2: Calculate the height of the normal distribution for each x value (NORM.DIST)

Step 3: Create the graph

Example Problem

You would like to graph the distribution of the cost of airplane tickets purchased by employees over the past 3 months. The average cost is \$485 and the standard deviation is \$260. You know that ticket prices are normally distributed based on historical data.

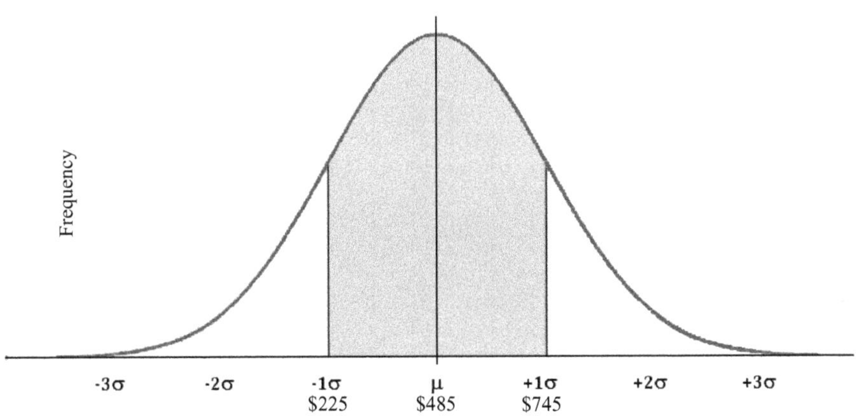

To do this in Excel will require several steps.

- Input the following ticket prices in column A

Cost of airplane tickets ($)
505
385
65
585
825
465
785
345
865
625
265
545
905
145
745
425
185
665
225
705
305
105

Step 1 Sort the data into sequential order using the SORT function.

Highlight the column of data to sort

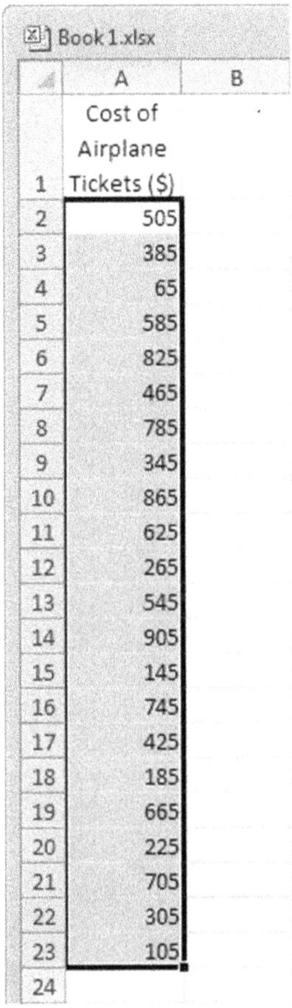

Book 1.xlsx

	A	B
1	Cost of Airplane Tickets ($)	
2	505	
3	385	
4	65	
5	585	
6	825	
7	465	
8	785	
9	345	
10	865	
11	625	
12	265	
13	545	
14	905	
15	145	
16	745	
17	425	
18	185	
19	665	
20	225	
21	705	
22	305	
23	105	
24		

  - If you want to preserve your original data, you need to copy and paste it into another column. When Excel sorts the data, it overrides the original order of the data with the sorted order of data.

- Right click anywhere in the column

- Select the **SORT** function from the menu

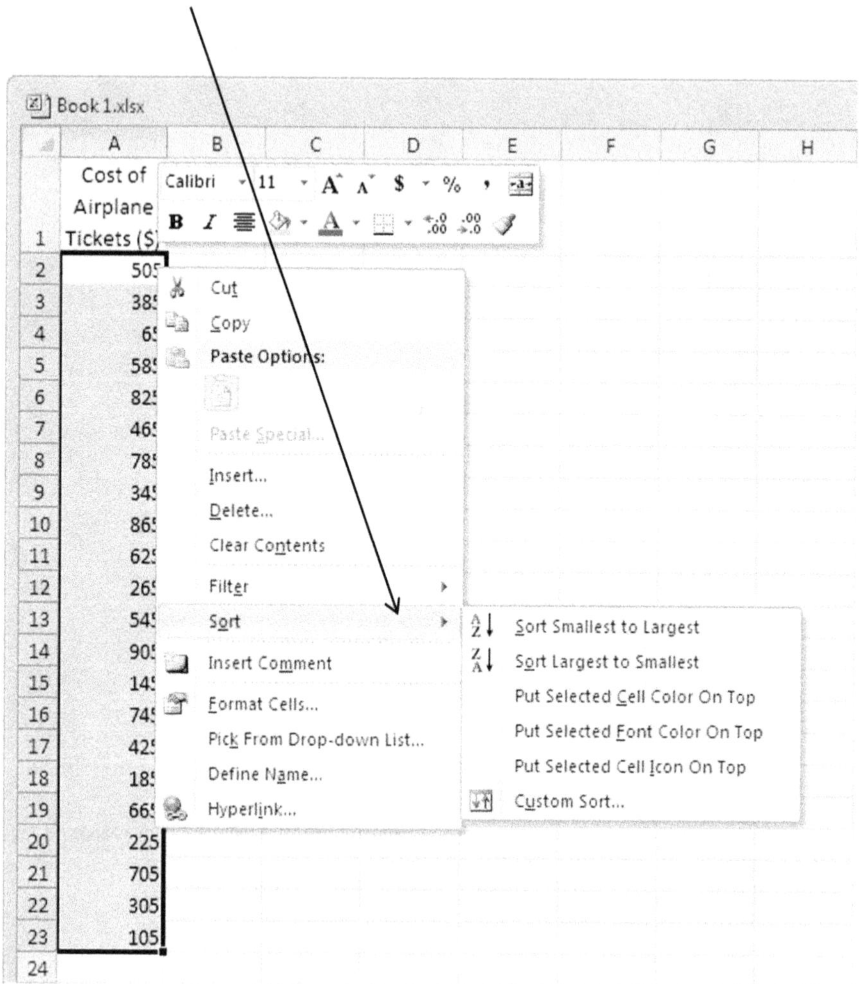

- Select which sort option you want. In this case, choose "**Smallest to Largest**"

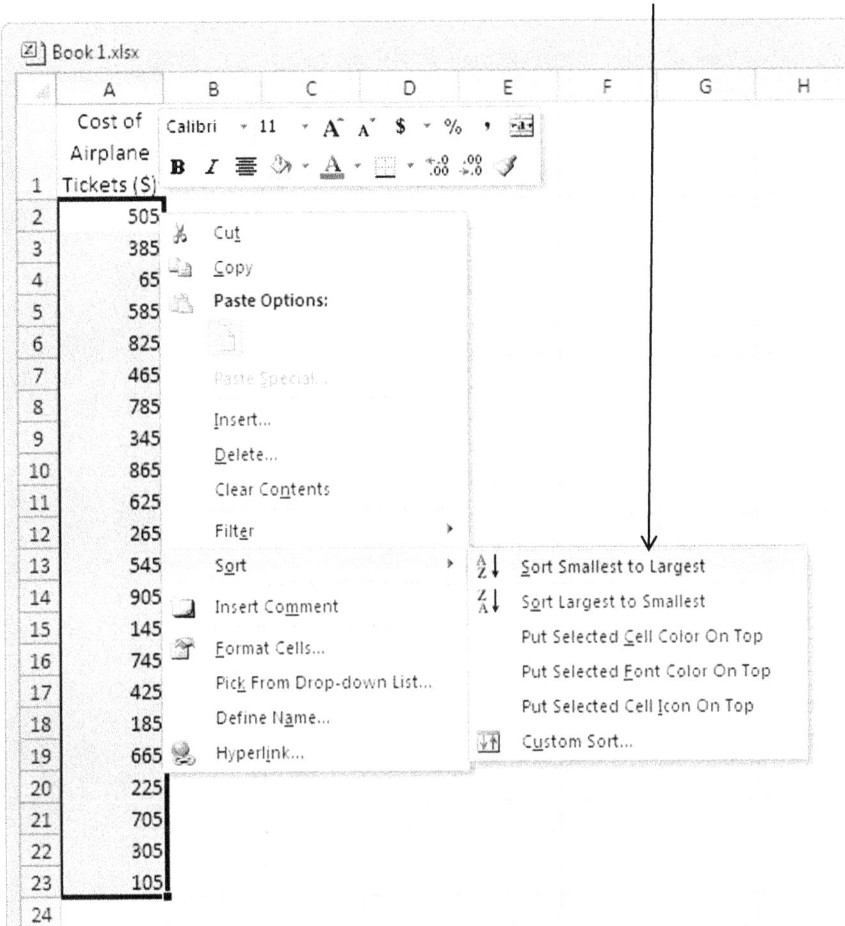

Step 2 Calculate the height of the normal distribution for each x value using the NORM.DIST function.

- Click on an empty cell where you want the output for the first value
- Click on the *fx* and select **NORM.DIST**

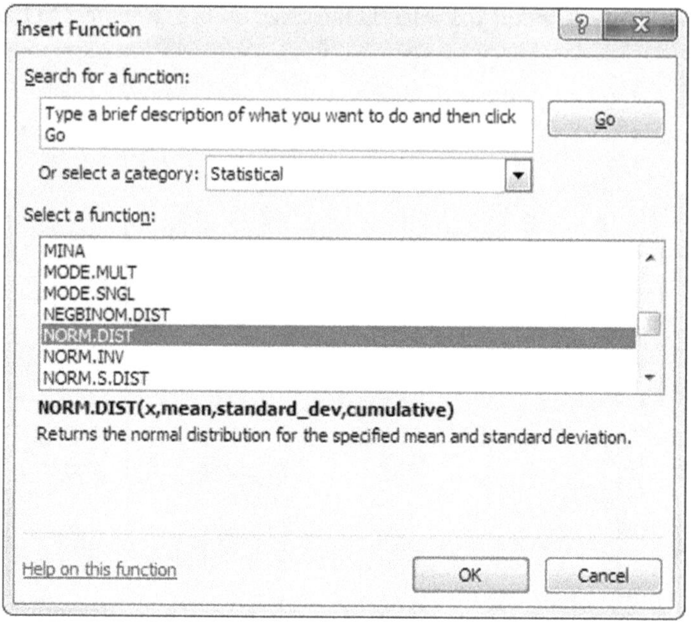

- Click **OK**
- Input the location (cell) of the first x value

Calculate the mean using the AVERAGE function

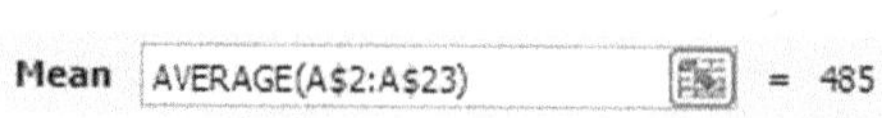

– Make sure to add "$" with each cell location to treat these cells as constant values when you cut and paste them for use with the other data values.

- Calculate the std deviation using the **STDEV** function

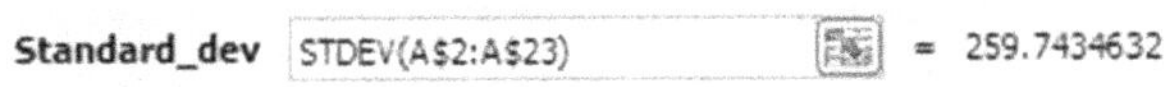

Examine the "Cell Reference" section in Chap. 2 for a refresher on "$" references.

- Input **0** (or **FALSE**) to calculate the height of the distribution for each ticket value in this normal distribution

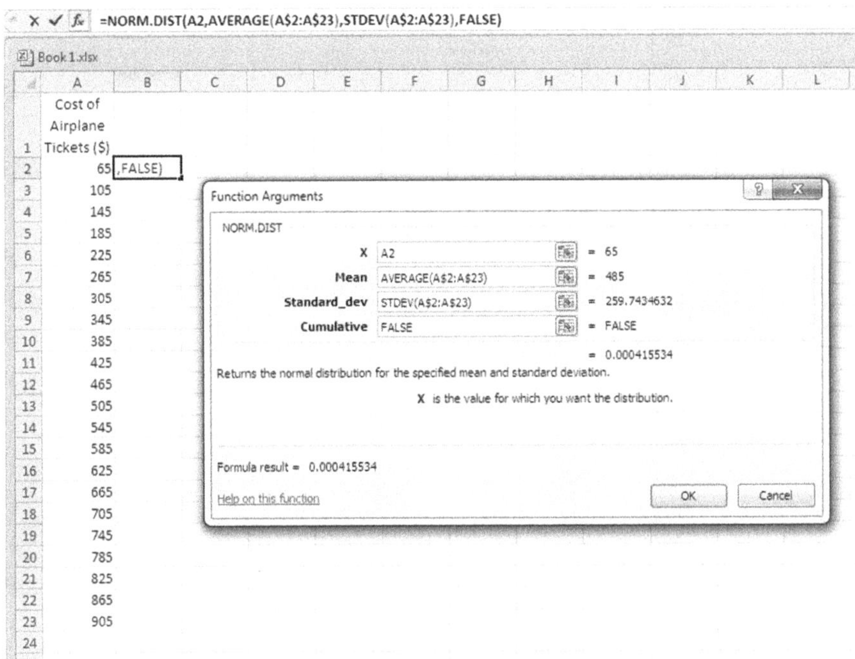

- Click **OK**
- Highlight cell B2 (or wherever your first calculation is located) and right click
- Select the **Copy** function

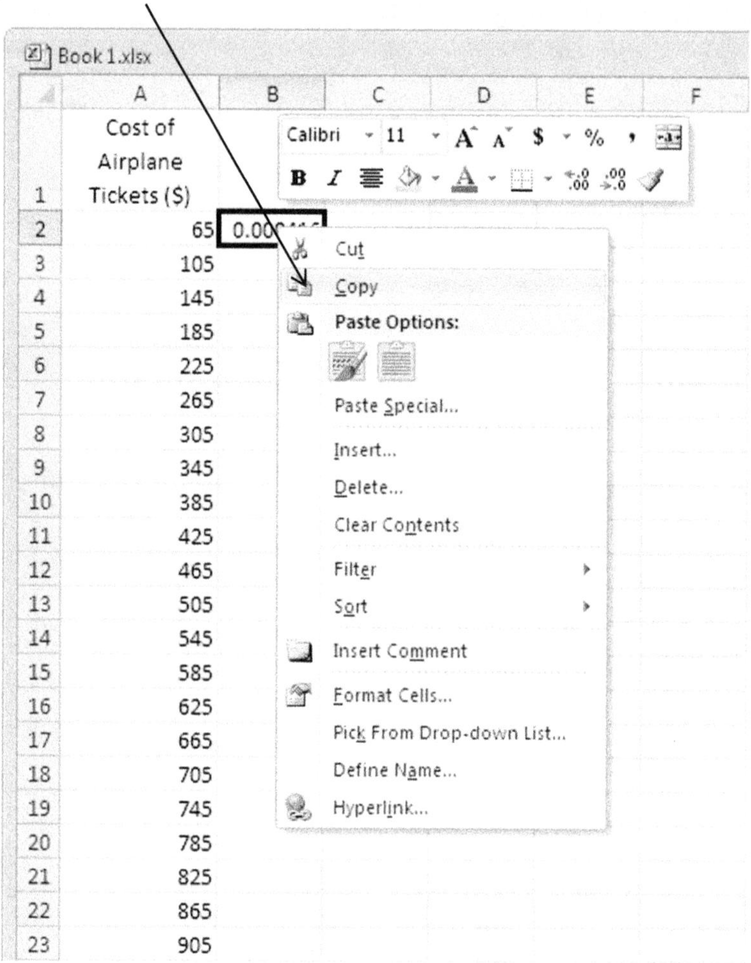

- Highlight the empty cells in this same column across from the other ticket prices, right click, and select the **Paste** function. Alternatively you can just double click with the left mouse button on the bottom right corner of the cell to have it automatically copy the formula for all the remaining values in the list.

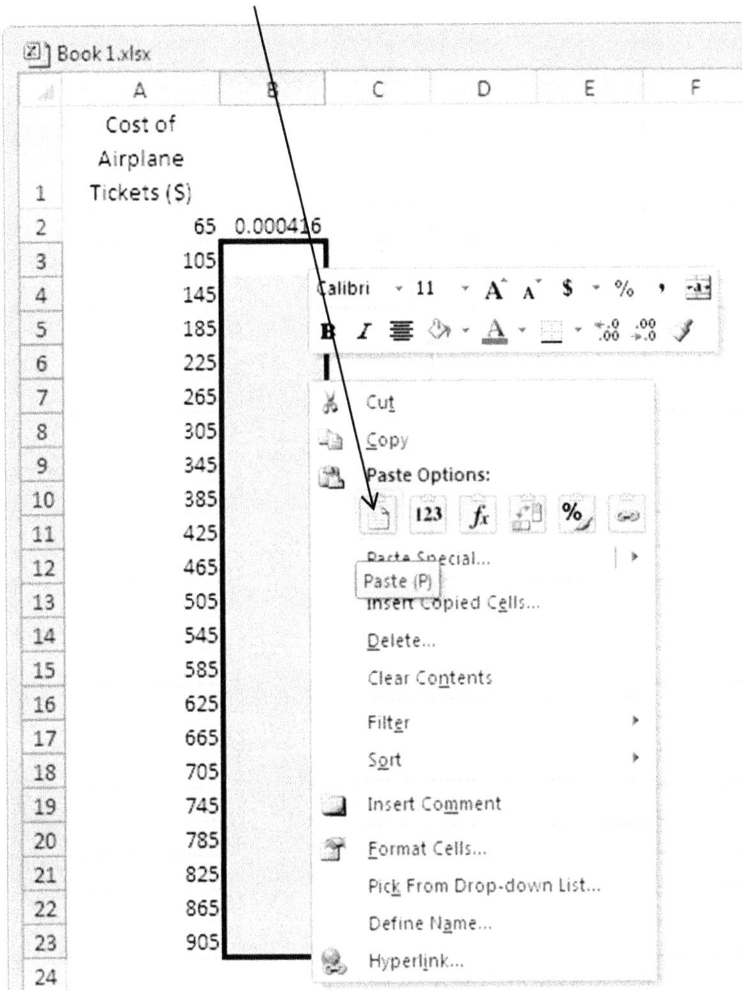

Step 3 Create the Graph.

- Highlight the second column of values

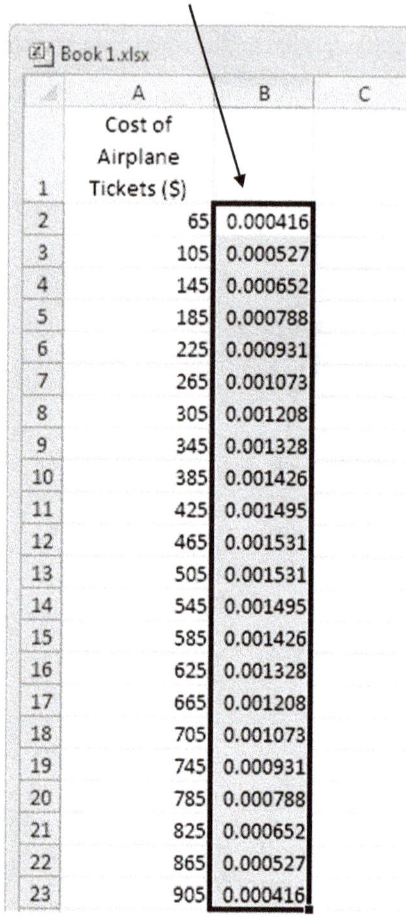

Book 1.xlsx

	A	B	C
1	Cost of Airplane Tickets ($)		
2	65	0.000416	
3	105	0.000527	
4	145	0.000652	
5	185	0.000788	
6	225	0.000931	
7	265	0.001073	
8	305	0.001208	
9	345	0.001328	
10	385	0.001426	
11	425	0.001495	
12	465	0.001531	
13	505	0.001531	
14	545	0.001495	
15	585	0.001426	
16	625	0.001328	
17	665	0.001208	
18	705	0.001073	
19	745	0.000931	
20	785	0.000788	
21	825	0.000652	
22	865	0.000527	
23	905	0.000416	

- Click the **Insert** tab
- Click the **Arrow** icon at bottom-right corner of the **Charts** tab

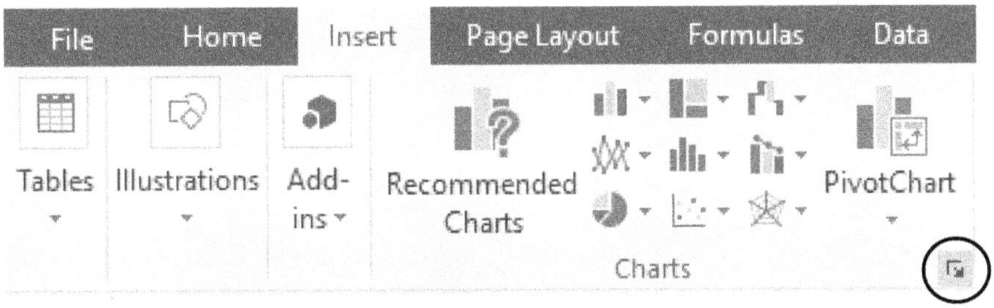

- Select the **Area** graph from the **All Charts** tab

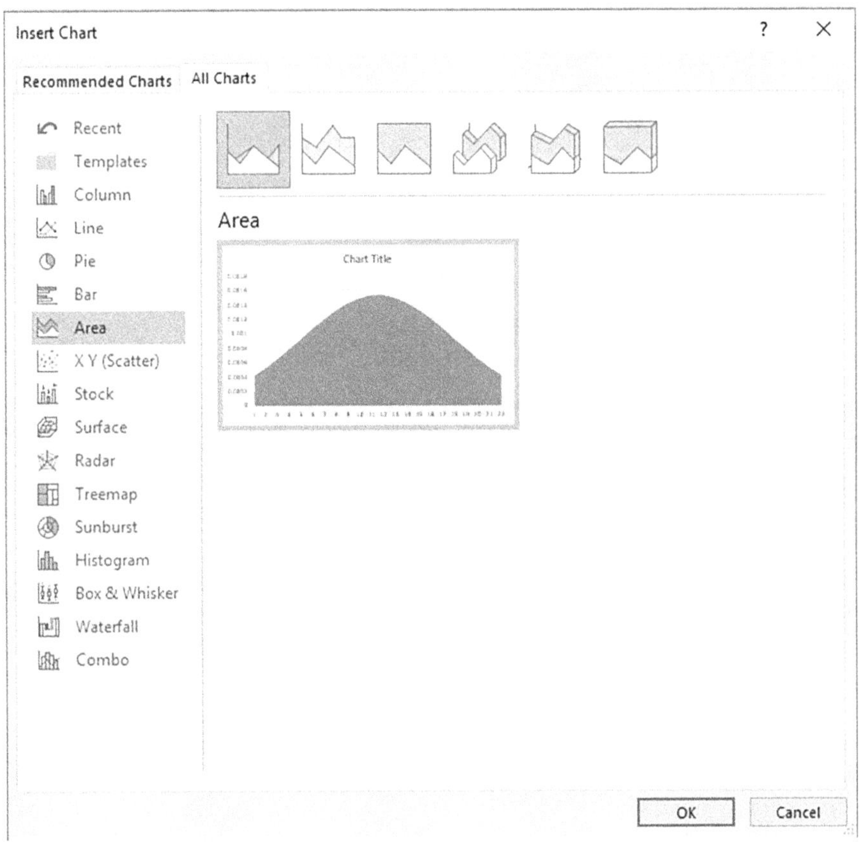

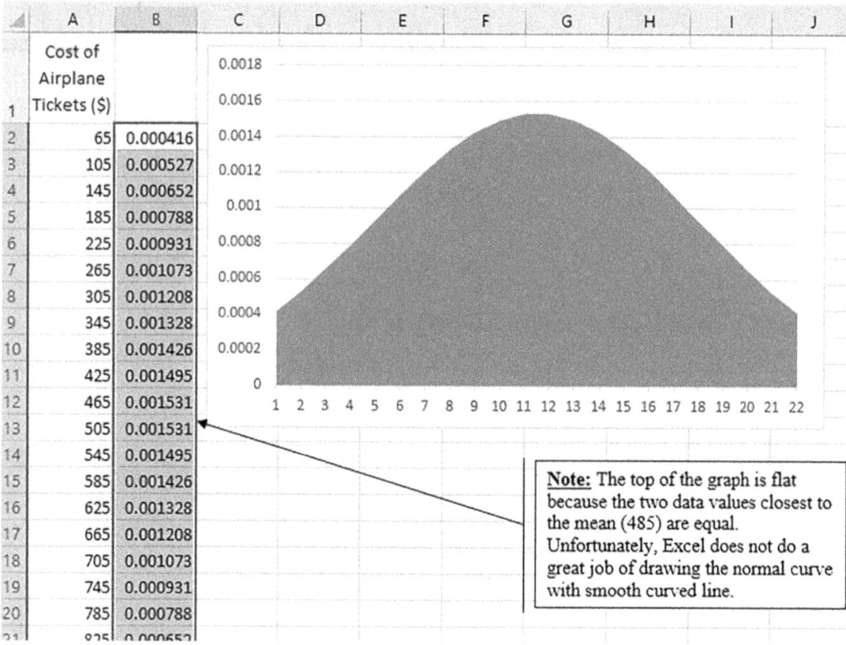

2. **Converting Percentiles to Measured Units (NORM.INV)**

Excel can covert percentile values in a normally distributed data set into measured units.

Example Problem

Your distribution company provides delivery service 7 days a week to stores selling synthetic fire logs. You need to determine how many trucks and drivers to have on hand to deliver the product. The data you have collected from last year indicates the number of stores requiring deliveries on a daily basis during the months of December and January. This data seems to follow a normal distribution. The average number of stores requiring deliveries on any given day during this time is 100. The standard deviation is 15 stores. Let's analyze the data.

Calculate the Number of Stores Corresponding with the 99th Percentile

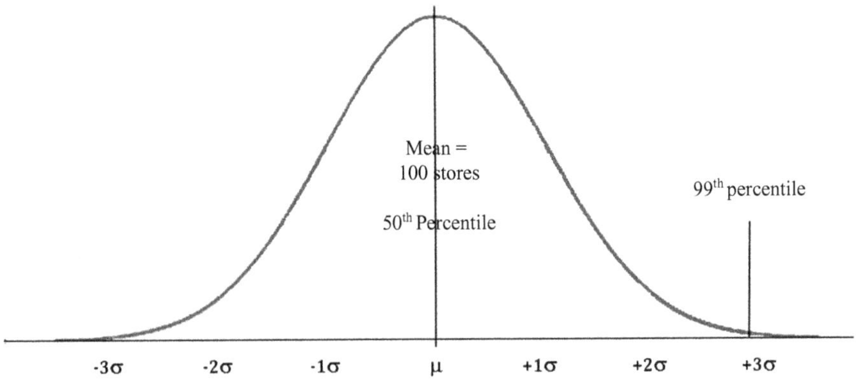

- Click on an empty cell where you want the answer
- Click on the ***fx*** function

- Type "=" in the white input box

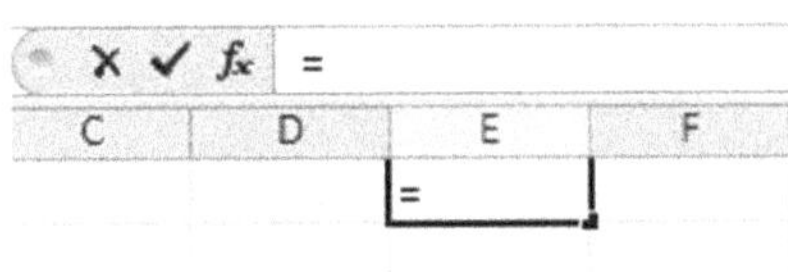

Type the function

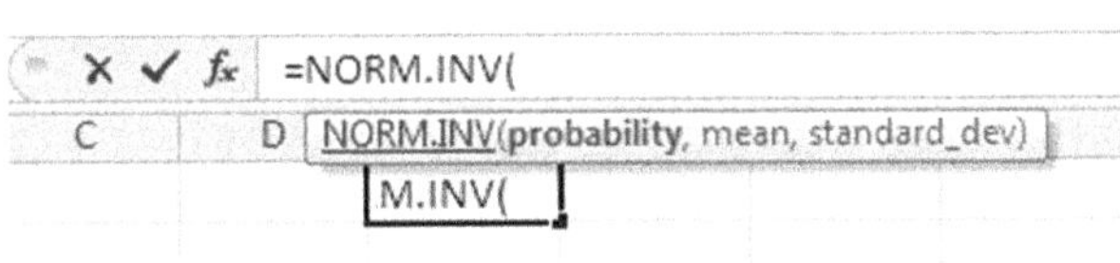

Or highlight the function from Insert Function window

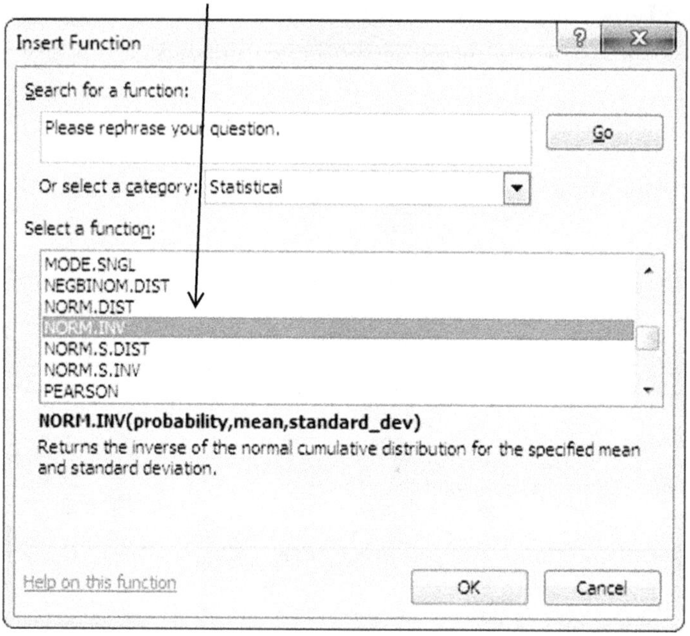

- Input the percentile value that you want to convert. **Input as a decimal value.** This is referred to as the probability by Excel.

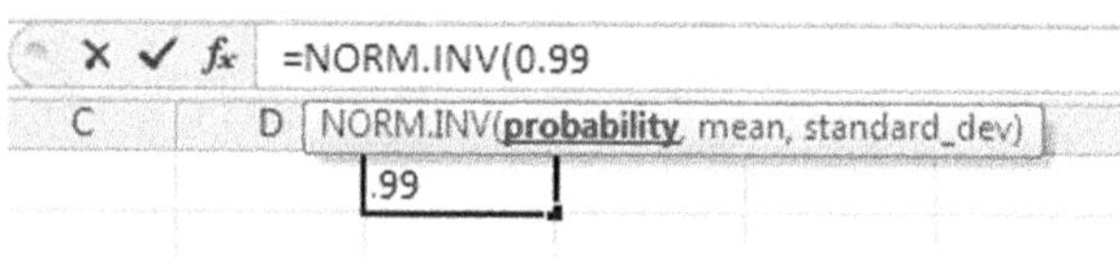

- Input the **mean** from your data set

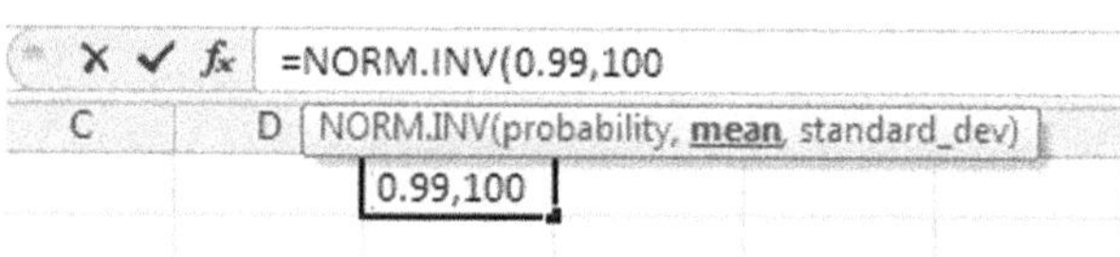

- Input the **standard deviation** from your data set

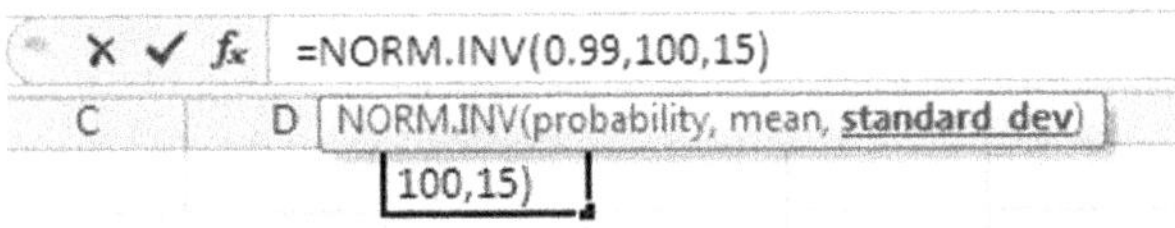

- Click the checkmark to the left of ***fx***

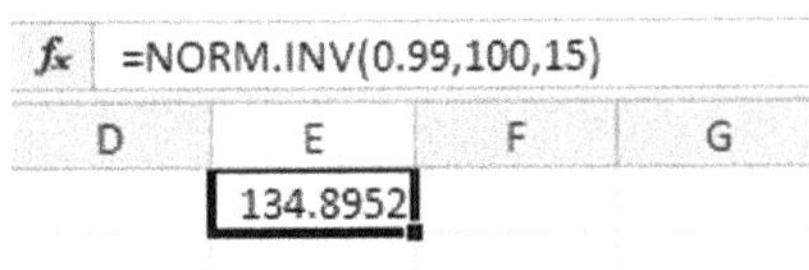

The answer is given as 134.8952, which means that the 99th percentile for this data set is about 135 stores. In other words, about 135 stores or less required deliveries on any given day 99% of the time during December and January of last year.

- No matter what the decimal indicates, we need to round up because we cannot have part of a store.

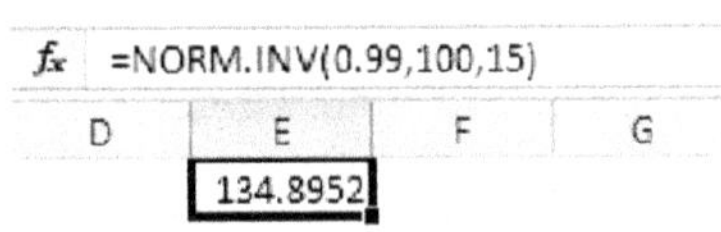

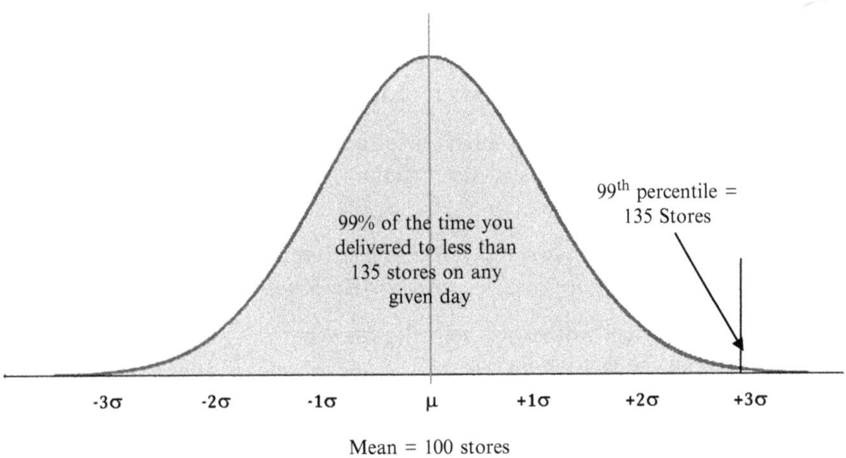

3. **Converting Measured Units to *z*-Scores (STANDARDIZE)**

Example Problem

Your distribution company provides delivery service 7 days a week to stores selling synthetic fire logs. You need to determine how many trucks and drivers to have on hand to deliver the product. The data you have collected from last year indicates the number of stores requiring deliveries on a daily basis during the months of December and January. This data seems to follow a normal distribution. The average number of stores requiring deliveries on any given day during this time is 100. The standard deviation is 15 stores.

Convert the Measured Value of 135 Stores to a z-Score

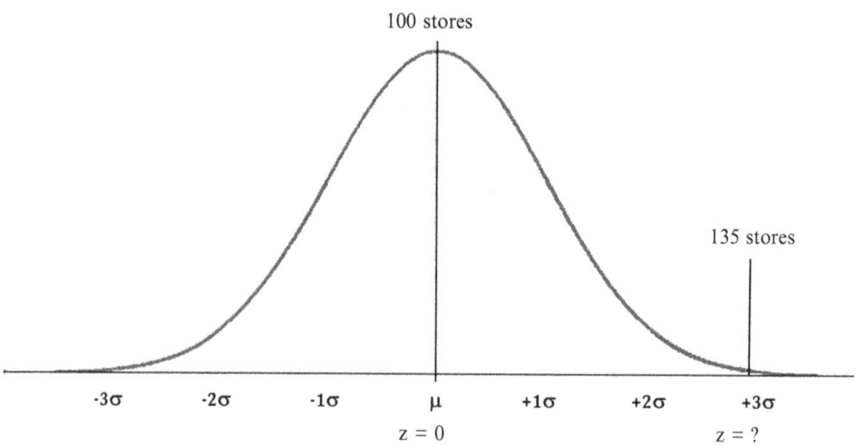

- Input in the **mean** number of stores as **100** into cell G5.
- Input the **standard deviation** as **15** stores into cell F5.
- Input the value that we want to convert of **135** stores into cell E5.
  - This problem could also be solved by directly typing in these required input values within the brackets of the function. The **STANDARDIZE** function uses the following format:

    STANDARDIZE (***observed value*** *in measured units to be converted to a z-score*, ***mean*** *of the data set*, ***standard deviation*** *of the data set)*

- Click on an empty cell where you want the answer
- Input "=" inside the ***fx*** formula box
- Type in the formula using cell locations for the observed value (x), mean, and standard deviation. The formula divides the z-score by the sd and adds it back to the mean.

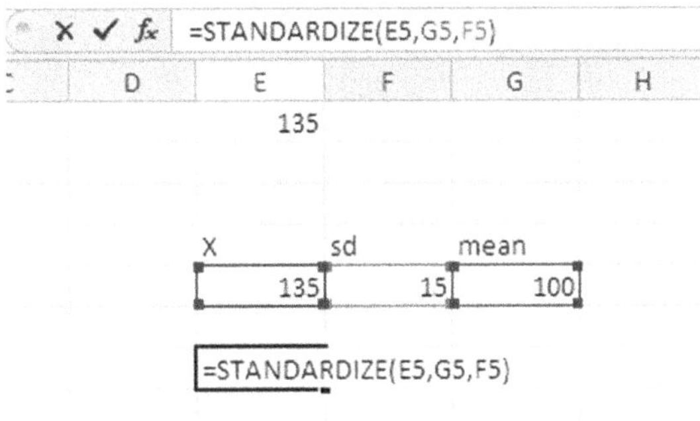

- Click on the checkmark

So in this case, 135 stores equate to a z-score of 2.326347 standard units. We should round the z-score to a value of approximately 2.33.

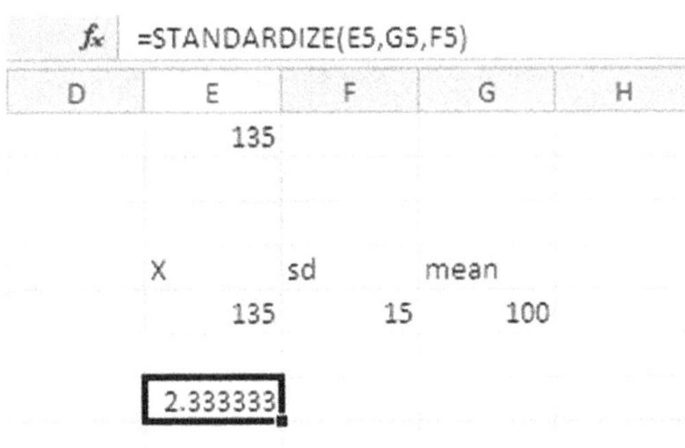

4. **Evaluating Normality**

Many continuous variables used in business closely follow a normal distribution. This section presents two approaches for determining whether a set of data can be approximated by the normal distribution:

1. Comparing the characteristics of the data with the theoretical properties of the normal distribution
2. Constructing a normal probability plot

1. **Comparing Data Characteristics to Theoretical Properties**

The normal distribution has several important theoretical properties:

- It is symmetrical; thus, the mean and median are equal. If the mean is greater than the median then the data is skewed to the right. If the mean is less than the median then the data is skewed to the left.
- It is bell-shaped; thus, the empirical rule applies. The empirical rule states that 68% of the data collected will fall between one standard deviation. It also states that 95% of the data falls between two standard deviations. Furthermore it states that 99.7% of the data should fall within 3 standard deviations.
- The skewness statistic is close to zero. Skewness is an evaluation of a given data set's asymmetry.
- The range is approximately equal to 6 standard deviations.

In actual practice, a continuous variable may have characteristics that approximate these theoretical properties. However, many continuous variables are neither normally distributed nor approximately normally distributed. For such variables, the descriptive characteristics of the data do not match well with the properties of a normal distribution. One approach to determining whether a data set follows a normal distribution is to compare the characteristics of the data with what would be expected if the data followed a normal distribution:

- Construct charts and observe their appearance. For small- or moderate-sized data sets, create a stem-and-leaf display or a boxplot. For large data sets, in addition, plot a histogram or polygon.
- Compute descriptive statistics and compare these statistics with the theoretical properties of the normal distribution. Compare the mean and median. Is the interquartile range approximately 1.33 times the standard deviation? Is the range approximately 6 times the standard deviation?
- Evaluate how the values are distributed. Determine whether approximately 68% of the values lie between the mean and ±1 standard deviation. Determine whether approximately 95% of values lie between the mean and ±2 standard deviations.

Figure 4.3 displays descriptive statistics for data that is not normally distributed. You can determine this by comparing several statistics:

- The mean (1.3169) does not equal the median (2.60); they would be equal if this was a perfect normal distribution.
- The skewness statistic is not close to zero. The negative sign indicates a left-hand skew.
- The range of 46.9 is equal to 6.13 standard deviation of the mean. In a normal distribution, the range is approximately 6 standard deviations, so this is very close.
- 76.69% of the returns are within ±1 standard deviation of the mean. (In a normal distribution, 68.26% of the values lie within ±1 standard deviation of the mean.)
- 93.89% of the returns are within ±2 standard deviations of the mean. (In a normal distribution, 95.44% of the values lie within ±2 standard deviations of the mean.)

Based on these statements, the data is highly left-skewed and has somewhat more values within ±1 standard deviation of the mean than expected. The range is

Fig. 4.3 Descriptive statistics for the data

2		
3	Mean	1.3169
4	Standard Error	0.5704
5	Median	2.6000
6	Mode	6.9000
7	Standard Deviation	7.6530
8	Sample Variance	58.5680
9	Kurtosis	1.7690
10	Skewness	-1.0445
11	Range	46.9000
12	Minimum	-31.9000
13	Maximum	15.0000
14	Sum	237.0500
15	Count	180.0000

approximately what would be expected in a normal distribution. Primarily because of the skewness you can conclude the characteristics of the data differ from the theoretical properties of a normal distribution.

2. **Constructing the Normal Probability Plot**

A normal probability plot is a visual display that helps you evaluate whether the data is normally distributed. One common plot is called the **quantile–quantile plot**. To create this plot, you first transform each ordered value to a z value. For example, if you have a sample of $n = 19$, the z value for the smallest value corresponds to a cumulative area of $1/(n + 1) = 1/(19 + 1) = 1/20 = 0.05$. The z value for the cumulative area of 0.05 (from z table) is -1.65. The table below illustrates the entire set of z values for a sample of $n = 19$.

Ordered value	z value	Ordered value	z value
1	−1.65	11	0.13
2	−1.28	12	0.25
3	−1.04	13	0.39
4	−0.84	14	0.52
5	−0.67	15	0.67
6	−0.52	16	0.84
7	−0.39	17	1.04
8	−0.25	18	1.28
9	−0.13	19	1.65
10	−0.00		

To construct the quantile–quantile plot, the z values are plotted on the x-axis, and the corresponding values of the variable are plotted on the y-axis. If the data is normally distributed, the values will plot along an approximately straight line.

Figure 4.4 illustrates the typical shape of normal probability plots for a left-skewed distribution (panel A), a normal distribution (panel B), and a right-skewed distribution (panel C). If the data is left-skewed, the curve will rise more rapidly at

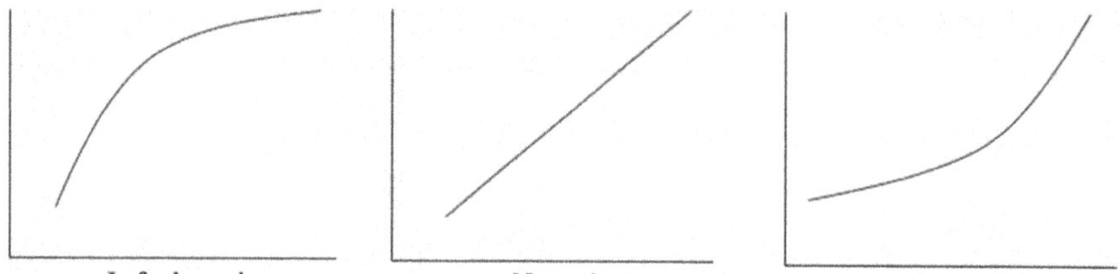

Fig. 4.4 Normal probability plots for a left-skewed distribution, a normal distribution, and a right-skewed distribution

first and then level off. If the data is right-skewed, the data will rise more slowly at first and then rise at a faster rate for higher values of the variable being plotted.

5. **Non-normal Distributions**

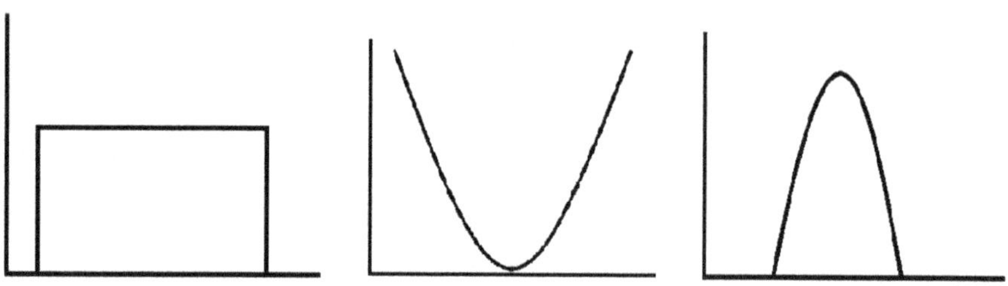

Not all distributions are normally distributed, or you simply may not know what the distribution looks like. To err on the side of conservative estimates, it is better to use the Chebyshev approximation in these cases. At least $(1-1/k^2)$ of the items in any data set will be within k standard deviations of the mean, where k is any value greater than 1.

- At least 75% of the items must be within $k = \pm 2$ standard deviations of the mean.
- At least 89% of the items must be within $k = \pm 3$ standard deviations of the mean.
- At least 94% of the items must be within $k = \pm 4$ standard deviations of the mean.

There are two steps in this type of problem:

Step 1: Convert measured units to standard units (STANDARDIZE)
Step 2: Use Chebyshev approximation (*fx*)

Example Problem
The data set of apartment rents has a mean of $490.80 and a standard deviation of $54.75. The data is not normally distributed.

Calculate what percentage of rents fall between $409 and $573

- With this approximation, we can only calculate the minimum amount of data that would fall in the range of interest.

Step 1: Convert the measured values to standard units

For the first boundary value of 409

- Click on an empty cell where you want the answer
- Input "=" inside the ***fx*** formula box

Type in the formula using cell locations for the observed value (x) of 409, mean, and standard deviation

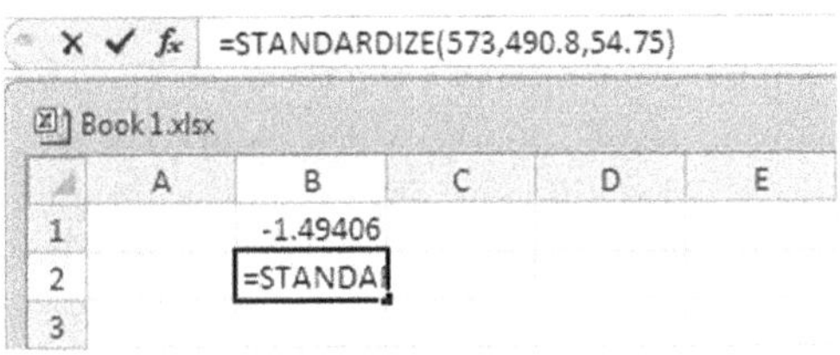

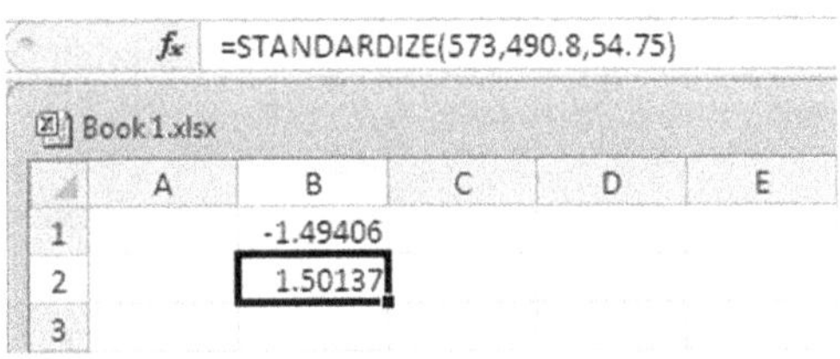

- The resulting standardized value for 409 is −1.49406 or approximately −1.5

For the second boundary value of 573

- Click on an empty cell where you want the answer
- Input "=" inside the ***fx*** formula box
- Type in the formula using cell locations for the observed value (x) of 573, mean, and standard deviation
- The resulting standardized value for 573 is 1.50137 or 1.5

Step 2: Use Chebyshev approximation = 1 – (1/(k)²), where k is the boundary value in standard units

- Click on an empty cell where you want the answer
- Input "=" inside the ***fx*** formula box
- Type in the formula using cell locations for the observed value (x) in standard units

In this example the formula $= \mathbf{1} - \left(\mathbf{1} / (\mathbf{k})^2\right)$ is input to Excel as

$$=1-\left(1/\left((-1.5)^2\right)\right)$$

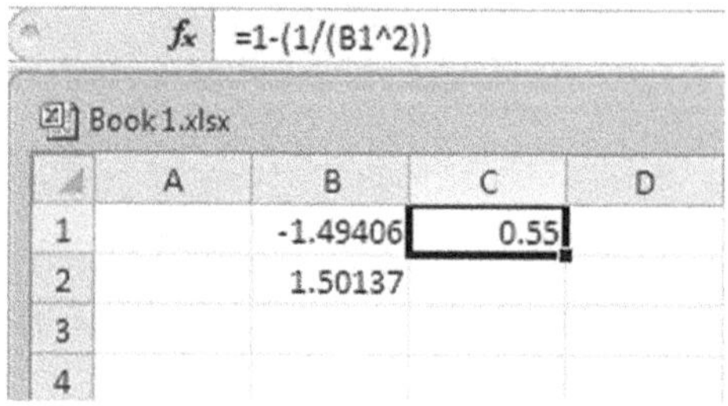

Answer: *At least* 55% of the rents fall between \$409 and \$573.

Note in this case the standard units were equal on either side of zero. For non-symmetric problems where the two *z*-scores are different distances away from the mean, you can perform a series of subtractions just as you would if the distribution was normal. For example, if the lower *z*-score was +1 and the upper *z*-score was +2, calculate how much data falls between ±1 using the Chebyshev formula and subtract that amount from what falls between ±2. Then divide by 2 so you have just the part greater than the mean. But this approximation only indicates the minimum amount of data that exists in this region.

Common Pitfalls

- Many of the Excel functions only differ by one letter, so make sure you have the right normal distribution function.
- Remember Excel gives percentages as decimals. Multiply by 100 to get a whole number percentage.
- Some versions of Excel may have a period in the function name, so make sure you include this to get the right function.
- Chebyshev approximation does not work with small standard deviations (e.g., the sd must be greater than 1).
- Remember the Chebyshev approximation only calculates the minimum amount of data (at least…).
- **NORM.INV** does not provide the percent of distributions nor frequency on the y-axis. It is used for graphing purposes only.

Final Thoughts and Activities

Practice Problems

1. Find the percentage of the area under the normal curve from −1.2 to 1.2 (in std units). In other words, find the area under the curve between these two *z*-scores. Use the table or the interactive graphic at the **mathisfun.com** website.

2. The heights of women who shop in your store average 63.5 inches with an sd of 3.0 inches. Use the normal curve to estimate the percentage of women with heights over 59 inches.
3. The data in the file **Amusementparks** contains the starting admission price (in $) for 1-day passes to 10 amusement parks in the United States.

   (a) Compute the mean, range, and standard deviation.
   (b) Compute the z-scores.
   (c) Based on the results from (a) and (b), what conclusions can you reach concerning the starting admission price for 1-day passes?

4. The data in the file **Candy** represents the cost per pound ($) for a sample of candy.

   (a) Compute the mean, range, and standard deviation.
   (b) Compute the z-scores.
   (c) Is the data skewed? If so, how?

5. You are an operations analyst for AT&T. The length of long-distance telephone calls may be normally distributed, but you are not certain. You select a random sample of 25 calls. The average length is 8 min and the SD is 2 min. What % of these calls should be between 4.8 and 11.2 min?
6. It has been reported that the average hotel check-in time, from curbside to delivery of bags into the room, is 12.1 min. Mary has just left the cab that brought her to her hotel. Assuming a normal distribution with a standard deviation of 2.0 min, she is interested in what percentage of guests get their bags to their rooms in a time. Complete using hand calculations and Excel:

   (a) Greater than 14.1 min?
   (b) Less than 8.1 min?
   (c) Between 10.1 and 14.1 min?
   (d) Between 10.1 and 16.1 min?

7. Toby's Trucking Company determined that the distance traveled per truck per year is normally distributed, with a mean of 50.0 thousand miles and a standard deviation of 12.0 thousand miles.

   (a) What proportion of trucks can be expected to travel between 34.0 and 50.0 thousand miles in a year?
   (b) What percentage of trucks can be expected to travel either below 30.0 or above 60.0 thousand miles in a year?

8. According to the American Society for Quality, a certified quality engineer (CQE) is a professional who understands the principles of product and service quality evaluation and control. In a survey, the mean salary of 1387 CQEs was $72,824, with a standard deviation of $18,796. Assume that the salaries of CQEs is approximately normally distributed. For a randomly selected CQE, what is the probability that he or she has a salary

(a) Below \$50,000?
(b) Above \$75,000?
(c) Above \$100,000?

8. An orange juice producer buys all his oranges from a large orange grove. The amount of juice squeezed from each of these oranges is approximately normally distributed, with a mean of 4.70 ounces and a standard deviation of 0.40 ounce.

   (a) What is the probability that a randomly selected orange will contain between 4.70 and 5.00 ounces of juice?
   (b) What is the probability that a randomly selected orange will contain between 5.00 and 5.50 ounces of juice?

10. A set of final examination grades in an introductory statistics course is normally distributed, with a mean of 73 and a standard deviation of 8.

   (a) What is the probability that a student scored below 91 on this exam?
   (b) What is the probability that a student scored between 65 and 89?
   (c) The probability is 5% that a student taking the test scores higher than what grade?

11. In 2019, the per capita consumption of coffee in Sweden was reported to be 8.2 kg, or 18.04 pounds. Assume that the per capita consumption of coffee in Sweden is approximately distributed as a normal random variable, with a mean of 18.04 pounds and a standard deviation of 5 pounds.

   (a) What is the probability that someone in Sweden consumed more than 10 pounds of coffee in 2019?
   (b) What is the probability that someone in Sweden consumed between 3 and 5 pounds of coffee in 2019?
   (c) What is the probability that someone in Sweden consumed less than 5 pounds of coffee in 2019?
   (d) 99% of the people in Sweden consumed less than how many pounds of coffee?

12. Given a distribution with a mean = 7 and a standard deviation = 2. Convert the following z-scores to measured units: −1.28, .5, 0

Discussion Boards

1. IQ scores are normally distributed with a mean of 100 and a standard deviation of 15. MENSA is an organization for people with high IQs; eligibility requires an IQ of about 131.5. *New York Times* reporter Trish Hall notes that people are scoring substantially higher on IQ tests than in the past. Discuss.

2. Discuss some of the cultural and political factors that could affect a normal distribution of voter ages in government election results in different countries around the world.

Group Activity

1. Queuing Theory is a branch of statistics that is very important to businesses such as McDonald's, Safeway, and Disney. Disney conducts extensive studies of lines at its amusement parks to keep patrons happy and to assist with expansion plans. Discuss how all of this relates to the normal distribution.
2. Ergonomics is the study of problems associated with people adjusting to their environments. A case in point was faced by the US Air Force in studying the cockpits of fighter jets. These were originally designed for men but of course now we also have women pilots. Research information on the web in conjunction with a normal distribution approach to address this issue.

Parting Thought

Smoking is one of the leading causes of statistics.

Problem Solutions

1. Find the percentage of the area under the normal curve from -1.2 to 1.2 (in std units). In other words, find the area under the curve between these two z-scores. Use the table or the interactive graphic at the mathisfun.com website.

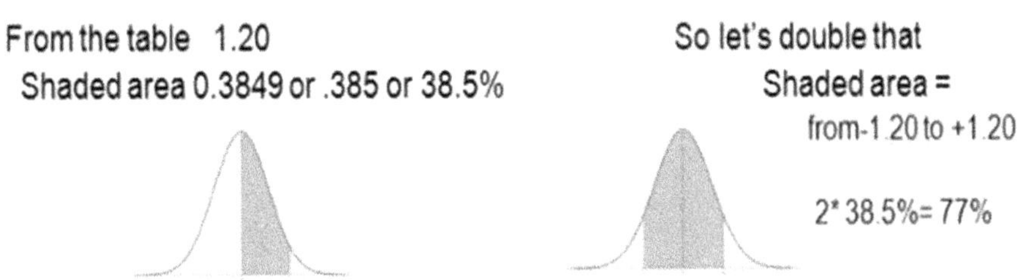

The area under the normal curve between $+1.2$ and -1.2 is 77%. In other words, 77% of the data falls between $+1.2$ and -1.2 std units… regardless of the measured units.

2. The heights of women who shop in your store average 63.5 inches with an sd of 3.0 inches. Use the normal curve to estimate the percentage of women with heights over 59 inches.

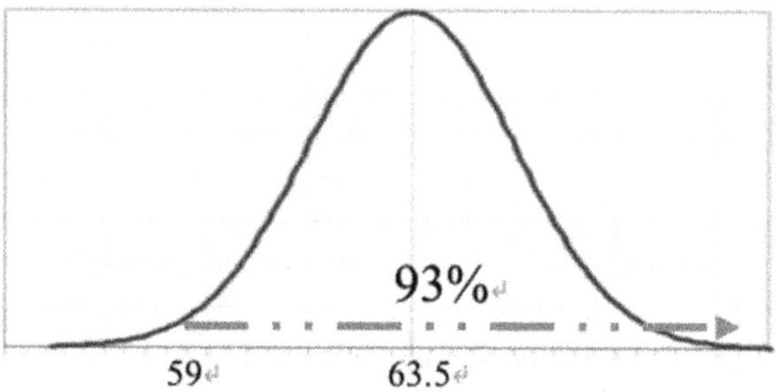

Convert 59 into su: (59 − 63.5)/3 = −1.5
If you use the z-table for 1.5: 86.64%
Calculate the tails: (100% − 86.54%) = 14% in both tails
The right-hand tail has: 14%/2 = 7%
Therefore, the area to the left of that tail has 100% − 7% = 93%
There are 93% of the women with heights over 59 inches.

3. The data in the file Amusementparks contains the starting admission price (in $) for 1-day passes to 10 amusement parks in the United States.
   (a) Compute the mean, median, range, and standard deviation.

Admission	
Mean	49.80
Median	45.50
Standard deviation	11.10
Sample variance	123.29
Range	34.00
Count	10.00

   (b) Compute the z-scores.

Location	Admission ($)	z-score
Boo's Gardens	61	1.01
Animal Adventures	66	1.46
Dolly's Wonderland	44	−0.52
Chocolate Heaven Park	45	−0.43
Bob's Children's Park	32	−1.60
Mouse's Great Adventure	53	0.29
Fairy Lake Park	65	1.37
Silver Lining Park	46	−0.34
Adventure City Rides	43	−0.61
Mountain Explorer's Park	43	−0.61

(c) Based on the results from (a) and (b), what conclusions can you reach concerning the starting admission price for 1-day passes?

1. Mean vs. Median: The mean (49.8) is higher than the median (45.5). This suggests the data is slightly skewed to the right (because a few higher values—65 and 66—pull the mean above the median).
2. Spread of Values (Range and Std. Dev.): The range is 34 (66–32), which is quite wide relative to the prices themselves. The standard deviation (about 11.3) indicates a moderate spread around the mean.
3. Mode: The most common price is $43, but it only appears twice, so the dataset isn't heavily concentrated around the mode.
4. Z-Scores: The largest positive z-score is about +1.43 (for $66). That's not extremely large (less than 2), so there aren't any extreme outliers on the high side. The lowest z-score is about −1.57(for $32). Also not extremely low.

4. The data in the file **Candy** represents the cost per pound of ($) for a sample of candy.

(a) Compute the mean, median, range, and standard deviation.

Cost ($)	
Mean	1.68
Median	1.63
Standard deviation	0.33
Sample variance	0.11
Range	0.96
Count	14.00

(b) Compute the z-scores. Are there any outliers? Explain

Cost ($)	z-score
1.43	−0.75
1.47	−0.63
1.67	−0.02
1.89	0.65
2.17	1.51
1.69	0.04
1.52	−0.48
1.32	−1.09
2.26	1.79
1.32	−1.09
1.30	−1.15
1.61	−0.20
2.16	1.48
1.65	−0.08

There are no outliers because none of the z-scores has an absolute value that is greater than 3.0.

(c) Is the data skewed? If so, how?

Based on the mean (1.68) being greater than the median (1.63), the distribution shows a slight right (positive) skew. In a right-skewed distribution, a few larger values pull the mean above the median. Since the difference (1.68 – 1.63 = 0.05) is relatively small compared to the standard deviation (0.33), the skew is likely modest rather than extreme.

5. You are an operations analyst for AT&T. The length of long-distance telephone calls may be normally distributed, but you are not certain. You select a random sample of 25 calls. The average length is 8 min and the SD is 2 min. What % of these calls should be between 4.8 and 11.2 min?

Given: avg = 8 min, sd = 2 min

- Convert 11.2 and 4.8 min to std units
- Enter "=STANDARDIZE(11.2,8,2)" in a cell
- Result: 1.6
- And enter "=STANDARDIZE(4.8,8,2)" into another cell
- Result: −1.6
- If $k = 1.6$ AT LEAST $\{1 - (1/k^2)\}$ will be within ± 1.6 standard deviations of the mean
- Enter "=1−(1/(1.6^2))" in one cell
- =1−(1/2.56)=Answer: .61 or AT LEAST 61% of the calls will be between 4.8 and 11.2 min

Book1.xlsx

	A	B	C	D
1		-1.6	0.61	
2		1.6	0.61	
3				

6. It has been reported that the average hotel check-in time, from curbside to delivery of bags into the room, is 12.1 min. Mary has just left the cab that brought her to her hotel. Assuming a normal distribution with a standard deviation of 2.0 min, she is interested in what percentage of guests get their bags to their rooms in a time. Complete using hand calculations and Excel:

(a) Greater than 14.1 min?

(b) Less than 8.1 min?

(c) Between 10.1 and 14.1 min?
(d) Between 10.1 and 16.1 min?

(a) Greater than 14.1 min:
Given in the problem:
$\mu = 12.1$ min, $\sigma = 2.0$ min
$P(x > 14.1) = P(z > 1.00) = .5 - .3413 = \boldsymbol{0.1587}$

$$z = \frac{x - \mu}{\sigma} = \frac{14.1 - 12.1}{2.0} = 1.00$$

Answer: 15.9%

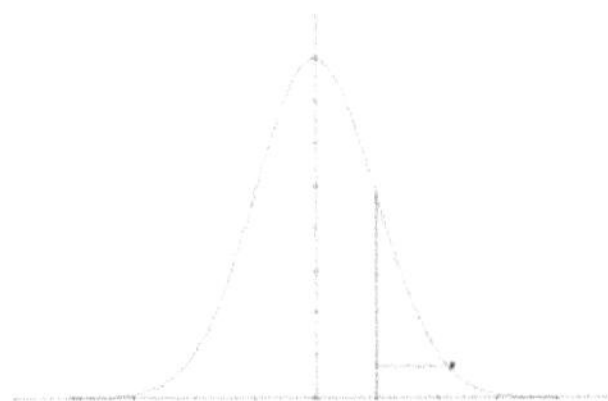

(b) Less than 8.1 min:
Given in the problem:
$\mu = 12.1$ min, $\sigma = 2.0$ min
$P(x < 8.1) = P(z < -2.00) = .5 - .4772 = \boldsymbol{0.0228}$

$$z = \frac{x - \mu}{\sigma} = \frac{8.1 - 12.1}{2.0} = -2.00$$

Answer: 2.2%

(c) Between 10.1 and 14.1 min
$\mu = 12.1$ min, $\sigma = 2.0$ min
$P(10.1 < x < 14.1) = P(-1.00 < z < 1.00) = 0.3413 + 0.3413 = \boldsymbol{0.6826}$
Answer: 68.3%

(d) Between 10.1 and 16.1 min
$\mu = 12.1$ min, $\sigma = 2.0$ min
$P(10.1 < x < 16.1) = P(-1.00 < z < 2.00) = 0.3413 + 0.4772 = \boldsymbol{0.8185}$
Answer: 81.9%

Excel Solution

Problem: What is the % of guests that receive their bags in delivery times:

(a) Greater than 14.1 min?
In a cell on an Excel worksheet, type
=1-NORMDIST(14.1,12.1,2,true) and you will see the answer: *0.1587*
(b) Less than 8.1 min?
In a cell on an Excel worksheet, type
=NORMDIST(8.1,12.1,2,true) and you will see the answer: *0.0228*
(c) Between 10.1 and 14.1 min?
In a cell on an Excel worksheet, type all on one line
=NORMDIST(14.1,12.1,2,true)-NORMDIST(10.1,12.1,2,true) and you will see the answer: *0.6826*
(d) Between 10.1 and 16.1 min?
In a cell on an Excel worksheet, type all on one line
=NORMDIST(16.1,12.1,2,true)-NORMDIST(10.1,12.1,2,true) and you will see the answer: *0.8185*

7. Toby's Trucking Company determined that the distance traveled per truck per year is normally distributed, with a mean of 50.0 thousand miles and a standard deviation of 12.0 thousand miles. (a) What proportion of trucks can be expected to travel between 34.0 and 50.0 thousand miles in a year? (b) What percentage of trucks can be expected to travel either below 30.0 or above 60.0 thousand miles in a year?
   (a) 0.9089
   (b) 0.0911
   (c) +1.96
   (d) −1.00 and +1.00
8. According to the American Society for Quality, a certified quality engineer (CQE) is a professional who understands the principles of product and service quality evaluation and control. In a survey, the mean salary of 1387 CQEs was $72,824, with a standard deviation of $18,796. Assume that the salaries of CQEs is approximately normally distributed. For a randomly selected CQE, what is the probability that he or she has a salary (a) Below $50,000? (b) Above $75,000? (c) Above $100,000?
   (a) 0.9599
   (b) 0.0228
   (c) 43.42
   (d) 46.64 and 53.36
9. An orange juice producer buys all his oranges from a large orange grove. The amount of juice squeezed from each of these oranges is approximately normally distributed, with a mean of 4.70 ounces and a standard deviation of 0.40 ounce. (a) What is the probability that a randomly selected orange will contain between 4.70 and 5.00 ounces of juice? (b) What is the probability that a randomly selected orange will contain between 5.00 and 5.50 ounces of juice?
   (a) $P(34 < X < 50) = P(-1.33 < Z < 0) = 0.4082$

(b) $P(X < 30) + P(X > 60) = P(Z < -1.67) + P(Z > 0.83) = 0.0475 + (1.0 - 0.7967) = 0.2508$

(c) $P(Z < -0.84) \approx 0.20$, $Z = -0.84 = \dfrac{X-50}{12}$, $X = 50 - 0.84 \times 12 = 39.92$ thousand miles, or 39,920 miles.

(d) The smaller standard deviation makes the Z values larger.

(a) $P(34 < X < 50) = P(-1.60 < Z < 0) = 0.4452$

(b) $$P(X<30)+P(X>60)=P(Z<-2.00)+P(Z>1.00) \\ =0.0228+(1.0-0.8413)=0.1815$$

(c) $X = 50 - 0.84(10) = 41.6$ thousand miles, or 41,600 miles.

10. According to the American Society for Quality, a certified quality engineer (CQE) is a professional who understands the principles of product and service quality evaluation and control. In a survey, the mean salary of 1387 CQEs was $72,824, with a standard deviation of $18,796. Assume that the salaries of CQEs is approximately normally distributed. For a randomly selected CQE, what is the probability that he or she has a salary (a) Below $50,000? (b) Above $75,000? (c) Above $100,000?
    (a) 0.1123
    (b) 0.4539
    (c) 0.0741
11. An orange juice producer buys all his oranges from a large orange grove. The amount of juice squeezed from each of these oranges is approximately normally distributed, with a mean of 4.70 ounces and a standard deviation of 0.40 ounce. (a) What is the probability that a randomly selected orange will contain between 4.70 and 5.00 ounces of juice? (b) What is the probability that a randomly selected orange will contain between 5.00 and 5.50 ounces of juice?
    (a) Five-number summary: 82 127 148.5 168 213; mean = 147.06, mode = 130, range = 131, interquartile range = 41, standard deviation = 31.69. The mean is very close to the median. The five-number summary suggests that the distribution is approximately symmetrical around the median. The interquartile range is very close to 1.33S. The range is about $50 below 6S. In general, the distribution of the data appears to closely resemble a normal distribution.
    (b) The normal probability plot confirms that the data appears to be approximately normally distributed.
12. A set of final examination grades in an introductory statistics course is normally distributed, with a mean of 73 and a standard deviation of 8. (a) What is the probability that a student scored below 91 on this exam? (b) What is the probability that a student scored between 65 and 89? (c) The probability is 5% that a student taking the test scores higher than what grade?
    (a) 0.9878
    (b) 0.8185
    (c) 86.16%

13. In 2019, the per capita consumption of coffee in Sweden was reported to be 8.2 kg, or 18.04 pounds. Assume that the per capita consumption of coffee in Sweden is approximately distributed as a normal random variable, with a mean of 18.04 pounds and a standard deviation of 5 pounds. (a) What is the probability that someone in Sweden consumed more than 10 pounds of coffee in 2019? (b) What is the probability that someone in Sweden consumed between 3 and 5 pounds of coffee in 2019? (c) What is the probability that someone in Sweden consumed less than 5 pounds of coffee in 2019? (d) 99% of the people in Sweden consumed less than how many pounds of coffee?
    (a) 0.9461
    (b) 0.0032
    (c) 0.0045
    (d) 29.6714

Chapter 5
Survey Design

Linda Herkenhoff and John Fogli

Case Study: Sampling for Potential Growth Areas
Infinity Auto Insurance is looking to expand their customer base, starting with large metropolitan areas across the United States. Because the Latino/Hispanic population is the largest growing population in the United States, Infinity Auto Insurance wishes to better understand this market segment and expand its advertising campaigns to include this population. They have decided to concentrate their strategic efforts to large metropolitan areas in four different states: California, Arizona, Texas, and Florida. For the initial study, they sampled 200 people in Los Angeles, California.

Supplementary Information The online version contains supplementary material available at https://doi.org/10.1007/978-3-031-46371-6_5.

L. Herkenhoff (✉)
Saint Mary's College, Moraga, CA, USA

J. Fogli
Diablo Valley College, Pleasant Hill, CA, USA

L. Herkenhoff, J. Fogli (eds.), *Applied Statistics for Business and Management using Microsoft Excel*, https://doi.org/10.1007/978-3-031-46371-6_5

Food for Thought

What other states/cities should they target?
What other demographic segments should they consider?
From what you understand of sampling, what sample size should they use to gather a statistically valid sample?

Possible Answers

What other states/cities should they target?
Houston, Texas; Phoenix, Arizona; Miami, Florida; Austin, Texas
What other demographic segments should they consider?
Lower income or less educated individuals

Over the years, Infinity Auto Insurance has increased their sample size from 200 to 400 or 600 depending on the depth and detail they wish to obtain from the sample. With a sample size of 400, Infinity Auto Insurance is able to statistically calculate valid information at a 95% level of confidence (±5 points); with 600, 95% level of confidence (±4 points).

Food for Thought

Would you suggest they gather a larger sample or is their current sample sufficient?
If you were working for Infinity Auto Insurance, how would you go about gathering a valid sample?
What issues would help or hinder you from creating a statistically valid sample?

Possible Answers

Would you suggest they gather a larger sample or is their current sample sufficient?
A sample size of 400 or 600 should be sufficient to capture the necessary data.
If you were working for Infinity Auto Insurance, how would you go about gathering a valid sample?
Use a random sample by gathering a database of randomly selected individuals. Contact the individuals in the database randomly to take the survey.
What issues would help or hinder you from creating a statistically valid sample?
Nonresponse, nonrandom sampling, etc.

Key Concepts

Bias, Bipolar scale, Extreme checking, Likert scale, Nonprobability sampling, Primary data, Probability sampling, Random sampling, Random sampling error, Response rates, Reversed items, Secondary data and Systematic error.

Discussion

This chapter provides a framework for effective survey design. All statistical analyses require high-quality input data. The first step in collecting unbiased, reliable, and valid survey data is having an unbiased, reliable, and valid survey instrument. Several effective examples are provided as well as some of the more common pitfalls to avoid.

Basic Concepts

Primary data is collected by the investigator conducting the research. **Secondary data** is collected by someone other than the investigator. The most common way to generate primary data is through surveys. Surveys provide quick, relatively inexpensive, and efficient means of assessing information about a population. The tasks of determining the survey distribution method, determining the list of questions, and designing the correct format of the survey are essential aspects of the development of a reliable/robust survey.

Most surveys are designed to collect information from a sample of the population. Sampling is the use of a subset of the population to represent the whole population. **Probability sampling**, or **random sampling**, is a sampling technique in which the probability of getting any particular respondent is equally as likely to occur. **Nonprobability sampling** does not meet this criterion and should be used with caution. Regardless of how the data is collected, it is important to minimize errors in the data collection process. This topic is covered in more detail in Chap. 6.

- When a selection procedure is biased, taking a larger sample does not help. This just creates the bias on a larger scale.
- To minimize bias, an impartial and objective probability method should be used to select the sample.

Although most surveys are conducted to quantify factual information, certain aspects of surveys may be qualitative. Although most surveys are descriptive, they can also be designed to provide causal explanations.

Survey Design

The survey designing process is a very important step to get your future survey produced. The first step is to start planning the purpose of the survey. Start by listing the survey objectives. As you start drafting questions, make sure that your survey items correspond to an initial objective listed. A common mistake is that researchers do not get the information needed because they do not ask the right questions in the survey. Finally, keep in mind that demographic questions can be critical to analyze results.

Scale

The **Likert scale** is one of the most common scales you will come across in surveys. This scale was introduced by Likert (England; 1932). Likert scaling is a unidimensional scaling method or "summative" scale.

One of the first questions in designing a survey is how many choices should be offered on the Likert scale. The most common are 5- or 7-point Disagree–Agree response scales. Examples are given in Fig. 5.1.

There are tradeoffs between choosing a 5-point scale and a 7-point scale. The 7-point scale provides more choice for the respondent and a more detailed response set for the analyst. However, more choices may lengthen the time to complete the survey, as respondents try to decide on their answer. Too detailed responses may require more time to complete: "Was I really somewhat satisfied or just satisfied the last time I responded to this survey?"

However, having too few choices, less than 5, can be frustrating to the respondents; none of the answers really fit. This smaller scale also limits a more detailed understanding of the responses. A 3-point scale is also given in Fig. 5.1.

The final score for the respondent on the scale is the sum of their ratings for all of the items (this is why it is sometimes called a "summated" scale). On some scales, you will have items that are reversed in meaning from the overall direction

1. = Strongly unfavorable to the concept
2. = Somewhat unfavorable to the concept
3. = Neutral
4. = Somewhat favorable to the concept
5. = Strongly favorable to the concept

or

1. = Extremely unfavorable to the concept
2. = Strongly unfavorable to the concept
3. = Somewhat unfavorable to the concept
4. = Neutral
5. = Somewhat favorable to the concept
6. = Strongly favorable to the concept
7. = Extremely favorable to the concept

or

1. Agree
2. Neutral
3. Disagree

Fig. 5.1 Likert scales

of the scale. These are called **reversed items**. You will need to reverse the response value for each of these items before summing for the total.

Bipolar questions offer positive, neutral, and negative choices. Whenever you are dealing with bipolar answers, it is important to have an odd number of choices no matter how large the scale. This becomes even more important when using a Likert scale with qualitative data. In this situation, the respondent is forced to decide whether they lean more toward the *agree* or *disagree* end of the scale for each item. Figure 5.2 illustrates bipolar and unidirectional Likert scales.

In Fig. 5.2, the first bipolar scale is designed to allow respondents to agree, disagree, or remain neutral about their level of satisfaction with the recent bonus check. The second unidirectional scale is designed to only capture positive responses. In other words, it is safe to say everyone would be positive to some degree to get a surprise bonus check for $5000. These scales are easily converted to quantitative scales if they are designed correctly, which allows for more rigorous analysis.

If the bipolar scale had only been a 4-point scale, we would not have been able to collect neutral responses. However, often less rigorous surveyors will simply average the responses to synthetically create a neutral response, as shown in Fig. 5.3. This is inaccurate data and should be avoided.

Types of Questions

Single Response/Select

Single select items are questions that ask the respondent to select one specific response to fulfill the question requirement. For example, most demographic questions are single response questions. The question asks for a specific characteristic of the person's demographic status (i.e., age, gender, household income, etc.), and the respondent is allowed to respond with one specific selection.

Example

1. What is your age?

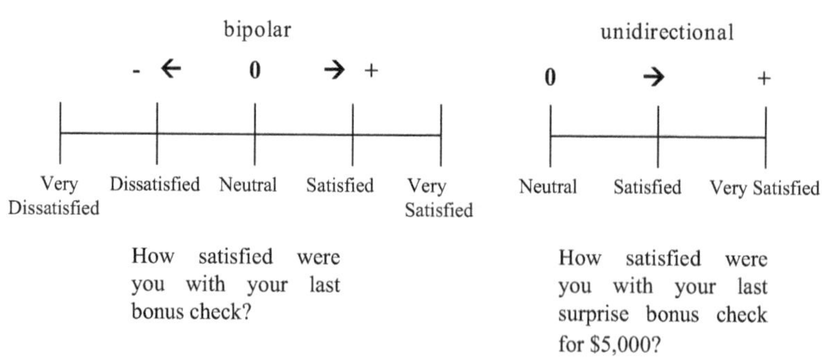

Fig. 5.2 Polarity with Likert scales

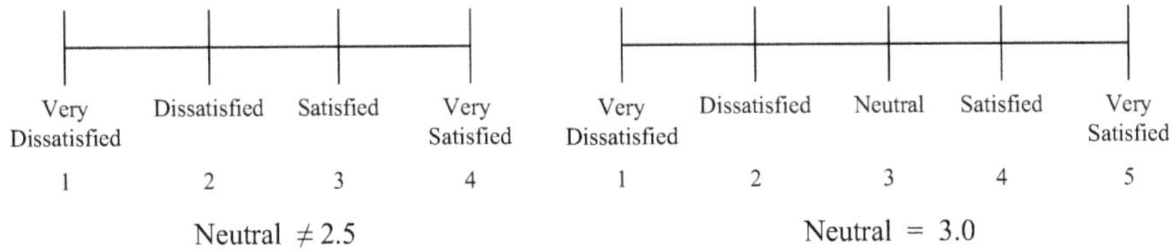

Fig. 5.3 Offering a neutral choice

- 18–24
- 25–34
- 35–44
- 45–54
- 55–64
- 65 or over

Because the respondent is a specific age, he/she will respond with only one item.

- For range items, such as age, make sure the numbers do not overlap and that the entire range is covered. Otherwise, this might cause confusion and inaccurate responses (i.e., age 18–25, age 25–30, age 30–40, etc.).

Multiple Response/Select

Multiple select questions are ones that ask the respondent to "select all that apply." A respondent can choose one or more responses. Sometimes the option "*None of the above*" is available when none of the other responses seem to fit.

Example

1. What are your favorite colors?

   - Red
   - Blue
   - Green
   - Yellow
   - Orange
   - None of the above

When a respondent chooses the different responses, he/she is allowed to choose as many responses as they deem fit. However, if the respondent chooses the "*None of the above*" option, it should deselect any previous answer selections for this question.

Structured Questions

Structured questions are questions that have a closed set of responses from which to choose. Structured questions are usually preferred because they make data collection/analysis much easier and they are quicker for the respondent to answer. Structured questions are recommended when you are not trying to capture new ideas or thoughts from the respondent. You need a good understanding of the responses so that you can develop the appropriate answer choices.

Example

Do you have an MBA?	Which international city where we have offices would you like to accept an assignment?
O Yes	O New York
O No	O Singapore
O In Progress	O Dubai
	O Hong Kong
	O Frankfurt
	O Stonetown

For structured questions, make sure that the list of answers includes every possible answer and that each of the answers is unique.

To ensure that the data you are collecting is indeed complete, you might want to include *"Other"* as the last answer choice. Including the answer choice *"Don't know"* to a response list provides a valid choice for respondents who are not capable of answering the question.

Do not include answer choices that are irrelevant to the question. You want to keep the survey as short and as easy to complete as possible.

Example of an Irrelevant Answer Choice
To which city should we relocate the head office?

- San Francisco
- Atlanta
- Jacksonville
- Detroit
- Disneyland

The last answer choice is not a real city and may be a very popular answer but is certainly not going to add much to your analysis. Although we might think it good to include humorous answers every once and awhile, keep in mind that when someone actually chooses these, you lose some real data.

All of the responses should be consistent so that no single response stands out from the other responses. Consistency ensures that you are not leading respondents to a particular answer by making that answer different from the others.

Example of an Inconsistent Answer Choice
To which city should we relocate the head office?

- San Francisco
- Atlanta
- Jacksonville
- Detroit
- Alaska

The last answer choice is not a city but a state. If we really did not want the respondent to choose a city in Alaska, this would be the way to ensure that. This confuses the respondent and distracts them from completing the survey in a valid manner.

Ranking and Rating

Surveys often gather opinion information by way of asking a person's opinion on a topic. Rating or ranking questions capture varying degrees of emotion about a subject. A rating question asks respondents to indicate the degree to which they feel about a certain topic. These are often referred to as Likert scales.

Example

How do you feel about the new flexible spending account?				
Unsatisfied	Somewhat satisfied	Satisfied	Very satisfied	Extremely satisfied
(1)	(2)	(3)	(4)	(5)

A **ranking** question asks respondents to explain how they feel about something by comparing it to other items in a list.

Example

Please rank the following health care benefits (1 being the highest rank and 4 being the lowest)
___Flexible health care savings account
___Employee assistance program
___Long-term insurance for retiree health care
___Exercise program subsidy

Nonstructured (Open-Ended) Questions

Nonstructured questions, or open-ended questions, are questions that have no list of answer choices from which to choose. Respondents write in their response to a question.

Example

What do you like best about our new CEO?
____
____

Use nonstructured questions when you are exploring new ideas and you don't really know what to expect from the respondents. This format gives them an opportunity to provide you with information that you may not have considered before. However, keep in mind that these types of questions take much longer to analyze.

Data

Labels

Another issue involves thinking you have transformed qualitative data into quantitative data because the labels have become numeric. By translating the label "satisfied" to a "3" does not make that data quantitative. Recall from our earlier chapters on descriptive statistics, we are much more limited in what type of analysis we can complete with qualitative data, regardless of whether the labels are numeric or not.

Demographic Data

To make sure the survey sample is representative of a wider population, we need to collect demographic information. Demographic items are typically categorical data. Some common demographic variables are presented in Fig. 5.4.

In addition to those variables listed in Fig. 5.4 and depending on the purpose of the survey, you may also want to collect demographic data on national culture and religion. Therefore, make sure you have included all the necessary demographic variables.

Response Rates

It is important to achieve a high response rate, otherwise we may have nonresponse bias. In other words, there might be some reason we would suspect that so many people did not respond. For example, if the survey was 15 pages long, people might not complete the survey because of how much time would be required. Those who do reply may also represent a bias. Perhaps, this group has more time on their hands, they are personal friends with the researcher, or they worked for the manager who initiated the survey. This type of bias weakens the usefulness of the data. Figure 5.5 summarizes some key design elements that may improve response rates.

Of course, the most influential factor is the subject matter of the survey. If you ask employees to complete a survey on their pay and benefits, there is usually a high response rate because it is of great interest/relevance to employees. However, if you ask employees to complete a survey about a new ad campaign there may be a very low response rate because they have been asked to do this every month for the past 6 months.

Editing: Data Quality

Before you start generating statistics from your data, you need to have a quick look at the raw data. Generate an Excel output of the raw data to check for unusual patterns in the data. Figure 5.6 gives you some examples.

In Fig. 5.6, there are seven questions and eight respondents. There are several patterns in these responses that may concern us.

- Respondent 1 has only checked 1 and 5 on a 5-point Likert scale. This type of response pattern is referred to as **extreme checking** and creates a bias in the data. Some respondents only really agree with something or really disagree; they just cannot take a neutral position.

Common Demographic Variables	
Gender	Level of education
Age	Job title
Marital status	Size of organization
Years of experience	Industry
Management level	

Fig. 5.4 Common demographic variables

1. Keep the survey as brief as possible.
2. Ensure all questions are clear and easy to understand. Avoid ambiguity ;it creates respondent frustration.
3. Maintain a consistent scale on short surveys; mix it up on longer surveys to keep it interesting for respondents.
4. Provide confidentiality and anonymity.
5. Don't ask the same question in different ways more than three times.
6. Explain what the survey results will help provide.

Fig. 5.5 Response rate factors

		Questions						
Respondent		**1**	**2**	**3**	**4**	**5**	**6**	**7**
1		5	1	5		1	1	5
2		3	2	4		3	4	6
3		3	3	3		5	2	3
4		1	1	1		1	1	1
5		3	1	2		5	5	5
6		3	2	4		3	1	5
7		3	1	5		2	5	2
8		3	3	1		5	2	3

Fig. 5.6 Patterns in response

- Question 1 seems to have an overwhelming "3" or neutral response on the 5-point Likert scale. This may mean the question was worded in a confusing way or there really was no way to disagree or agree with the question. Consider the question "*Over half of our employees were 'satisfied' in an earlier version of the survey and this is important. Do you agree or disagree?*" You may not have heard about an earlier version of the survey, let alone the survey results, so how can you agree or disagree. You don't know if the first part of the question is true or not. A remedy in this case would be to insert the word "*If*" at the start of the question.
- Question 4 has a large number of nonresponses. Again a source of potential bias.
- Respondent 2 shows a "6" as the response to question 7. This needs to be checked, since the possible answers change only from 1 to 5.
- Respondent 4 seems to have only selected "1" as a response to all questions. This respondent may not have bothered to read any of the questions. If you asked some of the questions with **reversed scales,** you can check if the answers are consistent. Consider the example.

How do you feel about adding a new cafeteria?

1	2	3	4	5
Strongly Disagree		Neutral		Strongly Agree

How do you feel about not adding a new cafeteria?

1	2	3	4	5
Strongly Disagree		Neutral		Strongly Agree

In this example, someone who strongly disagrees in the first question should strongly agree with not adding the cafeteria in the reversed question.

Coding

Coding is the process of identifying and classifying each answer with a numerical score or other character symbol. Before data can be tabulated, meaningful categories and character symbols must be established for groups of responses. In Fig. 5.7, the answers are coded as **1 for Yes**, **2 for No,** and **3 for Not sure**.

Open-ended questions pose more work in terms of coding as shown in Fig. 5.8. You need to complete a qualitative word categorization. There may be many comments concerning "*inequality*" or "*difficult to understand*." These categories can then be coded accordingly.

Be careful with **reversed scale** questions. As shown in Fig. 5.9, the answers for question 2 need to be reversed so the high agreement score still gets coded as a 1, rather than the 5.

Reversing the order of the codes for negative statements requires a simple data transformation, so the codes reflect the same direction and order of magnitude as the positive statements. Figure 5.10 shows the same questions with question 2 recoded appropriately.

The 0/1 coding is sometimes referred to as dummy variables or indicator variables. One of the most common dummy variables is gender, where we assign a 1 for

29. Do you belong to a union?

1 ☐ Yes
2 ☐ No
3 ☐ Not sure

Fig. 5.7 Basic coding of answers

30. Do you have any further comments on the new tax incentive?

____

Fig. 5.8 Coding open-ended questions

1. How do you feel about the new vacation policy?

1	2	3	4	5
Strongly Agree	Agree	Neutral	Disagree	Strongly Disagree

2. How do you feel about the new vacation guidelines?

1	2	3	4	5
Strongly Disagree	Disagree	Neutral	Agree	Strongly Agree

Fig. 5.9 Reversed scale questions without transformation

1. How do you feel about the new vacation policy?

1	2	3	4	5
Strongly Agree	Agree	Neutral	Disagree	Strongly Disagree

2. How do you feel about the new vacation guidelines?

5	4	3	2	1
Strongly Disagree	Disagree	Neutral	Agree	Strongly Agree

Fig. 5.10 Reversed scale questions with transformation

female and a 0 for male. In some situations, we may have more than two categories such as the quarters in the fiscal year or the seasons. In these cases, we use a number of indicator variables always starting with the number "0" as the first category.

Errors in Survey Question Creation

Loaded Questions

This type of question usually produces a socially desirable response or contains emotionally charged words. For example, "*Are you a cheater*?" These types of questions make assumptions about the respondent and tend to evoke certain responses from them. To avoid asking leading questions, use agree/disagree questions that are phrased with words that have neutral connotations and do not assume specific details about the respondent.

Leading Questions

This type of item phrases the question in a way that suggests to the respondent that he/she is supposed to answer in a specific manner (i.e., it leads the respondent to a particular answer). For example, "*Do you agree with Doctors worldwide that say obesity has become a chronic pandemic?*" To avoid leading questions, remove the leading words or phrases (e.g., the words that make a judgment or imply stupidity if the respondent disagrees). In this example, "*Doctors worldwide*" is leading the respondent to agree with "experts."

To avoid leading questions, phrase all questions in a neutral way. It is a common error for surveyors to word the questions in a manner that reflects their underlying opinion.

Example

Bad question: leading	Good question: neutral
Do you think that the new CEO who was hired from a completely different industry can effectively lead our company?	Do you think our new CEO can effectively lead our company?
O Yes	O Yes
O No	O No
O No opinion	O No opinion

The leading question hints that since the new CEO is from another industry this may influence his/her ability to lead. The neutral question presents a better way to phrase this question, and removes the bias.

Double-Barreled Questions

This type of question asks about more than one idea or issue in one question, which may lead to inaccuracies in the attitudes and opinions being measured. For example, "Do you agree or disagree that ice cream and cake are great?" The respondent may like ice cream but not cake, vice versa, both, or none. To avoid double-barreled questions, ask each idea separately.

- Make sure you are only asking one question at a time, this is different from multiple select type questions. These types of questions do not provide you with valid and reliable data.

Example

Bad question: double-barreled question	Good question
How have scientists and administrative staff at your research institute responded to the new parking policy?	How have *scientists* at your research institute responded to the new parking policy?
O Satisfied	O Satisfied
O Unsatisfied	O Unsatisfied
	How have *administrative staff* at your research institute responded to the new parking policy?
	O Satisfied
	O Unsatisfied

The double-barreled question is phrased in such an ambiguous way that you will not know what the respondent intended with their response (i.e., are the scientists, the administrative staff, or both satisfied/dissatisfied?). The key word (which differentiates the two questions) should be underlined to help ensure that the respondents are reading the questions correctly when the structures are so similar.

Errors in Survey Data Collection

Random Sampling Error

Even with probability sampling methods, statistical errors will occur because of chance error. **Random sampling errors** can be estimated and sample size can be increased to decrease the effect of this type of error in the data.

Systematic Error

This type of error results from some imperfect aspect of the survey or from a mistake in executing the survey. The two most common forms of **systematic error** are referred to as respondent error and nonresponse error. Nonresponse error occurs when certain groups are no longer represented in your survey because they did not respond. For example, if you were conducting a door-to-door poll in a neighborhood where typically both people work during the day, your sample will only include respondents such as retirees or people on sick leave or vacation.

If your nonresponse is high, you need to rethink your survey questions, format, and/or delivery method. Self-selection bias is another form of error that allows extreme positions to be overrepresented while other differing positions are underrepresented.

Response Bias

There are six specific categories of response bias: acquiescence, auspices, central tendency, extremity, interviewer, and social desirability. Take note that these biases are not mutually exclusive, so you can have several of these biases operating within the same survey. Surveyors should also be alert for culturally based response biases in international business research. One might expect more acquiescence bias within Japanese responses, as they have a more culturally homogeneous society and tend to not contradict their superiors.

- **Acquiescence bias**: This is a response bias due to the respondents' tendency to agree with questions when having doubt. Some respondents are very agreeable, whereas others may be very disagreeable throughout the survey. We typically find positive acquiescence when asking about a new product.
- **Auspices bias**: Answers may be influenced by the actual group doing the survey. If EarthWatch was conducting a survey about open pit mining, the respondent may be inclined to provide answers that align with the mission of the organization.

- **Central tendency bias**: This response bias usually occurs with respondents who wish to remain neutral or have no specific opinion. Often, participants who respond this way want to get through the survey quickly and don't care about the subject matter of the survey.
- **Extremity bias**: Some respondents will indicate extreme answers either in a positive and/or negative manner. These are the folks who only check the extreme ends of the scale, which may cause a bias in the data.
- **Interviewer bias**: There may be a bias introduced because of the interactions between the interviewer and the interviewee. Often, socially acceptable answers are given to impress the interviewer. The physical appearance of the interviewer may also affect the respondents. If the interviewer smiles after a particular answer is given, the respondent may tend to provide similar answers for further positive feedback. The gender of the interviewer may hamper candid responses on gender-specific issues. If the interviewer takes too long to ask the question, the respondent may feel like their time is being wasted and will start answering abruptly.
- **Social desirability bias**: The respondent may either consciously or unconsciously provide an answer to create a favorable impression. For example in asking if someone voted in the last election, you might create a social desirability bias as most respondents want to be considered responsible citizens and will report they voted. A more accurate question might be to ask if they know someone who did vote in the last election.

Survey Response Rates

After collecting your first round of data, you may be disappointed with the response rates. There are some common methods for trying to improve this situation. However, when attempting to improve survey response rates it is essential to prioritize ethical considerations to ensure the well-being and rights of participants.

- Personalized invitations: Addressing respondents by name and explaining the importance of their participation can increase response rates. Personalized invitations can be effective in creating a sense of importance, but may not work for larger sample sizes.
- Incentives: Offering small incentives (e.g., discounts, gift cards) can motivate respondents to complete the survey. Incentives can be effective in boosting response rates, but may attract respondents solely interested in the incentive rather than providing honest responses.
- Multiple reminders: Sending reminders at strategic intervals can improve response rates. Multiple reminders have been proven to be effective in reminding respondents, but excessive reminders may annoy or frustrate recipients.

By adhering to these ethical considerations researchers can foster trust, maintain participant engagement, and ensure the integrity of the survey process while striving to improve response rates. Ethical considerations in the use of surveys are crucial to ensure the well-being and rights of participants. The next few paragraphs will deal with the various roles of ethics in the survey process.

- Ethical considerations in survey design:
  1. Informed consent: Participants should provide informed consent before participating in a survey. This involves clearly explaining the purpose, potential risks/benefits, and their rights as participants. Example: Including a consent statement at the beginning of the survey explicitly stating that participation is voluntary and can be withdrawn at any time.
  2. Anonymity and confidentiality: Participants' identities and responses should be protected to maintain their privacy. Ensure that responses cannot be linked to individual participants unless explicitly permitted. Example: Using unique survey codes instead of personal identifiers to maintain anonymity.
  3. Voluntary participation: Participants should have the freedom to choose whether to participate in the survey without any coercion or pressure. Example: Clearly stating that participation is optional and unrelated to any rewards or consequences.
  4. Avoiding harm and deception: Surveys should not cause physical, emotional, or psychological harm to participants. Avoid using misleading or deceptive techniques that may compromise the integrity of the research. Example: Avoiding sensitive or triggering questions that could harm participants' well-being.
- Ethical considerations in survey conduct:
  1. Fair and balanced representation: Ensure that the survey questions are unbiased, impartial, and do not favor any particular group or viewpoint. Example: Avoiding leading or loaded questions that push participants toward a specific response.
  2. Data protection: Safeguard participants' personal information and ensure compliance with data protection regulations. Example: Storing survey data securely and using encryption when transmitting data.
  3. Respect for diversity and inclusion: Account for cultural, gender, and other relevant factors to ensure inclusivity in survey design and conduct. Example: Using gender-neutral language and offering response options that accommodate diverse perspectives.
  4. Professional integrity: Researchers should conduct themselves ethically, honestly, and transparently throughout the survey process. Example: Clearly stating the purpose of the research and disclosing any potential conflicts of interest.
- Ethical challenges in data collection:
  1. Informed consent: Ensuring participants fully understand the purpose, risks, and benefits of the survey, and providing them with the option to consent or withdraw their participation. Example: Failing to obtain informed consent when collecting sensitive information from participants.

2. Privacy and confidentiality: Protecting participants' identities and ensuring that their responses are kept confidential and used only for research purposes. Example: Sharing survey data with third parties without obtaining proper consent from participants.
3. Coercion and pressure: Ensuring that participants are not coerced or pressured to participate in the survey against their will. Example: Forcing employees to complete an internal satisfaction survey that may impact their employment or performance evaluations.

- Ethical challenges in data use and reporting:

  1. Misrepresentation or manipulation: Presenting survey data in a way that distorts or misrepresents the findings to fit a particular narrative or agenda. Example: Selectively reporting survey responses that support a preconceived hypothesis while ignoring contradictory data.
  2. Anonymity and confidentiality: Ensuring that participants' identities and responses remain anonymous and confidential in any reports or publications. Example: Accidentally revealing the identity of participants in a research report or academic publication.
  3. Data security and protection: Safeguarding survey data from unauthorized access, use, or disclosure to protect participants' privacy and prevent data breaches. Example: Failing to secure survey data, leading to a breach that exposes participants' personal information.
  4. Data ownership and intellectual property: Addressing issues related to data ownership, copyright, and intellectual property rights when using survey data collected from participants. Example: Using survey data without properly acknowledging the participants' contributions or seeking their permission for further use.

Navigating these ethical challenges requires researchers to establish clear protocols, obtain informed consent, maintain data privacy, and ensure the integrity of the data collection/reporting processes. Adhering to ethical guidelines promotes trust, protects participants' rights, and upholds the ethical standards expected in survey research.

Ethics Checklist

- **Clearly state your intentions with the research.**
  - At the top of your survey, write a brief statement explaining why you are collecting the information.
- **Reassure privacy of the data.**
  - Reassure each respondent that the information is entirely anonymous and confidential.

- **Include instructions with your survey.**
  - What may seem obvious may not be obvious to respondents; especially if the survey is being completed outside of your company. Include a short introductory set of instructions at the top of the survey and additional instructions for specific questions, as needed. Don't forget to include how the completed surveys should be submitted.
- **Don't ask for personal information unless you need it.**
  - Asking individuals to provide you with personal or demographic information may irritate some respondents and prevent them from completing your survey. (i.e., sexual orientation).
- **Keep the questions short and concise.**
  - The wording for survey questions should be short, concise, and easy to understand.
- **Order/group questions according to subject.**
  - A good way to minimize respondent distraction is to group questions together by subject. This way your respondents can focus their thoughts and answer a series of questions around these thoughts. However, be careful how you word and order your questions because order effects may also cause response bias.
- **Test the survey questionnaire.**
  - Once you have developed your survey, have a few colleagues or subject matter experts complete the survey to make sure there are no problems.

Excel

Below is an example of a survey written in Microsoft Word. With the example below, you will need to create a data dictionary that will enable you to properly code the data in Microsoft Excel. The actual data dictionary follows the survey.

- **Survey: Trends in Associating**

  Q1. How do you stay connected to others in your industry? (Select all that apply.)

  - Association/society conferences
  - Alumni networks
  - Social networking websites
  - Phone
  - Email
  - Blogs
  - Twitter

- Other: ___
- None of the above

Q2. How do you use social networking sites? (Select all that apply.)

- Staying connected with friends and family
- Communicating with others in my industry
- Communicating with clients
- Marketing my company
- Networking/finding professional opportunities
- Staying connected to a professional community (i.e., Facebook group)

Q3. Which of the following social networking sites do you use? (Select all that apply.)

- Facebook
- Twitter
- LinkedIn
- FourSquare
- Other______

Demographic Information
D1. What is your gender?

- Male
- Female

D2. What is your age?

D3. In which industry do you work?

O Arts/entertainment/recreation	O Automotive
O Construction	O Energy
O Health care	O Hospitality
O Information technology	O Manufacturing and distribution
O Marketing	O Miscellaneous
O Professional services	O Real estate
O Retail	O Telecommunications
O Transportation	O Other______

D4. How many years have you been in this industry?

A data dictionary assigns a value for each response in a question. For example, Question 1 (Q1) is a multiple select question and may cause the respondent to select multiple items. For this type of question, each response will be categorized as an individual question with the coding being 1, if the item is selected, and 0, if the item is not selected.

- **Data Dictionary**

Below is an example of how a data dictionary would appear in an Excel for the survey example.

ID. Respondent number		
[this is the number assigned to the respondent for record keeping]		
Q1. How do you stay connected to others in your industry?		
(a) Association/society conferences	1 = Selected	0 = Not Selected
(b) Alumni networks	1 = Selected	0 = Not Selected
(c) Social networking websites	1 = Selected	0 = Not Selected
(d) Phone	1 = Selected	0 = Not Selected
(e) Email	1 = Selected	0 = Not Selected
(f) Blogs	1 = Selected	0 = Not Selected
(g) Twitter	1 = Selected	0 = Not Selected
(h) Other	1 = Selected	0 = Not Selected
h_o.	[text response]	
(i) None of the above	1 = Selected	0 = Not Selected
Q2. How do you use social networking sites? (Select all that apply.)		
(a) Nonprofessionally (to stay connected with friends and family)	1 = Selected	0 = Not Selected
(b) Communicating with others in my industry	1 = Selected	0 = Not Selected
(c) Communicating with clients	1 = Selected	0 = Not Selected
(d) Marketing my company	1 = Selected	0 = Not Selected
(e) Networking/finding professional opportunities	1 = Selected	0 = Not Selected
(f) Staying connected to a professional community (i.e., Facebook group)	1 = Selected	0 = Not Selected
Q3. Which of the following social networking sites do you use? (Select all that apply.)		
(a) Facebook	1 = Selected	0 = Not Selected
(b) Twitter	1 = Selected	0 = Not Selected
(c) LinkedIn	1 = Selected	0 = Not Selected
(d) FourSquare	1 = Selected	0 = Not Selected
(e) Other	1 = Selected	0 = Not Selected
e_o.	[text response]	
D1. What is your gender?		
1 = Male		
2 = Female		
D2. What is your age?		
[text response]		

(continued)

D3. In which industry do you work?
1 = Arts/entertainment/recreation
2 = Automotive
3 = Construction
4 = Energy
5 = Health care
6 = Hospitality
7 = Information technology
8 = Manufacturing & distribution
9 = Marketing
10 = Miscellaneous
11 = Professional services
12 = Real estate
13 = Retail
14 = Telecommunications
15 = Transportation
16 = Other
D3o. Other:
[text response]
D4. How many years have you been in this industry?
[text response]

- **Note**: It is a standard practice to code your questions with a **Q** for the general survey questions, **o** for other, and a **D** for demographic questions.

After creating the data dictionary you can begin entering and coding your response data according to the variables you have established.

The following example includes an example of 10 responses to the previous survey. The question labels are in the first column and are read across each row. Each respondent is a separate column.

ID	1	2	3	4	5	6	7	8	9	10
Q1a	1	0	1	0	1	1	1	1	0	1
Q1b	0	0	0	0	0	0	0	0	1	0
Q1c	1	1	1	1	1	1	1	1	1	1
Q1d	0	1	0	0	0	1	0	0	0	1
Q1e	1	1	0	1	0	1	1	0	1	1
Q1f	0	0	0	0	1	0	0	0	0	1
Q1g	1	1	1	1	0	1	0	1	1	1
Q1h	0	0	0	0	0	0	0	0	0	0
Q1h_o										
Q1i	0	0	0	0	0	0	0	0	0	0
Q2a	1	1	0	1	1	1	1	1	1	1
Q2b	1	1	1	1	1	1	1	1	1	1
Q2c	1	0	0	0	1	0	0	1	0	0

(continued)

ID	1	2	3	4	5	6	7	8	9	10
Q2d	1	0	1	0	0	0	0	0	1	0
Q2e	0	1	0	1	0	0	0	0	0	1
Q2f	0	1	0	1	1	0	1	1	1	0
Q3a	1	1	1	1	1	1	1	1	1	1
Q3b	0	1	0	1	1	1	1	1	1	0
Q3c	1	1	1	1	1	1	1	1	1	1
Q3d	0	0	0	0	1	0	0	1	0	0
Q3e	0	0	0	0	0	0	0	0	0	0
Q3e_o										
D1	1	2	2	1	2	1	2	1	1	2
D2	25	32	56	45	22	60	38	42	54	35
D3	4	5	6	7	9	11	1	3	13	16
D3o										Tech
D4	2	5	7	15	2	40	10	24	30	6

After you have entered the data, you can analyze the numbers using descriptive statistics, frequency counts, and other methods in Excel.

Final Thoughts and Activities

Practice Problems and Case Studies

1. Below is an example of a survey. Take a look at the questions below and describe how you can improve them.

    1. Our in-store greeting

        Unacceptable—Bad—Good—Great—Excellent

    2. Store cleanliness

        Excellent—Great—Good—Bad—Unacceptable

    3. Our awesome in-store merchandise

        Not appropriate—Appropriate—Awesome

    4. Our overall store selection and bathrooms

        Unacceptable—Bad—Good—Great—Excellent

    5. Wait time at the checkout counter

        Too long—Just Right—Perfect

    6. Our sincere "Thank you" upon leaving our store

        Yes/No

7. Your overall store experience

   Awesome—Good—Not Good—Poor

2. We ask the survey question "What do you remember about advertising for the Ford pickup trucks during the Super Bowl?" How should the codes be structured for this question?
3. These are examples of leading questions. How might you correct them?
   - How great is our hardworking customer service team?
   - How awesome is the product?
   - What problems do you have with the design team?
4. The following questions are referred to as loaded questions. What are the underlying assumptions (required qualifying information) that make these questions loaded?
   - Loaded question: Where do you enjoy drinking beer?
   - Loaded question: How often do you exercise twice a day?
5. Which of the following statements about questionnaires is incorrect? Explain your answer.
   (a) Questionnaire design and development is straightforward as it does not require careful consideration and planning.
   (b) The key purpose of developing a survey questionnaire is to obtain raw information from participants to answer the research question.
   (c) A questionnaire can be very efficient in its design and structure but ineffective in answering the research question.
6. What are the advantages and disadvantages of using open-ended questions in survey design? Discuss with examples.
7. What are some common methods for increasing survey response rates? Discuss their effectiveness and potential drawbacks.
8. Discuss ethical challenges in data use and reporting.

Discussion Boards

1. How difficult is it to recruit households for the Nielsen panel? Why?
2. The use of Survey Monkey as an online tool has created some interesting ethical issues. Discuss.

Group Activity

1. Compile a list of at least five Census Bureau activities and explain what type of information is derived from these surveys.

2. Build your own survey around employee satisfaction and test it out on a small sample to get feedback.

Parting Thought

A statistician is a professional who diligently collects facts and data, then carefully draws confusions about them.

Problem Solutions

1. Answers may vary but here is an example of how the survey items can be improved

   How would you rate the following store experiences:

   1. Our in-store greeting

      Very Unacceptable—Unacceptable—Neutral—Acceptable—Very Acceptable

   2. Store cleanliness

      Very Unacceptable—Unacceptable—Neutral—Acceptable—Very Acceptable

   3. Our in-store merchandise (loaded question)

      Very Unacceptable—Unacceptable—Neutral—Acceptable—Very Acceptable

   4A. Our overall store selection (double-barreled question)

      Very Unacceptable—Unacceptable—Neutral—Acceptable—Very Acceptable

   4B. Our bathrooms

      Very Unacceptable—Unacceptable—Neutral—Acceptable—Very Acceptable

   5. Wait time at the checkout counter

      Very Unacceptable—Unacceptable—Neutral—Acceptable—Very Acceptable

   6. Our "Thank you" upon leaving our store (Leading loaded question)

      Very Unacceptable—Unacceptable—Neutral—Acceptable—Very Acceptable

7. Your overall store experience

   Very Unacceptable—Unacceptable—Neutral—Acceptable—Very Acceptable

Note: (1) The items all should have the same rating scale; should not switch between rating scale and yes/no question types. (2) All questions should have the same directional scales; should not switch between starting with very acceptable to starting with very unacceptable. Pick a direction and stick with it.

2. Answer:
   This question can be set as

   1. an open-ended question, thus the responses should be coded based on recurring themes and ideas. OR
   2. a nominal multiple response select question, thus the responses would be coded as 1 for remember an aspect of the commercial or 0 for no response to the item. OR
   3. a nominal single select question: the responses can be coded using numbers, but the numbers are not scalable and are only categorical labels.

3. These are examples of leading questions. How might you correct them?
   Answer:

   How great is our hardworking customer service team?
   Fixed: How would you describe your experience with the customer service team?
   How awesome is the product?
   Fixed: How would you rate this product?
   What problems do you have with the design team?
   Fixed: How likely are you to recommend working with the design team?

4. The following questions are referred to as loaded questions. What are the underlying assumptions (required qualifying information) that make these questions loaded?
   Answer:

   - Loaded question: Where do you enjoy drinking beer?
   - Required qualifying information: That the customer drinks beer
   - Loaded question: How often do you exercise twice a day?
   - Required qualifying information: That the customer exercises, and that they exercise twice a day

5. Which of the following statements about questionnaires is incorrect? Explain your answer:

   (a) Questionnaire design and development is straightforward as it does not require careful consideration and planning.

(b) The key purpose of developing a survey questionnaire is to obtain raw information from participants to answer the research question.
(c) A questionnaire can be very efficient in its design and structure but ineffective in answering the research question.

Answer:

The correct answer is (a). Questionnaire design and development is not straightforward and does require careful consideration and planning. Questionnaire design and development requires careful consideration of research objectives, target audience, question types, wording, sequence, and other factors. By following a systematic and thoughtful approach, you can create a questionnaire that effectively collects the desired information and provides reliable and meaningful results.

6. What are the advantages and disadvantages of using open-ended questions in survey design? Discuss with examples.
   Answer:

   Advantages:

   - Open-ended questions allow respondents to provide detailed and nuanced responses.
   - They can uncover unexpected insights and provide rich qualitative data.
   - They allow respondents to express their thoughts in their own words, giving a deeper understanding of their perspectives.
   - Example: "What are your thoughts on the company's recent product launch?"

   Disadvantages:

   - Open-ended questions can be time-consuming for respondents and may lead to survey fatigue.
   - Analyzing and interpreting open-ended responses can be challenging and time-consuming for researchers.
   - Responses may vary in length and quality, making it difficult to compare and quantify the data.
   - Example: "Please share your suggestions for improving customer service."

7. What are some common methods for increasing survey response rates? Discuss their effectiveness and potential drawbacks.
   Answer:

- Personalized invitations: Addressing respondents by name and explaining the importance of their participation can increase response rates. Personalized invitations can be effective in creating a sense of importance, but may not work for larger sample sizes.
- Incentives: Offering small incentives (e.g., discounts, gift cards) can motivate respondents to complete the survey. Incentives can be effective in boosting

response rates, but may attract respondents solely interested in the incentive rather than providing honest responses.
- Multiple reminders: Sending reminders at strategic intervals can improve response rates. Multiple reminders have been proven to be effective in reminding respondents, but excessive reminders may annoy or frustrate recipients.

8. Discuss ethical challenges in data use and reporting.

   (a) Misrepresentation or manipulation: Presenting survey data in a way that distorts or misrepresents the findings to fit a particular narrative or agenda. Example: Selectively reporting survey responses that support a preconceived hypothesis while ignoring contradictory data.
   (b) Anonymity and confidentiality: Ensuring that participants' identities and responses remain anonymous and confidential in any reports or publications. Example: Accidentally revealing the identity of participants in a research report or academic publication.
   (c) Data security and protection: Safeguarding survey data from unauthorized access, use, or disclosure to protect participants' privacy and prevent data breaches. Example: Failing to secure survey data, leading to a breach that exposes participants' personal information.
   (d) Data ownership and intellectual property: Addressing issues related to data ownership, copyright, and intellectual property rights when using survey data collected from participants. Example: Using survey data without properly acknowledging the participants' contributions or seeking their permission for further use.

Chapter 6
Sampling

Linda Herkenhoff and John Fogli

Key Concepts

Accuracy, Confidence level, Finite population, Infinite population, Proportion, Homogeneity, Population, Sample size, and Standard error.

Case Study: Customizing a Meaningful Report for Infinity Auto Insurance

Roughly 2.1 million Hispanics live in Houston, Texas. Infinity Auto Insurance has asked you to help them create a market research survey for the Houston, Texas, area. The Infinity marketing group wants to better understand the Hispanic automobile insurance market in this large metropolitan area. Infinity Auto Insurance has decided that a telephone survey would reach their target population (less acculturated Hispanics) better than any other data collection method. To get a representative sample, Infinity Auto Insurance has decided that 500 respondents would be sufficient to draw some conclusions.

Infinity Auto Insurance wants to better understand some of the following information; what percentage of Hispanics are insured, what companies are Hispanics insured with, how willing are Hispanics to buy or switch insurance companies, etc. Finally, Infinity Auto Insurance would like to know what attributes keep Hispanics buying auto insurance in the Houston area.

Supplementary Information The online version contains supplementary material available at https://doi.org/10.1007/978-3-031-46371-6_6.

L. Herkenhoff (✉)
Saint Mary's College, Moraga, CA, USA

J. Fogli
Diablo Valley College, Pleasant Hill, CA, USA

L. Herkenhoff, J. Fogli (eds.), *Applied Statistics for Business and Management using Microsoft Excel*, https://doi.org/10.1007/978-3-031-46371-6_6

Food for Thought

What types of questions would you ask?
What are some questions you would develop?
How would you ensure the questions were valid?
How long would your survey be?
What demographic attributes would you try to capture?

Possible Answers

What types of questions would you ask?
Likert questions
Single response questions
General demographic questions

What are some questions you would develop?
Do you own a vehicle?
Do you have auto insurance?
How likely or unlikely would you be to switch auto insurance companies?
Have you heard of [___, ___, Infinity Auto] Insurance?
Have you seen ads for [___, ___, Infinity Auto] Insurance?
How favorable or unfavorable would you rate [___, ___, Infinity Auto Insurance] based on their reputation?

How would you ensure the questions were valid?
Talk to Subject Matter Experts about questions
Check to make sure the questions are not double barreled
Check to make sure there are no leading questions
Check to make sure the questions do not have any biases
How long would your survey be?
Roughly 5–10 min long

What demographic attributes would you try to capture?
Age
Language spoken at home
Hispanic/non-Hispanic
Preferred Advertising Language
Highest Level of Education

INFINITY Auto Insurance

SAVE with Infinity!
(no coupon required)

Scan on your smartphone to learn how much you can save!

Dear

Do any of these describe you?

- You have a special filing system for coupons.
- You display receipts on your refrigerator like hunting trophies.
- You have a kitchen drawer devoted to bread ties.
- You think "Full Price" is the "Fool Price."
- You have an entire table setting of jelly jar drinking glasses.
- You've tried generic peanut butter.

Infinity salutes the Super Saver. Like you, we revel in the joy of saving. In fact, you're already prequalified to save up to $679* on car insurance from Infinity. Sound too good to be true? Well, it gets better. Respond before 4/30/2012, and get a FREE lifetime membership+ in Infinity DriverClub® roadside assistance. Simply visit InfinityAuto.com/Saver or call **855-727-2849** right now, and the Infinity DriverClub® membership is yours ... even if you don't switch to Infinity. No catches. No gimmicks. Think of it as our way of rewarding Super Savers.

We've built our reputation by saving drivers big money on their auto insurance. As one of the most respected companies in the business, we want you to know we're here for you. So, claim your FREE lifetime Infinity DriverClub® membership by calling **855-727-2849,** and you'll have access to valuable services including:

TOWING JUMP-START TIRE CHANGE LOCKOUT REFUELING

No contracts or membership fees! You only pay for roadside assistance when you need it. Call **855-727-2849** or visit InfinityAuto.com/Saver right now, and get a no-hassle quote that could save you hundreds on car insurance. Plus, request your FREE lifetime membership in the Infinity DriverClub® today. But hurry. We must hear from you before 4/30/2012. Did we mention it's FREE?

Save up to $679* on your Auto Insurance, and get a FREE lifetime Infinity DriverClub® membership!
SEE BACK FOR DETAILS.

Sincerely,

Joe Trevino
Chief Satisfaction Officer
InfinityAuto.com/Saver

** Amount of savings varies by state and is subject to the availability of discounts and dependent upon specific underwriting criteria and guidelines.*

+ Membership in Infinity DriverClub® is free and open to all U.S. drivers. Available online and at participating agencies only. At the free membership level, you have access to roadside assistance dispatching services 24 hours per day, 7 days per week; and, you pay only for the services you use, when you use them. Plus, you receive special member pricing. Infinity may discontinue the DriverClub program at anytime without further obligation to the participant.

Visit your local independent agent, and ask for INFINITY today!

You can choose to stop receiving "pre-screened" offers of insurance from this and other companies by calling toll-free 1-888-567-8688. See PRE-SCREEN & OPT-OUT NOTICE on other side for more information about pre-screened offers.

Flip for savings!

Discussion

Sampling is not a desired activity. We only sample when we have certain limitations such as limited access to the entire population of data and/or limited resources in terms of time and money.

We achieve more accurate results when we can avoid sampling and analyze the entire population of data. Sampling is a last resort as it introduces complexities in analyzing the data and introduces approximations in the final results.

This chapter will assist you in determining how large your sample should be to achieve the desired level of statistical rigor. But remember, the sample data is only ever as good as the population data it was derived from. The goal is to have a large enough sample to be representative of the entire population. The statistical values you calculate using just the sample data should also describe the population data if the correct sample size has been used.

If we undersampled the picture in Fig. 6.1, we might end up with an inaccurate understanding of what it actually depicts. Can you see both a face and an Eskimo? If samples are too small, they may not provide accurate generalizations of the entire population. If we had selected too small a sample of pixels from the image, we might see just the face or just the Eskimo.

The information to calculate the required sample size includes

1. **Homogeneity** of the population—proportion of the variable of interest (p)
2. **Accuracy** or the range of error you can live within your sample statistics ($\pm e$)

Fig. 6.1 Seeing both a face and an Eskimo in the samc imagc

3. **Confidence level** or chance that you have selected a representative sample; most common is 90.0%, 95.0%, 99.7% confidence level

To minimize bias, an impartial and objective probability method should be used to select the sample. The most common example of this is a simple random sample. Each data point in the sample has an equal chance of being selected.

Homogeneity refers to the diversity of the population. Consider a company of employees from ten different national cultures. The employees are surveyed about the acceptability of nepotism in the workplace. If the sample is too small, you may have certain cultures underrepresented or not represented at all. Since national culture may influence the answers, it is important to have a large enough sample to be representative across all ten cultures. If, on the other hand, our population data sample included only Americans, our sample would not have to be as large to remain representative of an American population. The higher the degree of homogeneity in the population, the smaller the sample required.

Accuracy of the data refers to the size of the standard error. The **standard error (SE)** is the likely size of a chance error and is often referred to as the "give or take" number. The idea is that the sample results are also the expected results for the population but will be off from this number by a certain amount. The SE indicates the likely size of that amount the number is off. This is described in more detail in Chap. 7.

In Chap. 3, we also defined **accuracy** as a measure of uncertainty around the mean. Accuracy answers the question: How close do you want your results to be to the true population? If we were willing to accept data with an associated standard error of 2, we could express 2 standard errors (SE) as a percent (%). To do this calculation, we convert 2 SE to measured units and express that amount as a percent of the population. It is generally stated as "plus or minus 'X' percent."

The **confidence level or confidence interval** can also be expressed as a percentage. A 95.0% confidence level means that if 100 samples are taken, 95 will contain the actual statistics of the population; in other words, "*would be representative of the population in question*." We can also express this by stating there is a 5% probability that the sample is not representative of the population. Most workplace data is analyzed with a 95.0% (within 2 SEs) or 99.7% (within 3 SEs) confidence level. Remember the higher the percentage value of the confidence, the wider the range of acceptable data. This may seem counterintuitive but to get more data (99.7% vs. 95.0%) you need to widen the range of data values in your accepted region.

Here are some examples in which these three criteria are combined.

- **Example 1**
  If a survey uses a plus or minus 3% *accuracy* and a *95.0% confidence level*, that means there is a *5 in 100 chance* that the survey results are *MORE than 3% away* from the results you would get if you had surveyed the entire population.
- **Example 2**
  If the average GPA of the sample is 3.2 ± 0.3, the actual population could have a GPA anywhere from 2.9 to 3.5. If we wanted a 95.0% confidence level, we would expect these results in 95 out of 100 samples.

- **Example 3**
  We can increase the confidence by adding observations to our sample (increase the sample size) or by widening the range (instead of 3.2 ± 0.3, we have 3.2 ± 0.5). To go from a 95.0% to a 99.7% confidence level, our range of data would increase from ±2 SEs to ±3 SEs.

Types of Problems

Mean Versus Proportion Problems Require a Slightly Different Treatment
This chapter will address calculating the appropriate sample size for **mean** and **proportion** problems. **Mean** problems involve collecting enough sample data to accurately solve for an average that represents the population. For example, you might want to collect enough data to solve for the average salary of a population.

Proportion problems involve collecting enough sample data to solve for the proportion of a variable. For example, you might want to collect enough data to solve for the percentage of women over 21 years old in a given population to explore female alcohol consumption.

Finite Versus Infinite Population Size Is Another Important Factor in Determining the Appropriate Sample Size
Finite population means you know the size of the population. However, if the sample size ends up being less than 5% of the known population size, then the population size would be considered infinite. But more generally, an **infinite** population means you do not know the size of the population. **In this chapter, we will assume we have an infinite population.** However, a finite population correction factor will also be provided for those who may want to fine-tune their estimates with known population size. The following formulas can be used for infinite population problems (Fig. 6.2).

Statistic	SD known	Formula
Mean	yes	$n = \frac{z^2\sigma^2}{e^2}$
	no	Use Range Rule to estimate σ
Proportion	NA	$n = \frac{z^2 p(1-p)}{e^2}$

Fig. 6.2 Sample size formula for infinite population problems
n = required sample size
z = z-value for desired level of confidence
p = estimated value of the population proportion
e = accuracy, the +/− error you can live with (chance error)
σ = standard deviation for the population (known or estimated)

Fig. 6.3 Estimating proportion

(p)	× (1-p)	=
.1	.9	.09
.2	.8	.16
.3	.7	.21
.4	.6	.24
.5	.5	.25
.6	.4	.24
.7	.3	.21
.8	.2	.16
.9	.1	.09

Range Rule for Estimating Standard Deviation

The **standard deviation estimate** may be useful in **mean** problems where the standard deviation is required. If the standard deviation of the population is unknown, you can estimate it by determining the largest expected observed value minus the smallest expected observed value. Some processes include a value called the **upper reasonable limit**, which could also be used as the maximum value.

Although there is not an explicit relationship between the range and standard deviation, there is a rule of thumb that can be useful in making a rough estimate of the standard deviation. This is referred to as the **Range Rule**:

$$\text{Standard deviation estimate} = \frac{\text{Largest possible value} - \text{Smallest possible value}}{4}$$

While this estimate is not as reliable as an estimate based on calculations over a large number of data points, it is often useful as a preliminary estimate. The denominator is based on the Gaussian Rule that says 95% of the data falls from two standard deviations below the mean to two standard deviations above the mean. Thus, nearly all of our normal distribution would stretch out over a line segment that is a total of four standard deviations long. Not all data is normally distributed and bell curve shaped. But most data is well behaved enough that going two standard deviations away from the mean on either side should capture nearly all of the data. We estimate and say that four standard deviations is approximately the size of the range, and so the range divided by four is a rough approximation of the standard deviation. Using the same logic, some textbooks use "σ" in the denominator.

Proportion problems require knowing the proportion of the attribute of interest in the population. If this is unknown, we assume it to be .5 (50%) to ensure we don't err on the side of underestimating the sample size. By using .5 for p in the formula will produce the largest sample size and we err on the side of having too large of a sample, rather than too small of a sample.

Proportion Rule for Estimating Proportions

Proportion problems require knowing the proportion of the attribute of interest in the population (p).

The largest value for sample size (n) occurs when $\rho \times (1 - \rho)$ is maximized (.25) as shown in Fig. 6.3

$$n = \frac{z^2 p(1-p)}{e^2}$$

Default for Confidence Level

If the confidence level is not stated, the default value is usually considered to be 95%. However, when we work with sample size formula, we need to convert this into a Z-value. The corresponding Z-value for 95% is 1.96.

Summary of Input Values for Sample Size Formula

Statistic	Description	Source
n	Sample size	Calculated from formula
z	Z-value converted from confidence level	Given or assumed to be 1.96 (95% confidence level)
P	Population proportion	Given or assumed to be .5
e	Accuracy	Given
Σ	Standard deviation	Given or use Range Rule (previous page)

Excel

Templates for Calculating Sample Size

Build your own templates so you can quickly calculate sample size for mean or proportion problems. Examples are provided below.

1. **Template Problem Type: Mean**

	A	B	C	D	E
1	**Sample Size Population Mean**				
2	**Finite Population**				
3					
4	Estimate for std. dev (σ)	0.5			
5	Maximum likelihood error (*e*)	0.03			
6					
7	Confidence Interval Desired	0.95			
8	Alpha (α)	0.05		=1-B7	
9	Corresponding *z* value	1.959964		=NORM.S.INV(0.5+(B7/2))	
10					
11	Required sample size is (n=)	1067.072		=((B9^2)*(B4^2))/(B5^2)	
12					

2. **Template Problem Type: Proportion**

Book1.xlsx

	A	B	C	D	E	F
1	**Sample Size Required for Estimating (α)**					
2	**Population Proportion:**					
3						
4	Estimate Population Proportion (*p*)	0.5		Population Proportion		
5	Maximum likelihood error (*e*)	0.03				
6						
7	Confidence Level Desired	0.95				
8	Alpha (α)	0.05		=1-B7		
9	Corresponding *z* value	1.96		=NORM.S.INV(0.5+(B7/2))		
10						
11	Required sample size is (n=)	1067.07		=((B9^2)*B4*(1-B4))/(B5^2)		
12						
13						

Problem Type: Infinite Mean

This is the formula for calculating a sample size for a problem in which you are calculating a mean for an infinite population.

Use this equation

$$n = \frac{z^2 \sigma^2}{e^2}$$

For example, let's suppose you want to estimate the mean of the population and the size of the population is infinite. We want to determine how many samples to take to ensure the mean of the sample matches the mean of the (infinite) population.

Example

Our highest cost for a bottle of water at a sports arena in the United States is \$5.00 and our lowest cost is \$3.00. We are willing to live with a \$0.03 error value and require a confidence level of 95%. Calculate the required sample size to estimate the average price of a bottle of water sold at this arena.

- Input estimated population standard deviation in cell B4. Remember if you don't have a clue about its size use the Range Rule from previous page.
  Range Rule

$$\text{Standard deviation estimate} = \frac{\text{Largest possible value} - \text{Smalest possible value}}{4}$$

Based on the data, the largest value is \$5.00 and the smallest possible value is \$3.00. This results in a standard deviation estimate of 0.5.

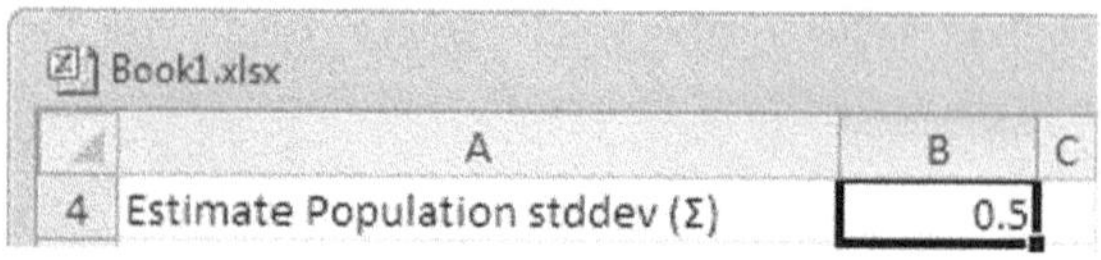

- Input maximum likelihood error (the ± error you can live with) in the given units of that error in cell B5

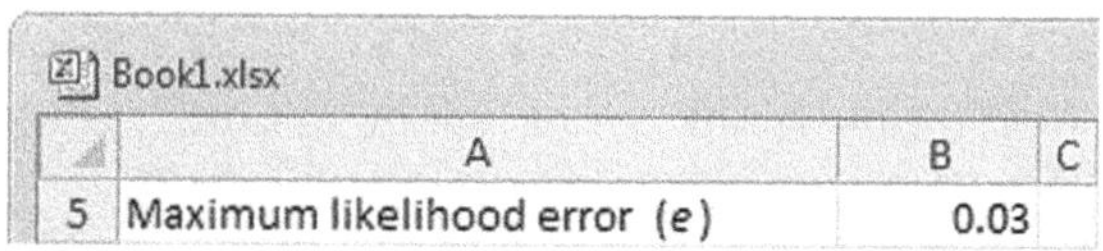

- Input the desired confidence level. This needs to be input as a decimal. Enter a **95%** confidence level as **0.95** in cell B7

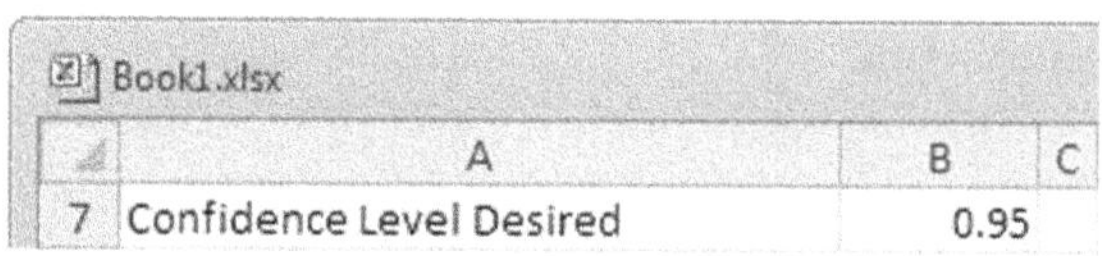

- Click on cell B8
- Click on the formula bar at the top of the page and type in
- $= \mathbf{1 - B7}$

  Alpha is the failure rate rather than the success rate (confidence level). It is the same information but the convention is to use alpha in the formula.

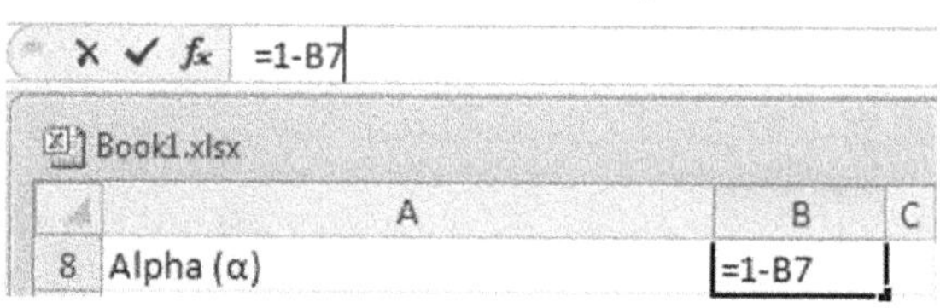

- Click the green checkmark

- Click on cell B9
- Click on the formula bar at the top of the page and type in
- $= \mathbf{NORM.S.INV}\left(\mathbf{0.5} + \left(\mathbf{B7} / \mathbf{2}\right)\right)$

  This is the z-score. Watch that "**S**" in the formula.

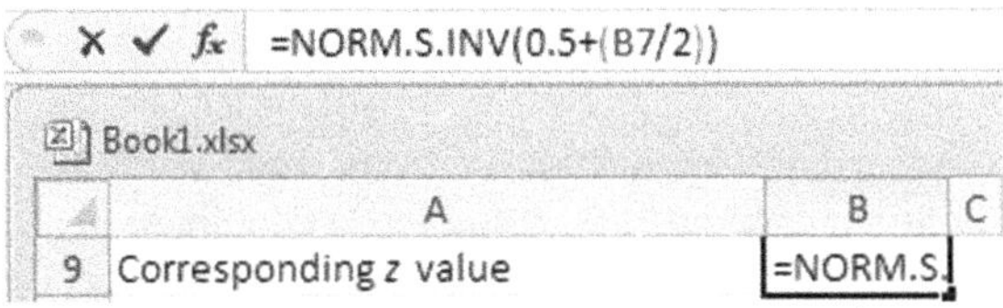

- Click the green checkmark

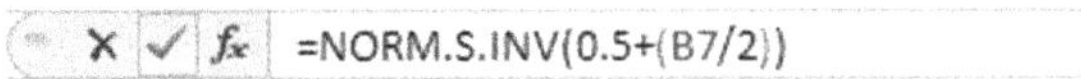

- Click on cell B11
- Click on the formula bar at the top of the page and type in

$$=\left((\mathbf{B9}\wedge 2)^{*}(\mathbf{B4}\wedge 2)\right)/(\mathbf{B5}\wedge 2)$$

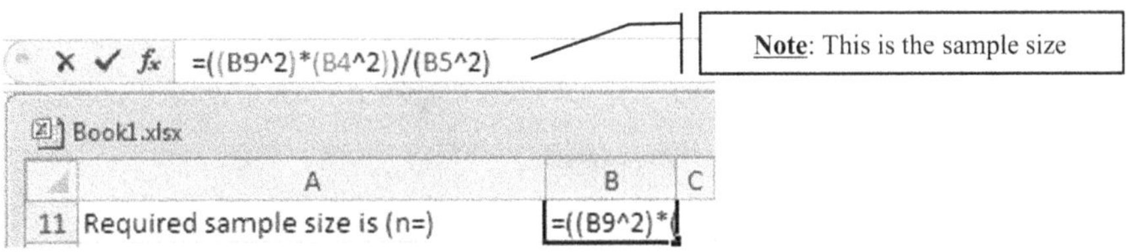

Click the green checkmark

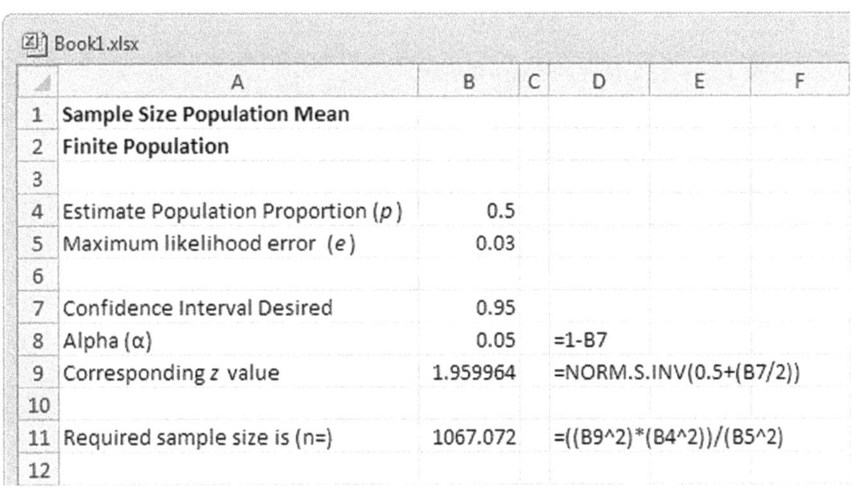

Book1.xlsx

	A	B	C	D	E	F
1	**Sample Size Population Mean**					
2	**Finite Population**					
3						
4	Estimate Population Proportion (*p*)	0.5				
5	Maximum likelihood error (*e*)	0.03				
6						
7	Confidence Interval Desired	0.95				
8	Alpha (α)	0.05		=1-B7		
9	Corresponding *z* value	1.959964		=NORM.S.INV(0.5+(B7/2))		
10						
11	Required sample size is (n=)	1067.072		=((B9^2)*(B4^2))/(B5^2)		
12						

Answer: We would round this answer up to 1068 bottles, as we cannot have part of a bottle.

Practice Problem for Infinite Mean

A company is concerned about employee overtime and has hired you to survey the employee records. You are an outside consultant and do not have access to the company population as this is a large multinational corporation. The estimated population standard deviation has been reported as 100 hours (h) of annual overtime per nonexempt employee (σ)(B4). How many employee records should you include in your sample to calculate the average annual overtime per employee with a desired confidence level of 95% (B7), within an accuracy of ±50 (B5) hours of overtime per employee (e)?

Excel			
B4	Estimate for std dev (σ)	100	Type this in from the problem
B5	Accuracy or maximum likelihood error (e)	50	Type this in from the problem
B7	Confidence level desired	0.95	The problem may state this as a % so convert to a decimal
B9	The corresponding z-value is	1.96	Excel will calculate this from the Confidence level
B11	The required sample size is n=	15.4	Excel will calculate

Answer: Always round up to be conservative on your sample size, so you need to survey 16 records.

Problem Type: Infinite Proportion

This is the formula for calculating a sample size for a problem in which you are calculating a proportion for an infinite population.

Use this equation

$$n = \frac{z^2 p(1-p)}{e^2}$$

For example, let's suppose you what to estimate the proportion of water bottles in the population that sold for more than \$4.00; the size of the population is infinite. We want to determine how many samples to take to ensure the proportion of the sample matches the proportion of the (infinite) population.

Example

Our highest cost for a bottle of water at a sports arena in the United States is \$5.00 and our lowest cost is \$3.00. We are willing to live with a \$0.03 error value and require a confidence level of 95%. Our standard deviation estimate (p) is .5 based on our previous example problem for calculating an estimate for an unknown standard deviation of the population. What is the required sample size to determine the proportion of bottles of water sold for more than $4.00 per bottle?

- Type in the population proportion in cell B4
  If you do not know this, use the most conservative estimate of 0.5.

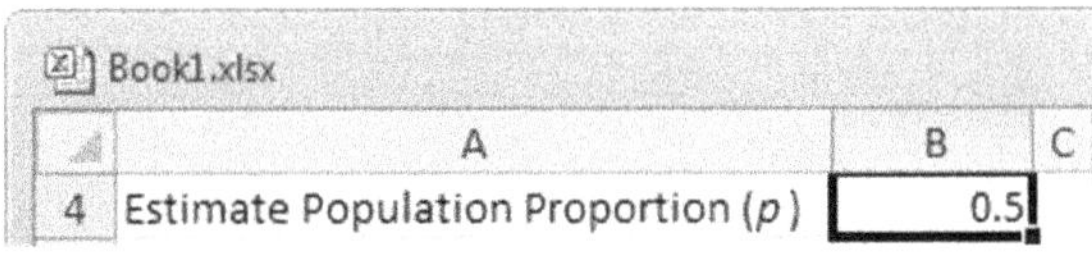

- Type in the maximum likelihood error (the ± error you can live with) in the given units of that error in cell B5

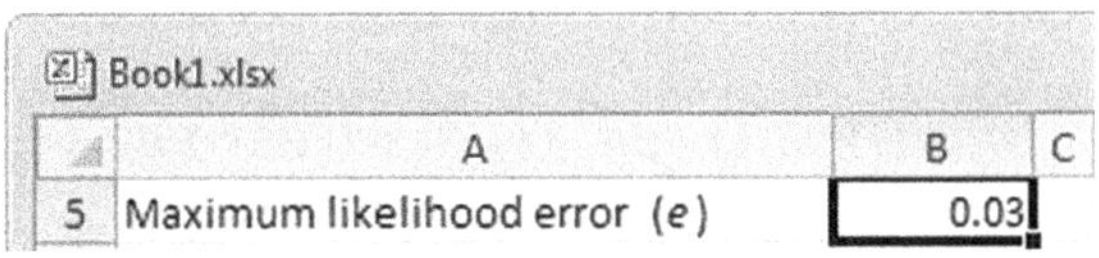

- Type in the desired confidence level desire. This needs to be input as a decimal. A **95.0%** confidence level would be entered as **0.95** in cell B7

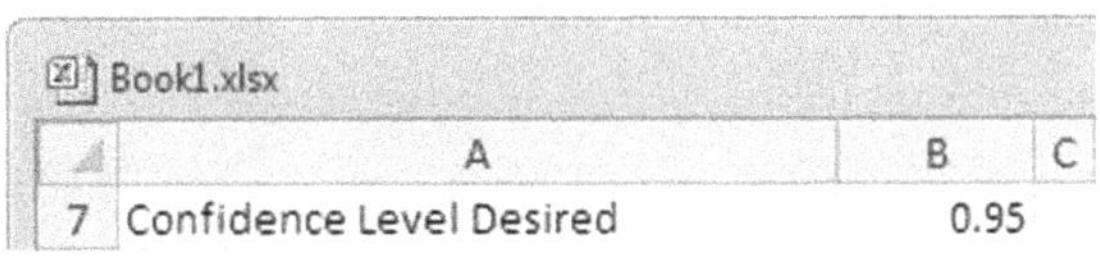

- Click on cell B8
- Click on the formula bar at the top of the page and type in

$$= \mathbf{1 - B7}$$

(This is the failure rate rather than the success rate)

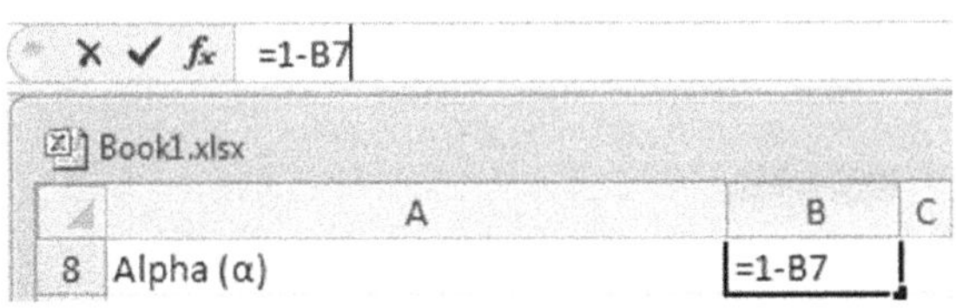

- Click the green checkmark

- Click on cell B9
- Click on the formula bar at the top of the page and type in

$$= \mathbf{NORM.S.INV}\left(\mathbf{0.5} + \left(\mathbf{B7/2}\right)\right)$$

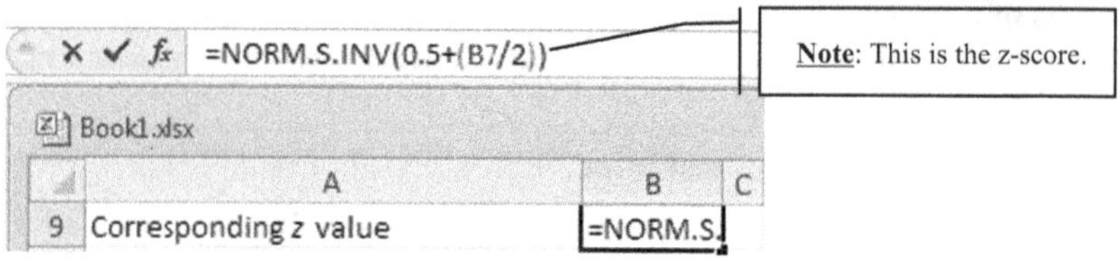

- Click the green checkmark

- Click on cell B11
- Click on the formula bar at the top of the page and type in

$$=\left(\left(\mathbf{B9}^\wedge\mathbf{2}\right)^{*}\left(\mathbf{B4}\right)^{*}\left(\mathbf{1}-\mathbf{B4}\right)\right)/\left(\mathbf{B5}^\wedge\mathbf{2}\right)$$

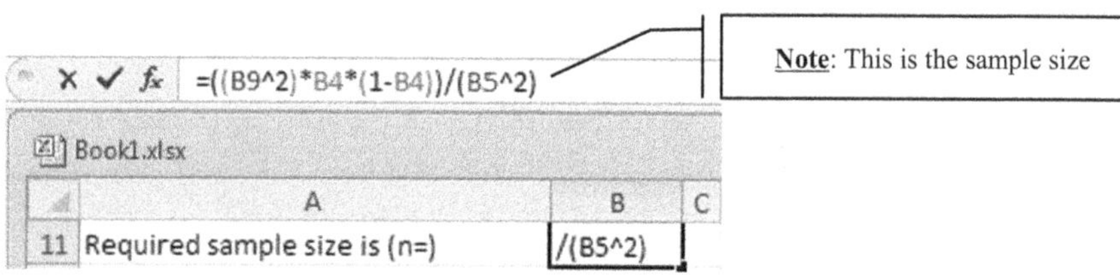

Book1.xlsx

	A	B	C	D	E	F
1	**Sample Size Required for Estimating (α)**					
2	**Population Proportion:**					
3						
4	Estimate Population Proportion (*p*)	0.5		Population Proportion		
5	Maximum likelihood error (*e*)	0.03				
6						
7	Confidence Level Desired	0.95				
8	Alpha (α)	0.05		=1-B7		
9	Corresponding *z* value	1.96		=NORM.S.INV(0.5+(B7/2))		
10						
11	Required sample size is (n=)	1067.07		=((B9^2)*B4*(1-B4))/(B5^2)		
12						
13						

Click the green checkmark

Answer: We would round this answer up to 1068.

Practice Problem for Infinite Proportion

You are an outside consultant. The Chinese company who hired you wants to know the proportion of full-time employees who might be willing to work overseas. The company population is unknown as records are not accurate about full-time versus part-time help. We have no idea how many employees may currently want to work overseas, so we assume 50% (B4) of the entire population would accept an overseas assignment. How large a sample do I need with a confidence level of 99.7% (B7) and accuracy of ±4% (B5)? My new survey needs to be a representative reflection of the entire population.

Excel			
B4	Pop proportion p	0.5	Type this in from the problem; if expressed as a % change to a decimal. If unknown use .5
B5	Maximum likely error, e:	0.04	Type this in from the problem (accuracy); this will be expressed as a % so convert to a decimal
B7	Confidence level desired:	0.997	Input from problem
B9	Corresponding z value:	2.97	Excel calculates this from the CI
B11	Required sample size:	1378.3	

Answer: Always round up to be conservative on your sample size, so we need to survey 1379 employees.

Finite Population Correction Factor (fpc)

This factor can be used to reduce the **sampling error** when you know the population size. The formula for the fpc follows

$$\text{fpc} = \sqrt{\frac{(N-n)}{(N-1)}}$$

where **N** is the population size and **n** is the sample size.

Let's assume we have a population of 1000 and a sample size calculated from our infinite sample size formula of 77. The fpc would equal 0.96 when n is very small relative to N; the fpc is close to 1 and has a small effect on the standard error. When n is large relative to N, the fpc can have a large effect on the standard error. (The fpc would equal 0.96 or as a percentage .096%. That means by using the fpc our sampling error will be reduced by almost 1%.)

The fpc can also be directly applied to the sample size by using it in the following format:

$$n = \frac{n_0 N}{\left(n_0 + (N-1)\right)}$$

where **N** is the population size, **n** is the sample size, and $\mathbf{n_o}$ is the sample size calculated using the infinite sample size formula.

Let's assume we have a population of 1000 (N) and a sample size calculated from our infinite sample size formula of 77 (n_0). The new sample size n from the formula would equal 71.56 rounded up to 72. By applying the fpc, we have reduced the required sample size from 77 to 72, which represents approximately 7% reduction in sample size.

The following table provides more examples of the impact of the FCF with a finite population of $N = 100$ and SD = 0.7.

Sample size	Standard error	Finite correction factor	SE with FCF
n	SD/sqrt n	Sqrt{(N – n)/(N – 1)}	[SD/sqrt n] * [Sqrt{(N – n)/(N – 1)}]
40	0.111	0.778	0.086
60	0.090	0.636	0.057
80	0.078	0.449	0,035
100	0.07	0	0

Final Thoughts and Activities

Practice Problems

1. There are several thousand machinery rebuilding and repairing companies in the United States. A tool manufacturer wishes to survey a simple random sample of these firms to find out what proportion of them is interested in a new tool design. If the tool manufacturer would like to be 95.0% confident that the sample proportion is within .04 of the actual population proportion, how many machinery repairing and rebuilding companies would he include in the sample?
2. A state politician would like to determine the average amount earned during summer employment by state teenagers during the past summer's vacation period. She wants a 95.0% confidence that the sample mean is within $50 of the actual population mean. Based on past studies, she has estimated the population std. deviation to be about $400. How large should the sample be?
3. There are numerous CEOs in the United States. A political scientist wants to estimate, with 99.7% confidence and within $3, what the average cost for their support staff is each year. Based on historical records, the population standard

deviation seems to be only \$10. How many CEOs should be interviewed about their support staff?
4. How would you respond to the statement that says that by increasing the sample size, the amount of sampling error will be decreased?
5. A company wants to conduct a customer satisfaction survey with a confidence level of 95% and a margin of error of ±4%. The population size is estimated to be 10,000. Calculate the required sample size for this survey.
6. Question: A marketing research firm is conducting a study to estimate the proportion of smartphone users who prefer Brand X over its competitors. They want to achieve a 99% confidence level with a margin of error of ±2%. The firm has no prior information about the proportion. What sample size should the firm use for this study?
7. A hospital wants to estimate the average waiting time for patients in its emergency department. The management decides to conduct a study with a 90% confidence level and a margin of error of ±5 minutes. The standard deviation of waiting times from a previous study is estimated to be 15 minutes. What sample size is required for this study?
8. What are the main reasons we sometimes have to work with sample data instead of population data?

Discussion Boards

1. What is RDD and what are the issues with using it?
2. Marine biologists often use the capture–recapture method for sampling. How does this work and how effective a sampling method is it?
3. What changes have occurred between the 1984 Gallup Survey and the most recent version?
4. Statistics as a discipline remains sharply divided on the fundamental definition of probability.

Group Activity

In 1936, the *Literary Digest* magazine predicted an overwhelming presidential victory for Alfred Landon who was running against Franklin Roosevelt. They predicted that Roosevelt would only get 43% of the vote based on a sample of 2.4 million people. However, Roosevelt won the election with a landslide vote of 62–38%. This is the largest error ever made by a major poll. George Gallup was just beginning his company. Using his own statistical methods, he predicted what the *Digest*'s predictions were going to be in advance of their publication with an error of only 1%. Find out where the *Digest* went wrong. Discuss how Gallup predicted the *Digest* results so well. After all he only had 36,000 people in his random sample.

Parting Thought

A frequentist is a person whose lifetime ambition is to be wrong 5% of the time. A Bayesian is one who, vaguely expecting a horse, and catching a glimpse of a donkey, strongly believes he has seen a mule …

Problem Solutions

1. **Problem type: Infinite Proportion**

$$n = \frac{z^2 p(1-p)}{e^2}$$

Substitute:

> p = assume 0.5 as our default when we don't know this number
> e = .04
> z = 1.96 (because the CI = 95.0%)
> Using our template, we end up with 600.23, which after rounding is 601 for our sample size.

Book1.xlsx

	A	B	C	D	E	F
1	**Sample Size Required for Estimating**					
2	**Population Proportion:**					
3						
4	Estimate Population Proportion (p)	0.5		Population Proportion		
5	Maximum likelihood error (e)	0.04				
6						
7	Confidence Level Desired	0.95				
8	Alpha (α)	0.05		=1-B7		
9	Corresponding z value	1.96		=NORM.S.INV(0.5+(B7/2))		
10						
11	Required sample size is (n=)	600.23		=((B9^2)*B4*(1-B4))/(B5^2)		
12						

2. **Problem type: Infinite Mean**

$$n = \frac{z^2 \sigma^2}{e^2}$$

Substitute:

$e = \$50$
$z = 1.96$ (because the CI = 95.0%)
$\sigma = \$400$
Using our template, we end up with 245.85, which after rounding is 246 for our sample size.

Book1.xlsx

	A	B	C	D	E	F
1	**Sample Size Population Mean**					
2	**Infinite Population**					
3						
4	Estimate Population stddev (*p*)	400				
5	Maximum likelihood error. (*e*)	50				
6						
7	Confidence Interval Desired	0.95				
8	Alpha (α)	0.05		=1-B7		
9	Corresponding *z* value	1.96		=NORM.S.INV(0.5+(B7/2))		
10						
11	Required sample size is (n=)	246		=((B9^2)*(B4^2))/(B5^2)		
12						

3. **Problem type: Infinite Mean**

$$n = \frac{z^2 \sigma^2}{e^2}$$

Substitute:

$e = \$3$
$z = 3.0$ (because the CI = 99.7%)
$\sigma = \$10$
Using our template, we end up with 100 for our sample size.

Book1.xlsx

	A	B	C	D	E	F
1	**Sample Size Population Mean**					
2	**Infinite Population**					
3						
4	Estimate Population stddev (σ)	10				
5	Maximum likelihood error (*e*)	3				
6						
7	Confidence Level Desired	0.997		Given in problem as .997		
8	Alpha (α)					
9	Corresponding *z* value	2.97		Given in problem		
10						
11	Required sample size is (n=)	100.00		=((B9^2)*(B4^2))/(B5^2)		
12						

In this case you type it in

4. How would you respond to the statement that says that by increasing the sample size, the amount of sampling error will be decreased?

Answer:

Increasing the sample size generally leads to a decrease in sampling error, but it is important to understand the underlying concepts. Sampling error refers to the discrepancy between a sample statistic (e.g., the mean or proportion) and the corresponding population parameter. It occurs because a sample only represents a subset of the entire population, and there is inherent variability in the data. By increasing the sample size, we can reduce the impact of sampling error for several reasons:

- Increased precision: A larger sample size provides more data points and reduces the random variation in the sample estimate. The larger the sample, the more likely it is to resemble the population characteristics, leading to more accurate estimates of the parameters.
- Decreased sampling variability: Sampling error arises from random fluctuations in the sample selection process. With a larger sample size, the impact of this random variation diminishes. The law of large numbers suggests that as the sample size approaches the population size, the sample mean becomes a better approximation of the population mean.
- Improved representation: A larger sample size allows for a more comprehensive representation of the population. It helps capture the diversity and heterogeneity present in the population, leading to more reliable estimates.

  However, it is important to note that while increasing the sample size reduces sampling error, it does not eliminate it entirely. Other sources of error, such as nonsampling errors (i.e., measurement errors or sampling bias), can still affect the accuracy of estimates, even with a large sample size. Therefore, it is crucial to consider various factors and use appropriate statistical techniques to minimize both sampling and nonsampling errors when conducting research or drawing conclusions from data.

5. A company wants to conduct a customer satisfaction survey with a confidence level of 95% and a margin of error of ±4%. The population size is estimated to be 10,000. Calculate the required sample size for this survey.

Answer:

The required sample size can be calculated using the formula

$$\mathbf{n} = \left[(\mathbf{z}^\wedge 2)^* (\mathbf{p})^* (\mathbf{p}-1) \right] / \mathbf{e}^\wedge 2$$

where z = z-value corresponding to the desired confidence level (e.g., $z = 1.96$ for 95% confidence level)

p = estimated proportion (0.5 can be used as a conservative estimate)

e = margin of error as a decimal.

Using the given values: n = (1.96^2 * 0.5 * 0.5) / 0.04^2

n = 600.25

The required sample size for this survey is approximately 601.

6. A marketing research firm is conducting a study to estimate the proportion of smartphone users who prefer Brand X over its competitors. They want to achieve a 99% confidence level with a margin of error of ±2%. The firm has no prior information about the proportion. What sample size should the firm use for this study?

Answer:

To calculate the sample size for estimating a proportion without prior information, the formula is

$$n = \left[(z^\wedge 2)^* (p)^* (p-1) \right] / e^\wedge 2$$

where z = z-value corresponding to the desired confidence level

p = estimated proportion (0.5 can be used as a conservative estimate)

e = margin of error as a decimal.

z = 2.58 for 99% confidence level and e = margin of error as a decimal

Using the given values: n = (2.58^2 * 0.25) / 0.02^2 n = 6640.5

The marketing research firm should use a sample size of at least 6641 for this study.

7. A hospital wants to estimate the average waiting time for patients in its emergency department. The management decides to conduct a study with a 90% confidence level and a margin of error of ±5 minutes. The standard deviation of waiting times from a previous study is estimated to be 15 minutes. What sample size is required for this study?

Answer:

The sample size for estimating a population mean can be calculated using the formula

$$n = \left(z^\wedge 2^* \sigma^\wedge 2\right) / e^\wedge 2$$

where z = z-value corresponding to the desired confidence level
σ = standard deviation of the population
e = margin of error as a decimal
z = 1.645 for 90% confidence level
Using the given values: n = (1.645^2 * 15^2) / 0.083^2 n = 397.07
The hospital should use a sample size of at least 398 for this study.

8. What are the main reasons we sometimes have to work with sample data instead of population data?

Answer:

There are several practical reasons why researchers often work with sample data instead of population data:

- Time and cost efficiency: Collecting data from an entire population can be time-consuming, expensive, and sometimes even logistically challenging. By working with a smaller sample, researchers can gather information more quickly and at a lower cost.
- Feasibility: In some cases, it may simply be impractical or impossible to access or measure the entire population. For example, when studying historical events or rare phenomena, it may not be feasible to obtain data from the entire population.
- Destructive testing: In certain scientific experiments or studies, the process of data collection may involve destructive testing or interventions that make it impossible to collect data from the entire population. In such cases, working with a sample is the only option.
- Inference: When researchers have access to a representative sample, they can use statistical techniques to make inferences about the population as a whole. With proper sampling methods, it is possible to draw valid conclusions about the population characteristics using sample data.
- Manageability: Working with sample data allows researchers to handle and analyze a smaller data set, making it easier to apply statistical techniques, identify patterns, and draw meaningful insights. Large population data sets can be overwhelming and computationally intensive to process.
- It is important to note that when working with sample data, researchers must ensure that the sample is representative of the population of interest. Using appropriate sampling techniques and considering potential biases are crucial steps to ensure the validity of any conclusions or generalizations made from the sample data.

Chapter 7
Inference

Linda Herkenhoff and John Fogli

Key Concepts

Confidence interval, Estimator, Expected value, Population, Sample, and Standard error.

Discussion

This chapter introduces the concept of inferring statistical information about a population from sample data. The **sample** is the group of selected data chosen from the entire set of possible data (**population**). An **estimator** is a formula or process for using sample data to estimate a population statistic. Inference is not a preferred statistical process but becomes a required process when you are working with only part of a larger data set; in other words, a sample of data. The first step is to assume an expected value, and the second step is to fine-tune that value using the standard error. The three basic problem types include inferring a population sum, a proportion from a sample data, and an average from the sample data. This chapter will include examples of all three of these problem types.

Supplementary Information The online version contains supplementary material available at https://doi.org/10.1007/978-3-031-46371-6_7.

L. Herkenhoff (✉)
Saint Mary's College, Moraga, CA, USA

J. Fogli
Diablo Valley College, Pleasant Hill, CA, USA

L. Herkenhoff, J. Fogli (eds.), *Applied Statistics for Business and Management using Microsoft Excel*, https://doi.org/10.1007/978-3-031-46371-6_7

The **expected value , E(x)**, pins down the center of the probability histogram and the **standard error** (SE) fixes its spread. Remember **standard deviation (SD)** is for a **list** of numbers, while **standard error** (SE) is for a chance **process**. The SD and SE are different. The expected value and standard error calculations vary based on the problem type.

1. The **expected value, E(x), for the sum** of selections made at random with replacement from a box is

$$\mathbf{E(x)\,for\,the\,sum} = (\mathbf{number\,of\,selections}) \times (\mathbf{average\,of\,box})$$
$$= (\mathbf{n}) \times (\bar{\boldsymbol{x}})$$

   But more generally, the **expected value** for the average or proportion of the population is the associated value calculated from the sample. The **standard error** fine-tunes that value.
2. The **expected value for the proportion** of the population is inferred to be the expected value of the sample proportion and likewise.

$$\text{Proportion} : E(x)_{\text{population}} = E(x)_{\text{sample}}$$

3. The **expected value for the average** of the population is inferred to be the expected value of the sample average.

$$\text{Average} : E(\bar{x})_{\text{population}} = E(\bar{x})_{\text{sample}}$$

- Be careful to use the correct formula for the **standard error** (SE). Check if the **expected value** is for the sum, average, or proportion.

  The **standard error (SE)** is the likely size of a chance error and is often referred to as the "give or take" number. The idea is that the sample results are also the expected results for the population but will be off from this number by a certain amount. The SE indicates the likely size of that amount the number is off. Keep in mind the SE is an approximation. The SE is always interpreted in the same way, but can take on slightly different forms depending on what the problem requires.

1. **The SE for the sum of numbers**

$$\boldsymbol{SE} = \sqrt{\textit{sample size}} \times \textit{SD of the population}$$

Example: Box Model SE for the Sum

0	2	3	4	6

Suppose we make 25 selections from this box with replacement and calculate the average of this box to be (0 + 2 + 3 + 4 + 6)/5 = 3, then the **expected value of the sum** of these tickets will be

$$E(x) = n \times \overline{x}$$
$$E(x) = 25 \times 3 = 75$$

To calculate the **standard error,** we need the standard deviation of the box. To get the standard deviation, you can use the STDEV function in Excel or just calculate by hand (see Chap. 3).

$$\begin{aligned} \text{SD} &= 2 \\ \text{SE for sum} &= \{\text{SQRT } (n)\} * \text{SD} \\ &= \{\text{SQRT} 25)\} * 2 \\ &= 10 \end{aligned}$$

This means the expected sum of the 25 selections is likely to be 75 give or take 10. We would expect any sum of 25 numbers randomly selected from this box to fall within the range of 65–85.

2. **The SE for the average of numbers**

$$\textbf{SE for the average} = \frac{\textbf{\textit{SE for the sum}}}{\textbf{\textit{sample size}}}$$

But this is more commonly written as

$$\textbf{SE} = \frac{\boldsymbol{SD}}{\sqrt{\boldsymbol{n}}}$$

Here is how we derive that formula, from the previous **SE for the Sum** for those who have a mathematical interest… others can skip over this section.

$$\textbf{SE for the average} = \frac{SE\ for\ the\ sum}{n} \times \frac{(\sqrt{n})}{(\sqrt{n})} = \frac{\sqrt{n} \times SD}{n} \times \frac{(\sqrt{n})}{(\sqrt{n})}$$

$$= \frac{n \times SD}{n(\sqrt{n})} = \frac{SD}{\sqrt{n}}$$

Example: SE for the Average

Find the SE for the average daily salary in the population when the sample size is 2500, the standard deviation, as calculated from the sample data, is \$25, and the expected average value is \$250.00.

$$\textbf{SE for the average} = \frac{SD}{\sqrt{n}} = \frac{25}{50} = 0.5$$

This means the expected average of the 2500 daily salaries is likely to be $250.00 give or take $0.50.

3. **The SE for proportion percent**

   This SE needs to be reported as a percentage

$$\textbf{SE} = \left(\frac{SD}{\sqrt{n}}\right) \times 100\%$$

Example: SE for the Proportion Percent

Find the SE for the proportion of female MBA students in the population when the sample size is 100 and the standard deviation, as calculated from the sample data, is 2 people (female MBA students) and the expected proportion from the sample is 40 people or 40%.

$$\textbf{SE} = \frac{SD}{\sqrt{n}} \times 100\% = \frac{2}{10} \times 100\% = 20\%$$

This means the expected sum of the 100 MBA student selections is likely to be 40% female MBA students give or take 20%.

It is important when working with inference statistics and sample data to include confidence intervals. Confidence intervals are a bit difficult to understand because they involve thinking about the actual sample as well as about other samples that could have been drawn. Confidence intervals indicate how often you will get the same results with different samples. For example, a 95% confidence interval indicates the same results will occur in 95 out of 100 samples from the same population. The following synonymous terms may be used interchangeably: confidence interval, confidence level, degree of confidence.

A **confidence interval (CI)** is a range of values used to estimate the true value of a population statistic. The definition for degree of confidence uses **"α"** (**alpha**) to represent a probability or area. The value of **α** is the complement to the **degree of confidence.** For a 95% (.95) degree of confidence, $\alpha = 0.05$. For a 0.99 (or 99%) degree of confidence, $\alpha = 0.01$.

The confidence interval is determined by the **standard error.**

- The sample percentage interval of ±1 SE corresponds to a 68% **confidence interval** for the population percentage.
- The sample percentage interval of ±2 SEs corresponds to a 95% **confidence interval** for the population percentage.
- The sample percentage interval of ±3 SEs corresponds to a 99.7% **confidence interval** for the population percentage.

Confidence statistics are often stated as being "about" so much. That is because the **standard errors** have been estimated from the data and the normal "approximation" has been used. We are not dealing with an exact number. We are dealing with an approximation. It may seem counterintuitive that an increase in **standard error** causes an increase in the **confidence interval** for the population percentage. However, when one is increasing the amount of **standard error** they are essentially increasing the range of acceptable measurements. For example, if the standard error for a given average measurement is ±1 SE and the average is 50, then all measurements between 49 and 51 are acceptable. Now if you increase the amount of **standard error** to ±3 SEs then all measurements between 47 and 53 are acceptable, thus giving one more confidence that a given random measurement will agree with the previously measured average.

Example Confidence Interval

The following diagram depicts the concept of a 95% CI for a crate of oranges. Twenty survey organizations are asked to estimate the percentage of bad oranges in this shipment. But they don't know that the actual percentage of bad oranges was measured as 35% of the crate population of oranges. Each organization takes a simple random sample of 2500 oranges from the crate and computes a 95% CI for the percentage of bad oranges in the crate, which corresponds ± 2 SEs. The percentage of bad oranges and the corresponding SE is different for each sample. Some of the samples cover the correct percentage of bad oranges, others fail.

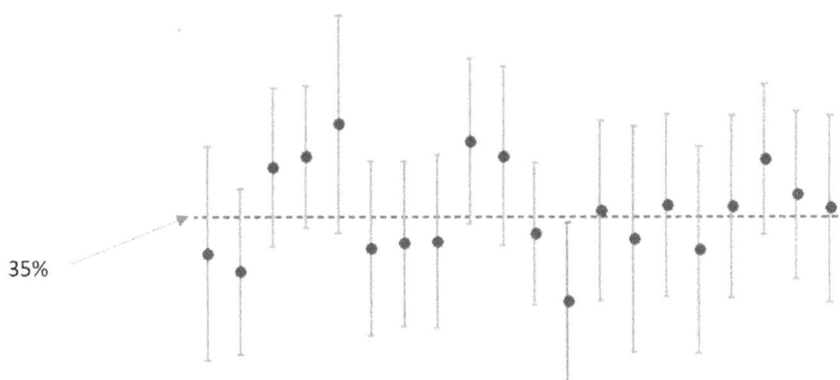

- The central horizontal red dashed line is the correct % (35%) from the population.
- The shorter vertical lines represent each sample. The dot on the line is the percentage of bad oranges and the length of the lines is the +/− SE for the sample.

Out of the 20 samples, we see 19 cover the actual population value of 35%; only one does not cross the main horizontal line at any point (solid red line). Therefore, 19 are representative of the population and 1 would give an erroneous answer. Also, 19 out of 20 equates to 95/100 = 95%… This represents a 95% confidence interval.

Remember the confidence interval tells us in a sample of 100 samples how many will have the population statistic and how many are not representative samples. But the issue is that we have no way of knowing which is which.

Inferring Proportions

Example Problem

There is an election for the President of the Academy of Management this year. There are 100,000 eligible voting members around the world, but you only want to enter if you have a good chance of winning. You take a simple random sample of 2500 members through some of the regional chapters that you have access. In the sample, 1328 members favor you. What should you expect on voting day when the entire population of 100,000 voting members participates?

In this problem, we have sample data and want to infer something about a proportion (percentage) for the entire population. In statistical terms, we want to know the **expected value** for the proportion of voters who will support the candidate and calculate how good an estimate that is using the **standard error** for the sum of the voting members.

When setting up the box model, it is mandatory to assign the "1" to the variable of interest; the variable you are interested in counting (summing). In this case, it is the number of votes favoring the candidates.

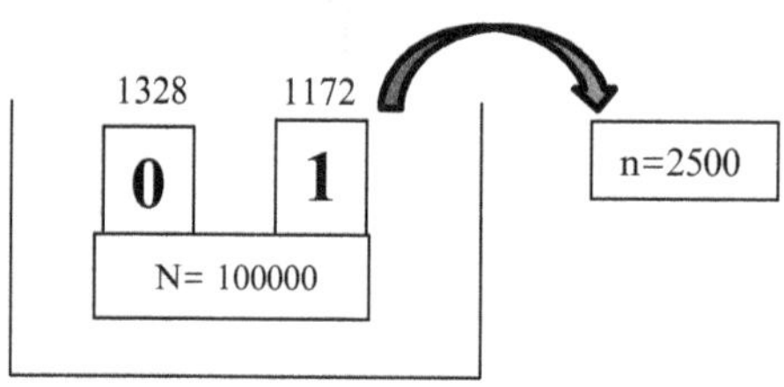

Excel

As we start these problems, let's keep in mind the shortcut formula for standard deviation when solving proportion-related questions (zero/one box).

$$\sigma = \text{sqrt}\left(\left(\frac{x_1}{x_2}\right) * \left(1 - \left(\frac{x_1}{x_2}\right)\right)\right)$$

where

x_1 = the count of zeros in the box (votes not favoring the candidate)
x_2 = the count of ones in the box (votes in favor of candidate)

- Input in A1 **Sample for the Candidate** and in A2 **1328**

Note: This number is given from number of positive votes for the candidate in the sample.

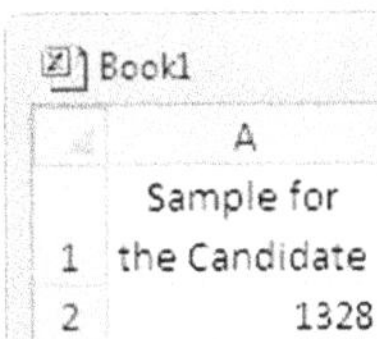

- Input in B1 **Sample Size** and in B2 **2500**

Note: given in problem as the actual sample size.

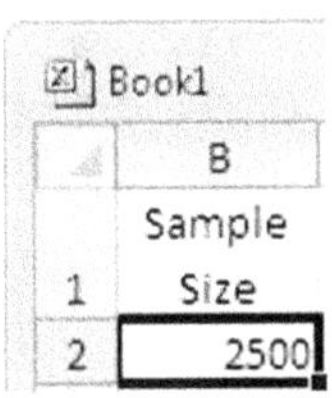

- Input in A3 = **(A2/B2)** * **100,** resulting in 53.12

Note: This formula calculates a percentage of positive votes in the sample.

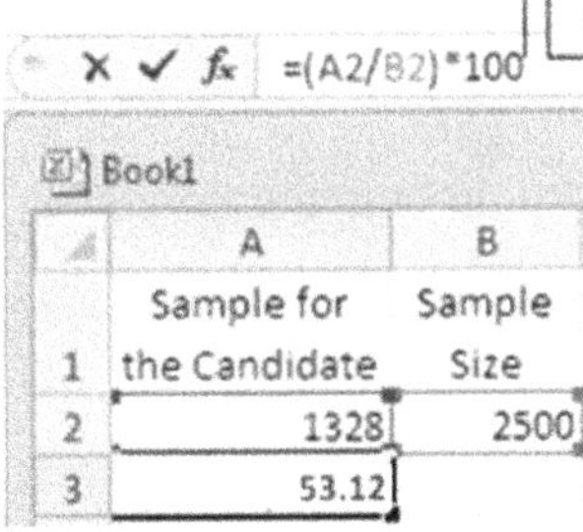

- Input in C1 **Population** and in C2 **100,000** (i.e., entire population size)
- Input in D1 **Standard Deviation**, and in D2 input = **SQRT((A2/B2)*(1–(A2/B2)))**
- After typing in the formula, click on the green checkmark

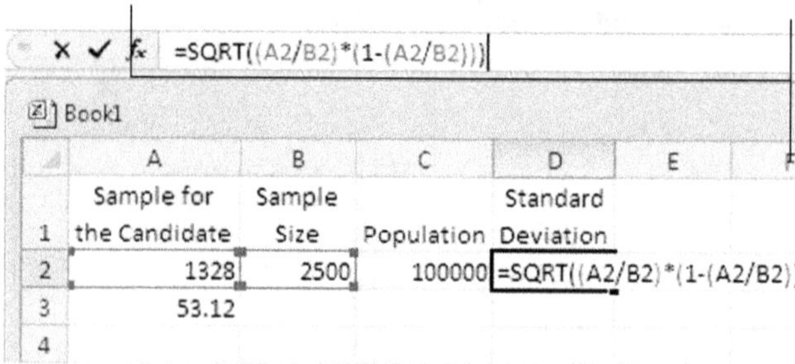

Note: This formula can be used to calculate SD when you have a zero, one box.

- After you have typed in this formula, click the checkmark

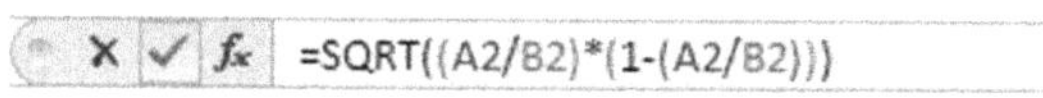

- Input in E1 **Expected Value** and in E2 input = **(A3*C2)/100**

Note: You assume the same percentage of votes in the population as you calculated for the sample data. Referred to as the Bootstrap Method.

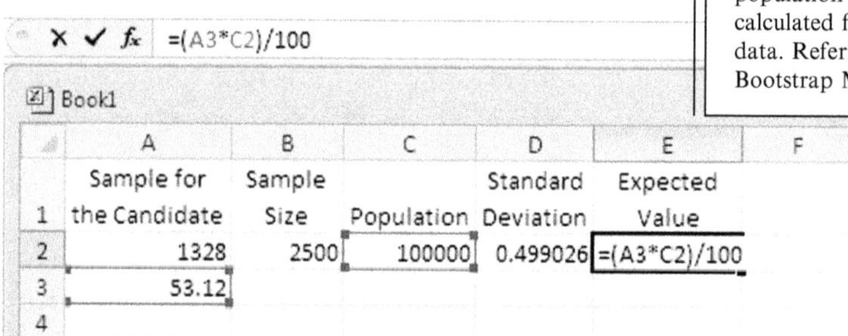

- After you have typed in the formula, click the checkmark

- Input in E3 = **A3**

Note: This is just showing the same percentage in the population as you calculated in the sample. You only show here for completeness. The information in this cell is not actually used in any of the calculations

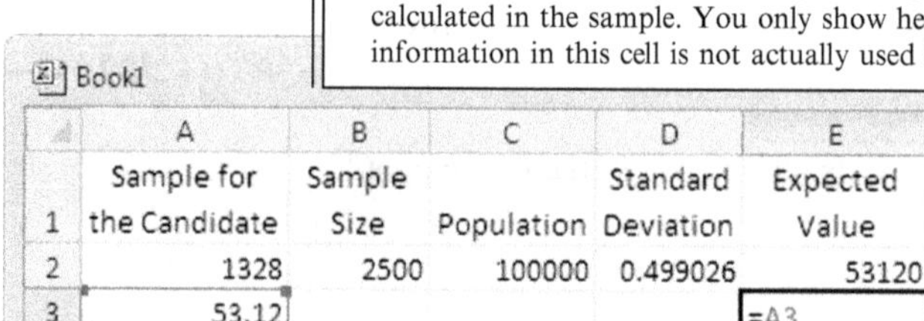

- Click on the green checkmark
- Input in F1 **Standard Error for Population**
- Input in F2 = **(D2/SQRT (B2))*100**

Note: This is the SE for the population as %

Book1.xlsx

	B	C	D	E	F	G
1	Sample Size	Population	Standard Deviation	Expected Value	Standard Error for Population	
2	2500	100000	0.499026	53120	=(D2/SQRT(B2))*100	
3				53.12		
4						

- Input in F3 = **F2*C2/100**

Note: This is the SE for the population as an actual headcount

Book1.xlsx

	B	C	D	E	F
1	Sample Size	Population	Standard Deviation	Expected Value	Standard Error for Population
2	2500	100000	0.499026	53120	0.998051221
3				53.12	=F2*C2/100
4					

For this problem, you should expect to receive 53,120 votes (E2) if all 100,000 votes were submitted, or 53% of the vote (E3). But, this is only an estimate, and we know there is an error due to the process of sampling. That error turns out to be about 1% (F2) or 998 votes (F3).

Book1.xlsx

	A	B	C	D	E	F
1	Sample for the Candidate	Sample Size	Population	Standard Deviation	Expected Value	Standard Error for Population
2	1328	2500	100000	0.499026	53120	0.998051221
3	53.12				53.12	998.0512211
4						

So you could expect to receive 53,120 ± 998, which is between 52,122 and 54,118 votes. Sounds like a winning proposition!

- Be careful how many decimal places you carry in these estimation calculations because you can't have a fraction of a vote.

Inferring Averages

Example Problem

Suppose a city manager wants to know the average income of the 25,000 families living in his town. He hires a survey organization to take a simple random sample of 1000 families. The total of all the sample incomes is \$32,396,714. Therefore, the average sample income for the 1000 families in the sample must be \$32,396,714/1000 = \$32,397. It turns out that the standard deviation for the sum of the sample of salaries is \$19,000. What is the average income for all 25,000 families likely to be, plus or minus a chance error? (This is the SE for an average.)

Excel

- Input in A1 **Sample Size** and in A2 **1000**

<u>Note</u>: given in problem as the sample size.

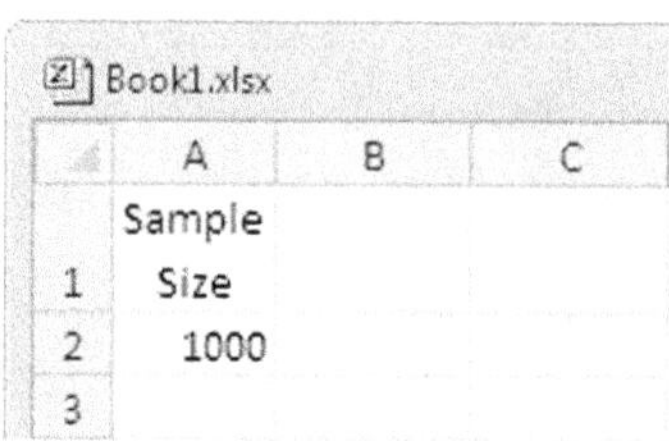

- Input in B1 **Sample Total Income** and in B2 **32,396,714**

<u>Note</u>: given in problem as the sum of the incomes of the sample families.

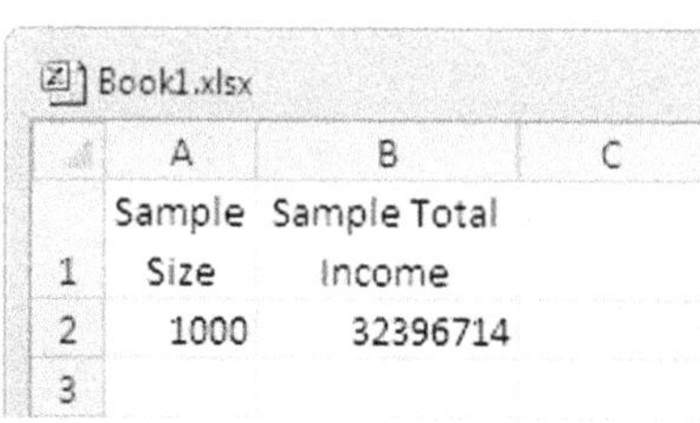

- Input in C1 **Sample Avg Income** and input in C2 **= B2/A2**
- After you have typed in the formula, click the checkmark

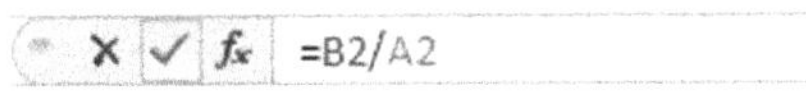

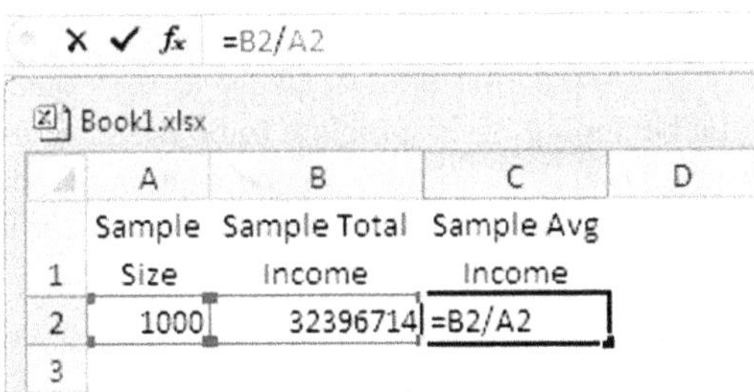

Note: Calculating the average income for the sample data. This may or may not be given in the problem. We infer this is also the average income for the population.

- Input in D1 **Sample SD** and in D2 **19,000**

Note: given in problem as the standard deviation for the sum of the sample salaries.

- Input in E1 **SE for the average income** and input in E2 **= D2/SQRT(A2)**

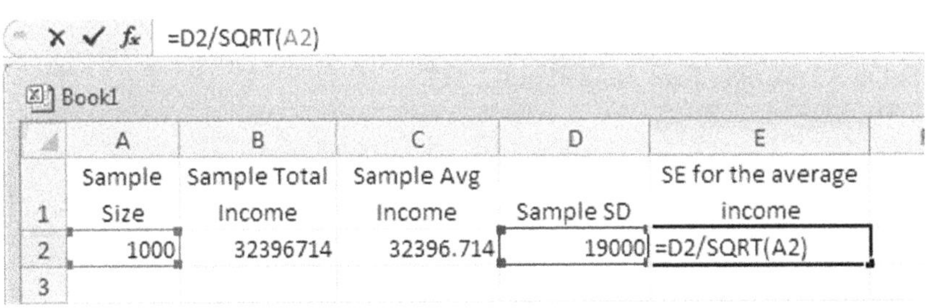

- After you have typed in the formula, click the checkmark

=D2/SQRT(A2)

Book1

	A	B	C	D	E
1	Sample Size	Sample Total Income	Sample Avg Income	Sample SD	SE for the average income
2	1000	32396714	32396.714	19000	600.8327554
3					

Therefore, the average income of the population is most likely to be around $32,400 (C2), which is inferred at the beginning of the problem to be the same as the sample average income, give or take $600 (E2), which is the population SE for the average income. The $600 represents the standard error or the margin of error for this estimate.

Confidence Intervals with Proportion Inference

Example Problem

Your company is considering opening a new location in a nearby town with a population of 25,000 people, and you need more female workers for your plant because they are more interested in the female products produced. You hire a consultant to complete a simple random sample of 1600 persons from the town. It turns out 917 people in the sample are females. Find a 95% confidence interval for the proportion of females among all 25,000 residents.

Excel

- Input in A1 **Sample Females** and in A2 **917**

Note: given in problem as the number of females in the sample.

Book1.xlsx

	A
1	Sample Females
2	917

- Input in B1 **Sample Size** and in B2 **1600**

Note: given in problem as the sample size.

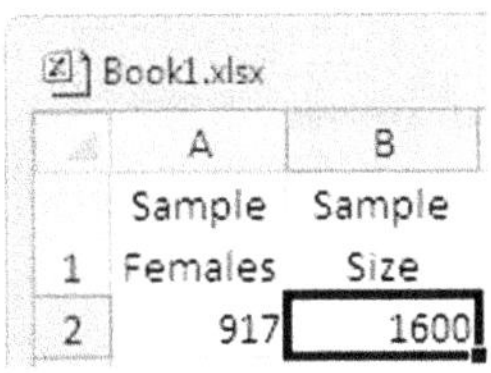

- Input in A3 = **(A2/B2)*100**

Note: Convert to a percentage of females in the sample.

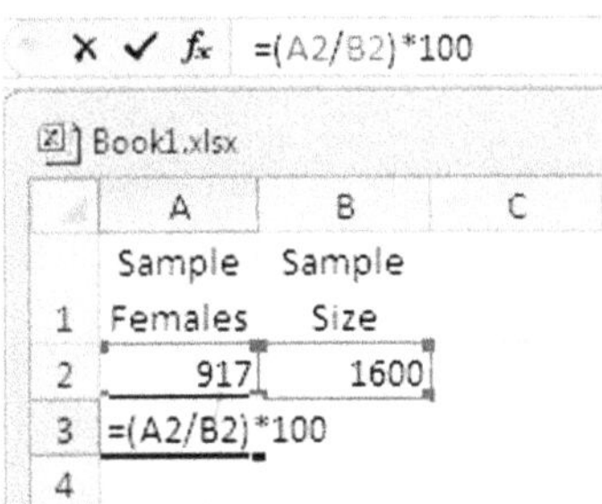

- After you have typed in the formula, click the checkmark and get the answer **57.3125**

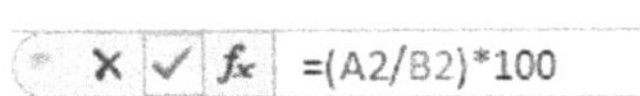

- Input in C1 **Population** and in C2 **25,000**

Note: given in problem as the population size.

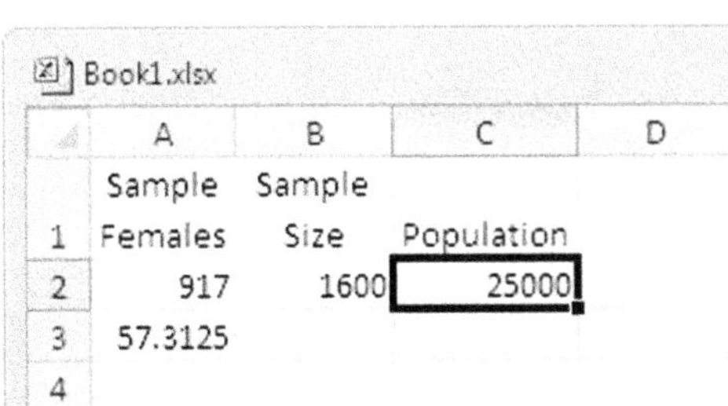

- Input in D1 **Standard Deviation** and in D2 input = **SQRT((A2/B2)*(1–(A2/B2)))**

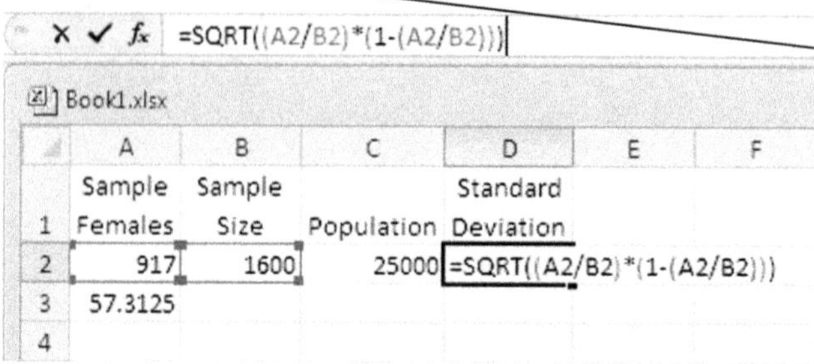

	A	B	C	D	E	F
1	Sample Females	Sample Size	Population	Standard Deviation		
2	917	1600	25000	=SQRT((A2/B2)*(1-(A2/B2)))		
3	57.3125					
4						

Note: This formula can be used to calculate SD when you have a zero, one box.

- After you have typed in the formula, click the checkmark, resulting in **0.494624**

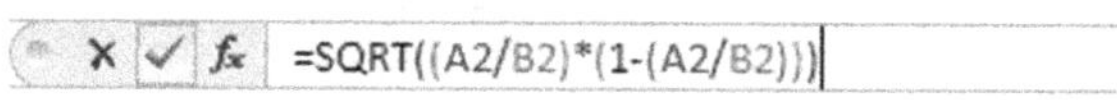

- Input in E1 **Expected Value** and input in E2 = **(A3*C2)/100**

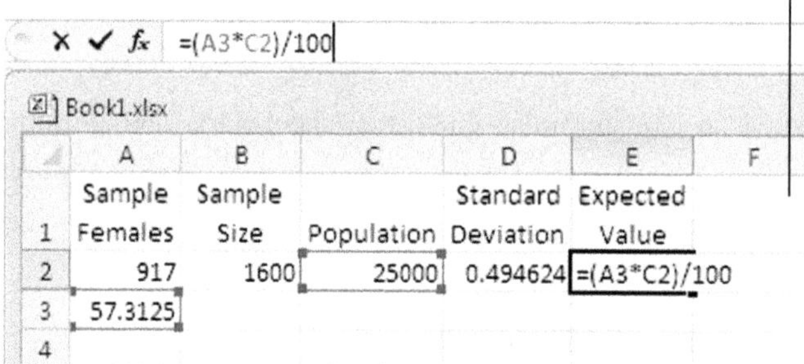

	A	B	C	D	E	F
1	Sample Females	Sample Size	Population	Standard Deviation	Expected Value	
2	917	1600	25000	0.494624	=(A3*C2)/100	
3	57.3125					
4						

Note: You assume the same percentage of females in the population as you calculated for the sample data. This is referred to as the Bootstrap Method.

- After you have typed in the formula, click the checkmark, resulting in **14,328.125**

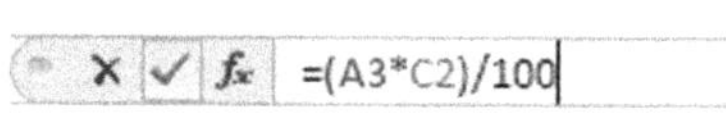

- Input in E3 = **A3**

Note: This is showing the same percentage in the population as you calculated in the sample.

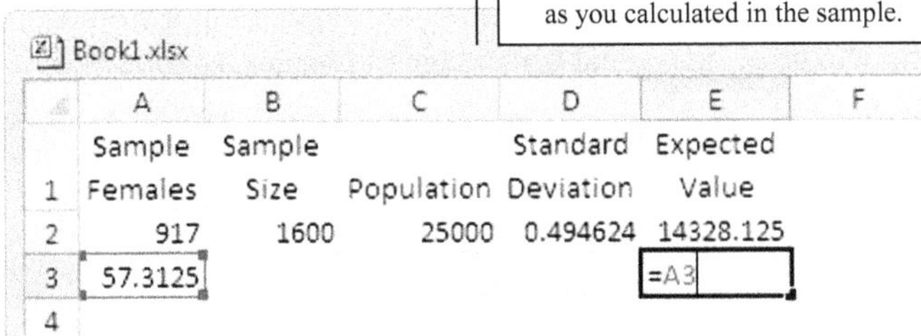

- Input in F1 **Standard Error for Percent**
- Input in F2 = **D2/SQRT(B2)*100**

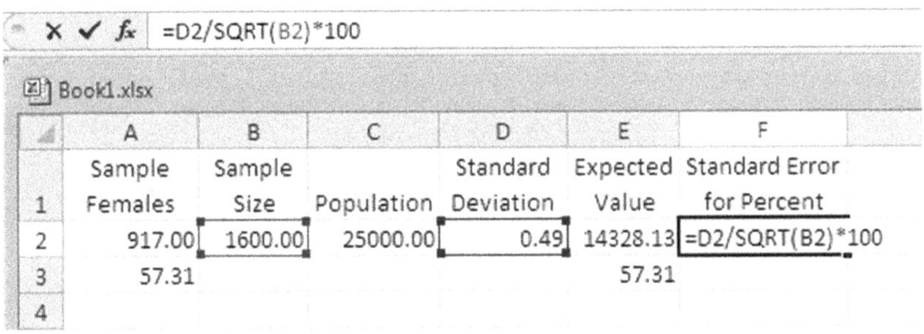

- After you have typed in the formula, click the checkmark, resulting in **1.24**

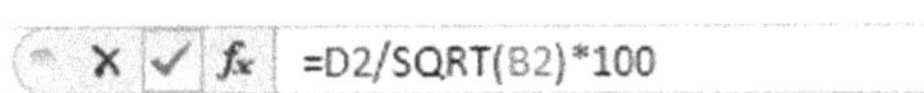

- Input in G1 **Standard Error for Population**
- Input in G2 = **(F2*C2/100)**

Note: This is the population standard error as an actual headcount. This is just calculated for completeness

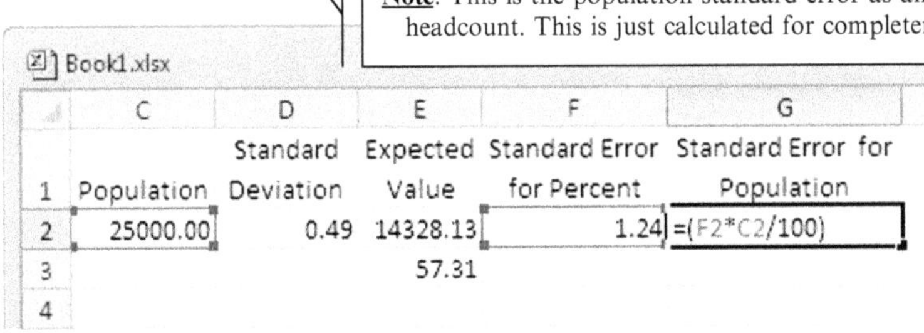

After you have typed in the formula, click the checkmark, resulting in 310

- The expected value for the percentage of females in the population of workers is 57% (E3) and is likely to be off the actual percentage of females by about 1.24% (F2) or so. Now, we need to consider the confidence interval calculations.
- Input in A5 **Confidence Intervals**

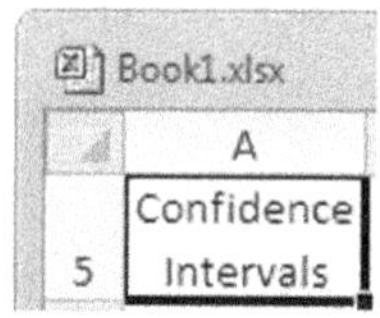

- Input in B5 **1SE (68%)**

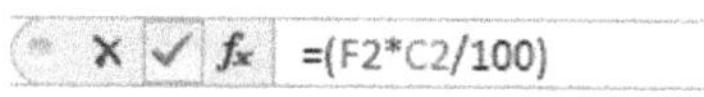

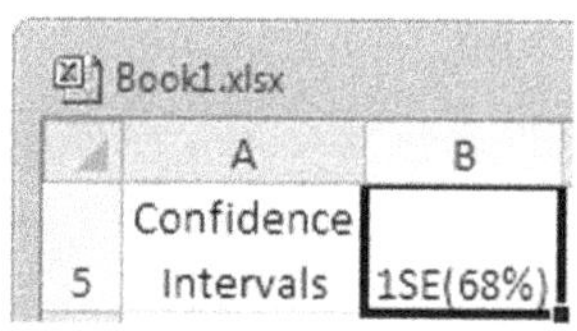

- Input in C5 **2SE (95%)**

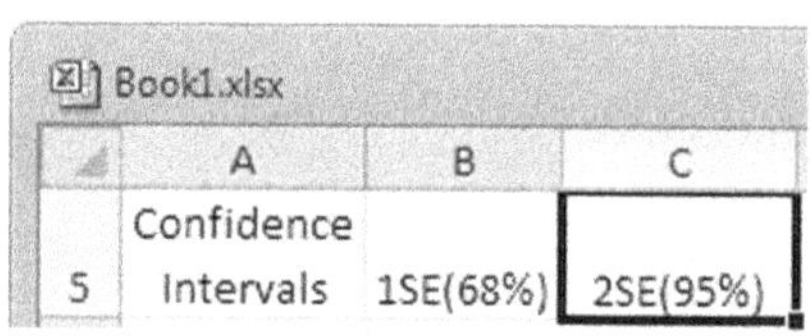

- Input in D5 **3SE (99.7%)**

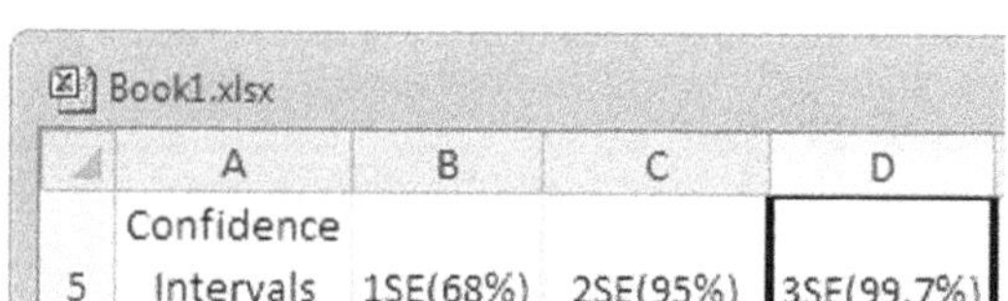

- Input in B6 = **F2**

Note: This is the population Standard Error as a %

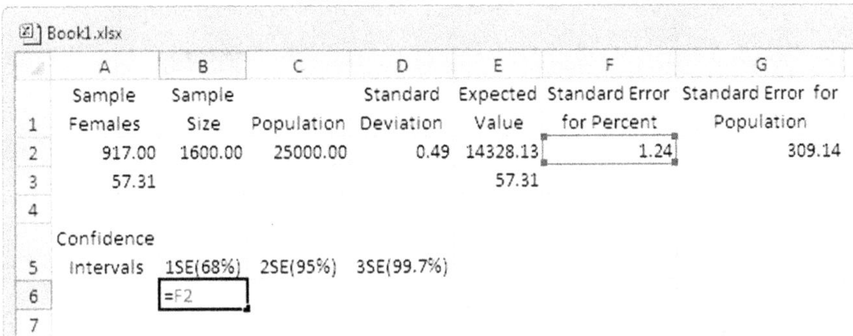

Book1.xlsx

	A	B	C	D	E	F	G
1	Sample Females	Sample Size	Population	Standard Deviation	Expected Value	Standard Error for Percent	Standard Error for Population
2	917.00	1600.00	25000.00	0.49	14328.13	1.24	309.14
3	57.31				57.31		
4							
5	Confidence Intervals	1SE(68%)	2SE(95%)	3SE(99.7%)			
6		=F2					
7							

- Input in C6 = **F2*2**

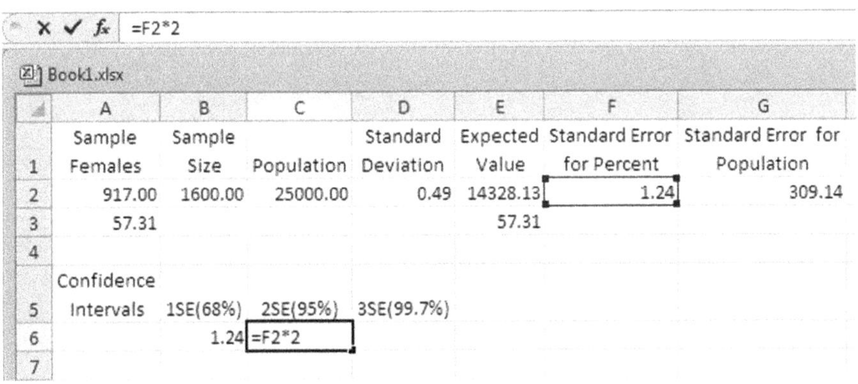

=F2*2

Book1.xlsx

	A	B	C	D	E	F	G
1	Sample Females	Sample Size	Population	Standard Deviation	Expected Value	Standard Error for Percent	Standard Error for Population
2	917.00	1600.00	25000.00	0.49	14328.13	1.24	309.14
3	57.31				57.31		
4							
5	Confidence Intervals	1SE(68%)	2SE(95%)	3SE(99.7%)			
6		1.24	=F2*2				
7							

- After you have typed in the formula, click the checkmark, resulting in **2.47**

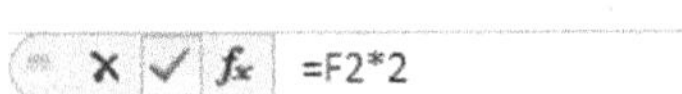

- Input in D6 = **F2*3**

× ✓ f_x =F2*3

Book1.xlsx

	A	B	C	D	E	F	G
1	Sample Females	Sample Size	Population	Standard Deviation	Expected Value	Standard Error for Percent	Standard Error for Population
2	917.00	1600.00	25000.00	0.49	14328.13	1.24	309.14
3	57.31				57.31		
4							
5	Confidence Intervals	1SE(68%)	2SE(95%)	3SE(99.7%)			
6		1.24	2.47	=F2*3			
7							

After you have typed in the formula, click the checkmark, resulting in 3.71

× ✓ f_x =F2*3

A 95% confidence interval for the percentage of females among the 25,000 residents corresponds to 2 SE or, in other words, ±2.47% (C6).

Book1.xlsx

	A	B	C	D	E	F	G
1	Sample Females	Sample Size	Population	Standard Deviation	Expected Value	Standard Error for Percent	Standard Error for Population
2	917.00	1600.00	25000.00	0.49	14328.13	1.24	309.14
3	57.31				57.31		
4							
5	Confidence Intervals	1SE(68%)	2SE(95%)	3SE(99.7%)			
6		1.24	2.47	3.71			
7							

The expected value for the percentage of females in the population of workers is 57% (E3) and is likely to be off the actual percentage of females by about 1.24% (F2) or so. We can be 95% confident that between 54.53% and 59.47% of the residents in this town are female. That sounds like enough females for us to consider this as a viable town for opening up our new plant.

Confidence Intervals with Average Inference

Example Problem

Suppose a city manager wants to know the average income of the 25,000 families living in his town. He hires a survey organization to take a simple random sample of 1000 families. The total of all the sample incomes is \$32,396,714. Therefore, the average sample income for the 1000 families in the sample must be \$32,396,714/1000 = \$32,397. It turns out that the standard deviation for the sum of the sample of salaries is \$19,000. What is the average income for all 25,000 families likely within a 99.7% confidence interval?

NOTE: this is the same problem we solved for inferring averages.

Now we are asked to compute the inferred average for the population within a 99.7% confidence interval. In other words, these results should be the average income +/3 SE.

We already know the average income should be around \$34,400 and the standard error is \$600. Therefore, we expect the population average income to be about \$34,400 plus or minus \$1800 within a 99.7% confidence level. In other words, we expect to get these sample results 99.7 times out of 100 samples.

Final Thoughts and Activities

Practice Problems and Case Studies

1. Examine a current magazine and take a sample of 30 pages. Record how many of those pages have some form of advertising on them. Based on the results, construct a 95% confidence interval estimate of the percentage of pages that you would expect to have advertising for the entire magazine.
2. In Florida, members of the citrus industry make extensive use of statistical methods. One issue involves how growers are paid for oranges used to make orange juice. A sample is taken from each truckload of oranges. This sample is weighed, squeezed, and the amount of sugar in the juice is measured. Based on these sample results, an estimate is made of the total sugar in the truckload. Payment is based on the % of sugar. In a similar vein, how do customs officers decide how many boxes to inspect in a company's shipments at the docks? How about the security checks at airports? Any concerns with using samples and inference in these situations?
3. A real estate company has been provided with a portfolio of townhomes and single-family homes at Tahoe Keys in California. The company surveys a nearby affluent town of 25,000 households to determine which form of residence is preferred as a vacation home, so they can target their marketing accordingly. The company took a sample of 500 of these households. It turns out only 79 households preferred single-family residences as a vacation home over townhomes.

   (a) Calculate the proportion of households in the town who prefer townhomes as a vacation home.
   (b) About how far off do you expect this estimate to be off?
   (c) Calculate the 99% confidence interval for the expected value.

4. The HR director of a large corporation wants to study absenteeism among administrative workers at the corporate office during the year. A random sample of 25 administrative workers reveals that the average number of days absent was 12 days. Construct a 95% confidence interval for the population average of absences for the clerical workers during the year given $\sigma = .495$.
5. In a random sample of 40 houses, the average house price is \$426,000. Construct a 99% confidence interval for the population average of house prices using a SD of \$49,500.
6. Why does central limit theory work even with non-normal distributions?
7. Calculate the SE for the population average from a sample of 50 travelers. The variable of interest is how much an average traveler spends on food at the airport per trip. The sample standard deviation is \$10.
8. In inference, we need to use the confidence intervals to estimate the statistically relevant range of the inferred value. Calculate the confidence intervals for 68%, 95%, and 99%. The SE is .466 units.

Discussion Boards

1. Daniel Yankelovich states, "Warning labels about sampling error say nothing about whether or not the public is conflict-ridden or has given a subject much thought. This is the most serious source of opinion poll misinterpretation."
2. Studies show that if children took the 1932 IQ test today, half would score above 120 on the 1932 scale. Currently IQ scores have a mean of 100, a SD of 15, and are normally distributed. If today's American children took the same IQ test used in 1932, the mean score would be 120 instead of 100 as it was in 1932. The net effect is that today's IQ scores are 20 points higher than in 1932. Any concerns?
3. "When you or anyone else attempts to tell me that a sample of 1,223 persons account for our opinions and tastes, here in America, I get mad as hell! This seems astounding and unfair and should be outlawed!" This is a case in which a sample poll of 1,223 people was projected to represent 120 million. Do you agree with the concern?
4. About 30% of Americans aged 19–28 claim that they have used an illicit drug other than marijuana. If a random sample of 40 Americans aged 19–28 finds 21 who claim to have used an illicit drug other than marijuana, would you be surprised? Explain these statistics.

Group Activity

1. Sherlock Holmes often used inference. He often discussed his process of "reasoning backward." Choose one of his cases and explain how inference was successfully employed.
2. Research how the *Current Population Survey* is used to estimate US unemployment rates.
3. Research the type of data and statistical analysis that form the foundation of Neilson Media Research.
4. Using surveys to estimate rare events typically leads to overestimating. For example, the National Rifle Association reports three million dues-paying members or about 1.5% of American adults. In national random telephone surveys, however, 4–10% of respondents claim that they are dues-paying NRA members. Do your own web research. Discuss possible reasons for bias and methods of minimizing the bias.

Parting Thought

You know how dumb the average person is? Well, by definition, half of the population is even dumber than that …

Problem Solutions

1. Examine a current magazine and take a sample of 30 pages. Record how many of those pages have some form of advertising on them. Based on the results, construct a 95% confidence interval estimate of the percentage of pages that you would expect to have advertising for the entire magazine.

   Answer:

   This is a proportion problem.
   For this problem, we need to know the total number of pages in the magazine. These numbers will vary based on which magazine you select. Here is an example answer:
   $N = 86$ (for the selected magazine)
   $n = 30$
   Number of pages in the sample with ads = 13 or 13/30 = .433 or 43%
   sd = we can use the short cut because either the pages have ads or they don't (zero, one model)

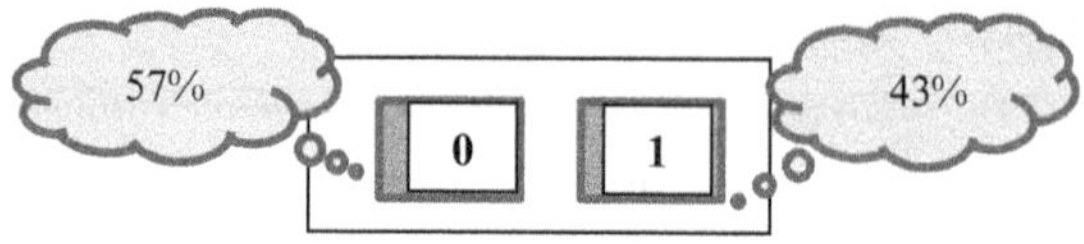

SD = SQRT (0.57 * 0.43) = .5
Use the formula SE for the proportion:
SE = (sd/sqrt n) × 100%
SE = (.5/SQRT 30) × 100% = (.5/5.47) × 100% = 9.1%
95% CI = 2*9.1 = 18.2% of the pages or, in other words, approximately 6 pages.
If we took more 30-page samples, we would still expect to find about 6 pages with advertising in 95 out of 100 samples.
As far as the population (all the pages), we would expect about 43% would have ads, plus or minus 18%.

2. In Florida, members of the citrus industry make extensive use of statistical methods. One way includes how growers are paid for oranges used to make orange juice. A sample is taken from each truckload of oranges. This sample is weighed, squeezed, and the amount of sugar in the juice is measured. Based on these sample results, an estimate is made of the total sugar in the truckload. Payment is based on the % of sugar. In a similar vein, how do customs officers decide how many boxes to inspect in a company's shipments at the docks? How about the security checks at airports? Any concerns?

   Answer:

   In all three situations it is only reasonable to collect samples. The results would be more accurate if we looked at every part of the population but that would be too time intensive.
   Based on the sample the farmers may be underpaid, but on the other hand they may be overpaid. So everyone would like a high confidence interval (CI) to ensure the results are as accurate as possible without sampling the entire shipment. The CI and the standard error would be part of what gets negotiated between the farmers and the buyers. But, you can imagine understanding this data is difficult for anyone who is not up to speed with their statistics.
   Consider that the truck has 1000 oranges and we only sample 100 with an average sugar content of 8 g. A sugar value of 12 g per medium size orange is more desirable, so the farmers get paid less than the premium price. The sample size corresponds to only a 68% CI. This would mean if we took another 100 different samples from this truck we would expect the same sugar content in only 68 oranges, but in 32 of the samples we might have a higher or lower sugar content.
   Similarly customs officers and TSA officials cannot examine every box and every person in great detail, so there is a tradeoff in risk versus cost. The lowest level of SE risk occurs when you survey the entire population (CI = 100%).

3. A real estate company has been provided with a portfolio of townhomes and single-family homes at Tahoe Keys in California. The company surveys a nearby affluent town of 25,000 households to determine which form of residence is preferred as a vacation home, so they can target their marketing accordingly. The company took a sample of 500 of these households. It turns out only 79 households preferred single-family residences as a vacation home over townhomes.

   (a) Calculate the proportion of households in the town who prefer townhomes as a vacation home.
   (b) About how far off do you expect this estimate to be off?
   (c) Calculate the 99% confidence interval for the expected value.

Answer:

$N = 25{,}000$
$n = 500$
CI = 99%
Number preferring single-family residents = 79 or 79/500 = .158 or 15.8%
sd = we can use the short cut because either they like townhomes (0) or single-family residents (1)

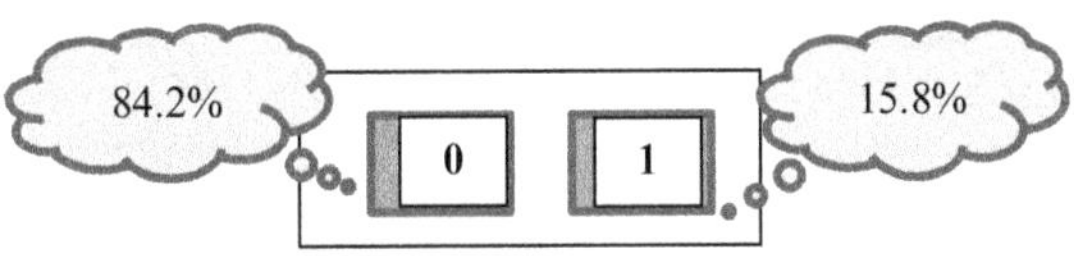

SD = sqrt (.84 * .16) = .37
Use the formula SE for the proportion
SE = (sd/sqrt (n)) × 100%
SE = (.37/ [SQRT 500]) × 100% = (.37/22.36) × 100% = 1.65%
99% CI = 3 × 1.65 = 4.95% of the households prefer single-family vacation homes. In other words, about 25 households prefer single-family vacation homes

A. 84.2% prefer townhomes
B. .0165 × 500 = 8.25 households. As a percent 8.25/500 = .0165 or approximately 1.7%
C. 99% CI = 3 * 1.7 = 5.1%

4. The HR director of a large cooperation wants to study absenteeism among administrative workers at the corporate office during the year. A random sample of 25 administrative workers reveals that the average number of days absent was 12 days. Construct a 95% confidence interval for the population average of absences for the clerical workers during the year given $\sigma = .495$.

   Answer:

   $n = 25$
   $\overline{x} = 12$
   CI = 95%

The average number of days that administrative workers were absent is 12 days.
SD = .495
Use the formula SE for the average:
SE = (sd/sqrt (n))
SE = .495/ (SQRT 25) = 0.099
95% CI = 2 * 0.099 = .198 days of absenteeism
Workers will be absent on average 12 days plus or minus 0.20 days.

5. In a random sample of 40 houses, the average house price is $426,000. Construct a 99% confidence interval for the population average of house prices using a SD of $49,500.
   Answer:

   n = 40
   avg house price = $426,000
   CI = 99%
   SD= $49,500
   Use the formula SE for the average:
   SE = (sd/sqrt (n))
   SE = 49,500/ (SQRT 40) = $7826
   99% CI = 3 * 7826 = $23,478

6. Why does central limit theory work even with non-normal distributions?
   Answer:
   The Central Limit Theorem (CLT) is a fundamental concept in statistics that states that regardless of the shape of the population distribution, the sampling distribution of the sample mean tends to follow a normal distribution as the sample size increases. This remarkable property of the CLT allows us to make certain assumptions and perform statistical inference even when the population distribution is non-normal.

   The reasons why the Central Limit Theorem holds true for non-normal distributions are

   1. Averaging effect: The CLT relies on the fact that when we take repeated random samples from any population, the individual sample means tend to cluster around the population mean. As the sample size increases, the effect of individual observations on the sample mean decreases and the average tends to stabilize. This averaging effect helps normalize the sampling distribution, even if the population distribution is not normal.
   2. Sum of random variables: The Central Limit Theorem is based on the idea that a sample mean is the sum of a large number of random variables. According to mathematical theory, the sum of independent and identically distributed (i.i.d.) random variables, regardless of their underlying distribution, tends to be normally distributed. This is often referred to as the "sum of random variables" property.

3. Law of large numbers: The Central Limit Theorem is related to the law of large numbers which states that as the sample size increases, the sample mean converges to the population mean. The CLT builds upon this concept, indicating that the distribution of the sample mean approaches normality as the sample size becomes larger.

It is important to note that the Central Limit Theorem applies particularly well when the sample size is large (typically above 30) due to the mathematical assumptions underlying the theorem. However, even with smaller sample sizes the CLT approximation can still be reasonably accurate, especially if the population distribution is not heavily skewed or has extreme outliers.

7. Calculate the SE for the population average from a sample of 50 travelers. The variable of interest is how much an average traveler spends on food at the airport per trip. The sample standard deviation is $10.

   Answer:

   SD = $10
   Use the formula SE for the average:
   SE = (sd/sqrt (*n*))
   SE = 10/ (SQRT 50) = $1.41

8. In inference, we need to use the confidence intervals to estimate the statistically relevant range of the inferred value. Calculate the confidence intervals for 68%, 95%, and 99%. The SE is .466 units.

   Answer:
   68%, 1 * .466 = .466 units
   95%, 2 * .466 = .932 units
   99%, 3 * .466 = 1.398 units

Chapter 8
Probability

Linda Herkenhoff and John Fogli

Key Concepts

Addition rule, Binomial distribution, Complement rule, Combinations, Factorial function, Conditional probability, Dependence, Multiplication rule, Mutual exclusivity, Permutations, and Probability, Probability density function, Random variable, With and Without replacement.

Discussion

This chapter will explore the world of probabilities. It is very important to remember that probability is an estimation tool describing expected outcomes. Probability statistics explain the fraction of time an event is expected to happen in the long run, if the situation is repeated over and over again. In this area of statistics, we have specific rules for how to combine probabilities and we also have special symbols to show these processes. In some cases, the historical data fits a theoretical distribution, allowing us to answer questions about the likelihood of an event occurring.

It is important to differentiate between discrete and continuous probability distributions. In the discrete world, you have a few distinct values that are often integers.

Supplementary Information The online version contains supplementary material available at https://doi.org/10.1007/978-3-031-46371-6_8.

L. Herkenhoff (✉)
Saint Mary's College, Moraga, CA, USA

J. Fogli
Diablo Valley College, Pleasant Hill, CA, USA

L. Herkenhoff, J. Fogli (eds.), *Applied Statistics for Business and Management using Microsoft Excel*, https://doi.org/10.1007/978-3-031-46371-6_8

Continuous distributions describe a random process with continuous valued outcomes. Excel tools for handling both of these distributions will be explained.

Probability is the chance or likelihood that an event will happen. It is defined as the number of ways an outcome can occur, divided by the total number of outcomes that are possible for discrete events. There are some basic rules associated with probability:

1. Probabilities are between 0% and 100%.
2. The probability of an event happening is 100% minus the chance that it does not happen. If the chance of rain is 30%, then the chance of no rain is 70%. This is often referred to as the **Complement Rule** because the event and its opposite (complement) add up to 100%.
3. When you draw at random, all items have the same probability of being selected.
4. When drawing at random with replacement, the draws are independent. Without replacement, the draws are dependent.

Addition rule: This rule is used to combine probabilities one **or** the other may occur. When at least one of the events happens, we combine the events by addition. In mathematical terms, we often refer to this as the **union (U)** of two events. To show the probability of the union of two events, it is written as

$$P(A \cup B) = P(A) + P(B)$$

Mutual exclusivity: Two events are **mutually exclusive** when the occurrence of one prevents the other event from occurring. This concept is important when we add probabilities.

$$\textbf{Event A} = \text{being over 21year old} \qquad \textbf{Event B} = \text{being 21year old or younger}$$

Events A and B are **mutually exclusive.** It is impossible for both to occur at the same time. If Event A is true, then Event B must be false.

Multiplication rule: This rule is used to combine probabilities when one event **and** the other may occur. When both of the events happen, we combine the events by multiplication. In mathematical terms, we often refer to this as the **intersection** (∩) of two events. To show the probability of the intersection of two events, it is written as

$$P(A \cap B) = P(A) \times P(B)$$

Dependence: Events are **dependent** when the occurrence of one event affects the probability of occurrence of the other. Two events are **independent** if the chances for the second event stay the same, no matter how the first one turns out. This concept is important when we multiply probabilities.

Event A = being under 25 year old **Event B** = being a CEO

In this example, consider that Events A and B are dependent. The probability of being a CEO may be 15%, but if Event A is true (the person is under 25 years old), that probability may drop to less than 1%. Event A has affected the probability of Event B.

Dependence is commonly referred to as "**without replacement,**" while **independence** is referred to as "**with replacement.**" This analogy is based on reaching a hypothetical box and selecting a card. After the first draw if the card is placed back in the box, then the second draw is independent of the first outcome because the second draw has been completed "with replacement." This is illustrated in Fig. 8.1.

Conditional probability: $P(A \mid B)$ means the probability of A given that B has already happened. There is an implied temporal or sequential relationship: Event B has already occurred and that event is then followed by Event A.

Probability distributions: Discrete probability distributions may be displayed with a bar chart in which the height of each bar is equal to the probability of the event as shown in Fig. 8.2. To find the probability of a group of events, we add up the individual probabilities of the events of the group.

Continuous probability distributions consider the area under the curve as the probability associated with a range of values. Continuous distributions are usually described by a **probability density function (pdf)**, where the area under the density function for a specified range of values is the probability.

Normal distribution: One of the most common continuous probability distributions in business is the **Normal Distribution.** Excel includes functions for both the

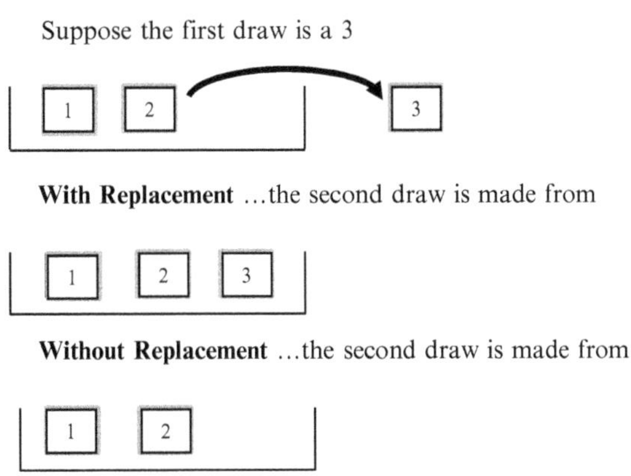

Fig. 8.1 The difference in drawing with and without replacement

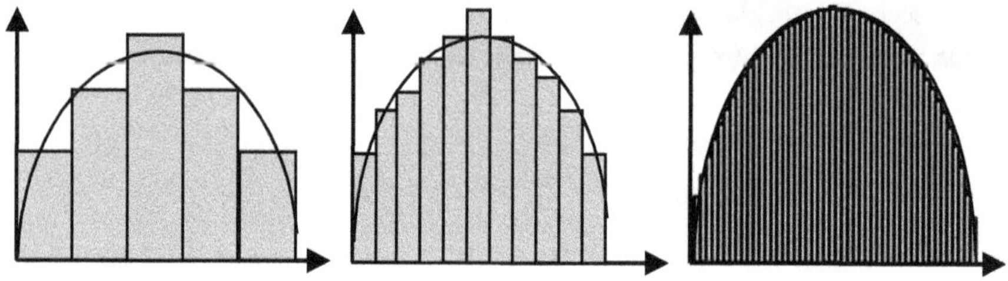

Fig. 8.2 Discreet and continuous distributions

standard normal distribution (mean = 0 and standard deviation = 1) and a general normal distribution (mean other than 0 and/or standard deviation other than 1). To obtain probability values for a standard normal distribution, the Excel function **NORM.S.DIST** is used. Otherwise, for the general normal distribution, the Excel function is named **NORM.DIST** (note: there is not an **S** between the **M** and the **D**).

Random variable: A **random variable** is a variable whose value is subject to variations due to chance. Conceptually it does not have a single, fixed value; it can take on a set of possible different values, each with an associated probability.

Combinations and permutations:

We can use combinations and permutations to assist in counting the number of possible outcomes, which allows us to determine the probability of the event occurring. Sometimes it may be difficult to decide whether you should use a combination or a permutation. The following key words apply to situations involving combinations where order is not important: subsets, committees, work teams, work groups, and delegations. If sequencing or order is important, use permutations.

Combination: order of like items is not important	Permutation: order of like items is important
$C\binom{n}{x} = \frac{n!}{x!(n-x)!}$	$P\binom{n}{x} = \frac{n!}{(n-x)!}$

n = total number of items and x = number of items of a like kind to be chosen

The large brackets include the information that asks how many ways can we choose x items from n total items. In **combinations (C),** the order of the items is not important but in **permutations (P)**, order is important.

The **factorial function** (symbol: **!**) just means to multiply a series of sequentially descending whole numbers. Examples:

- **4!** $= 4 \times 3 \times 2 \times 1 = 24$
- **7!** $= 7 \times 6 \times 5 \times 4 \times 3 \times 2 \times 1 = 5040$
- **1!** $= 1$
- **0!** $= 1$

> Note: This is just one you have to remember: 0! will always equal 1.

Example 1

Consider the letters A, B, C. What is the probability of choosing the following selection (A, B, C) from this group without replacement? You must first calculate the total number of possible outcomes from this group when

Order is not important	Order is important
$n = 3, x = 3$	$n = 3, x = 3$
$C\binom{3}{3} = \frac{3!}{3!(3-3)!} = \frac{3!}{3!(0)!} = 1$	$P\binom{3}{3} = \frac{3!}{(3-3)!} = \frac{3^*2^*1}{(0)!} = 6$
ABC = BCA = CBA, etc., so there is only one possible outcome, you will always have an A, B, and C no matter what	ABC, BCA, ACB, BAC, CAB, and CBA all meet the criteria. So there are six ways to get the desired selection

The next step is to calculate the actual probability. Remember probability is defined as the number of ways an outcome can occur divided by the total number of outcomes. The number of ways an outcome can occur can be thought of as the number of ways of achieving **success** out of the total number of ways possible. We just calculated one possible outcome for the combination and six possible outcomes for the permutation.

Combination: There is only one pattern we are interested in (A, B, C) or there is only one way to be successful because *ABC* is the same as *BCA*, is the same as *CBA* so $x = 1$, and we just calculated the total number of outcomes as being only 1. The total number of ways of being successful divided by the total number of possible outcomes is (1/1) = 1 or 100% probability you will be successful and select (A, B, C).

Permutation: There is only one pattern we are interested in (A, B, C) or there is only one way to be successful so $x = 1$, and we just calculated the total number of outcomes as 6. The total number of ways of being successful divided by the total number of possible outcomes is (1/6). So, there is a 16.7% probability that you will be successful in selecting (A, B, C), in that specific order.

Example 2

Consider choosing three employees from a group of five employees without replacement: (Tom, Sue, Bob, Ann, Mary)

What is the probability that we end up with **(Ann, Mary, Tom)**? First, we need to know how many outcomes are possible.

Order is not important	Order is important
$C\begin{pmatrix}5\\3\end{pmatrix}=\frac{5!}{3!2!}=\frac{5^*4^*3^*2^*1}{(3^*2^*1)(2^*1)}=10$	Because we are filling a slate of positions, where team leader is the first slot, facilitator the second slot, and scribe the third slot. We want to know how many different slates of candidates are possible
Tom, Sue, Bob = Bob, Sue, Tom	$P\begin{pmatrix}5\\3\end{pmatrix}=\frac{5!}{2!}=\frac{5^*4^*3^*2^*1}{2^*1}=60$
It is the same three people, so we don't want to count this more than once. Here are the possible outcomes	Because:
Tom, Bob, Ann Tom, Sue, Mary	Tom, Sue, Bob is a different slate from Sue, Tom, Bob, etc.
Tom, Bob, Sue Bob, Ann, Sue	
Tom, Bob, Mary Bob, Ann, Mary	
Tom, Ann, Sue Bob, Sue, Mary	
Tom, Ann, Mary Ann, Sue, Mary	

The next step is to calculate the probability.

Combination: There is only one pattern we are interested in **(Ann, Mary, Tom)** because *Ann, Mary, Tom* is the same as *Tom, Ann, Mary*, etc., and we just calculated the total number of outcomes as being 10. The total number of ways of being successful divided by the total number of possible outcomes is (1/10) or a 10% probability that you will be successful in selecting (Ann, Mary, Tom) in any order.

Permutation: There is only one pattern we are interested in **(Ann, Mary, Tom)** because we want Ann as the team leader, Mary as the facilitator, and Tom as the scribe. We just calculated the total number of outcomes as 60. The total number of ways of being successful divided by the total number of possible outcomes is (1/60) or a 1.6% probability that you will be successful in selecting Ann as the team leader, Mary as the facilitator, and Tom as the scribe.

Binomial distribution: This approach to calculating probabilities should be used when there are only two possible outcomes: success and failure.

Use the binomial distribution if and only if the situation satisfies the following conditions

1. There is a fixed number of trials.
2. Each trial is independent of one another.
3. There are only two possible outcomes (a Success or a Failure) for each trial.
4. The probability of success, p, is the same for every trial.

An example of an experiment that has a binomial distribution would be a coin toss.

1. You would toss the coin n (a fixed number) times.
2. The result of a previous toss does not affect the present toss (trials are independent).
3. There are only two outcomes (Heads or Tails).

The probability of success (whether a head is considered a success or a tail is considered a success) is constant at 50%.

Decision Trees

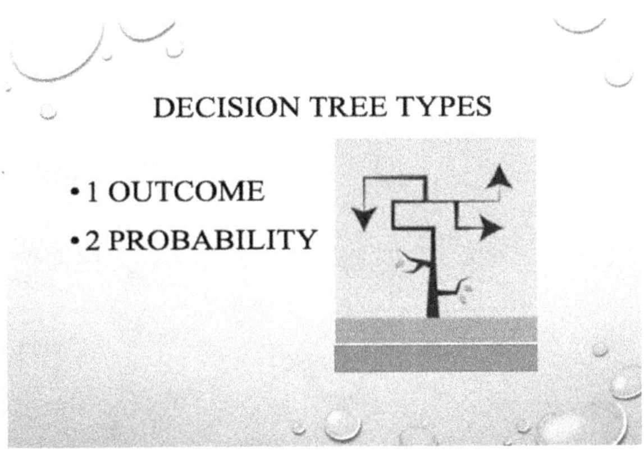

A decision tree is a decision support tree that uses a tree-like model of decision and/or outcome as well as the possible consequences. They are a powerful visualization tool that allows the analyst to see the overall picture. These diagrams typically include outcomes (outcome tree; Fig. 8.3) or chance event probabilities (probability tree; Fig. 8.4). Decision trees are commonly used in operation management to help identify strategy most likely to reach a goal.

Templates for drawing decision trees can be easily found online (Fig. 8.5).

All decision trees have nodes connecting the branches. Nodes represent decision points. These are commonly shown as circles or boxes.

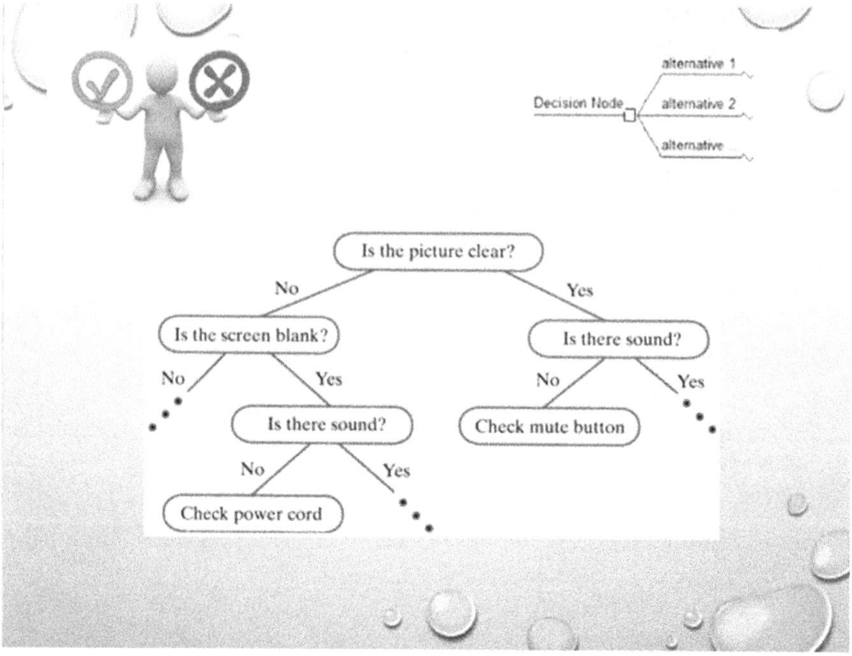

Fig. 8.3 Outcome decision tree

Fig. 8.4 Probability decision tree

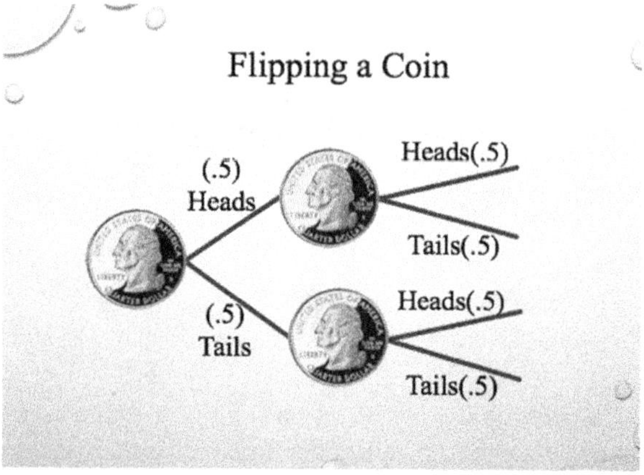

Outcome Decision Trees

Bradley has invested in two stocks, Mesa Oil and Collins Mining. Bradley has determined that the possible outcomes of these investments 3 months from now are as follows.

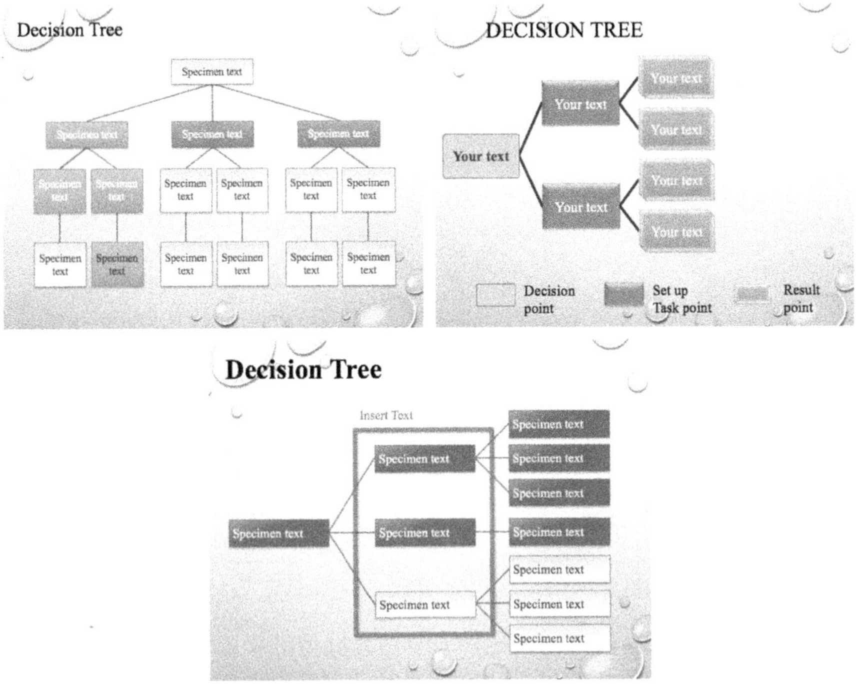

Fig. 8.5 Examples of decision tree templates

Investment Gain or Loss in 3 Months (in $000)	
Mesa Oil	Collins Mining
10	8
5	−2
0	
−20	

Bradley Investments can be viewed as a two-step experiment; it involves two stocks, each with a set of experimental outcomes.

Mesa Oil:	$n_1 = 4$
Collins Mining:	$n_2 = 2$
Total number of possible outcomes:	
$n_1 n_2 = (4) \times (2) = 8$	

This data can be easily shown in an outcome decision tree (Fig. 8.6). The first step is to identify the two stages. The overall outcomes are shown at the end of each branch.

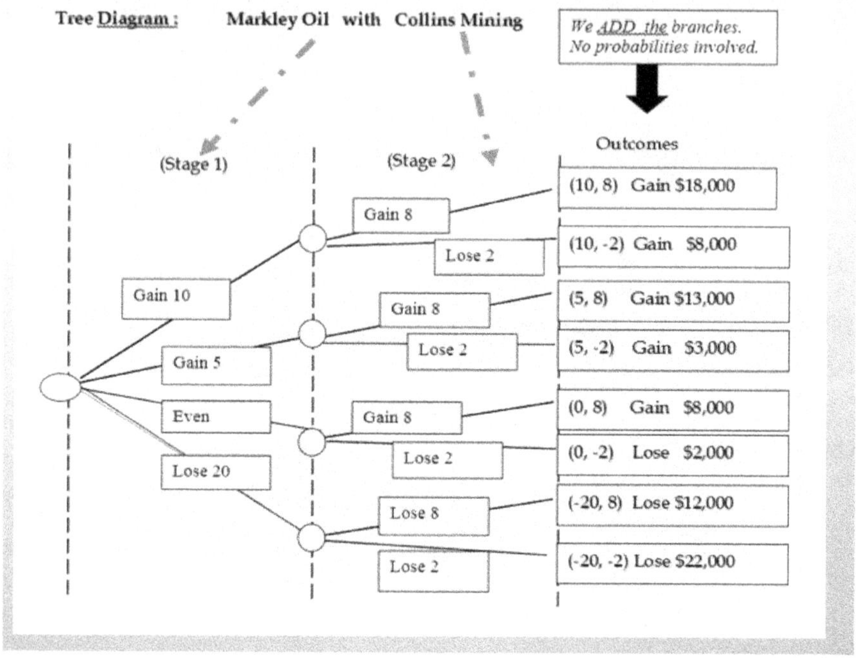

Fig. 8.6 Outcome decision tree example

Probability Decision Trees

The Royal Banks is concerned about a low retention rate for employees. On the basis of past experience, management has seen an annual employee turnover of 10%. For any employee chosen at random, management estimates a probability of 0.1 that the person will not be with the company next year. Choosing three hourly employees at random, what is the probability that one of them will leave the company next year?

Step 1: Draw the probability decision tree diagram

In this problem, it is important to correctly identify the three stages. Stage 1 deals with the first worker who can stay or leave. Stage 2 deals with the second worker who can also stay or leave. Remember each new rear branch needs to connect with the previous stage branches. Stage 3 deals with the third worker staying or leaving.

Show the associated probabilities that are all independent of each other (Fig. 8.7).

Step 2: Calculate the outcomes

The variable we are concerned with is how many of the three employees are leaving. Let's refer to this as "x." Show the associated x value at the end of each branch.

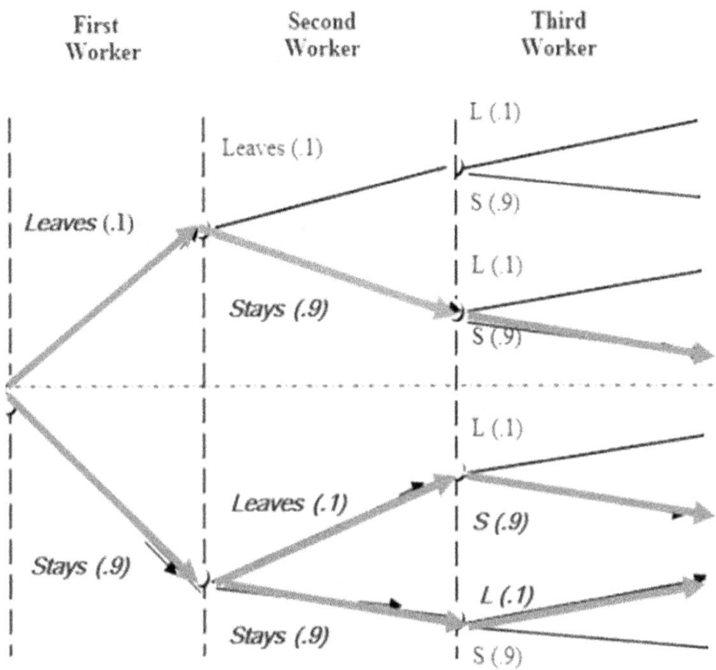

Fig. 8.7 Probability decision tree example

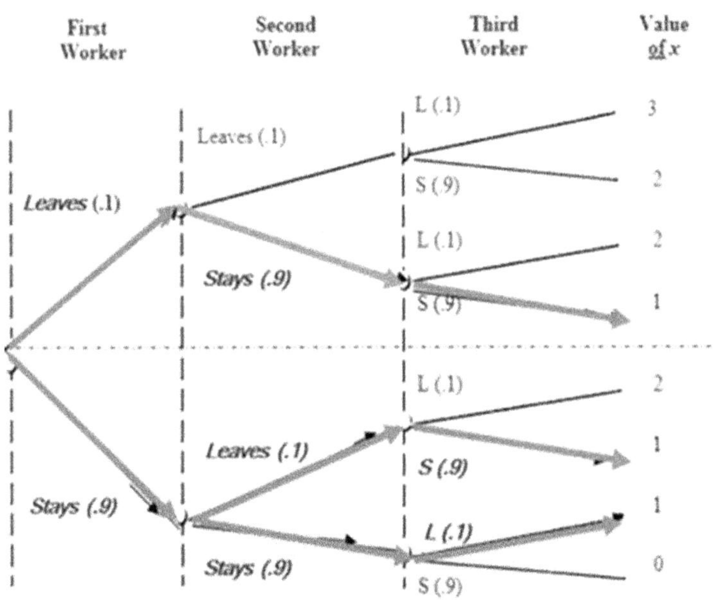

Step 3: Calculate the final probabilities

Since these probabilities are from independent events, in other words, whether one employee stays or leaves does not affect the probability of other employees staying or leaving.

We multiply each of the probabilities as we follow along a branch of the decision tree. For example, consider the upper branch of the tree: leave (.1) and leave (.1) and leave (.1). Therefore, the combined probability for all three employees leaving is .1 × .1 × .1 = .0010

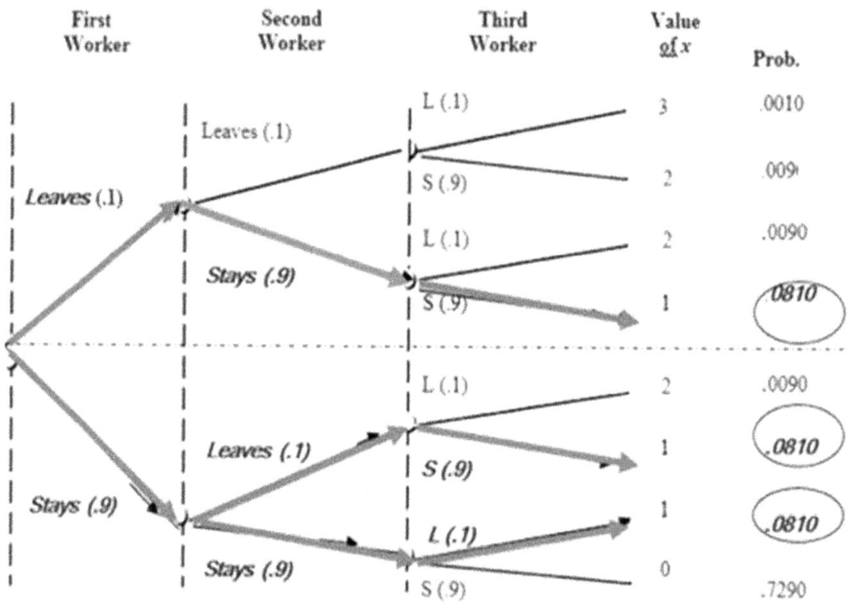

Step 4: What is the overall probability that one employee will leave during the next year?

This requires combining all the probabilities that correspond to branches with only one employee leaving. The question is concerning the events when employee 1 leaves or employee 2 leaves or employee 3 leaves. This is an "or" problem, so add the probabilities corresponding to only one employee leaving: .0810 + .0810 + .0810 = .2430. In other words, there is a 24.3% chance that one employee out of three will leave within the year.

Excel

NORM.DIST is the function that returns the cumulative left-tail probability that the normal random variable is less than or equal to x.

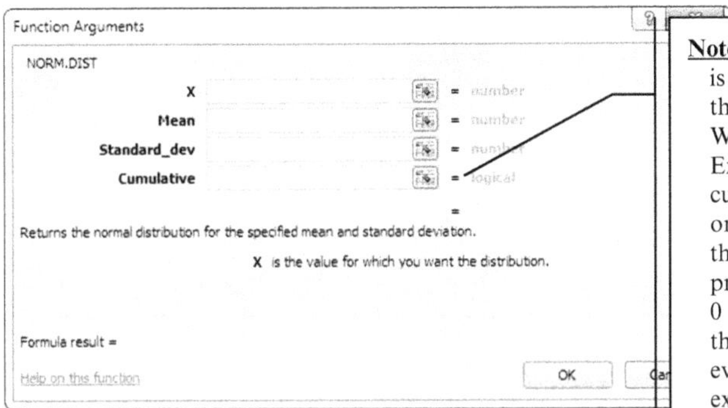

Note: When a Cumulative box is part of the required input there are two possible inputs. When TRUE or 1 is selected, Excel calculates the cumulative probability of "X or fewer" events of interest; this is a cumulative probability. When FALSE or 0 is selected, Excel calculates the probability of "exactly X" events of interest as in this example.

Finding Probabilities Using Normal Distributions

In this example, we have data on the number of cars sold each week at local dealerships over the past 6 months. Consider a normal random variable x (number of cars sold per week) with a mean (μ) of 100 cars sold per week and standard deviation (σ) of 15 cars per week.

What Is the Probability that a Dealership Will Sell 90 Cars or Less ($x \leq 90$) per Week?

- Highlight the cell where you want the answer to go
- Click on the ***fx*** button by the function input bar

- Select **Statistical** under **Category**

- Select **NORM.DIST** under **Function**

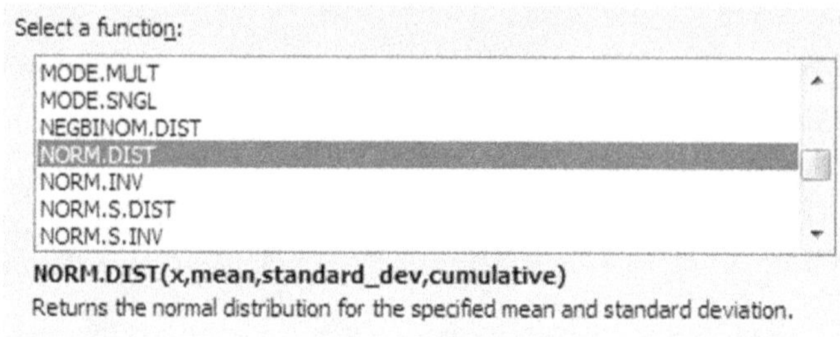

- Click **OK**

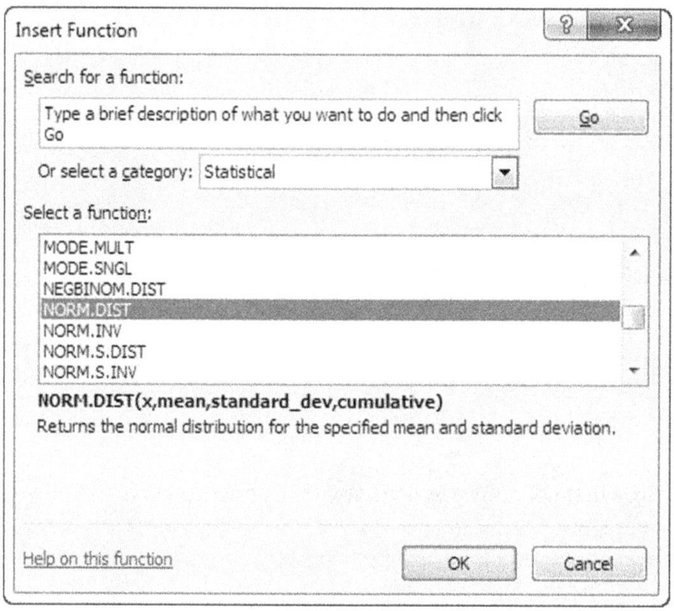

- Input **90** for **X**

- Input **100** for the **Mean**

- Input **15** for the **Standard_dev**

- Input a value for **Cumulative** of **1** or **True**

Click OK

Excel provides an answer of 0.2525, which can be converted to 25.25% probability that a dealership will sell 90 or less cars per week.

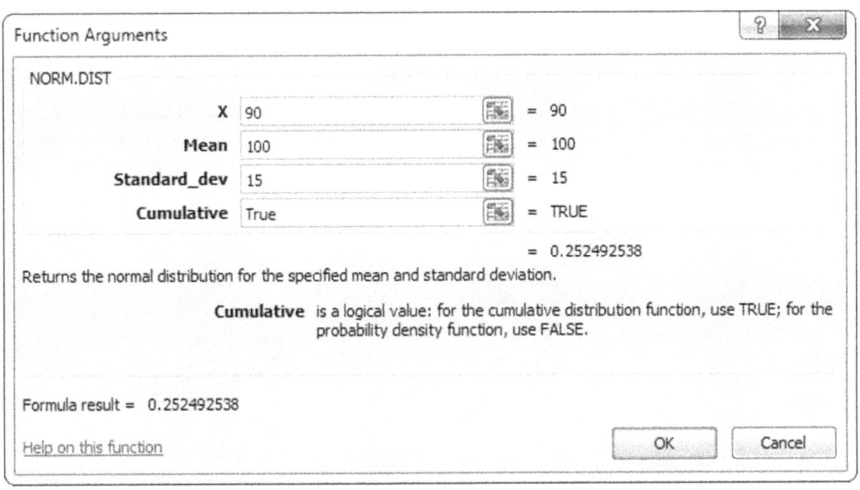

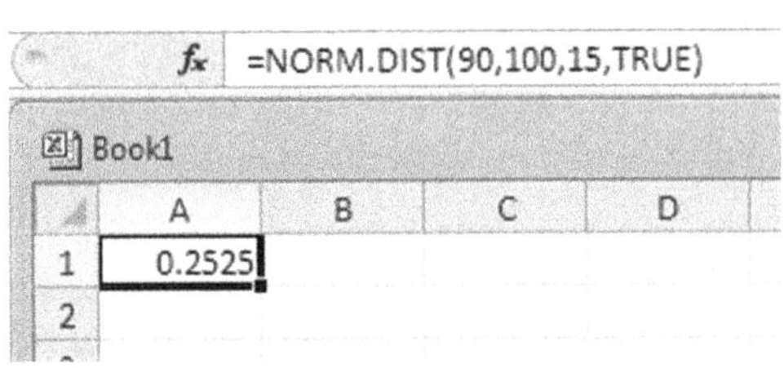

What Is the Probability that a Car Dealership Will Sell at least 130 ($x \geq 130$) Cars per Week?

- Highlight the cell where you want the answer to go
- Click on the ***fx*** button by the function input bar

- Select **Statistical** under **Category**
- Select **NORM.DIST** under **Function**
- Click **OK**
- Input **130** for **X**

- Input **100** for the **Mean**

- Input **15** for the **Standard_dev**

- Input a value for **Cumulative** of **1** or **True**

<u>Note</u>: The NORM.DIST function with True (1) provides the probability the car dealership sells less than 130 cars.

- Click **OK**

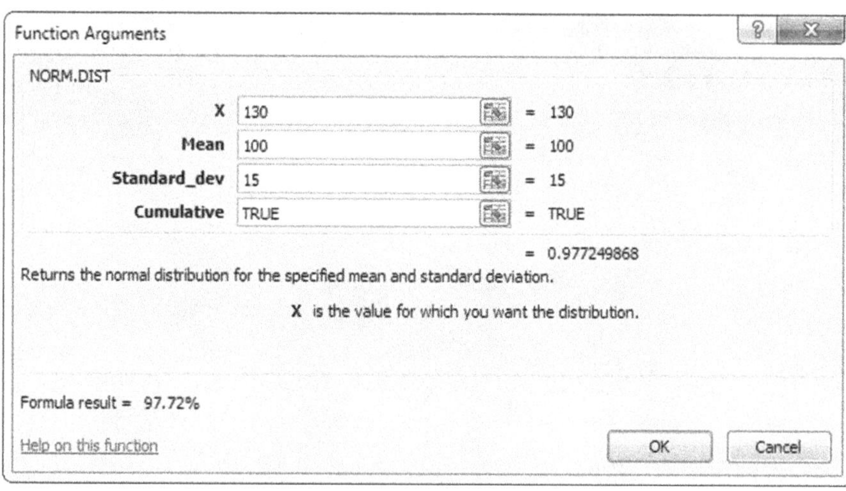

- Double click on the cell to edit the formula
- Subtract this amount from 1 in the function input box

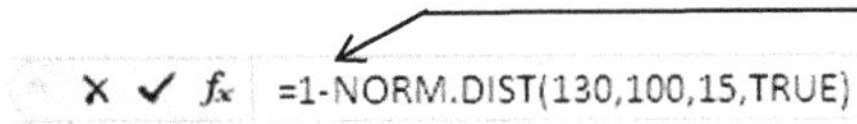

Note: To calculate the probability of at least 130 cars, you must subtract this number from 1 (as two probabilities must sum to 1).

Click the green check

Excel provides an answer of 0.0228 (2.28%) or there is a probability of about 2.3% that dealerships will sell 130 or more cars per week.

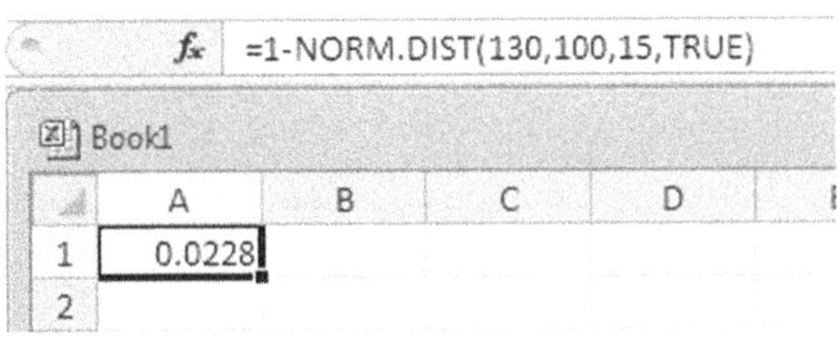

What Is the Probability that a Car Dealership Will Sell Between 90 and 120 Cars per Week?

- Highlight the cell where you want the answer to go
- Click on the ***fx*** button by the function input bar
- Select **Statistical** under **Category**
- Select **NORM.DIST** under **Function**
- Input **120** for **X**

- Input **100** for the **Mean**

- Input **15** for the **Standard_dev**

- Input **1** or **True** for **Cumulative**

- Click **OK**

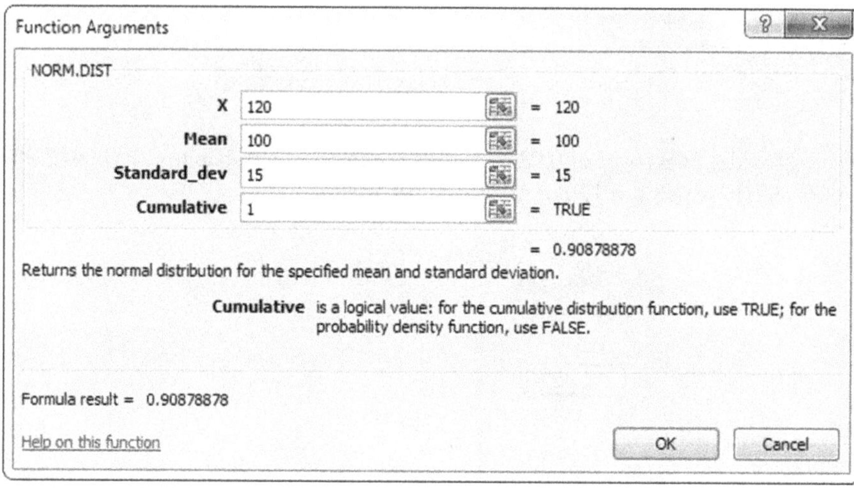

- Double click the cell to edit
- Inside the formula bar input you will need to type in the rest of the equation

$$-\textbf{NORM.DIST}(\textbf{90}, \textbf{100}, \textbf{15}, \textbf{1})$$

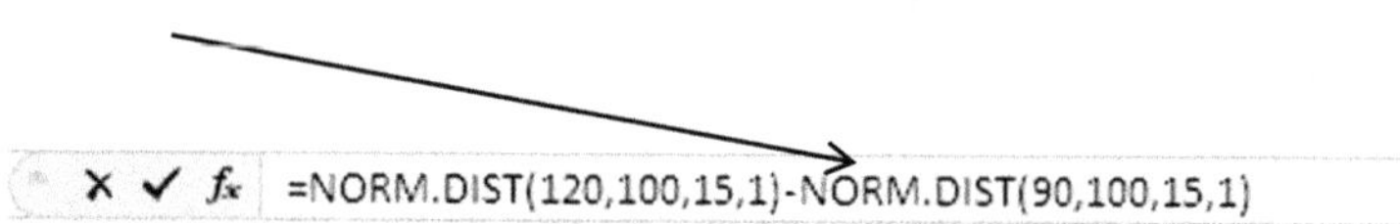

Click the green check

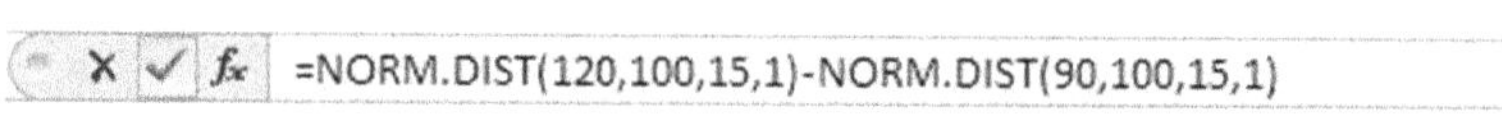

Excel provides an answer of 0.656296 or there is a probability of about 65% that the dealership will sell between 90 and 120 cars per week.

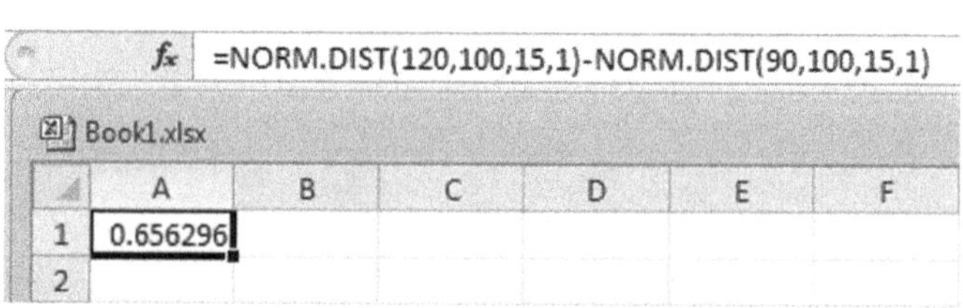

Calculating Combinations and Permutations

The main technical problems in your company involve laptops, monitors, speaker setups, software issues, mouse connectivity, and Internet connectivity. It is important to forecast utilization of your IT professionals and understand the probability of the type of issues they might deal with on a day-to-day basis. On a typical day, they can handle four of these problem areas.

Permutation

- **Order is important** as the IT staff work shift work and not all employees on each shift have the same expertise. What is the probability that the daily roster of problems will be **(Connectivity, Software, Laptops, Monitors)**? In this case, $x = 4$ and $n = 6$.
  - Highlight the cell where you want the answer to go
  - Click on the ***fx*** button by the function input bar
  - Select **Statistical** under **Category**
  - Select **PERMUT** under **Select a function:**
  - Click **OK**

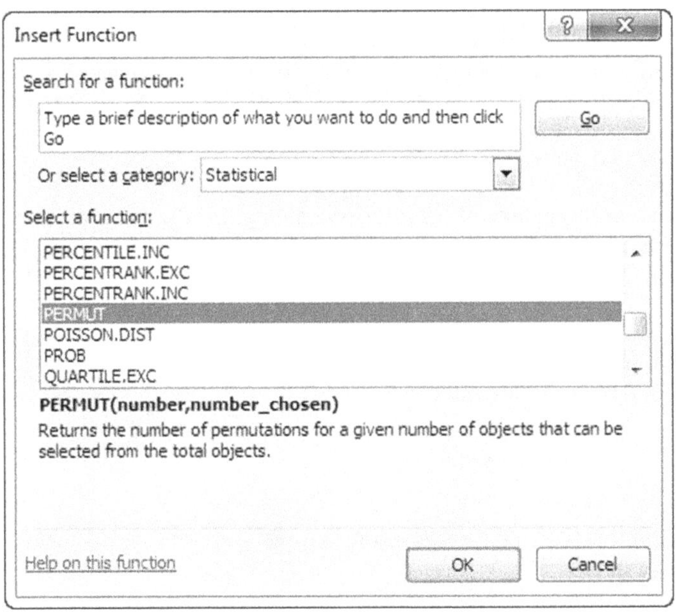

  - Input the **6** for **Number** (n) in the first entry box followed by **4** for **Number_chosen** (x) in the second input box, then click **OK**

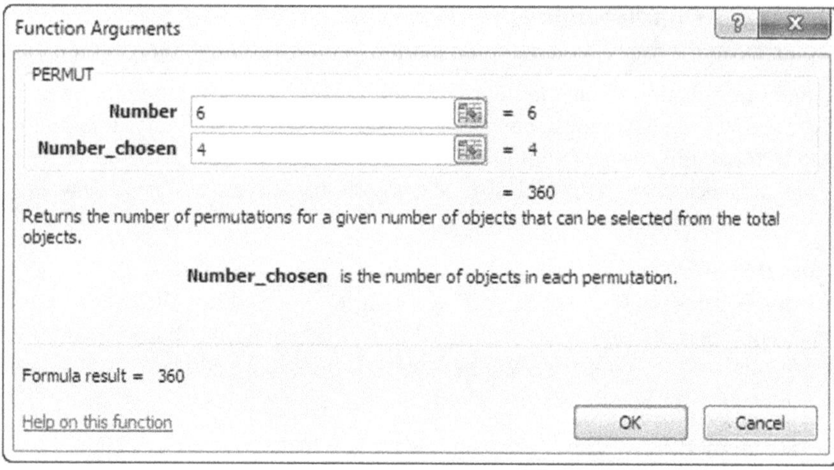

Excel will provide the answer that *360 permutations are possible*, or, in other words, there are 360 possible outcomes.

The total number of ways of being successful divided by the total number of possible outcomes is (1/360) or a .3% probability that you will be successful in having these four items in that order on your roster. There is a very low probability that your IT staff will be faced with connectivity, software, laptop, monitor problems in that specific order on any given day. You might have to look at a different kind of analysis to achieve more accurate forecasting results.

Combination

- **Order is not important.** What is the probability that the daily roster of problems will include these four areas **(Connectivity, Software, Laptops, Monitors)**? Remember in this case $x = 4$ and $n = 6$.
  - Highlight the cell where you want the answer to go
  - Click on the ***fx*** button by the function input bar
  - Select **All** under **Category**

  - Select **COMBIN** under **Select a function:**
  - Click **OK**

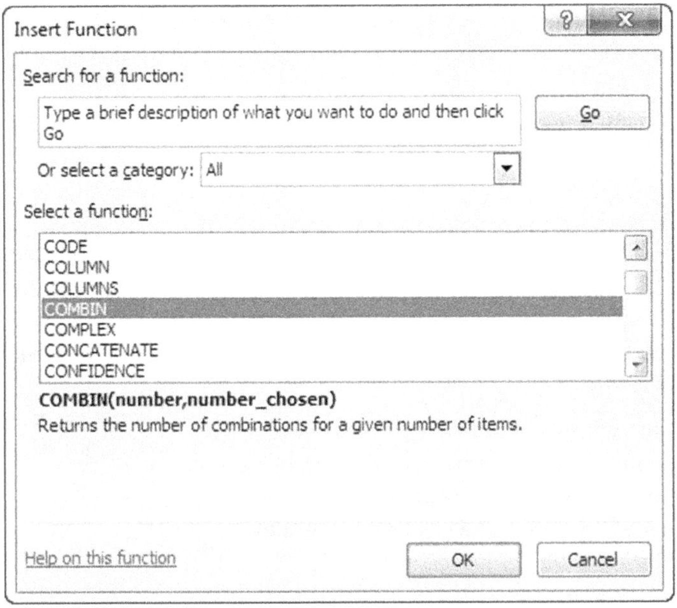

– Input the **6** for **Number** (n) in the first entry box followed by **4** for **Number_chosen** (x) in the second input box, then click **OK**

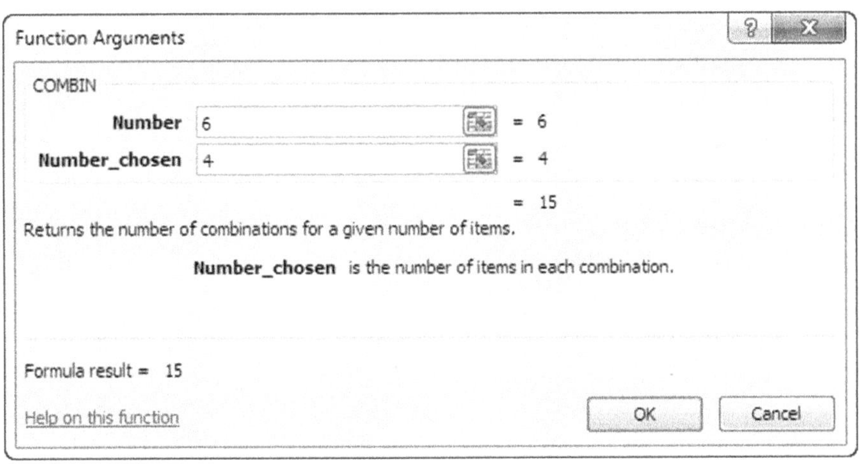

Excel will provide the answer that *15 combinations are possible*, or, in other words, there are 15 possible outcomes.

The total number of ways of being successful divided by the total number of possible outcomes is (1/15) or a 6.7% probability that these IT issues will occur. There is about a 7% chance that your IT staff will have to deal with connectivity, software, laptop, and monitor problems on any given day.

Finding Probabilities Using the Binomial Distribution

Royal Bank Retention Problem

The Royal Bank is concerned about a low retention rate for employees. On the basis of past experience, management has seen an annual employee turnover of 10%. For any employee chosen at random, management estimates a probability of 0.1 that the person will not be with the company next year.

Choosing four hourly employees at random, what is the probability that 0 of them will leave the company next year, everyone stays? For this problem: Number of employees that leave is the number of successes ($x = 0$), the number of employees in the group is the number of trials ($n = 4$), and the probability of turnover is the probability ($p = 10\%$).

- Highlight the cell where you want the answer to go
- Click on the ***fx*** button by the function input bar
- Select **Statistical** under **Category**
- Select **BINOM.DIST** under **Select a function:**
- Click **OK**

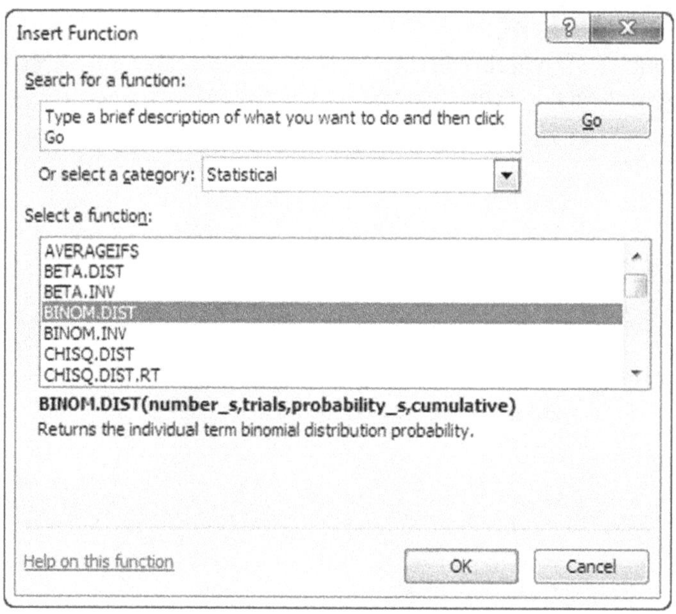

- Input the number of successes in **Number_s**; in this case, the number of ways of being successful (x) is 0, but we have typed it into cell A2, so we can input A2

- Input the number of trials in the **Trials**; in this case, the number of trials (n) is 4

- Input the probability in **Probability_s**; in this case, the percentage of turnover (p) is 10%

- Type in **FALSE** or **0** (exact probability)

Note: When a **Cumulative** box is part of the required input there are two possible inputs. **TRUE** or **1** is selected calculates the probability of "**x or fewer**" events of interest; this is a cumulative probability. **FALSE** or **0** calculates the probability of "**exactly x**" events of interest as in this example.

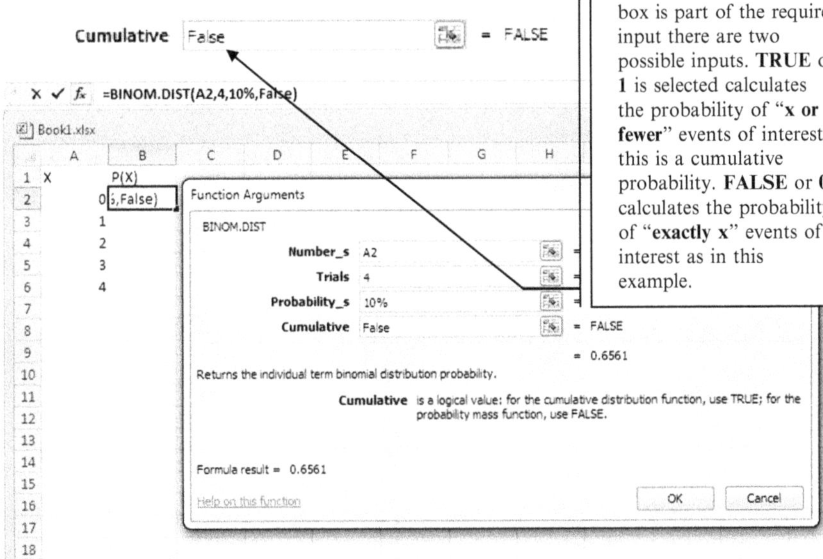

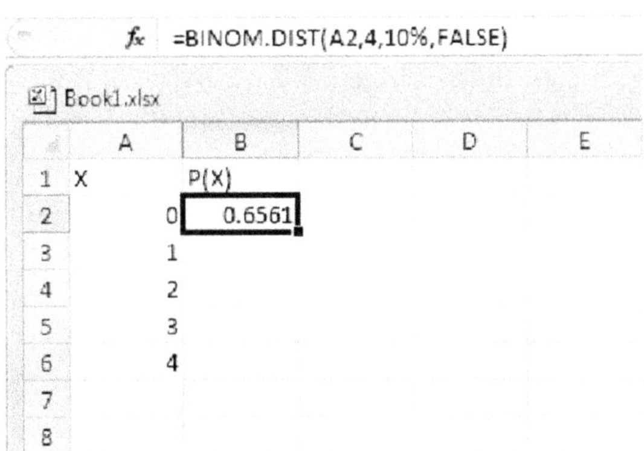

Excel will provide an answer of .6561 or about a 65.6% chance that zero out of four employees will leave the company next year. In other words, there is about a 66% chance that everyone will stay or conversely a 34% that someone will leave.

Common Excel Pitfalls

- Do not use probabilities that exceed 100%.
- Be careful if you need to take into account the order of elements in counting the number of possible outcomes. If so, then use a permutation not a combination.
- In earlier versions of Excel, there may be a period in the statistical function name, so make sure you include this to get the right function.
- If events are mutually exclusive, then by definition they are dependent on one another.
- If the distribution function requires a **cumulative** box to be filled in, remember that **False (0)** calculates the probability of **exactly x** events and **True (1)** is used for **x or fewer** events.
- Do not forget to include the "%" sign when calculating Probability_s for the BINOM.DIST; otherwise, make sure to convert the percentage into decimal format.

Final Thoughts and Activities

Practice Problems

1. From four registered nurses (RN) and three licensed vocational nurses (LVN), find the number of committees that can be formed consisting of two RNs and one LVN, if order is irrelevant.
2. How many ways can five open bank teller positions be filled with nine qualified candidates? Note that order is important because each of the teller positions has a different pay level and different title.
3. Suppose three friends decide to go to Hamburger Hut. In the last month, 86.1% of the orders were filled correctly. What is the probability that

   (a) All three of the orders of the three friends will be filled correctly?
   (b) None of the three will be filled correctly?
   (c) At least two of the three will be filled correctly?

4. A project at Kentucky Power and Light Company (KPL) is designed to increase the generating capacity of one of KPL's plants in northern Kentucky. The project is divided into two sequential stages: stage 1 is the design phase and stage 2 is the construction phase. While each stage will be closely scheduled and con-

trolled, management cannot predict beforehand the exact time required to complete each stage of the project. An analysis of similar construction projects has shown completion times for the design stage as **2, 3, or 4** months and completion times for the construction stage as **6, 7, or 8** months. Because of the critical need for additional electrical power, management has set a goal of *10 months or less* for the completion of the entire project. Since there are three possible completion times for the design stage (step 1) and three possible completion times for the construction stage (step 2), the counting rule for multiple step experiments can be applied. Therefore, the total possible outcomes is of $3 \times 3 = 9$. Show these using a tree diagram. Which scenarios will meet the 10-month requirement?

5. An airplane manufacturer called Aerosmith Aeronautics has their planes manufactured at three factories A, B, and C, which produce 50%, 25%, and 25%, respectively, of Aerosmith's planes. 70% of all of the Aerosmith planes produced in factory A are passenger airplanes, 25% of those produced in factory B are passenger airplanes, and 25% of the airplanes produced for Aerosmith in factory C are passenger airplanes. Remember there are lots of other types of planes such as military carriers, commercial carriers, private jets, etc. If an Aerosmith plane is selected at random at San Francisco airport, calculate the probability the airplane will be a passenger plane.
6. In a sample of 100 retail buyers, 50 purchased brown bread and 40 purchased peanut butter. However, 30 buyers purchased both brown bread and peanut butter. If a retail buyer selected at random purchased brown bread, what is the probability that he also purchased peanut butter? Also, determine the probability that a randomly chosen buyer has purchased brown bread, given that he also purchased peanut butter.

	A	B
6		
7	Probability	Value
8	Purchase of Peanut Butter P(A)	0.40
9	Purchase of Brown Bread P(B)	0.50
10	Purchase of Peanut Butter and Brown Bread at the same time P(A ∩ B)	0.30

7. Your company target market section includes teenagers between the ages of 15 and 18 years old. Calculate the following conditional probability:
   - The probability of a boy owning an iPhone
   - The probability of a girl owning an iPhone
   - The probability of a boy not owning an iPhone
   - The probability of a girl not owning an iPhone

8		
9	Probability of a Boy, P(A)	0.49
10	Probability of a Girl, P(B)	0.51
11	Probability of Having iPhone, P(C)	0.71
12	Probability of Not Having iPhone, P(D)	0.29

8. A manufacturing company produces two types of products: Product A and Product B. The probability of a randomly selected product being defective is 0.15 for Product A and 0.08 for Product B. The company randomly selects a product from its inventory.
   (a) What is the probability that the selected product is defective?
   (b) If the company randomly selects two products consecutively without replacement, what is the probability that both products are defective?

Discussion Boards

1. Reliability of systems can often be improved with redundancy of critical components. Airplanes have two independent electrical systems and typically two separate radios. If one component has a 0.001 probability of failure, the probability of the two independent components both failing is only 0.000001 using the multiplication rule. Discuss where else redundancy like this is important.
2. In a Chicago study involving the use of olestra as a substitute for fat in potato chips, Proctor and Gamble researchers reported that on a nine-point taste scale, olestra chips got a mean of 5.6 where regular chips got a mean of 6.4. When Proctor and Gamble were asked to provide the raw data, they refused. Is this ethical?

Group Activity

1. What role does probability play in sabermetrics? Research this "Billy Bean" approach in professional baseball.
2. In the movie *Twenty Four*, several MIT students worked with their professor in developing a method of counting cards in Las Vegas to change their odds of winning in the game of blackjack. In the end research, why the system failed and the impact it had on the gaming industry.
3. We use probability to influence our daily activities and decisions. For example, we may use several variables in determining which checkout lane will have the

highest probability of being the quickest at our local grocery store. What are some other daily activities in which you use basic probability in making decisions?

Parting Thought

Did you hear the one about the statistician? Probably …

Problem Solutions

1. From four registered nurses (RN) and three licensed vocational nurses (LVN), find the number of committees that can be formed consisting of two RNs and one LVN, if order is irrelevant.

   In this case for the RN, $x = 4$ and $n = 2$. For the LVN, $x = 3$ and $n = 1$. Order doesn't matter. We multiply because the committee is for RNs **AND** LVN.

$$\begin{aligned}\text{In Excel, use equation} :&= \text{COMBIN}(4,2)*\text{COMBIN}(3,1)\\ &= 6^*3\\ &= 18\end{aligned}$$

$$\text{Mathematically} = \left(\frac{4!}{2!(4-2)!}\right)*\left(\frac{3!}{1!(3-1)!}\right) = \left(\frac{4!}{2!*2!}\right)*\left(\frac{3!}{1!*2!}\right)$$

$$= \left(\frac{4^*3^*2^*1}{2^*1^*2^*1}\right)*\left(\frac{3^*2^*1}{1^*2^*1}\right) = \left(\frac{24}{4}\right)*\left(\frac{6}{2}\right) = 6^*3 = 18$$

2. How many ways can five open bank teller positions be filled with the nine qualified candidates?

   Answer:

   Order **does** matter.

$$\begin{aligned}\text{In Excel, use equation} :&= \text{PERMUT}(9,5)\\ &= 15{,}120\end{aligned}$$

$$= \left(\frac{9!}{(9-5)!}\right) = \left(\frac{9!}{4!}\right) = \left(\frac{9^*8^*7^*6^*5^*4^*3^*2^*1}{4^*3^*2^*1}\right) = \left(\frac{362{,}880}{24}\right)$$

$$= 15{,}120$$

3. Suppose three friends decide to go to Hamburger Hut. In the last month, 86.1% of the orders were filled correctly. What is the probability that

   (a) All three orders will be filled correctly?

$$\text{Use Equation} := \text{BINOM.DIST}\left(3,3,86.1\%,\text{FALSE}\right)$$

Answer: 63.83%

   (b) None of the three will be filled correctly?

$$\text{Use Equation} := \text{BINOM.DIST}\left(0,3,86.1\%,\text{FALSE}\right)$$

Answer: 0.27%

   (c) At least two of the three will be filled correctly?

$$\text{Use Equation} := \text{BINOM.DIST}\left(2,3,86.1\%,\text{FALSE}\right) + \text{BINOM.DIST}\left(3,3,86.1\%,\text{FALSE}\right)$$

Answer: 94.74%

You add the two probabilities because the question asks for the probability of at least two of the three will be filled correctly; you add the chance of having two correct and the chance of having three correct.

4. A project at Kentucky Power and Light Company (KPL) is designed to increase the generating capacity of one of KPL's plants in northern Kentucky. The project is divided into two sequential stages: stage 1 is the design phase and stage 2 is the construction phase. While each stage will be closely scheduled and controlled, management cannot predict beforehand the exact time required to complete each stage of the project. An analysis of similar construction projects has shown completion times for the design stage as **2, 3, or 4** months and completion times for the construction stage as **6, 7, or 8** months. Because of the critical need for additional electrical power, management has set a goal of *10 months or less* for the completion of the entire project. Since there are three possible completion times for the design stage (step 1) and three possible completion times for the construction stage (step 2), the counting rule for multiple step experiments can be applied. Therefore, the total possible outcomes is of $3 \times 3 = 9$. Show these using a tree diagram. Which scenarios will meet the 10-month requirement?

   Answer:

   Step 1: Draw decision tree diagram

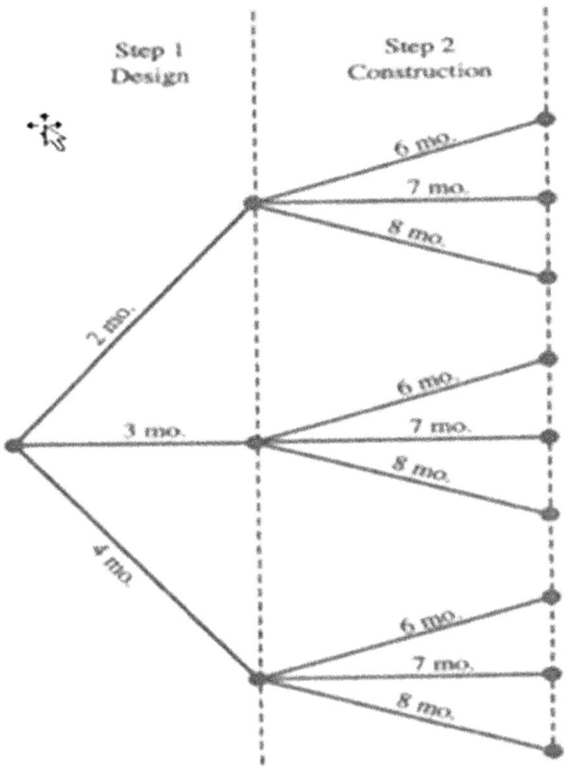

Step 2: Calculate outcomes

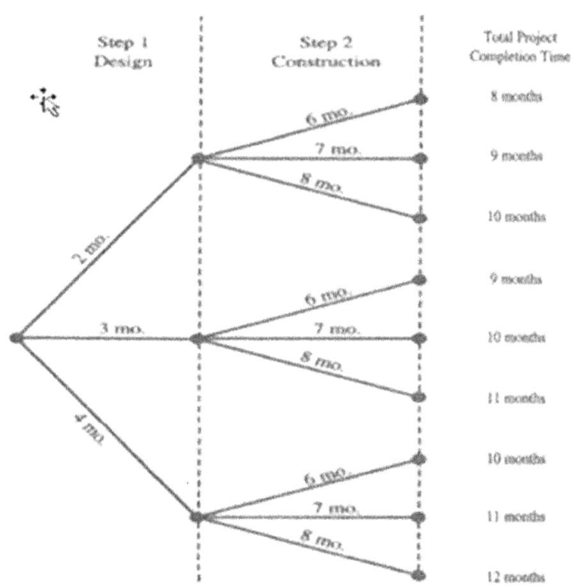

Step 3: Which scenarios are within a 10-month outcome?

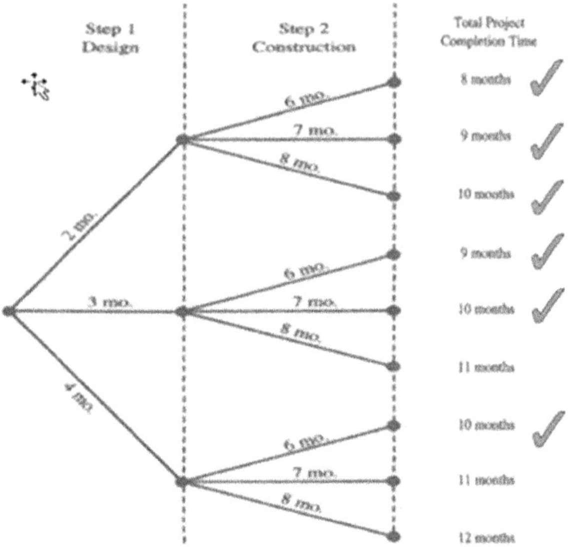

5. An airplane manufacturer called Aerosmith Aeronautics has their planes manufactured at three factories A, B, and C, which produce 50%, 25%, and 25%, respectively, of Aerosmith's planes. 70% of all of the Aerosmith planes produced in factory A are passenger airplanes, 25% of those produced in factory B are passenger airplanes, and 25% of the airplanes produced for Aerosmith in factory C are passenger airplanes. Remember there are lots of other types of planes such as military carriers, commercial carriers, private jets, etc. If an Aerosmith plane is selected at random at San Francisco airport, calculate the probability the airplane will be a passenger plane.

   Answer:

   Step 1: Draw a probability decision tree diagram

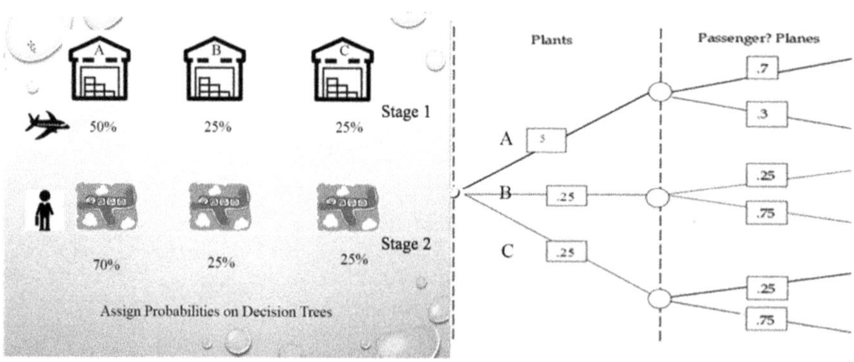

Remember that the probabilities are assigned based on the given frequencies.

Step 2: Calculate the outcomes

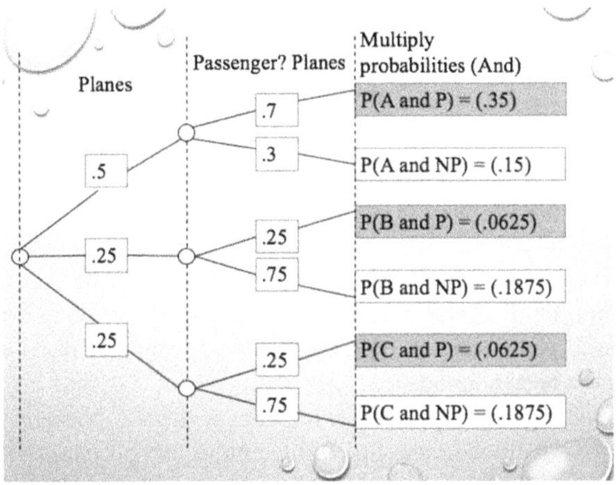

As we follow along the branches, we can multiply the probabilities. This is an "and" problem and the probabilities are independent. Let's follow along the upper branch. The probability of the plane produced at Plant A is .5 and the probability of the plane being a passenger plane is .7. The combined probability is .7 × .5 = .35. The probability is 35%.
Step 3: Calculate the final probabilities

What is the chance the selected plane is a passenger plane? The plane could be from Plant A or Plant B or Plant C. Let's add these probabilities to get the final answer:

$$.35 + .0625 + .0625 = .475$$

There is a 47.5% chance that the plane selected will be a passenger plane.

6. In a sample of 100 retail buyers, 50 purchased brown bread and 40 purchased peanut butter. However, 30 buyers purchased both brown bread and peanut butter.
   (a) If a retail buyer selected at random purchased brown bread, what is the probability that he also purchased peanut butter?
   (b) Determine the probability that a randomly chosen buyer has purchased brown bread, given that he also purchased peanut butter.

Probability	Value
Purchase of Peanut Butter P(A)	0.40
Purchase of Brown Bread P(B)	0.50
Purchase of Peanut Butter and Brown Bread at the same time P(A ∩ B)	0.30

Answer:

(a) Conditional probability = 0.30 / 0.50 = 0.60
(b) Conditional probability = 0.30 / 0.40 = 0.75

Therefore, if a buyer chosen at random is found to have purchased brown bread, then there is a 60% chance that he has also purchased peanut butter. On the other hand, if a buyer chosen at random has purchased peanut butter, then there is a 75% chance that he has also purchased brown bread.

7. Your company target market section includes teenagers between the ages of 15 and 18 years old. Calculate the following conditional probability:

(a) The probability of a boy owning an iPhone
(b) The probability of a girl owning an iPhone
(c) The randomly chosen person doesn't own an iPhone, that in boys
(d) The randomly chosen person doesn't own an iPhone, that in girls

	Have iPhone	Don't Have iPhone	Total
Boys	0.37	0.12	0.49
Girls	0.34	0.17	0.51
Total	0.71	0.29	1

Probability of a Boy, P(A)	0.49
Probability of a Girl, P(B)	0.51
Probability of Having iPhone, P(C)	0.71
Probability of Not Having iPhone, P(D)	0.29
Probability of boy Having iPhone, P(A ∩ C)	0.37
Probability of girl Having iPhone, P(B ∩ C)	0.34
Probability of boy Not Having iPhone, P(A ∩ D)	0.12
Probability of girl Not Having iPhone, P(B ∩ D)	0.17

Answer:

Conditional probability is calculated using the formula

$$P(A|B) = P(A \cap B) / P(B)$$

Step 1: Firstly, determine the probability of occurrence of the first Event B.
Step 2: Next, determine the probability of both Events A and B happening together simultaneously. This is the joint probability of Events A and B.
Step 3: We can derive the formula for the conditional probability of Event A, given that Event B has already occurred, by dividing the joint probability of Events A and B (step 2) by the probability of Event B (step 1), as shown below.

(a) Conditional probability = 0.37/0.71 = 0.52
The probability of a boy owning an iPhone = 0.52
(b) Conditional probability = 0.34/0.71
The probability of a girl owning an iPhone = 0.48
(c) Conditional probability = 0.12/0.49
The probability of a boy not owning an iPhone = 0.24
(d) Conditional probability = .17/0.51
The probability of a girl not owning an iPhone = 0.33

8. A manufacturing company produces two types of products: Product A and Product B. The probability of a randomly selected product being defective is 0.15 for Product A and 0.08 for Product B. The company randomly selects a product from its inventory.

   (a) What is the probability that the selected product is defective?

Answer:

(a) To find the probability that the selected product is defective, let D represent the event of selecting a defective product.

$P(D) = (.15 \times .5) + (.08 + .05)$, $P(D) = .75 + .04$, $P(D) = .115$ or 11.5%

$$P(D) = P(D|A)P(A) + P(D|B)P(B) = (0.15)(P(A)) + (0.08)(P(B))$$

$$P(\text{both defective}) = P(D)^* P(D|\text{not } D) = P(D)^* P(D-1) = (P(D))(P(D-1))$$

The calculation of $P(D - 1)$ depends on the outcome of the first selection. If the first selection is defective, the probability of selecting another defective product will differ from the case where the first selection is nondefective.

Chapter 9
Correlation

Linda Herkenhoff and John Fogli

Key Concepts

Association, Correlation coefficient, Correlation matrix, Causation, Direct relationship, Inverse relationship, Pearson correlation coefficient, and Positive relationship.

Discussion

This chapter explores the linear relationship between variables. The correlation statistics quantify the magnitude and direction of the relationship. Correlational analysis can be applied between *x*s as well as between the *x*s and *y*s. This chapter will describe several ways to generate these statistics in Excel.

Correlation Correlation is a powerful measure of the **association** between two variables. This is not the same as causation. **Causation** infers that one variable causes an effect in the other. A change in one variable creates a change in the other.

Statistic The **Pearson correlation coefficient** (***r***) is commonly used to measure the degree of correlation between data sets.

Supplementary Information The online version contains supplementary material available at https://doi.org/10.1007/978-3-031-46371-6_9.

L. Herkenhoff (✉)
Saint Mary's College, Moraga, CA, USA

J. Fogli
Diablo Valley College, Pleasant Hill, CA, USA

L. Herkenhoff, J. Fogli (eds.), *Applied Statistics for Business and Management using Microsoft Excel*, https://doi.org/10.1007/978-3-031-46371-6_9

This coefficient can range from +1 to −1. The correlation coefficient provides two important pieces of information: the strength of the relationship between variables and the type of relationship between the data sets. If both data sets increase or both decrease together, we refer to that as a **direct relationship** and the r statistic will be **positive**. If the values for a data set increase, while the other values for another data set decrease, we refer to this as an **inverse relationship** and the r statistic will be **negative**.

Positive $r \rightarrow x$ and y change in the same direction

ex. The higher employee satisfaction, the higher productivity.

Negative $r \rightarrow x$ and y change in opposite directions

ex. The more children I have, the fewer expensive vacations I take.

Figure 9.1 provides some graphical examples. The lower left graph shows no association between the x data set and the y data set; as x increases, y shows no tendency to increase or decrease. Therefore, the r value is zero with no sign. The lower right graph shows a slight linear pattern to the data sloping upward to the right. This slight association between the variables results in an r value of .4. As the variable on the x-axis increases, the variable on the y-axis also increases, resulting in a positive sign for the correlation coefficient. The two upper graphs show strong linear patterns with an r of .9. When data slopes downward to the right the sign of r becomes negative (upper right graph). In other words, as the x data increases, the y data decreases, indicating an inverse relationship between the two data sets.

Weaker associations are common in social science data. For example, consider a university entrance application that associates test scores to college success. The behavioral relationships of people are not always strongly correlated, so we often see r values between 0.3 and 0.7. In physics and engineering, we might seek higher r values. In designing the space shuttle, we would want a stronger r value, perhaps as high as .99 in analyzing the relationship between heat transference and the amount of a particular alloy used in designing the outer shell of the shuttle. A "rule of thumb" translation guide is provided in Fig. 9.2.

Note that $r = 0.40$ does not mean that 40% of the data is tightly clustered around a line. It also does not mean that the data displays twice as much linearity as data with an $r = 0.20$.

- The correlation **coefficient** can only measure linear associations.

Because r is a pure number, meaning it is computed using standard units, it is not affected by a change in scale of the measured units. This means r is not affected

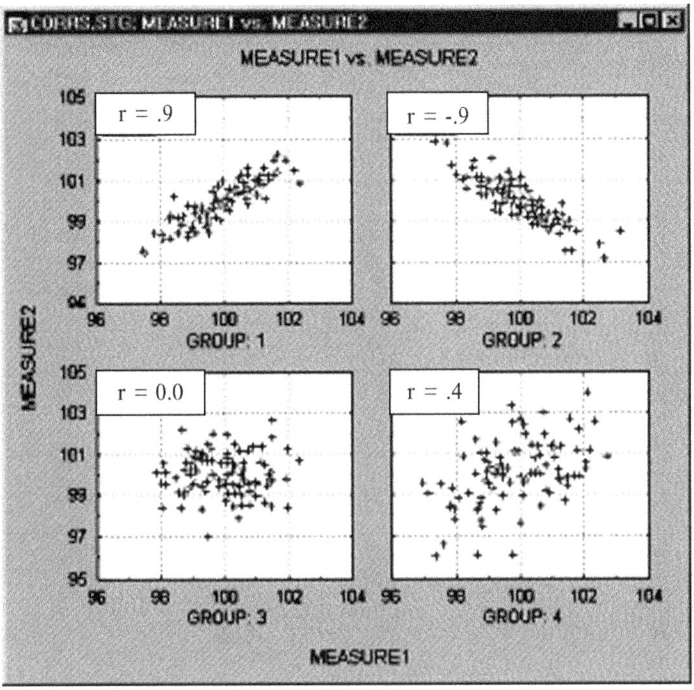

Fig. 9.1 Graphical examples of *r*

when you multiply each value of one variable by the same number. It is not affected if you add the same number to all values of one variable. Finally, it is not affected by interchanging the two variables ($x \leftrightarrow y$). These properties are shown in Figs. 9.3 and 9.4.

The correlation coefficient is sensitive to **outliers** . Remember the general definition of outliers is data that exceeds ±3 standard deviations. When outliers are present, the r value can be misleading. Consider Fig. 9.5.

Since outliers tend to reduce the correlation coefficient, careful consideration should be given in determining whether they remain in the calculation of *r*. It is also important to define what an outlier would be for a given experiment before collecting the data so that one isn't artificially grooming a data set to produce a greater *r* value.

Magnitude of r (applies to both possitive and negative values)	Associated Terminology
0.00	No association
.01-.20	Very weak
.21-.40	Weak to moderate
.41-.60	Medium to substantial
.61-.80	Very strong
.81-.99	Extremely strong
1.00	Perfect correlation

Fig. 9.2 General terminology in the magnitude of *r*

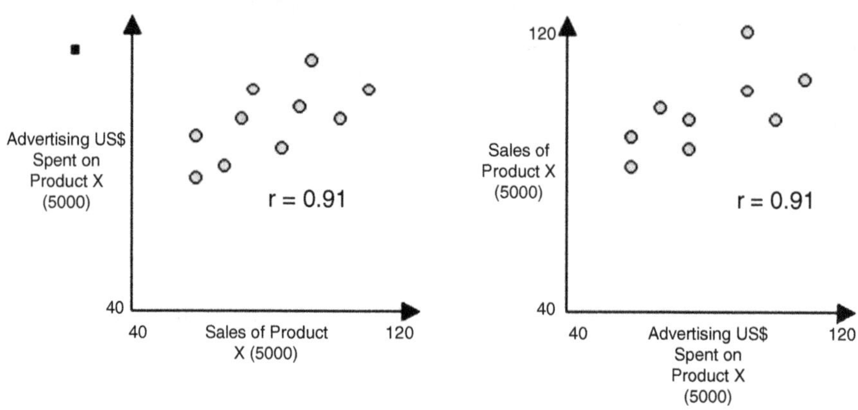

Fig. 9.3 Interchanging the variables ($x \leftrightarrow y$)

Nonlinear Data Caution

There are several cautions in working with correlation coefficients. The first is that the correlation coefficient can only measure "linear" associations. Consider the nonlinear data presented in Fig. 9.6.

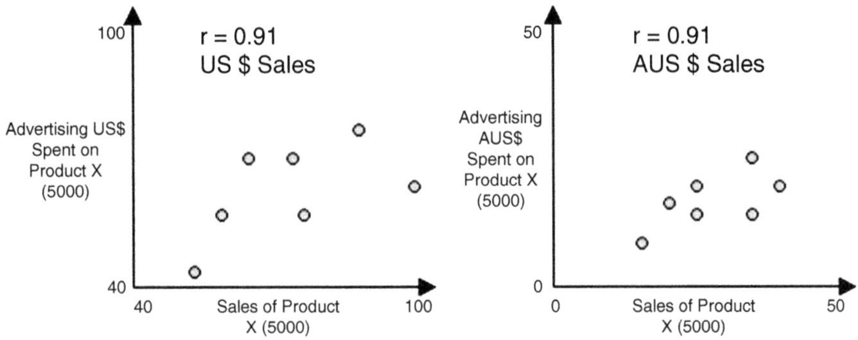

Fig. 9.4 Adding and multiplying the same number(s) to all variables. Converting from US $ to Australian $ with 1 month's worth of recorded daily averages shows the same correlation coefficient $r = .91$

Fig. 9.5 The effect of outliers on r

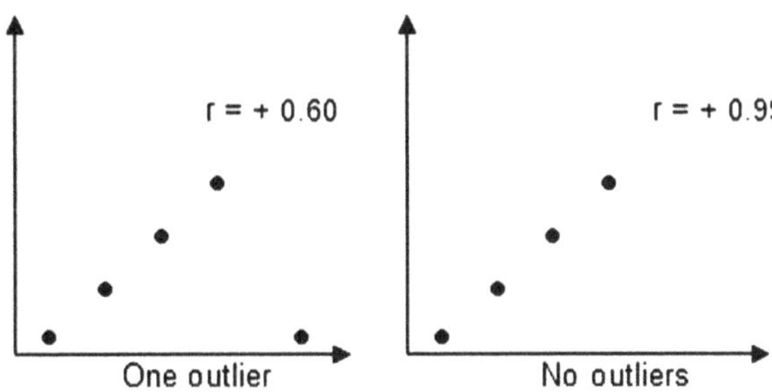

Figure 9.6 suggests there is a strong positive relationship between the amount of caffeine consumed and how many pages someone can read up to a certain point. After that point, no matter how much more caffeine you consume, there is little to no change in the number of pages you can read. Then you reach a certain level of caffeine consumption that starts reducing the number of pages you can read. (Perhaps you are too jittery to hold the book still.)

Using a single value of r cannot clearly define this nonlinear relationship. In this case, you could calculate three separate r values in the three different (x, y) relationships; this is shown in Fig. 9.7.

Average Data Caution

The second issue involves working with average data. Aggregated or averaged data is data that has been collected at the individual level and combined in some way into group data. Correlations based on averages can be misleading.

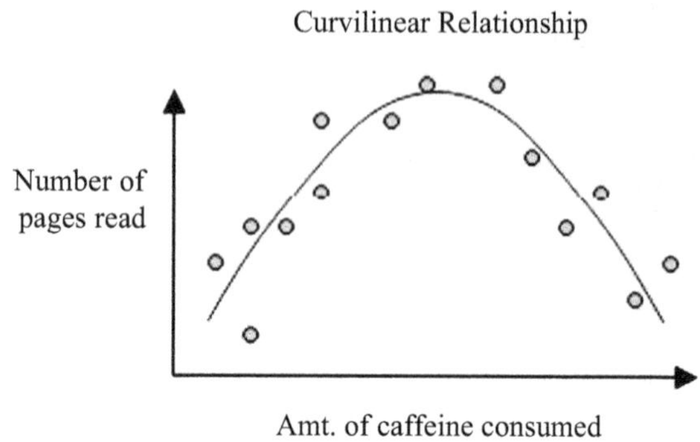

Fig. 9.6 Nonlinear data: the relationship between the amount of caffeine consumed and number of pages read

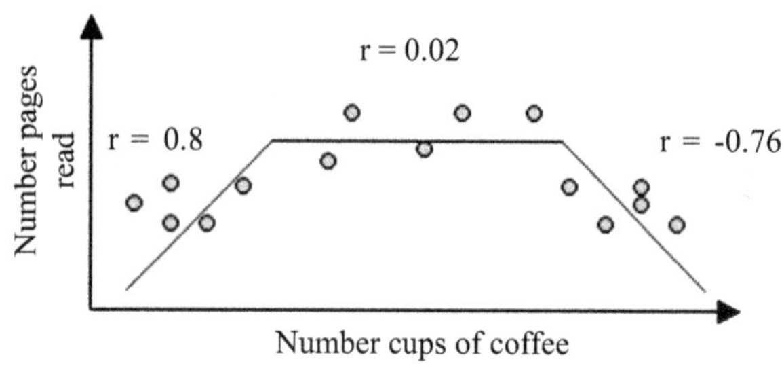

Fig. 9.7 Multiple *r* values for multiple relationships

In individual-level data, there may be a lot of spread around the averages. If you average that data into group data, the spreads are reduced and you generate an inflated *r* value. Consider the example in Fig. 9.8.

In Fig. 9.8, the panel on the left indicates the individual-level data generated by measuring annual salary and job satisfaction among accountants, earth scientists, and financial analysts within one company. The resulting coefficient is $r = 0.71$.

The second panel shows the average salary and job satisfaction values for each of the three professional groups. By averaging the data, we have reduced the spread and produced a stronger linear association. The *r* value in this case has increased to 0.92. It is less accurate to use the correlation coefficient based on average data to predict relationships at the individual level. These correlations tend to overstate the strength of association, so use caution.

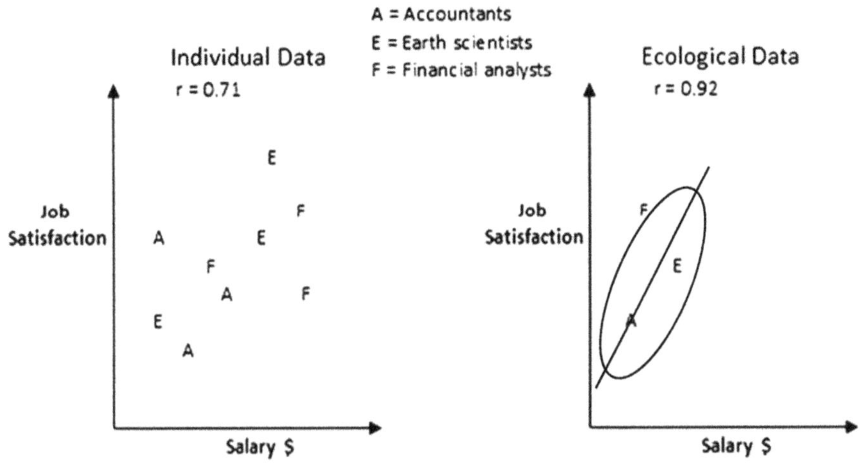

Fig. 9.8 Correlations of averaged or ecological job satisfaction data produces inflated *r* values

Excel

Correlation: One r *Value or Correlation Matrix*

There are two common ways to access the correlation functionality inside of Excel. They are both equally easy and vary by how many data sets (columns of data) you want to correlate simultaneously.

Method 1: Two or More Data Sets (Matrix)
Enter the data below:

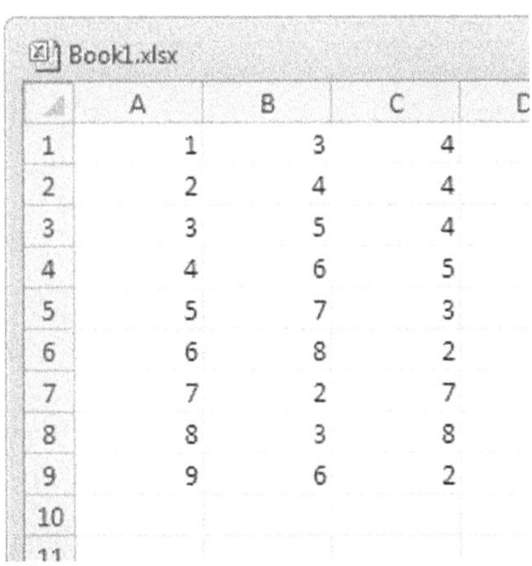

Book1.xlsx

	A	B	C
1	1	3	4
2	2	4	4
3	3	5	4
4	4	6	5
5	5	7	3
6	6	8	2
7	7	2	7
8	8	3	8
9	9	6	2
10			

- Select the **Data** tab
- Select the **Data Analysis** function
- Choose **Correlation** from the list of Analysis Tools

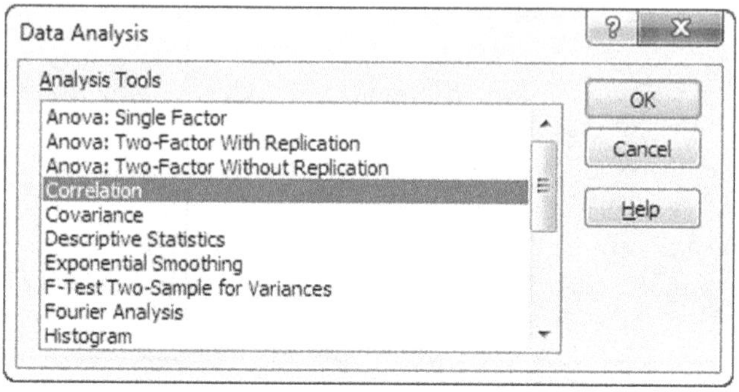

- Click **OK**
- Input the data you want to correlate in the **Input Range**
  - You need a minimum of two columns of data but you can input as many columns as you want to analyze.

- Click the box next to **Labels in first row**, if you have labels. However in this example, no labels are used so we leave the box unchecked.
  - It is highly recommended you input labels to help keep track of the variables.

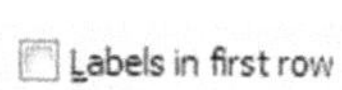

- Click on the **Output Range** and input **Cell B11**
- Click **OK**

The output should appear as a matrix starting from cell B11 as shown in the screenshot below. The output matrix shows just the lower half, as the upper half is a mirror image of the lower half. This is because the correlation between *x* and *y* is the same value as the correlation between *y* and *x*. The diagonal of 1s indicates the perfect correlation each variable has with itself. The diagonal divides the upper half correlation values from the lower half correlation values.

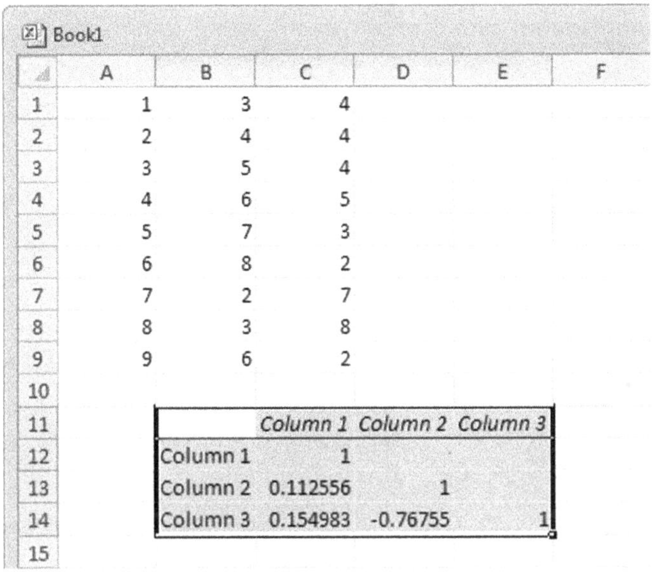

Method 2: Only Two Data Sets
For this method you don't even have to be within the data analysis tab, just start from your Excel Home tab.

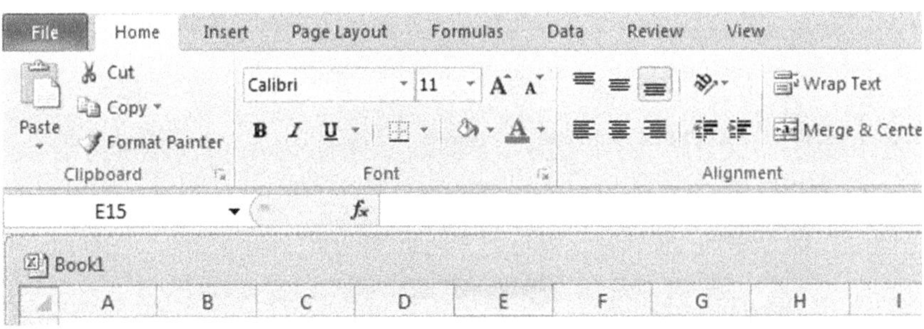

- Click the *fx* button on the Formula Bar
- Select **Statistical** from the **Or select a category:** drop-down menu

- Select **CORREL** or **PEARSON** (these two functions produce the same answer) from the **Select a function:** list

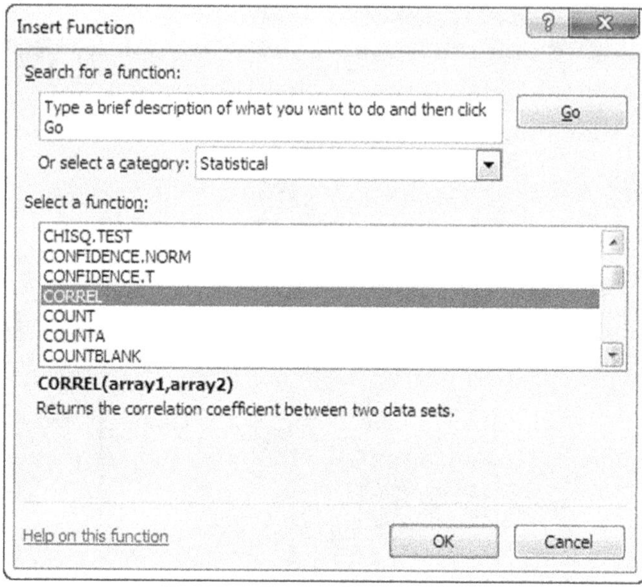
Insert Function
Search for a function:
Type a brief description of what you want to do and then click Go
Go
Or select a category: Statistical
Select a function:
CHISQ.TEST
CONFIDENCE.NORM
CONFIDENCE.T
CORREL
COUNT
COUNTA
COUNTBLANK
CORREL(array1,array2)
Returns the correlation coefficient between two data sets.
Help on this function
OK
Cancel

OR

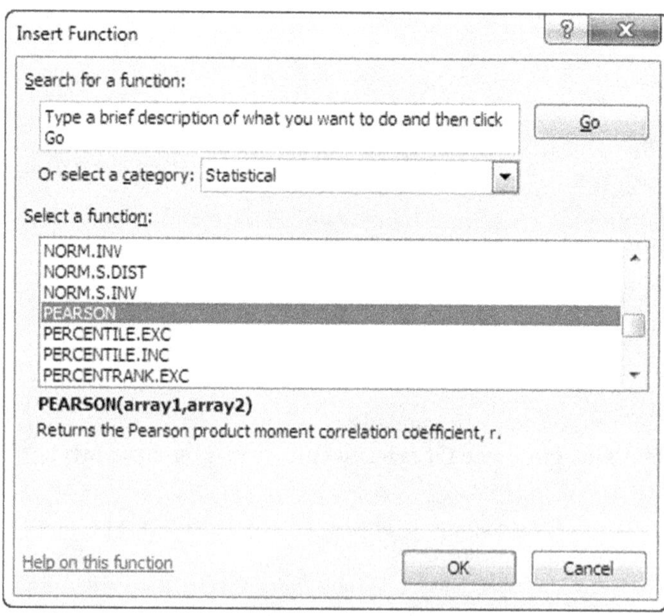
Insert Function
Search for a function:
Type a brief description of what you want to do and then click Go
Go
Or select a category: Statistical
Select a function:
NORM.INV
NORM.S.DIST
NORM.S.INV
PEARSON
PERCENTILE.EXC
PERCENTILE.INC
PERCENTRANK.EXC
PEARSON(array1,array2)
Returns the Pearson product moment correlation coefficient, r.
Help on this function
OK
Cancel

- Click **OK**
- Input the range for the two data sets you are correlating
- Click **OK**

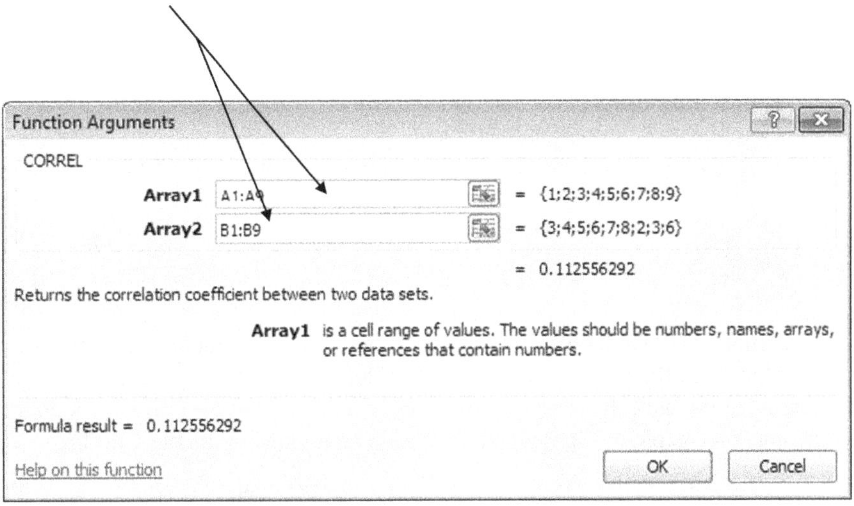

The results will be the Pearson correlation coefficient (r). If using an older version of excel (any version before Excel 2003) then it is better to use the CORREL function as it will have less rounding errors compared to the PEARSON function. However, when using a version of excel that is from 2003 or later then both functions should produce identical results.

Common Excel Pitfalls

- Make sure to click the box indicating you are including variable labels if your data has variable labels.
- Make sure to click inside the Array 1 input box before you highlight the input data field. Then click inside the Array 2 data input box before highlighting the next input data field. If you don't click inside Array 2 data input box, you will just overwrite the input you selected for Array 1.
- By default, Excel will display as many decimals as it can to fill a cell. To reduce the number of digits displayed after the decimal format the cell to display no more than four decimal places.

Final Thoughts and Activities

Practice Problems

1. The following is a set of data from a sample of $n = 12$ items:

   (a) Create a scatter plot of the variables X and Y. What relationship, if any, appears to exist between the two variables?
   (b) Compute the correlation coefficient.
   (c) How strong is the relationship between the two variables? Explain.

X	3	7	5	15	12	4	9	8	10	11	21	2
Y	6	14	10	28	24	8	18	16	20	22	40	4

2. Customers who made online purchases from an Internet retailer were randomly sampled from the retailer's database. The dollar value of each customer's purchases, along with the time the customer spent shopping the company's online catalog was recorded. The sample results are contained in the file **Catalog**.

   (a) Create a scatter plot of the variables, ***Time (x)*** and ***Purchases (y)***. What relationship, if any, appears to exist between the two variables?
   (b) Compute the correlation coefficient for this sample data.
   (c) What inferences can you make about the relationship between time spent on the company's online catalog and the dollars spent shopping at the company's site?

3. InTech has created a new technological widget. In the first few months InTech produces very few widgets, but as time progresses more widgets are available on the market. The data in the **InTech** file provides the number of items in stock at the local warehouse and the cost per item.

   (a) Create a scatter plot of the variables, ***# of Items in Stock (x)*** and ***Price per Item (y)***. What relationship, if any, appears to exist between the two variables?
   (b) Compute the correlation coefficient for this sample data.
   (c) What inferences can you make about the supply of widgets and the cost per widget?

4. Different countries sell a variety of drugs at different prices. For each of the 10 most popular prescription drugs, the file contains a list of the retail price (US $) for the drug in different countries. Use the data in **DrugPrices** to calculate the answers to the problem.

   (a) Compute the correlation coefficient for this sample data.
   (b) What inferences can you make about the 10 most popular prescription drugs and the different costs per country?

5. The following is a set of data from a sample of $n = 11$ items:

X	7	5	8	3	6	10	12	4	9	15	18
Y	21	15	24	9	18	30	36	12	27	45	54

   (a) Compute the correlation coefficient.
   (b) How strong is the relationship between X and Y? Explain.

6. The cereal table below lists the calories and sugar, in grams, for one serving of seven breakfast cereals:

Cereal	Calories	Sugar
Indulge Bran Nuggets	80	6
Indulge Corn Flakes	100	2
Wheatoes	100	4
Indulge Organic Multigrain Flakes	110	4
Indulge Rice Flakes	130	4
Indulge Shredded Wheat Vanilla Almond	190	11
Indulge Mini Wheats	200	10

   (a) Compute the correlation coefficient.
   (b) Based on (a), what conclusions can you reach about the relationship between calories and sugar?

7. Movie companies need to predict the gross receipts of individual movies once a movie has debuted. The following results are the first weekend gross, the US gross, and the worldwide gross (in millions of $) for the eight Harry Potter movies:

Cereal	First Weekend	US Gross	Worldwide Gross
Awesome Stone	90.295	317.558	976.458
Chamber of Gods	88.357	261.988	878.988
Prisoner of Alcatraz	93.687	249.539	795.539
Goblet of Wind	102.335	290.013	896.013
Order of the Birds	77.108	292.005	938.469
Half-Blood King	77.836	301.460	934.601
Deathly Crows—Part 1	125.017	295.001	955.417
Deathly Crows—Part 2	169.189	381.011	1328.111

   (a) Compute the coefficient of correlation between first weekend gross versus US gross, first weekend gross versus worldwide gross, and US gross versus worldwide gross.
   (b) Based on (a), what conclusions can you reach about the relationship between first weekend gross, US gross, and worldwide gross?

8. These three graphs show correlation relationships, with x corresponding to the age of the driver and y corresponding to the retail price of the most recent car purchase. Interpret each of the graphs.

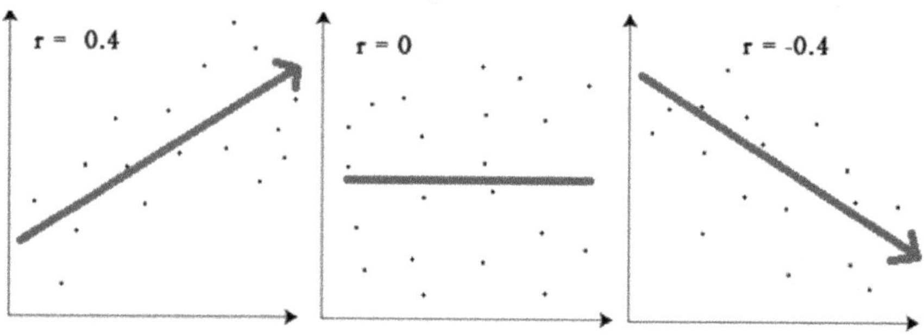

Discussion Boards

1. Unemployment is related to violent crime.
2. Baseball players with higher batting averages receive higher salaries. If I am a professional baseball player, raising my batting average will get me a higher salary.
3. Many studies have found an association between smoking and heart disease. One study found an association between coffee drinking and heart disease. As part of your company's wellness program, should you conclude coffee drinking causes heart disease and ban coffee in the workplace?
4. Think of some relationships and associations that you might find at your organization. For example, how are wellness programs related to job satisfaction or company child care services associated with attendance?

Group Activity

In designing a profitable airline, the number of seats in a plane is an important variable. Likewise, customer comfort is also important. Leonardo da Vinci provided some of the earliest data on human measurements. He also developed correlational relationships between various measurements. Choose one of his correlational relationships that might still be valid in designing a profitable aircraft.

Parting Thought

I have had my results for a long time, but I do not know yet how I am to arrive at them. . .

Problem Solutions

1. The following is a set of data from a sample of $n = 12$ items:

X	3	7	5	15	12	4	9	8	10	11	21	2
Y	6	14	10	28	24	8	18	16	20	22	40	4

(a) Create a scatter plot of the variables X and Y. What relationship, if any, appears to exist between the two variables?
Answer:

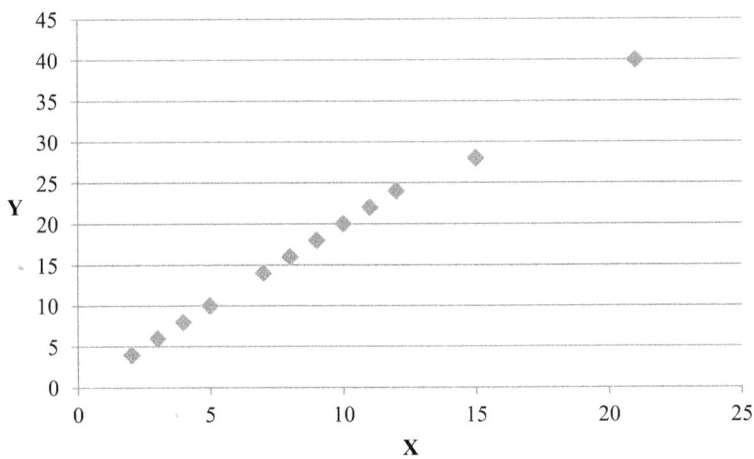

(b) Compute the correlation coefficient.
Answer: $r = 0.9989$

(c) How strong is the relationship between the two variables? Explain.
Answer: There is an almost perfect positive relationship between X and Y. Almost all of the points fall into a straight line with a positive slope.

2. Customers who made online purchases from an Internet retailer were randomly sampled from the retailer's database. The dollar value of each customer's purchases, along with the time the customer spent shopping in the company's online catalog was recorded. The sample results are contained in the file **Catalog**.

(a) Create a scatter plot of the variables **Time (x)** and **Purchases (y)**. What relationship, if any, appears to exist between the two variables? Answer: A very strong positive correlation between online purchases and time spent shopping online.

Answer:

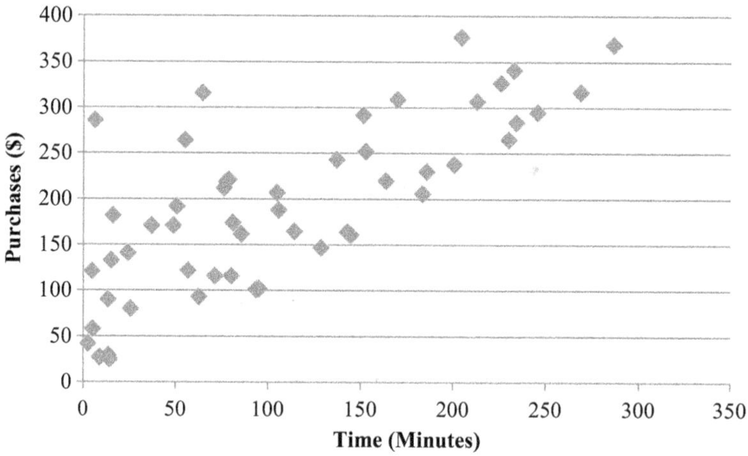

(b) Compute the correlation coefficient for this sample data.

Answer: $r = 0.7564$

(c) What inferences can you make about the relationship between time spent on the company's online catalog and the dollars spent shopping at the company's site?

Answer: The correlation of 0.7564 implies a strong positive relationship between time spent looking at the company's catalog and the amount of dollars spent shopping at the company's site. Customers who spend a longer period of time looking at the company's online catalog are more likely to spend more money shopping at the company's site.

3. InTech has created a new technological widget. In the first few months InTech produces very few widgets, but as time progresses more widgets are available on the market. The data in **InTech** provides the number of items in stock at the local warehouse and the cost per item.

(a) Create a scatter plot of the variables, **# of Items in Stock (x)** and **Price per Item (y)**. What relationship, if any, appears to exist between the two variables?

Answer:

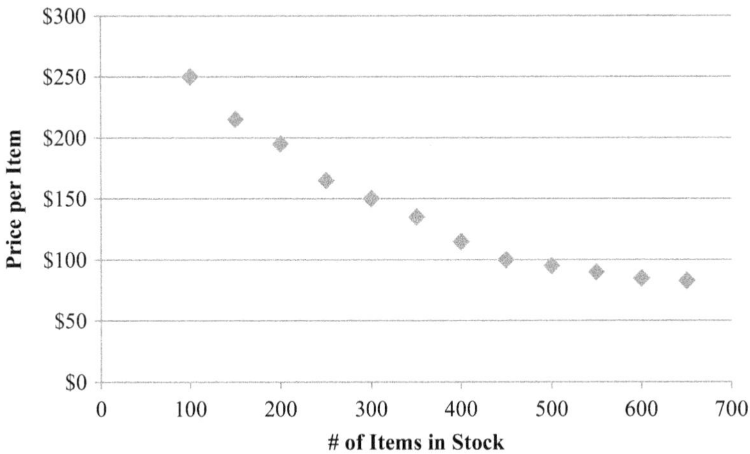

(b) Compute the correlation coefficient for this sample data.
Answer: Cannot calculate, not linear.

(c) What inferences can you make about the supply of widgets and the cost per widget?
Answer: In general, there seems to be a trend that indicates that when there are more widgets in stock, the price per item is lowered. But, you need to handle this nonlinear data set with a different model before any conclusions can be statistically substantiated.

4. Different countries sell a variety of drugs at different prices. For each of the 10 most popular prescription drugs, the file contains a list of the retail price (US $) for the drug in different countries. Use the data in **DrugPrices** to calculate the answers to the problem.

(a) Compute the correlation coefficient for this sample data.
Answer:

	US Price	*Canada Price*	*Great Britain Price*	*Australia Price*
US Price	1			
Canada Price	0.9983	1		
Great Britain Price	0.9953	0.9980	1	
Australia Price	0.9959	0.9985	0.9994	1

(b) What inferences can you make about the 10 most popular prescription drugs and the different costs per country?

Answer: The correlations imply a strong positive relationship between each of the countries' prescription drug costs. Prescription drugs that tend to cost more in the United States tend to cost more in Canada, Great Britain, and Australia. Prescription drugs that tend to cost less in the United States tend to cost less in Canada, Great Britain, and Australia.

5. The following is a set of data from a sample of $n = 11$ items:

X	7	5	8	3	6	10	12	4	9	15	18
Y	21	15	24	9	18	30	36	12	27	45	54

   (a) Compute the correlation coefficient.
   Answer: 1.0
   (b) How strong is the relationship between X and Y? Explain.
   Answer: There is a perfect positive linear relationship between X and Y. The increase of Y is three times the increase of X.

6. The cereal table below lists the calories and sugar, in grams, for one serving of seven breakfast cereals:

Cereal	Calories	Sugar
Indulge Bran Nuggets	80	6
Indulge Corn Flakes	100	2
Wheatoes	100	4
Indulge Organic Multigrain Flakes	110	4
Indulge Rice Flakes	130	4
Indulge Shredded Wheat Vanilla Almond	190	11
Indulge Mini Wheats	200	10

   (a) Compute the correlation coefficient.
   Answer: 0.8391
   (b) Based on (a), what conclusions can you reach about the relationship between calories and sugar?
   Answer: There is an extremely strong positive linear relationship between calories and sugar.

7. Movie companies need to predict the gross receipts of individual movies once a movie has debuted. The following results are the first weekend gross, the US gross, and the worldwide gross (in millions of \$) for the eight Harry Potter movies:

Cereal	First Weekend	US Gross	Worldwide Gross
Awesome Stone	90.295	317.558	976.458
Chamber of Gods	88.357	261.988	878.988
Prisoner of Alcatraz	93.687	249.539	795.539
Goblet of Wind	102.335	290.013	896.013
Order of the Birds	77.108	292.005	938.469
Half-Blood King	77.836	301.460	934.601
Deathly Crows—Part 1	125.017	295.001	955.417
Deathly Crows—Part 2	169.189	381.011	1328.111

(a) Compute the coefficient of correlation between first weekend gross versus US gross, first weekend gross versus worldwide gross, and US gross versus worldwide gross.

Answer:

	First Weekend	US Gross	Worldwide Gross
First Weekend	1		
US Gross	0.728417481	1	
Worldwide Gross	0.823319135	0.96419656	1

(b) Based on (a), what conclusions can you reach about the relationship between first weekend gross, US gross, and worldwide gross?

Answer: Worldwide gross has an extremely strong positive linear relationship with both first weekend gross and US gross. First weekend gross has a very strong positive linear relationship with US gross.

8. These three graphs show correlation relationships, with x corresponding to the age of the driver and y corresponding to the retail price of the most recent car purchase. Interpret each of the graphs.

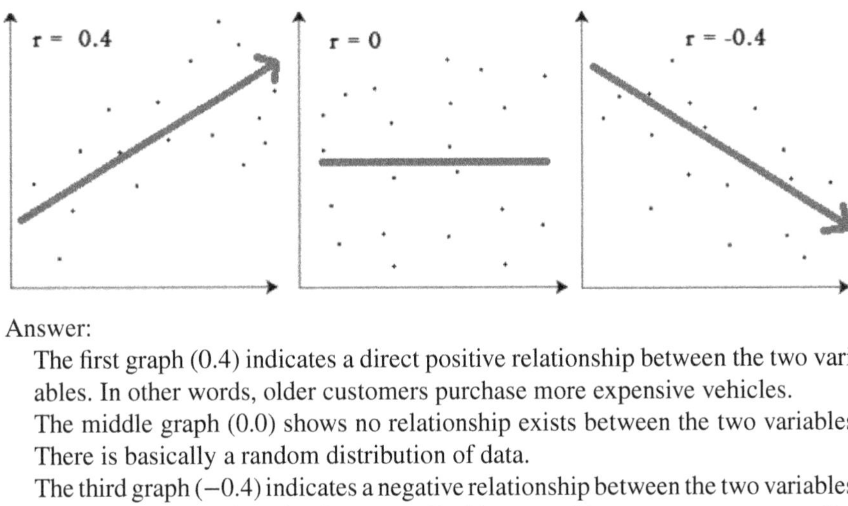

Answer:

The first graph (0.4) indicates a direct positive relationship between the two variables. In other words, older customers purchase more expensive vehicles.

The middle graph (0.0) shows no relationship exists between the two variables. There is basically a random distribution of data.

The third graph (−0.4) indicates a negative relationship between the two variables. As one decreases, the other increases. In this case, older customers are spending less on their most recent vehicle purchase compared to younger customers.

Chapter 10
Simple Linear Regression

Linda Herkenhoff and John Fogli

Key Concepts

Coefficient of determination, Method of least squares, Market model, Regression, Residual, Residual plots, Slope, Standardized residuals, y- intercept, $\overline{y}$, y_i, and $\hat{y}$.

Discussion

This chapter will introduce the simple linear regression technique that involves only one x and one y. This is a technique that creates a linear fit to the data set. In the simple linear regression technique, the (x, y) data generates a line of the form $y = mx + b$. Figure 10.1 illustrates the linear relationship between temperature in degrees C (x) and the number of flower buds (y). Excel creates the linear trendline that best fits the data. To understand this statistical technique, several important concepts are defined throughout this chapter. A step-by-step process flow is developed in Excel to complete the creation and testing of a simple linear regression model.

Simple linear regression This is a technique that models the linear relationship in your data set. The *simple* refers to using only one x to predict y. The *linear* reminds

---

Supplementary Information The online version contains supplementary material available at https://doi.org/10.1007/978-3-031-46371-6_10.

L. Herkenhoff (✉)
Saint Mary's College, Moraga, CA, USA

J. Fogli
Diablo Valley College, Pleasant Hill, CA, USA

L. Herkenhoff, J. Fogli (eds.), *Applied Statistics for Business and Management using Microsoft Excel*, https://doi.org/10.1007/978-3-031-46371-6_10

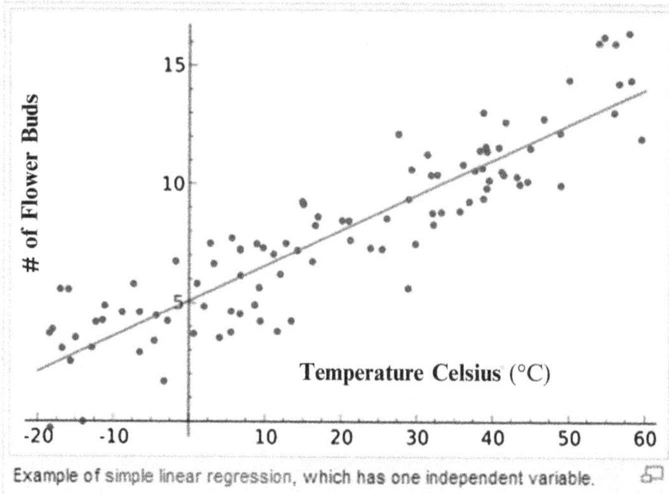

Fig. 10.1 Simple linear regression

you that we are fitting a straight line through the data in this approach. The *regression* means pulling the data points back (regressing) to a line that tries to go through the "middle" of the data distribution. The (x, y) data generates a line of the form $\boldsymbol{y = mx + b}$ that best describes how changes in y can be predicted from changes in x. This will allow us to estimate the y value for any given x value.

Statistics The **coefficient of determination** (R^2 or r^2) indicates how well the line fits your data. It tells you how much of your y can be explained by the x you used in the model. It is most easily understood when expressed as a %.

The regression line does not usually describe the data perfectly. It is a useful summary of the main trend, but it does not capture the random variation of the data points about the line. But how useful is the regression line? The answer is based on the **coefficient of determination** or r^2. The "r squared" tells you how much of the variability in y is explained by x. Larger values of r^2 are considered better because they indicate a stronger relationship between x and y. Often the **coefficient of determination** appears in its capitalized form R^2. It ranges in value from 0 to 1. An r^2 of 1 would indicate that the data all fall perfectly along a straight line.

Although the **correlation coefficient** is restricted to a linear relationship between two variables, the **coefficient of determination** can be used for nonlinear relationships and for relationships that have two or more independent variables. In that sense, the **coefficient of determination** has a wider range of applicability.

The ***y*-intercept (b)** is the y value when x equals zero. In other words, it is where the regression line cuts the y-axis. If the line stops before it hits the y-axis, then you can extrapolate the line back to where it would cut the axis. If this is not a real value on the line, it may not have any real meaning. For example, graphing the number of employees (x) versus number of units produced (y) when no employees are working may not make sense. Let's assume the y-intercept is 100 when x equals zero. This would not make any sense. You would not expect to make 100 units with no employees. So be careful if you extrapolate the line to create a y-intercept, it may not make any sense in the real world! In Fig. 10.2, the y-intercept is 2.0. When there is no wage inflation, there is still a price inflation of 2%. That intercept makes sense.

The **slope (*m*)** of the regression line is just the increase in y for a unit increase in x. It will be positive if x and y move in the same direction, that is, both increase or both decrease. It will be negative if they move in opposite directions, that is, as x increases then y decreases or vice versa. In Fig. 10.2, the slope is 0.5547. The resulting equation from this data set is $\mathbf{y = 0.5547\,x + 2.0107}$. For each increase of 1(%) of x, this means y will increase about 0.55(%). In other words, a 1% increase in wage inflation will result in a 0.55% increase in price inflation.

In the basic model, the (x, y) data generates a line of the form $y = mx + b$ that best describes how changes in y can be predicted from changes in x. This allows you to estimate the y value for any given x value.

The **market model** is an important application of simple linear regression. In this model, we assume that the rate of return on a stock is linearly related to the rate of return on the overall market. This can be defined as

$$\begin{array}{lcccc} & R = & \beta_0 + & \beta_1 & R_m \\ \textbf{compare} & \updownarrow & \updownarrow & \updownarrow & \updownarrow \\ & y = & b + & m & x \end{array}$$

where

R = return on a particular stock (the dependent variable)
β_0 = y-intercept

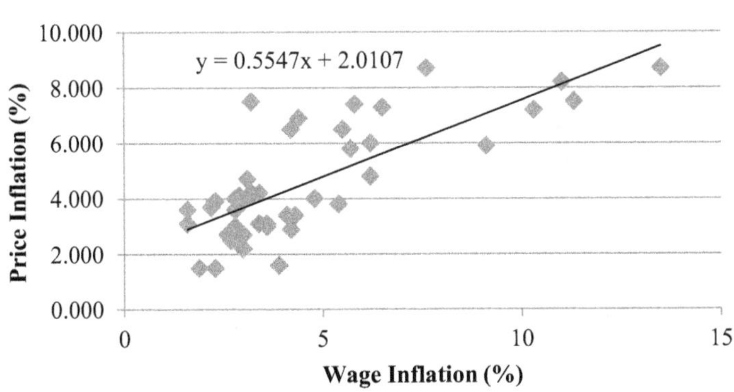

Fig. 10.2 Price versus wage inflation

β_1 = slope or the beta coefficient that measures how sensitive the stock's rate of return is to change in the level of the overall market
R_m = return on some major stock index

When β is greater than 1, the stock's rate of return is more sensitive to changes in the level of the "overall" market than is the average stock. In other words, a stock with β greater than 1 will be more volatile than the overall market.

In this model, the slope is a measure of the stock's systematic risk because it measures the volatility of the stock's price in comparison to the volatility across the system (overall market).

However, there is also nonsystematic risk that is associated with the listed stock company. Advertising effectiveness, a recent merger or if the company has filed Chap. 11 (bankruptcy) can be variables that also affect the volatility of the stock. This type of risk is minimized by creating a stock portfolio. However, the systematic or market-related risk remains.

The portfolio's β_1 is estimated by calculating the mean of the entire individual stock betas that are in the portfolio. Depending on an investor's risk aversion, different β values will be desirable. Risk-averse investors, who predict a fall in the market, will seek portfolios with β_1 less than 1. Likewise, those who believe the market is about to increase will seek a portfolio with β_1 greater than 1, which represents higher volatility.

A common approach in determining the best-fitting line utilizes the **method of least squares**. The following symbols are important in understanding this part of the discussion:

Symbols

x: independent variable
y: dependent variable
y_i: ith value of the measured or observed y
$\hat{y}$: predicted or estimated value of y
$\overline{y}$: average or mean value of y
ε: error or residual term
b: y-intercept
m: slope of the line

The **Method of Least Squares** uses the data to provide those values of the y-intercept and the slope that minimize the sum of the squares of the differences between the observed values (y_i) of the dependent variable and the estimated (predicted) values of the dependent value ($\hat{y}$). This difference is referred to as the residual or error term (ε). The **residual** is simply the difference between the measured value of y and the predicted value of y. The predicted value of y is generated from the least-squares regression line. The real value of y is the data you collected.

Figure 10.3 includes data from a model in which we want to predict quarterly sales for restaurants by knowing the student populations nearby. Residuals are shown in the last column. The residual is the difference between the actual observed

Restaurant	Student Pop (1000s)	Quarterly Sales ($US 1000s)	Excel Predicted Quarterly Sales ($US 1000s)	Residuals ($US 1000s)
1	2	58	70	-12
2	6	105	90	15
3	8	88	100	-12
4	8	118	100	18
5	12	117	120	-3
6	16	137	140	-3
7	20	157	160	-3
8	20	169	160	9
9	22	149	170	-21
10	26	202	190	12

Fig. 10.3 Residuals

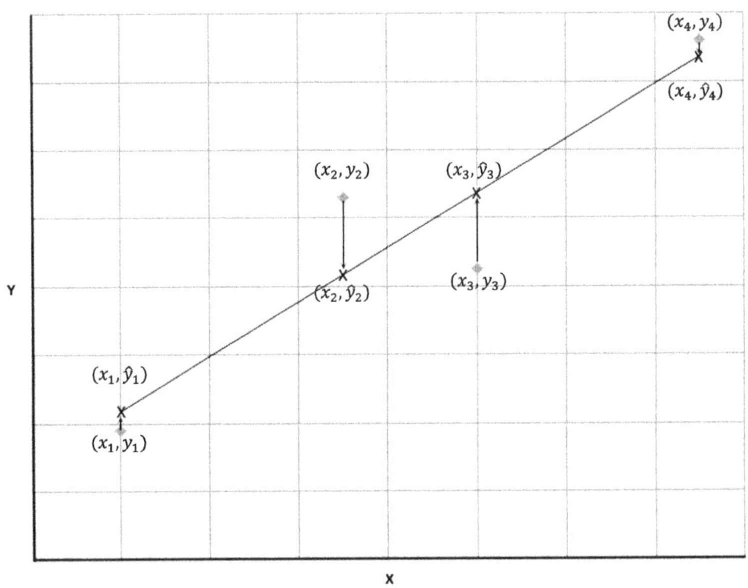

Fig. 10.4 Observed values (y_i) and estimated values $(\hat{y}_i)$ for the dependent variable

sales data in the third column, minus the sales data predicted by the regression line in the fourth column.

In Fig. 10.4, the values of the independent variable (x) are plotted along the horizontal axis. The vertical axis is used to plot both the observed values of y and the Excel estimated (predicted) values of y for each and every value of x. This graph shows the difference between the observed values (y_i) of Quarterly Sales and the Excel predicted values $(\hat{y}_i)$ of Quarterly Sales as vertical lines.

Residuals: Tests for Linearity—Constant Variance of Residuals

To accurately apply the linear regression model, it is important to verify that our data really does follow a linear relationship. One way to do this is to make sure we have a relatively constant variance of residuals.

Assume we have the same data set where the independent variable is student population and the dependent variable is quarterly sales at local pizza restaurants. Remember in this model we are interested in predicting sales from the student population data.

Figure 10.5 depicts a residual plot, with the residuals ($y - \hat{y}$) on the vertical axis and the independent variable (x) along the horizontal axis. The symbol for residuals is the lowercase Greek letter epsilon (ε).

Residual plots support a linear model when they show a random scatter of points forming an approximately constant width (variance) around the identity line. A residual plot is a graph that shows the residuals on the vertical axis and the independent variable on the horizontal axis. If the points in a residual plot are randomly dispersed around the horizontal axis, a linear regression model is appropriate for the data; otherwise, a nonlinear model is more appropriate.

Residual scatter plots provide a visual examination of the assumption homoscedasticity (points at widely varying differences from the line), between the predicted dependent variable scores and the errors of prediction. The primary benefit is that the assumption can be viewed and analyzed with one glance; therefore, any violation can be determined quickly and easily.

The plot in Fig. 10.5 shows a random distribution of residuals equally dispersed around the horizontal axis, which is what we want if the data is truly linear in nature. In other words, we do not want to see a pattern that would allow us to feel comfortable in predicting the next residual value. A pattern that raises a red flag suggests that we really do not have a distribution of data that supports using a linear model. Figures 10.6 and 10.7 are examples of residual plots that raise a red flag.

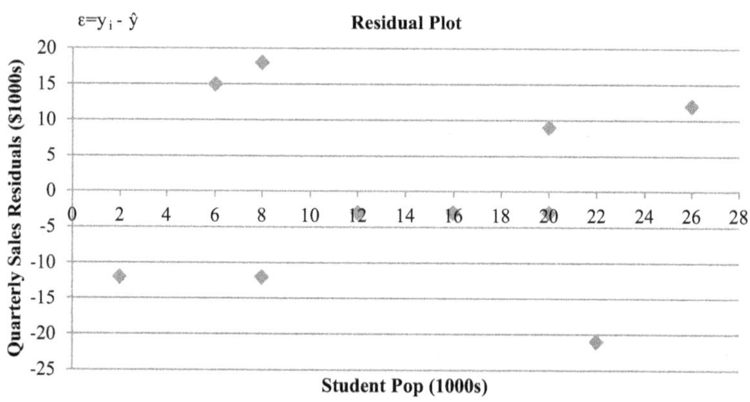

Fig. 10.5 Residual plots

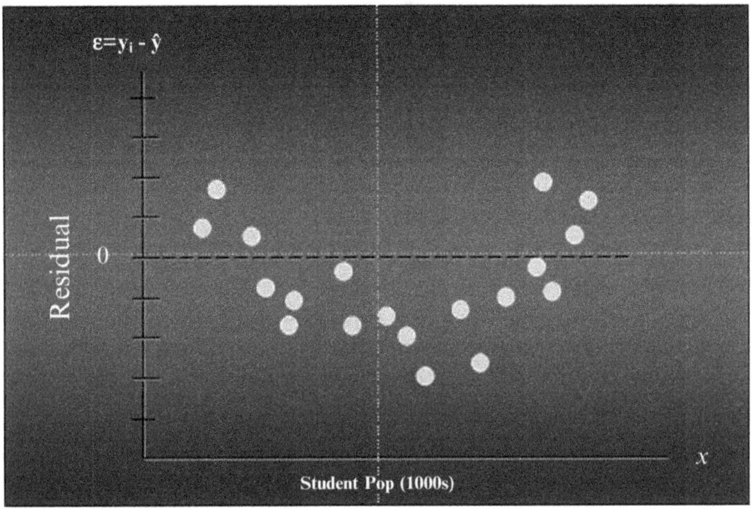

Fig. 10.6 Residual plots

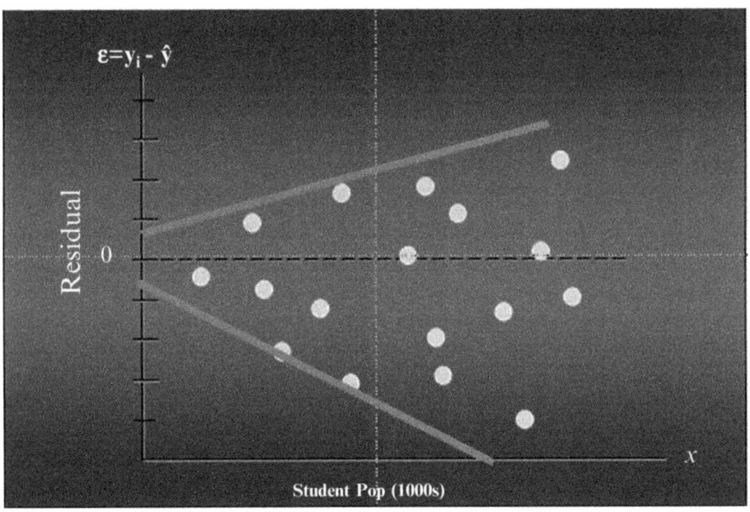

Fig. 10.7 Residual plots

Figure 10.6 shows a curvilinear distribution of residuals; not good. Figure 10.7 shows small values or residuals for small values of x and progressively larger residuals for larger values of x in a funnel distribution; also, not good. Neither of these data sets suggests that a simple linear model would make sense to use for analysis and prediction.

Standardized Residuals: Tests for Outliers and Linearity

When residuals are converted into standard units, we refer to them as **standardized residuals.** This data is very important in checking for linearity and outliers.

1. **Tests for Linearity**

Histogram of Standardized Residuals
A robust model will include a normal distribution of the standardized residuals. Normality of the residuals is an assumption of running a linear model. One of the simplest ways to identify the distribution is to run a histogram on the standard residuals. So, if your standardized residuals are normal, it means that your linear assumption is valid and the model inference (confidence intervals, model predictions) should also be valid.

Normal Probability Plot
Normal Probability Plots (also called Normal Test Plots or Normal Quartile Plots) are another way to investigate whether the residuals exhibit the standard normal "bell curve" distribution. First, the x-axis is transformed so that a cumulative normal density function will plot in a straight line. Then, using the mean and standard deviation that are calculated from the data, the data is transformed to the standard normal values, that is, where the mean is zero and the standard deviation is one. The data points are plotted along the fitted normal line.

The nice thing is that you don't have to understand all the transformations. All you have to do is look at the plotted points and see how well they fit the normal line. If they fit well, you can safely assume that your process data is normally distributed. That gives you confidence in your process capability indices, and in % of nonconforming parts projections. If the data values fall along a roughly straight line at a 45-degree angle, then the data is normally distributed. If the line is skewed to the left or right, it means that you do not have normally distributed data.

2. **Outliers**

Residual analysis in standard units can help us determine whether outliers are present in the data. For normally distributed **standardized residuals**, we would expect about 99% of our residuals to fall between ±3 standard units. Check if this is true by eyeballing the data. If you have values that are less than −3 or greater than +3, these may be outliers or they may be extreme values that should still be included. The first step is always to check if the measurement was recorded accurately. If after investigation the value is valid, you could run the model with and without the outlier to determine the impact and then decide if it is OK to exclude the value from your model (Fig. 10.8).

None of these standardized residuals violate our rule, so we will assume we do not have any outliers in this data set.

Regression Summary
The next section will provide detailed Excel instructions on completing each of the steps shown below. A thorough discussion of significance tests associated with

Standardized Residuals			
Observation	Predicted Y	Residuals	Standard Residuals
1	70	-12	-0.920357987
2	90	15	1.150447483
3	100	-12	-0.920357987
4	100	18	1.38053698
5	120	-3	-0.230089497
6	140	-3	-0.230089497
7	160	-3	-0.230089497
8	160	9	0.69026849
9	170	-21	-1.610626477
10	190	12	0.920357987

Fig. 10.8 Standardized residuals

regression is described in the next two chapters rather than inserting it here. But all model analysis should also complete significance testing in addition to the six steps indicated.

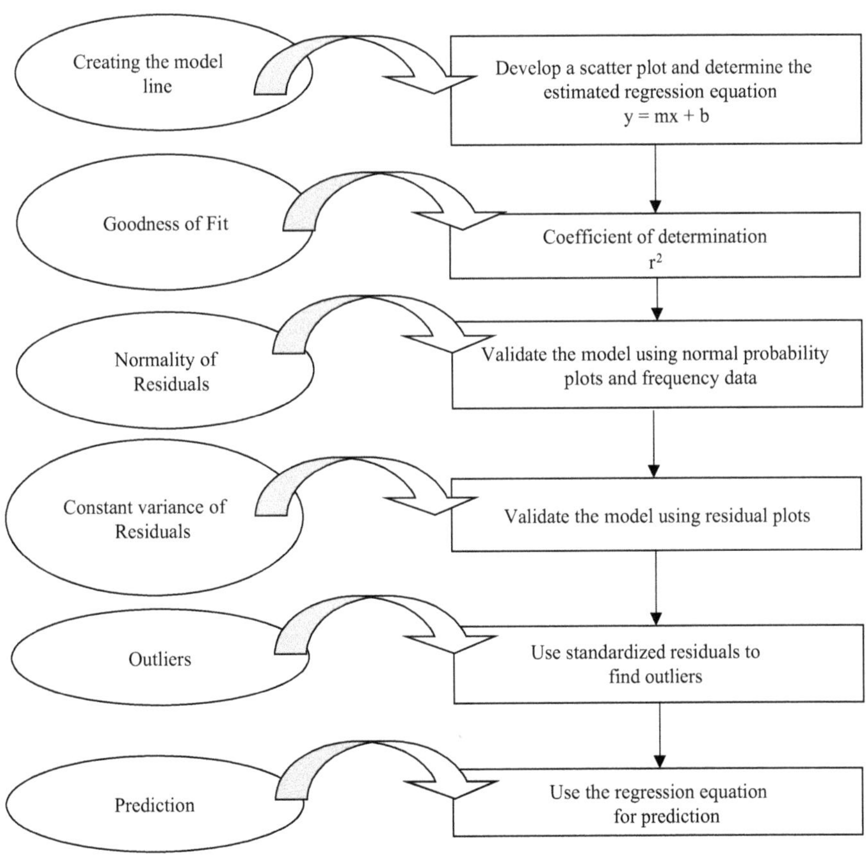

- Remember to complete significance testing as described in Chaps. 11 and 12.

Excel

Regression tools include analysis completed through use of the scatter plot and through use of the Regression function. Both of these techniques will be described in detail in this next section. The topics covered in this section will include

1. Computing the regression line and the coefficient of determination using the **Scatterplot** function (**Create model, R^2 Goodness of fit**)
2. Developing a regression model using the **Regression** function (**Create model**)
3. Compute residual plots using the **Regression** function
4. Testing for normality of the distribution of the residuals (**Normality of residuals**)
5. Testing for constant variance of the residuals (**Constant variance of residuals**)
6. Testing for outliers

After completing the analyses using these EXCEL tools, you are now in a position to use the model for prediction.

Scatter Plot: Compute the Regression Line and the Coefficient of Determination

A prospective buyer of a pizza restaurant is interested in knowing if the student population near the restaurant may help him estimate what the expected sales should be. Data was collected such that the independent variable is student population at universities near 10 pizza restaurants and the dependent variable is quarterly sales at those local pizza restaurants.

Check your data is set up correctly so the x (independent) data is in the first column and the y (dependent) data is in the second column. Remember the y variable is your variable that you are interested in predicting, the x variable helps you do that.

Input Data for Example Problem:

- Input the data from columns two and three. Include labels
- Create a scatter plot using the **Scatter Chart** function in **Insert** tab

Student population (1,000s)	Quarterly sales ($US 1,000s)
2	58
6	105
8	88
8	118
12	117
16	137
20	157
20	169
22	149
26	202

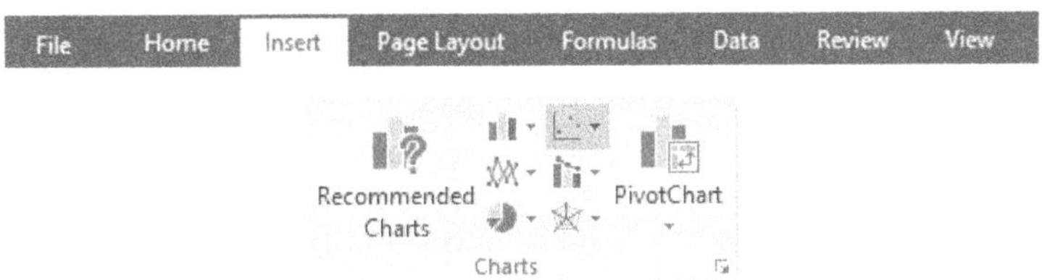

- Position the mouse over any data point in on the scatter graph and right click to display a list of options

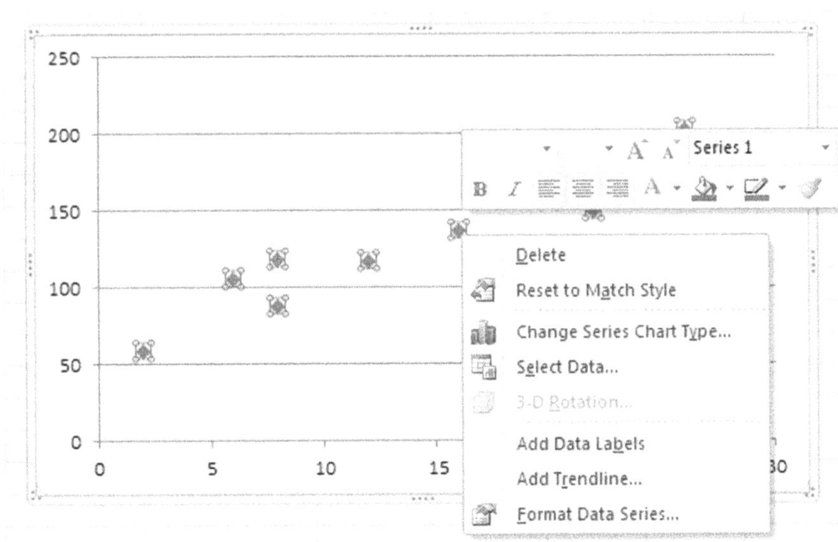

- Select the **Add Trendline** option

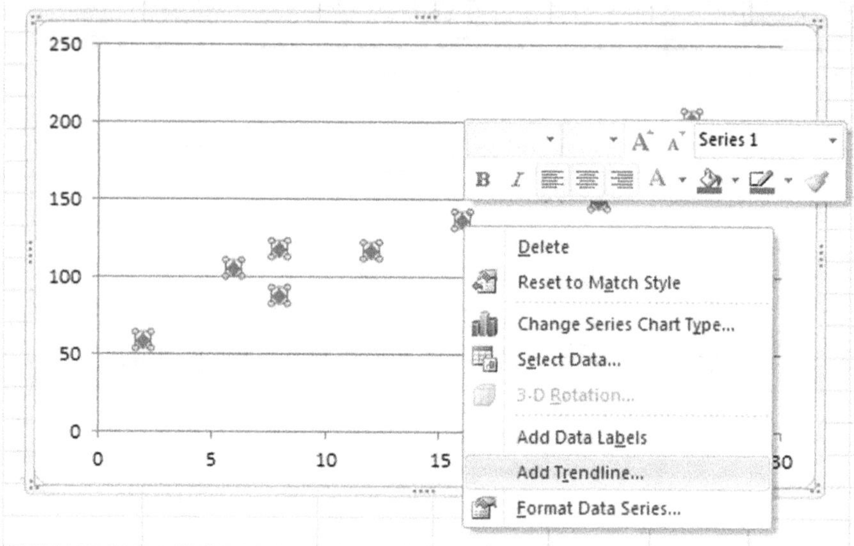

x must be column 1, *y* must be column 2

- Select **Linear**

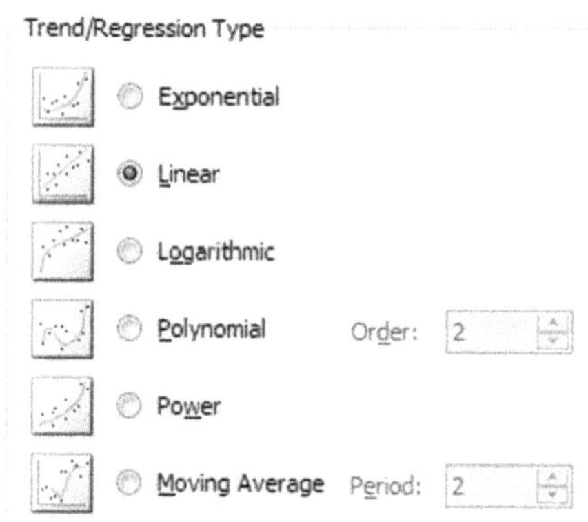

- Choose **Display R-squared value on chart** option

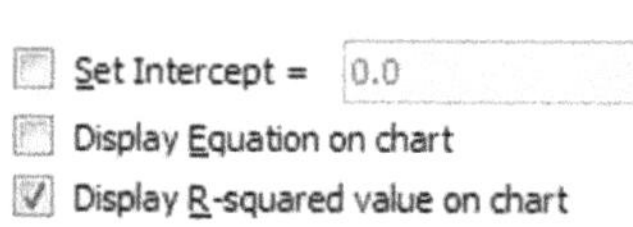

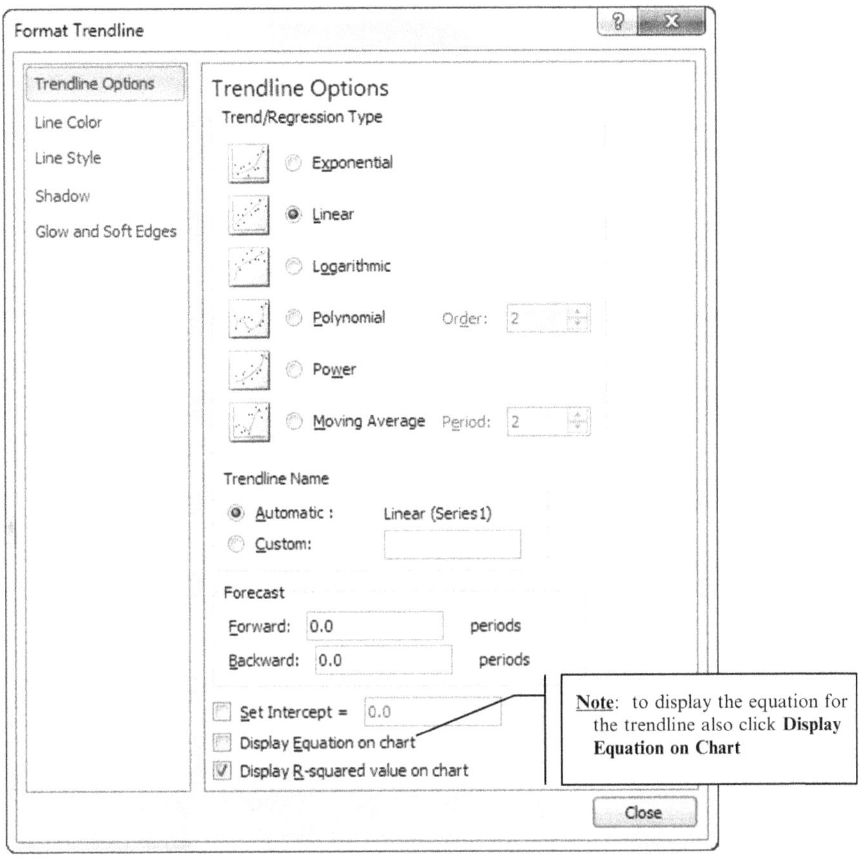
Format Trendline
Trendline Options
Line Color
Line Style
Shadow
Glow and Soft Edges
Trendline Options
Trend/Regression Type
Exponential
Linear
Logarithmic
Polynomial
Order: 2
Power
Moving Average
Period: 2
Trendline Name
Automatic : Linear (Series1)
Custom:
Forecast
Forward: 0.0 periods
Backward: 0.0 periods
Set Intercept = 0.0
Display Equation on chart
Display R-squared value on chart
Close
Note: to display the equation for the trendline also click **Display Equation on Chart**

- Click **Close**

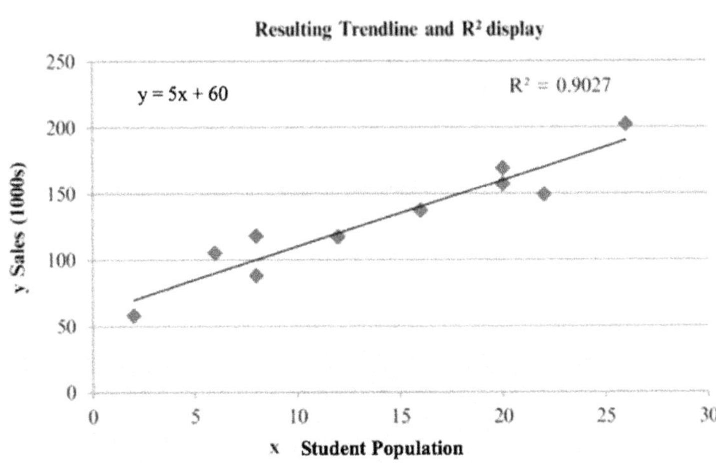
Resulting Trendline and R² display
y = 5x + 60
R² = 0.9027
y Sales (1000s)
x Student Population
0
50
100
150
200
250
0
5
10
15
20
25
30

- Click anywhere on graph
- Click **Design**
- Click **Add chart element** to add labels

If you want to change the font or size of the labels for the trendline, just right click on them. You can select **Edit Text** or **Font** to allow you to edit as desired.

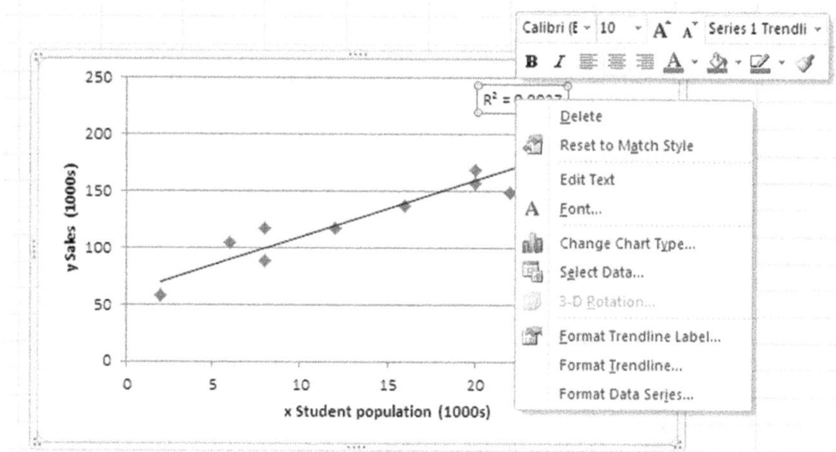

If you want to move the labels to a different location on the graph, just left click on one of the corners of the text box and drag it to another location.

Regression Function: Compute the Regression Model

- Select the **Data** tab
- Choose the **Data Analysis** function
- Choose the **Regression** from the list of Analysis Tools

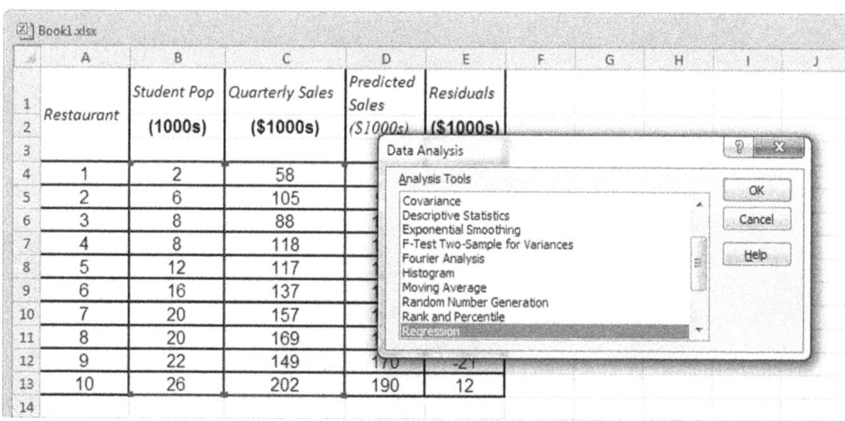

- Click **OK**
- Enter the y data in the **Input Y Range** box

Note when you input a column or row of data into an Excel function that Excel will automatically add $ signs in the input range boxes. Remember the concept of the $ sign allows you to copy the formula in Excel without changing references. Using the $ sign before the row and column coordinates makes an absolute cell reference.

- Enter the x data in the **Input X Range** box

- Select **Labels** if you have them (a good idea!)

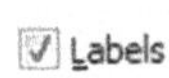

- Select **Output Range**. Enter the cell to identify the upper left corner of the section of the worksheet where the output will appear

- Click **OK**

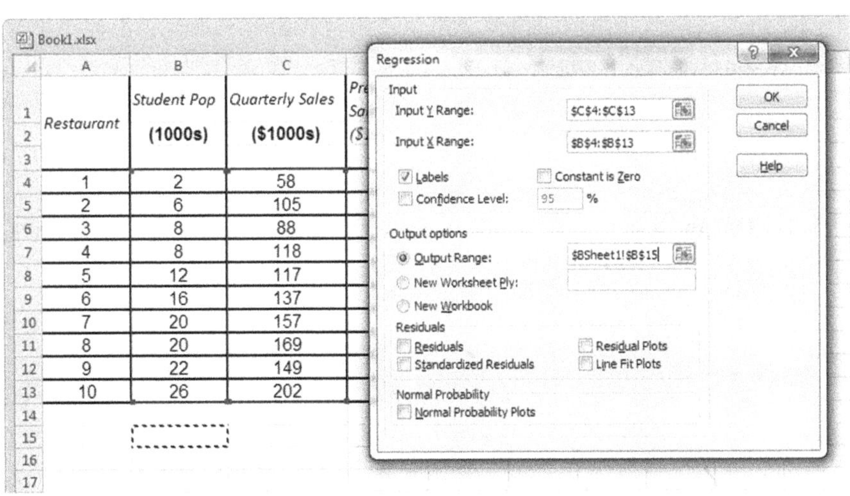

See Fig. 10.9 for the actual output.

Compute Residual Plots Using the Regression Function

- Select the **Data** tab
- Choose the **Data Analysis** function
- Choose the **Regression** function from the list of Analysis Tools

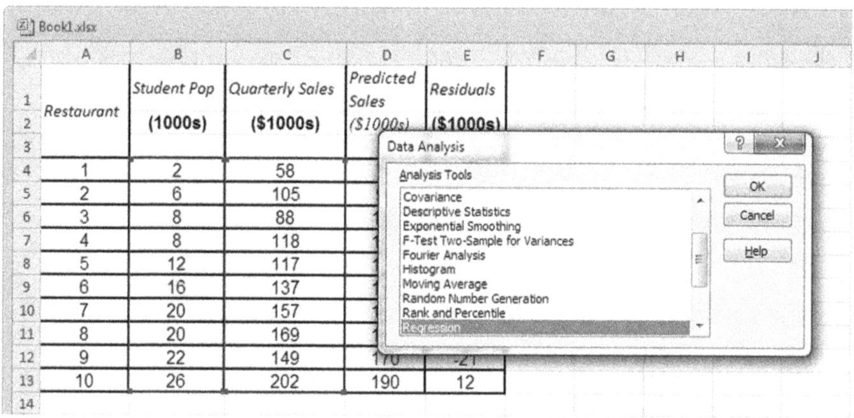

When the Regression dialog box appears:

- Enter your **Input Y Range** for your data

- Enter your **Input X Range** for your data

- Select **Output Range** and include upper-left corner of the section of the worksheet where the output will be printed

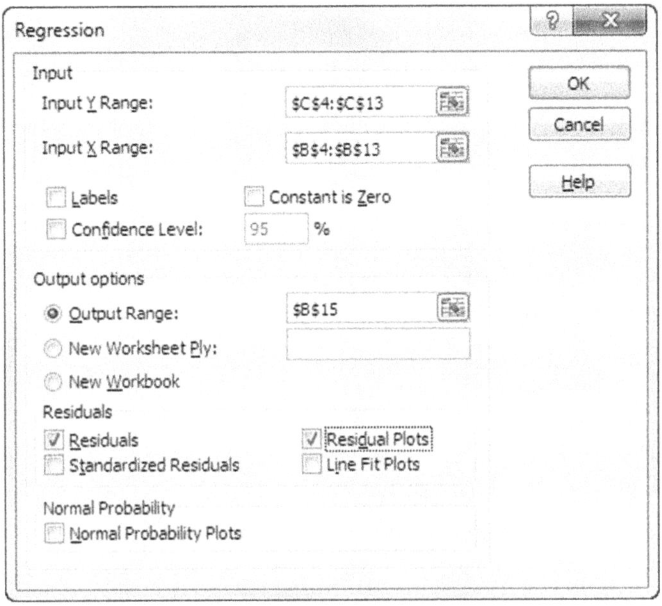

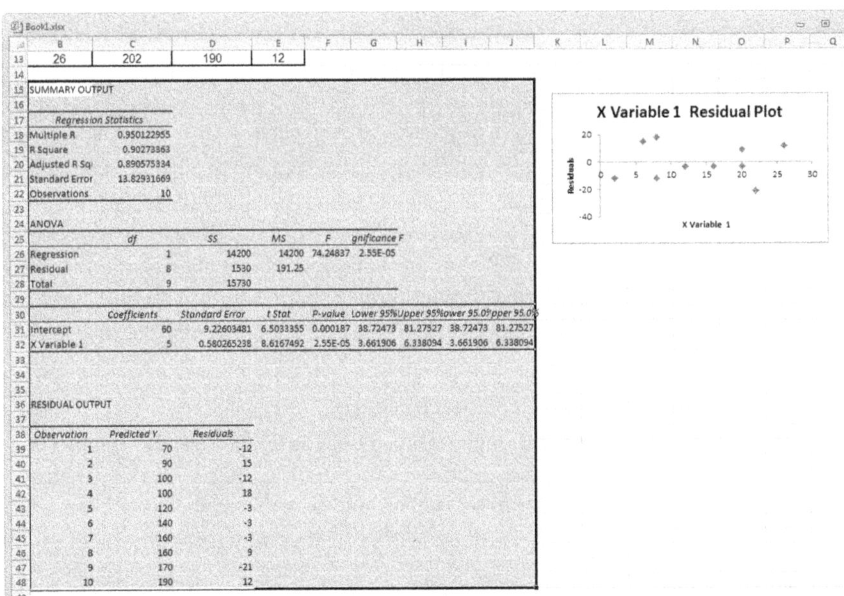

26	202	190	12

SUMMARY OUTPUT

Regression Statistics	
Multiple R	0.950122955
R Square	0.90273363
Adjusted R Sq	0.890575334
Standard Error	13.82931669
Observations	10

ANOVA

	df	*SS*	*MS*	*F*	*gnificance F*
Regression	1	14200	14200	74.24837	2.55E-05
Residual	8	1530	191.25		
Total	9	15730			

	Coefficients	*Standard Error*	*t Stat*	*P-value*	*Lower 95%*	*Upper 95%*	*ower 95.0%*	*pper 95.0%*
Intercept	60	9.22603481	6.5033355	0.000187	38.72473	81.27527	38.72473	81.27527
X Variable 1	5	0.580265238	8.6167492	2.55E-05	3.661906	6.338094	3.661906	6.338094

RESIDUAL OUTPUT

Observation	*Predicted Y*	*Residuals*
1	70	-12
2	90	15
3	100	-12
4	100	18
5	120	-3
6	140	-3
7	160	-3
8	160	9
9	170	-21
10	190	12

Fig. 10.9 Output chart and residual plot

The resulting output will show a plot of the residuals ($y_i - \hat{y}$) against the independent variable (x). If you also selected **Residuals** as well as **Residual Plots**, Excel will provide a table of all the observed y values and the associated predicted y values. Remember the numerical difference between these two y values is defined as the residual.

Normality Test for Data: Using Excel's Regression Tool to Test for Normality of the Distribution of Residuals

A robust model will include a normal distribution of residuals. Normality of the residuals is an assumption of running a linear model. So, if your residuals are normal, it means that your assumption is valid and model inference (confidence intervals, model predictions) should also be valid.

Method 1: Normal Probability Plot
You follow the same steps as you used to construct a residual plot but now you should also select **Normal Probability Plot**. Excel will provide you with a normal probability plot. The normalized values of the residuals are plotted on the y-axis and the independent variable remains on the x-axis. If the points approximate a straight line, the residuals could have come from a normal distribution, which is what you want. If the line is skewed to the left or right, it means that you do not have normally distributed data.

First, the x-axis is transformed so that a cumulative normal density function will plot in a straight line. Then, using the mean and standard deviation (sigma) that are calculated from the data, the data is transformed to the standard normal values, that is, where the mean is zero and the standard deviation is one. Then the data points are plotted along the fitted normal line. The nice thing is that you don't have to understand all the transformations. All you have to do is look at the plotted points, and see how well they fit the normal line. If they fit well, you can safely assume that your process data is normally distributed. If your plotted points don't fit the line well, but significantly curve away from it in places, you may have a non-normal distribution (Fig. 10.10).

Method 2: Histogram of Standardized Residuals
In addition to running a normal probability plot, you could simply create a frequency plot of the standardized residuals as shown in Fig. 10.11. Build your histograms selecting bins between −3 to +3, which will become your x-axis. Frequency will be plotted along the y-axis. As long as this histogram approximates a normal distribution, you are good to go.

In this case, both tests for normality of the residual distribution are acceptable, which means that your assumption is valid and model inference (confidence intervals, model predictions) should also be valid.

Using Excel's Regression Tool to Test for Constant Variance of Residuals

One of the key assumptions of linear regression is that the residuals have constant variance at every level of the predictor variable(s). If this assumption is not met, the residuals are said to suffer from heteroscedasticity. When this occurs, the estimates for the model coefficients become unreliable.

Construct a residual plot by selecting **Residuals,** and it is still OK to also select the **Standardized Residuals**. Excel will provide you with a table of both residuals and standardized residuals for each *y* value, but the residual plots use residual values, not the standardized residual values.

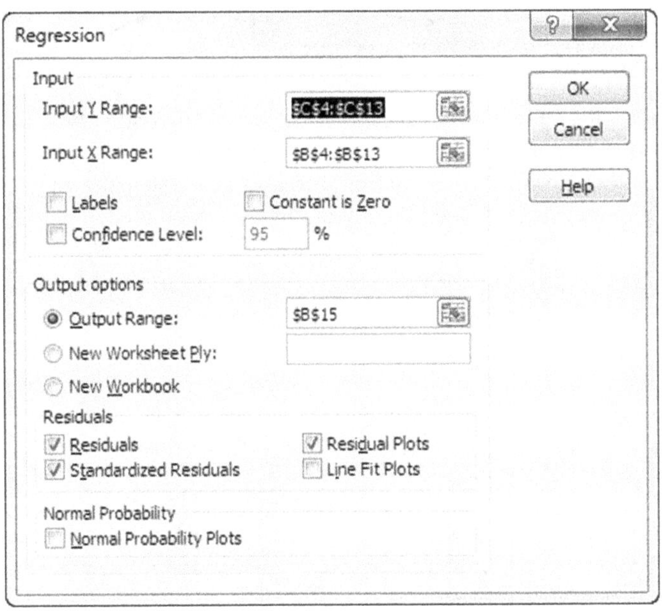

Standard residual plots are often quite small and need to be enlarged to identify any patterns. To do this just, grab a corner and pull to the size you desire.

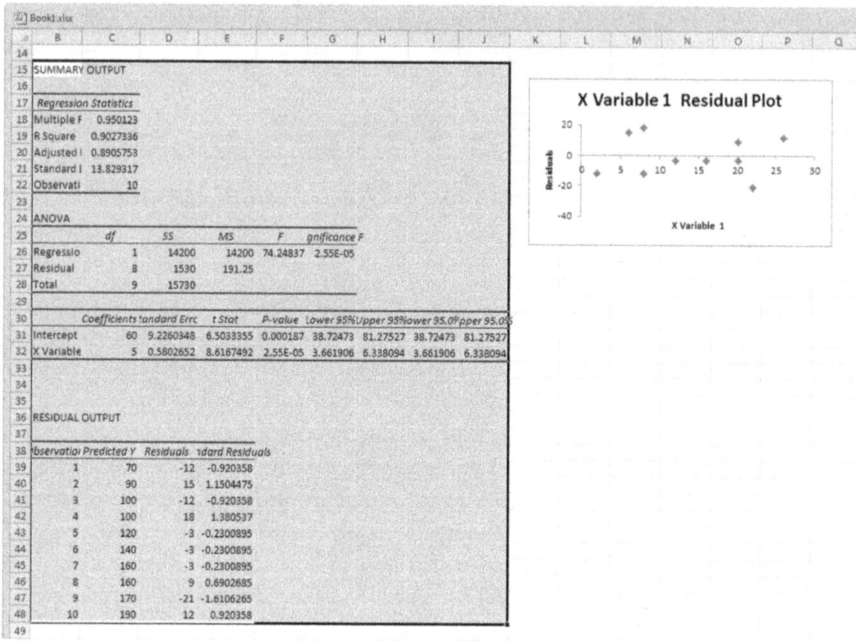

SUMMARY OUTPUT

Regression Statistics	
Multiple F	0.950123
R Square	0.9027336
Adjusted I	0.8905753
Standard I	13.829317
Observati	10

ANOVA

	df	*SS*	*MS*	*F*	*gnificance F*
Regressio	1	14200	14200	74.24837	2.55E-05
Residual	8	1530	191.25		
Total	9	15730			

	Coefficients	*tandard Errc*	*t Stat*	*P-value*	*Lower 95%*	*Upper 95%*	*ower 95.0%*	*pper 95.0%*
Intercept	60	9.2260348	6.5033355	0.000187	38.72473	81.27527	38.72473	81.27527
X Variable	5	0.5802652	8.6167492	2.55E-05	3.661906	6.338094	3.661906	6.338094

RESIDUAL OUTPUT

bservatio	*Predicted Y*	*Residuals*	*ndard Residuals*
1	70	-12	-0.920358
2	90	15	1.1504475
3	100	-12	-0.920358
4	100	18	1.380537
5	120	-3	-0.2300895
6	140	-3	-0.2300895
7	160	-3	-0.2300895
8	160	9	0.6902685
9	170	-21	-1.6106265
10	190	12	0.920358

Remember in a "good" residual plot the residuals are randomly scattered about zero with no systematic increase or decrease in variance. In a "bad" residual plot, the variance of the residuals increases or decreases in a systematic way. In this case, we have a plot that supports the linearity of our regression model.

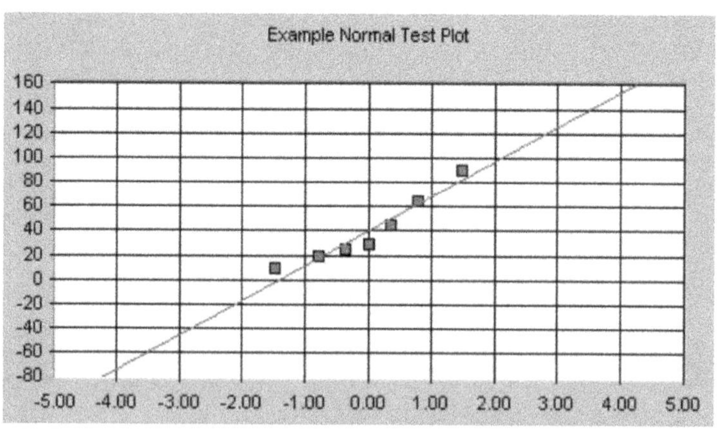

Fig. 10.10 Normal probability plot

Fig. 10.11 Frequency plot of residuals

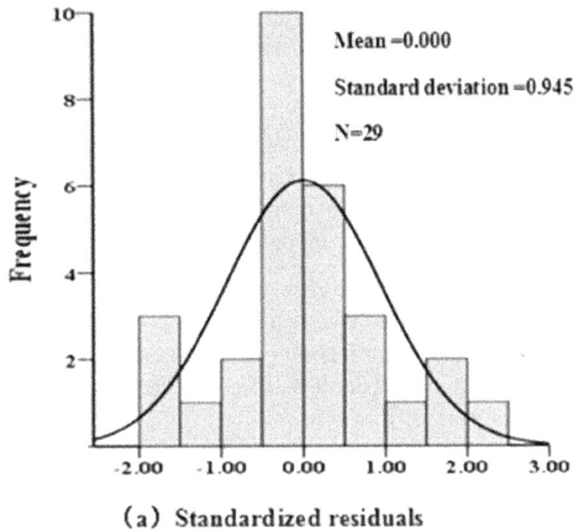

Using Excel's Regression Tool to Test for Outliers

The only analysis tool you require is a list of the standardized residuals. Remember an outlier is defined as any value that does fall between the range of −3 to +3.

Summary of Regression Analysis Process

Step 1: Develop a model that is based on realistic relationships. For the dependent variable, find an independent variable that makes sense as a variable that can generate a linear relationship. For example, using how much caffeine you have consumed to predict how many pages in a managerial report you can read doesn't make sense as a linear relationship. However, how many times your office assistant interrupts you may have a linear relationship with number of pages in a report that you can read. We know that most likely the relationship won't be perfect, so we need to recognize the presence of an error term (residuals). (Create model)

Step 2: Gather x, y data pairs. (Create model)

Step 3: Create a scatter plot to decide if a linear model is appropriate. At this point, outliers may already be apparent; you need to investigate these before going forward. Remember an outlier can greatly influence all your regression statistics. (Outliers)

Step 4: Use Excel to calculate the regression equation. It is usually helpful to have the equation printed on the scatter plot. (Create model)

Step 5: Use Excel to generate the coefficient of determination (r^2). This gives you an idea of how much of the variance in y can be explained by the variance in x. This value can range from 0 to 1. The closer the r^2 is to zero, the weaker the fit. (Goodness of Fit)

Step 6: Validate the linear regression model that can be used to represent this data. The validation tests have to do with the residuals because our entire modeling technique is based on minimizing the residual. You can also think of it as running checks on "ε" in our model equation.

The **first test** is to check that the distribution of frequency data of the residuals is normal. We have two ways to do this: (1) A histogram of the residuals and (2) a normal probability plot. Remember, in a normal distribution of residuals, we would expect 99% of our residuals to fall between +/−3 standard units. (Normality of residuals)

The **second test** is a residual plot that compares residuals to the x values. Residuals should be randomly distributed around zero; otherwise, a clearly visible pattern in this plot indicates a red flag. Without relatively constant variance (change in the residuals), we have a problem and another model should be used. In other words, the residual plots should not indicate any strong pattern that would allow you to predict the next residual value. (Constant variance of residuals)

Step 7: Check for outliers using the standardized residuals. An outlier is any value that does not fall in the range of between ±3 standard units.

Step 8: If all tests were good in Steps 6 and 7, it is now appropriate to use the regression equation to predict values of the dependent variable (y) for given values of (x). (Prediction)

Common Excel Pitfalls

- Make sure to click the box indicating you are including variable labels.
- Make sure to click inside the **Input Y** box before you highlight the input data field. Then, click inside the **Input X** box before highlighting the next input data field. If you don't click inside the **Input X** box, you will just overwrite the input you selected for **Input Y**.
- Look at the data for a linear model before you run a linear regression; don't just cram data in a model for the sake of generating a model.
- Be consistent with your units throughout the analysis including on the graphs.
- Pay attention to outliers.
- Make sure to label your independent and dependent variables, x being your independent variable and y being your dependent variable.

Final Thoughts and Activities

Practice Problems

1. The marketing manager of a supermarket chain is trying to see if shelf space can predict sales for breakfast cereal. A random sample of 12 equal-sized stores is selected with the following results.

Number of shelves	Sales ($U.S. 1000s)	Aisle location
5	160	0
5	220	1
5	140	0
10	190	0
10	240	0
10	260	1
15	230	0
15	270	0
15	280	1
20	260	0
20	290	0
20	310	1

   (a) Construct a scatter plot.
   (b) Determine the coefficient of determination, R^2, and interpret its meaning.
   (c) How useful do you think this regression model is for predicting sales?

2. The director of human resources at Jaunk-It moving company wants to predict the number of cubic feet of goods moved, based on labor hours (stored in **Jaunkit** file). In other words, if he pays his workers for 12 h of work, about how many cubic feet of goods are they moving into the storage unit? Use the data to help the director understand this relationship.

   (a) Construct a scatter plot.
   (b) Determine the coefficient of determination, R^2, and interpret its meaning.
   (c) How would you explain this relationship to the director?

3. Nine's Wine Vineyard makes a selection of wines that goes through independent taste testing every year. In the file **Wine**, the different varieties of wine are ranked by taste and number of bottles sold. Use the data to help the Nine's Wine Vineyard understand the relationship between its wine and sales.

   (a) Construct a scatter plot.
   (b) Determine the coefficient of determination, R^2, and interpret its meaning.
   (c) How would you use this regression model for predicting sales?

4. A company that holds the streaming service distribution rights to movies previously released only in theaters has the business objective of developing estimates of the sales revenue of the streaming service. Toward this goal, a company analyst plans to use box office gross to predict streaming service sales revenue. For 22 movies, the analyst collects the box office gross (in $millions) in the year that they were released and the streaming service revenue (in $millions) in the following year. For these data,

Gross:	109.92	177.02	166.82	93.78	106.66	90.22	85.54	66.70	87.34	101.40	33.26
	144.13	60.15	46.69	60.73	56.75	79.30	40.68	39.32	41.80	39.26	17.74
DVD	81.60	107.54	44.41	60.21	62.82	58.81	48.79	38.78	44.67	34.88	31.05
Revenue:	33.14	27.12	28.16	26.43	26.10	42.04	25.03	21.45	17.51	21.08	18.78

(a) Construct a scatter plot.
(b) Assuming a linear relationship, use the least-squares method to determine the slope and y-intercept.
(c) Interpret its meaning of the slope in this problem.
(d) Predict the sales revenue for a streaming movie that had a box office gross of $75 million.

5. In the previous problem, you used box office gross to predict DVD revenue. Using the results of that problem,

(a) Determine the coefficient of determination, r^2, and interpret its meaning.
(b) How useful do you think this regression model is for predicting DVD revenue?
(c) Can you think of other variables that might explain the variation in DVD revenue?

6. The following results show the X values, residuals, and a residual plot from a regression analysis:

X	Residuals
1	0.70
2	1.58
3	1.03
4	0.33
5	-0.39
6	-0.67
7	-0.56
8	-0.65
9	-1.19
10	-0.84
11	-0.29
12	-1.28
13	-0.21
14	-0.37
15	0.22
16	-0.16
17	0.82

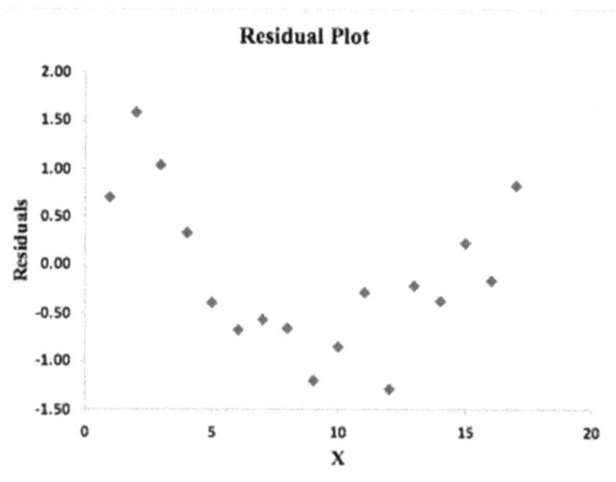

Is there any evidence of a pattern in the residuals? What does that mean?

7. A mail-order catalog business that sells personal computer supplies, software, and hardware maintains a centralized warehouse for the distribution of products ordered. Management is currently examining the process of distribution from the warehouse and has the business objective of determining the factors that affect warehouse distribution costs. Currently, a small handling fee is added to the order, regardless of the amount of the order. Data that indicates the warehouse distribution costs and the number of orders received has been collected over the past 24 months. The results are

Months	Distribution Cost ($ Thousands)	Number of Orders
1	52.95	4015
2	71.66	3806
3	85.58	5309
4	63.69	4262
5	72.81	4296
6	68.44	4097
7	52.46	3213
8	70.77	4809
9	82.03	5237
10	74.39	4732
11	76.84	4413
12	54.06	2921
13	62.98	3977
14	72.3	4428
15	58.99	3964
16	79.38	4582
17	94.44	5582
18	59.74	3450
19	90.5	5079
20	93.24	5735
21	69.33	4269
22	53.71	3708
23	89.18	5387
24	66.8	4365

(a) Assuming a linear relationship, use the least-squares method to find the regression equation.
(b) Predict the monthly warehouse distribution costs when the number of orders is 4500.
(c) Plot the residuals versus the time period.
(d) Based on the results of (c), is there reason to question the validity of the model?

8. This graph is part of a report that you need to review. What data is missing before you are able to fully assess these results?

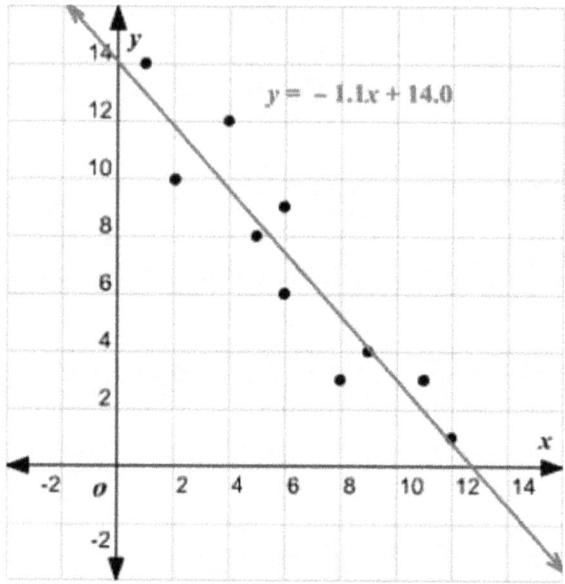

Discussion Boards

1. Research at the University of Texas indicates a negative correlation between economic growth and the percentage of lawyers in the workforce. Predatory litigation lowered GNP by 10% below its potential level during the 1980s; discuss this relative to the past 10 years.
2. Monthly charges for cell phones have decreased as worldwide subscribers have increased. Is it reasonable to use worldwide subscriptions to predict your monthly budget for your company cell phone?
3. CEOs have been under serious criticism from organized labor for the fat paychecks they receive. Many CEOs claim they put in longer hours, endure more stress, etc. In fact, there is a linear relationship between CEO salaries and financial success of a company. Discuss this using current data.

Group Activities

1. Provide data from the web to develop a regression relationship between calories and grams of fat for pizza. As a conscientious employee, you want to be able to predict the calorie content of the new pizza you are featuring in the cafeteria that has 10.5 g of fat.

2. Discuss the beta values of your favorite stock and favorite mutual fund. Have they changed much over the last 5 years?
3. Your company is getting ready to move into a new office space, but painting will be required. The toughness of paint is usually a good indication of how long it will last. Consumer Reports rates paints on their toughness and also provides an overall rating for each paint. The paint that has been selected for your office has an overall rating of 65, what toughness would you expect? Build a linear model from the data provided on the Consumer Reports website.
4. Moving people into cubicles increases our profit. Choose a position (pro or con) on this statement. Collect data from the web to support your position. Include some regression analysis.

Parting Thought

Statisticians must stay away from children's toys because they regress so easily. and.. .the last few available graves in the cemetery are called residual plots.

Problem Solutions

1. The marketing manager of a supermarket chain is trying to see if shelf space can predict sales for breakfast cereal. A random sample of 12 equal-sized stores is selected with the following results.

Number of shelves	Sales ($U.S. 1000s)	Aisle location
5	160	0
5	220	1
5	140	0
10	190	0
10	240	0
10	260	1
15	230	0
15	270	0
15	280	1
20	260	0
20	290	0
20	310	1

Note in the solution for this problem we don't actually need to use the aisle location data. Be careful to identify the key variables for each regression model. The independent variable is the number of shelves and our variable of interest is the dependent variable, sales.

(a) Construct a scatter plot

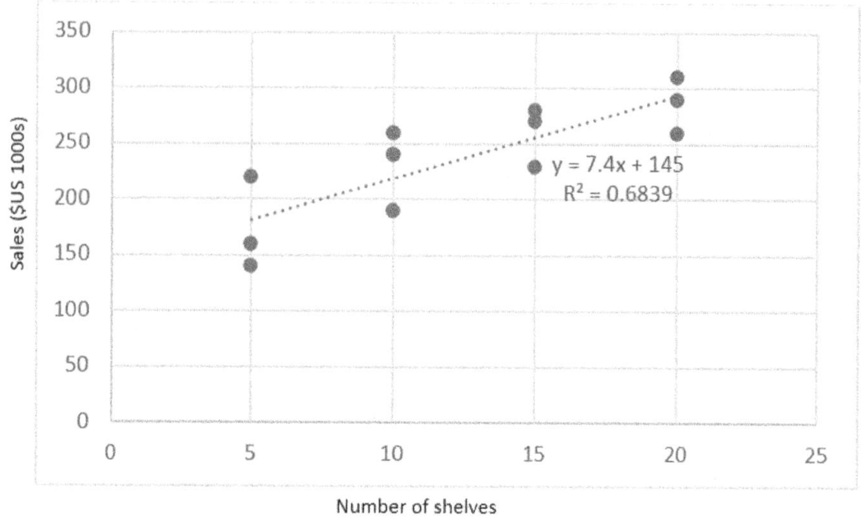

(b) Determine the coefficient of determination, R^2, and interpret its meaning.
Answer: $R^2 = 0.6839$
Therefore, 68% of sales can be explained by number of shelves.

(c) How useful do you think this regression model is for predicting sales?
This regression is helpful in predicting sales because it helps the marketing manager understand that for each increase in shelf space of an additional foot, weekly sales are estimated to increase by $7.40 (from the slope).

2. The director of human resources at Jaunk-It moving company wants to predict the number of cubic feet of goods moved, based on labor hours (stored in **Jaunkit** file). In other words, if he pays his workers for 12 h of work, about how many cubic feet of goods are they moving into the storage unit? Use the data to help the director understand this relationship.

The independent variable is the number of hours and our variable of interest is the dependent variable, cubic feet of goods being moved.

(a) Construct a scatter plot

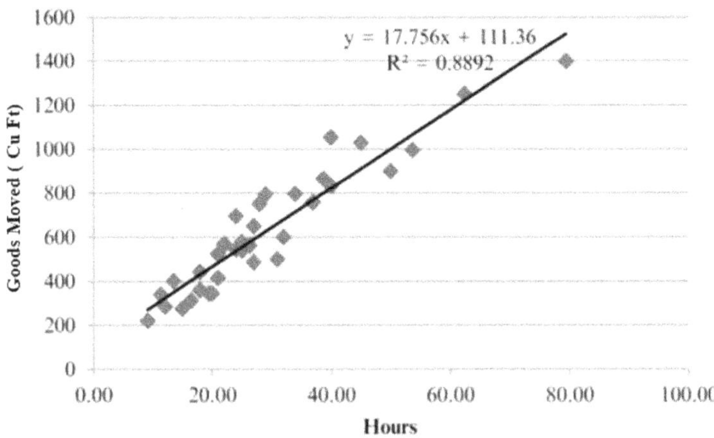

(b) Determine the coefficient of determination, R^2, and interpret its meaning.

Answer: $R^2 = 0.8892$

Therefore, 89% of cubic feet moved can be explained by the labor hours worked.

(c) How would you explain this relationship to the director?

There is a strong linear relationship between number of hours and cubic feet moved. This means that the more cubic feet on a property that requires moving, the more hours it will take for workers to move these things. Knowing how many hours he has paid his employees should help him predict how many cubic feet of stored goods have been moved. If he pays his employees for 30 h of moving time, he should expect **17.76 (30) + 111.36 = 644.16**, or approximately 644 cubic feet of storage space to be filled with moved goods. Once this relationship is established, he can flag lower-performing employees. Higher-performing employees may warrant bonuses.

3. Nine's Wine Vineyard makes a selection of wines that goes through independent taste testing every year. In the file **Wine**, the different varieties of wine are ranked by taste and number of bottles sold. Use the data to help the Nine's Wine Vineyard understand the relationship between its wine and sales.

The independent variable is the taste ranking and our variable of interest is the dependent variable, number of bottles sold.

(a) Construct a scatter plot.

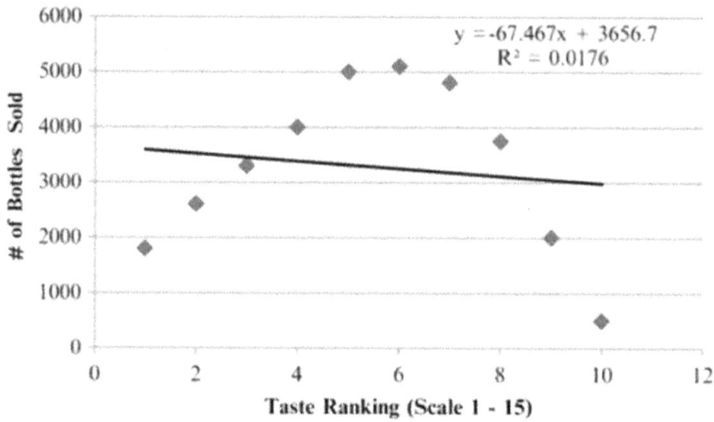

(b) Determine the coefficient of determination, R^2, and interpret its meaning.
Answer: $R^2 = 0.017$
Therefore, only about 2% of bottles of wine sales can be explained by the taste ranking.

(c) How would you use this regression model for predicting sales?
This regression model is a poor predictor of sales because it is not a linear relationship, but a curvilinear relationship. **You should not use this model for any type of prediction.**

4. A company that holds the streaming service distribution rights to movies previously released only in theaters has the business objective of developing estimates of the sales revenue of streaming services. Toward this goal, a company analyst plans to use box office gross to predict streaming service sales revenue. For 22 movies, the analyst collects the box office gross (in \$millions) in the year that they were released and the streaming service revenue (in \$millions) in the following year. For these data,

Gross	Revenue
109.92	81.60
177.02	107.54
166.82	44.41
93.78	60.21
106.66	62.82
90.22	58.81
85.54	48.79
66.70	38.78
87.34	44.67
101.40	34.88
33.26	31.05
144.13	33.14
60.15	27.12
46.69	28.16
60.73	26.43

(continued)

Gross	Revenue
56.75	26.10
79.30	42.04
40.68	25.03
39.32	21.45
41.80	17.51
39.26	21.08
17.74	18.78

(a) Construct a scatter plot.

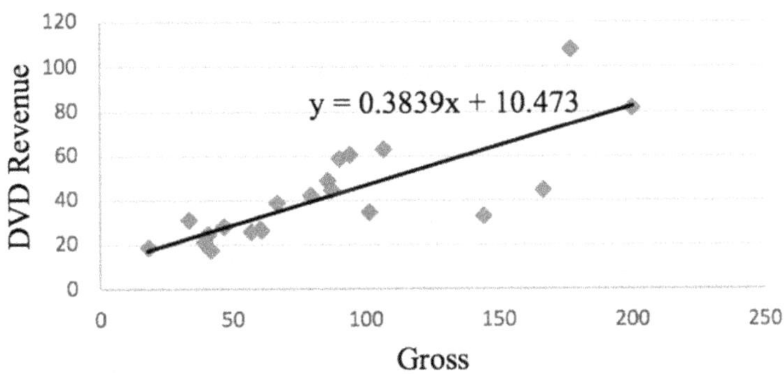

(b) Assuming a linear relationship, use the least-squares method to determine the slope and y-intercept.

Answer: $b = 10.473$, $m = 0.3839$

(c) Interpret its meaning of the slope in this problem.

For each increase of 1 million dollars of box office gross, the predicted DVD revenue is estimated to increase by \$0.3839 million.

(d) Predict the sales revenue for a movie DVD that had a box office gross of \$75 million.

Answer: $y = mx + b$

$y = 0.3839(75) + 10.473 = \$39.2658$ million.

5. In the previous problem, you used box office gross to predict DVD revenue. Using the results of that problem,

(a) Determine the coefficient of determination, r^2, and interpret its meaning.

Answer: $r^2 = 0.5452$

54.52% of the variation in DVD revenue can be explained by the variation in box office gross.

(b) How useful do you think this regression model is for predicting DVD revenue?

Based on (a), the model is useful for predicting DVD revenue.

(c) Can you think of other variables that might explain the variation in DVD revenue?

Other variables that might explain the variation in DVD revenue could be the amount spent on advertising, the timing of the release of the DVDs, and the type of movie.

6. The following results show the x values, residuals, and a residual plot from a regression analysis:

X	Residuals
1	0.70
2	1.58
3	1.03
4	0.33
5	-0.39
6	-0.67
7	-0.56
8	-0.65
9	-1.19
10	-0.84
11	-0.29
12	-1.28
13	-0.21
14	-0.37
15	0.22
16	-0.16
17	0.82

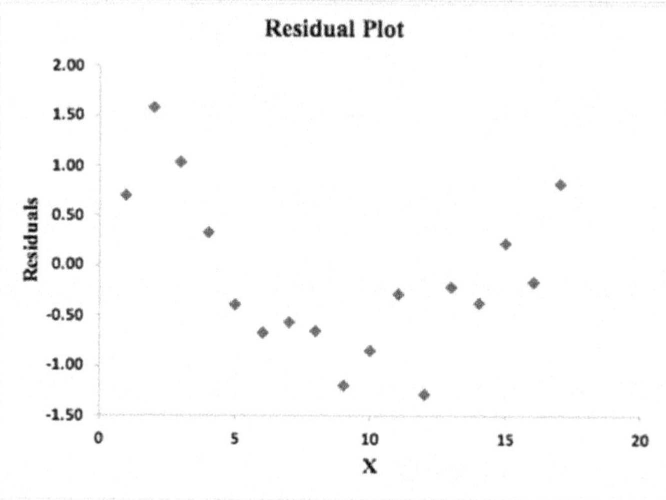

Is there any evidence of a pattern in the residuals?

A residual analysis of the data indicates a pattern, with sizable clusters of consecutive residuals that are either all positive or all negative. This pattern indicates a violation of the assumption of linearity. A curvilinear model should be investigated.

7. A mail-order catalog business that sells personal computer supplies, software, and hardware maintains a centralized warehouse for the distribution of products ordered. Management is currently examining the process of distribution from the warehouse and has the business objective of determining the factors that affect warehouse distribution costs. Currently, a small handling fee is added to the order, regardless of the amount of the order. Data that indicates the warehouse distribution costs and the number of orders received has been collected over the past 24 months. The results are

Months	Distribution cost ($ Thousands)	Number of orders
1	52.95	4015
2	71.66	3806
3	85.58	5309
4	63.69	4262
5	72.81	4296
6	68.44	4097
7	52.46	3213
8	70.77	4809
9	82.03	5237

(continued)

Months	Distribution cost ($ Thousands)	Number of orders
10	74.39	4732
11	76.84	4413
12	54.06	2921
13	62.98	3977
14	72.3	4428
15	58.99	3964
16	79.38	4582
17	94.44	5582
18	59.74	3450
19	90.5	5079
20	93.24	5735
21	69.33	4269
22	53.71	3708
23	89.18	5387
24	66.8	4365

(a) Assuming a linear relationship, use the least-squares method to find the regression equation.

Answer: Highlight column 2 and column 3. Insert scattergram. Right click on any data point. Add trendline and R^2 to graph

$y = 52.378x + 660.46$

(b) Predict the monthly warehouse distribution costs when the number of orders is 4500.

Answer:

$y = 660.46 + 52.378x = 660.46 + 52.378(4500) = 2366{,}361.46$ or 2366362

(c) Plot the residuals versus the time period.

There is no evidence of a pattern in the residuals over time.

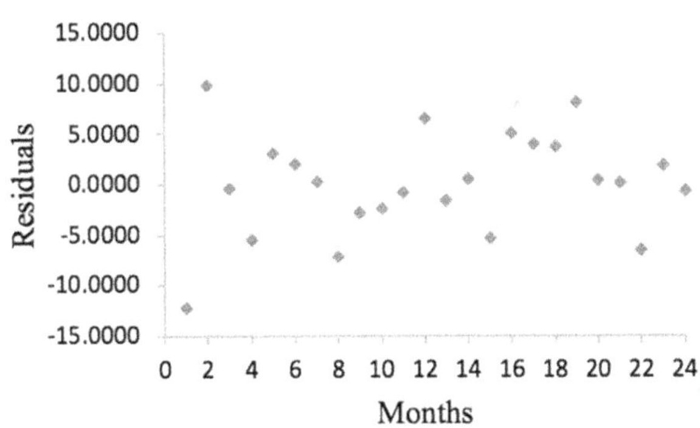

(d) Based on the results of (c), is there reason to question the validity of the model?

Based on a residual analysis, the model appears to be adequate.

8. This graph is part of a report that you need to review. What data is missing before you are able to fully assess these results?

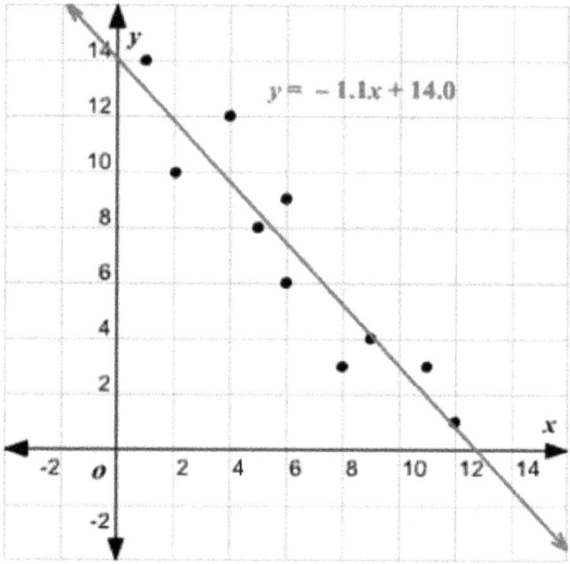

- The graph is missing a title.
- Both x and y axes are missing labels and units.
- You need to R sq value to assess how well the data fits the regression model.
- You need to see the residual plots to check for linearity.
- You need significance statistic (p value) for the x variable (needs to be less than 0.05).
- Ask if all the data were included—any outliers that were removed? If so, explain.

Chapter 11
Significance Tests Part 1

Linda Herkenhoff and John Fogli

Key Concepts

Alternate hypothesis, Chi-square test, Confidence level, Critical values, Cronbach's alpha, Degrees of freedom, F-test, Hypothesis, Null hypothesis, Precision, Reliability, Statistical significance, t-test, Validity, X^2-test, and z-test.

Discussion

The next two chapters help you test your sample results to ensure they are real and not due to chance. Remember none of these tests are necessary if you have access to the entire population. In Part 1 of this chapter, we will discuss the ***F*-test** and the ***t*-test**. In Part 2 Chap. 12, we will discuss the **X^2-test** and the ***z*-test**. Note there are many more exotic significance tests, beyond the basic four tests presented here.

Every test of significance examines whether the observed difference is real or just due to chance variation. We use **hypotheses** or statements of expected outcome. There are **null hypotheses** and **alternative hypotheses**. Each hypothesis represents one side of the discussion. The **null hypothesis** (**H_0**) expresses the idea that an observed difference is due to chance. The **alternative hypothesis** (**H_α**) expresses

Supplementary Information The online version contains supplementary material available at https://doi.org/10.1007/978-3-031-46371-6_11.

L. Herkenhoff (✉)
Saint Mary's College, Moraga, CA, USA

J. Fogli
Diablo Valley College, Pleasant Hill, CA, USA

L. Herkenhoff, J. Fogli (eds.), *Applied Statistics for Business and Management using Microsoft Excel*, https://doi.org/10.1007/978-3-031-46371-6_11

the idea that an observed difference is real and not due to chance. The **null hypothesis** is the one that is always tested. The **alternative hypothesis** is set up as the opposite of the **null hypothesis**. So if the **null hypothesis** is not true, we can accept the **alternative hypothesis**.

Remember **statistical significance** does not mean "important." **Statistical significance** is measuring the probability that our results are real and not due to chance error. The significance level is a critical probability in choosing between the null hypothesis and the alternative hypothesis.

Basic Concepts

Degrees of freedom is the number of values in the final calculation of a statistic that are free to vary; calculated by the number of terms minus one. In other words if you have five variables in a relationship and you select values for four of them, there is no freedom in choosing the fifth. The fifth can only be one value. In this case the number of degrees of freedom would be four.

Validity indicates the extent to which our test is measuring what it is supposed to be measuring. For example, many critics claim that although IQ tests provide consistent results they don't really measure intelligence. They argue that IQ tests have a low **validity**.

Reliability refers to the consistency with which results occur. Since classical IQ tests provide consistent results we say they have a high **reliability**.

The following analogy depicts both of these concepts. A customer in a pub is given a set of six darts to throw at a dartboard on the nearby wall. However, the dartboard has a white sheet covering it and the area around it. The customer has a general idea where the target is and proceeds to throw the six darts. Figure 11.1 shows the pattern of dart hits on the sheet from the first customer.

Second and third customers also participate. Their results are shown in Figs. 11.2 and 11.3.

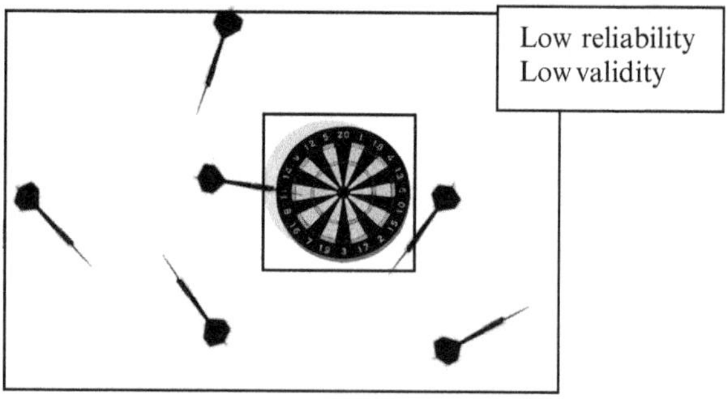

Fig. 11.1 Customer 1 dart hits

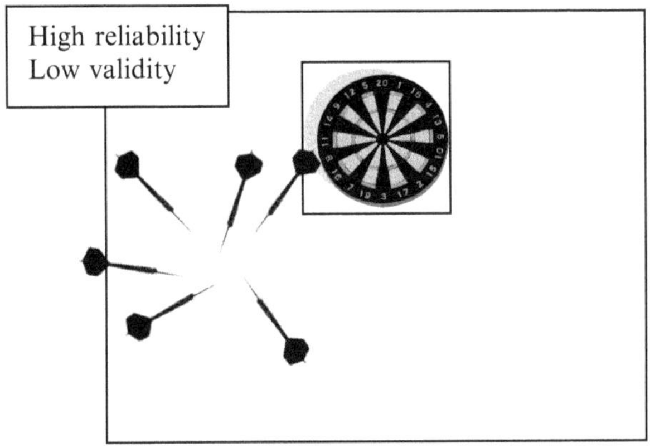

Fig. 11.2 Customer 2 dart hits

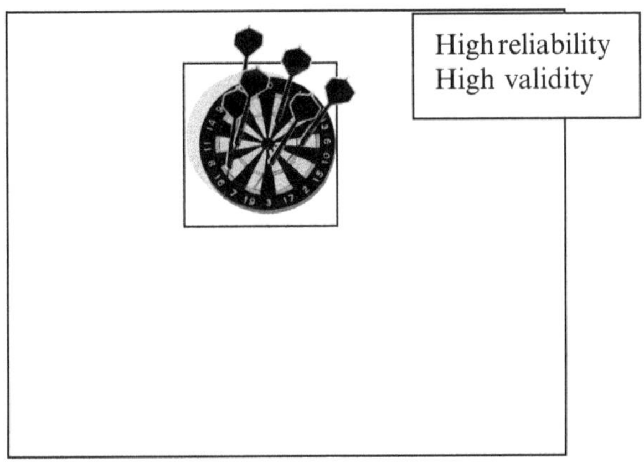

Fig. 11.3 Customer 3 dart hits

Customer 2 and Customer 3 demonstrate high **reliability**. Dart after dart lands in nearly the same place. In other words, you can count on these results being generated no matter how many darts are thrown. In significance testing terms, you can count on these results being generated no matter how many samples are chosen.

Proximity to the target is the analogy of **validity** as indicated by Customer 3's results. Close grouping (high reliability) of the dart hits doesn't prove too useful, unless they hit the desired target. In significance testing the desired target is "hitting" the real population statistic. We can also think about **reliability** in terms of **precision**. Therefore, any **standard error** implies an estimation of **reliability**.

To summarize these relationships, see Fig. 11.4.

One of the most common **reliability** measures is **Cronbach's alpha** (**α** or **R_α** or **r_α**). This **reliability** coefficient can hold values from −1 to +1. The generally agreed upon lower limit is ±0.70, although it may be much less in exploratory research. We typically think of R_α in absolute terms.

Fig. 11.4 Validity and reliability

If you have...	You can have...
High validity	High reliability only
Low validity	High or low reliability
Low reliability	Low validity only
High reliability	High or low validity

Validity can be measured empirically by the correlation between defined sets of variables. The three most widely accepted forms of validity are **convergent**, **discriminant**, and **nomological validity**. **Convergent validity** assesses the degree to which two measures of the same concept are correlated. **Discriminant validity** is the degree to which two conceptually similar ideas are distinct. The empirical test is correlation among measurements, but this time the summated scale (combining several individual variables into a single composite measurement) is correlated with a similar but conceptually distinct scale. The correlation should be low demonstrating the two scales are sufficiently different. Finally, the **nomological validity** refers to the degree that the summated scale makes accurate predictions of other concepts. These types of validity can be assessed in several different ways such as multi-trial, multi-method matrices, and structural equation modeling. However, these various methods of validity assessment are beyond the scope of this book.

Assessing Reliability

The reliability of a measurement procedure is the technical term for its consistency. It is well known that no measurement procedure—whether it is the use of a bathroom scale or the administration of an achievement test—is perfectly consistent. If you were to step on and off your bathroom scale five times in a row, chances are you would observe three or four different measurements of your weight. Depending on the quality of the scale or whether you placed your feet in exactly the same spot on the scale, the scale's indication of your weight might fluctuate by three or four pounds across the five measurements. A more reliable bathroom scale will generate more consistent measurements of weight. If the scale were to indicate exactly the same weight for you every time you stepped on and off (a highly unlikely occurrence), it would be totally consistent. We would then say that it was perfectly "reliable" *even though it might indicate the wrong weight every time* and thus give you an invalid weight. (In the *everyday* sense of "reliable" we would include validity too, but in the technical sense validity is a completely separate issue.)

When you observe the weight indicated by your bathroom scale you are reading what measurement theorists call an **observed score**. This observed score can be thought of as consisting of two parts: one is your true weight (the **true score**) and the other is the result of a variety of factors that have nothing to do with your true weight (the **error component**). These other factors might include the particular position of your feet on the scale (step too far forward and the scale might read low; step too far backward it might read high), the fact that you placed the scale on a thick rug instead of on the floor, and the tendency of the scale's springs to fatigue/recover depending on the rates at which you climb on and off the scale. In every measurement procedure, regardless of what is being measured or how it is being measured, the observed score that results from taking a measurement is equal to the sum of the true score that results from taking a measurement and the error component. The reliability of the measurement procedure depends on the relative sizes of the true score and the error component. The larger the error component, the lower the reliability; the smaller the error component, the higher the reliability.

Measurement reliability is usually expressed as an index that can take on values between 0 and 1 much like a correlation coefficient. However, unlike a correlation coefficient, the reliability of a measurement procedure can never be a negative number. A reliability of 0 means that the observed scores consist entirely of error components. A reliability of 1 would mean that the observed scores consist entirely of true scores, however in reality the observed scores are composed of both true scores and error components. Furthermore, observed scores often have their reliabilities as less than 1.

Another statistic that is closely associated with reliability is called the **standard error of measurement**. The standard error of measurement is equal to the *standard deviation of the error components* in a set of observed scores. If the reliability is equal to one, there are no error components and the standard error of measurement is equal to zero. If the reliability is equal to zero, the observed scores consist totally of error components and the standard error of measurement is equal to the standard deviation of observed scores.

In many research/evaluation reports that include assessments of measurement quality, standard errors of measurement are provided together with estimates of the reliabilities of measurement procedures. These standard errors of measurement can be interpreted by assuming that errors follow a normal distribution as is commonly the case. For example, if a 50-item test were administered to a group of students and the standard error of measurement were reported to equal two points, you could draw the following conclusions from your knowledge of the normal distribution: sixty-eight percent of the individuals tested would have true scores that were within one standard error of measurement (two points) of their observed scores. Ninety-five percent of the individuals tested would have true scores that were within two standard errors of measurement (four points) of their observed scores. Alternatively, suppose a given individual had an observed score of 30 on the test. You could then say that the chances were 68% that this person's true score was between 28 (one standard error of measurement below 30) and 32 (one standard error of measurement above 30). Also the chances would be 95% that this person's true score was between 26 and 34.

Although reliable measurement procedures are always desirable in research/evaluation, your need for a particularly high level of reliability might depend on the nature of your research since high reliability may be expensive and unnecessary. Of course, reliability is not at stake when a true score is changing on its own. For example, if you were measuring weights using a bathroom scale, you would want the measurement procedure to provide consistent results at a particular time on a particular day. But, you wouldn't be distressed if the scale showed different weights for people on Friday than it did on Monday. Since individuals' true weights change over a period of 5 days, the scale's indications of their weights should also change.

In contrast, suppose that you administered an IQ test to a group of adults on Monday and then administered the same test to them on Friday. Ignoring the possibility that they might recall the items and hence do better because they've had practice, you would expect the intelligence levels of these individuals to be exactly the same on Friday as they were on Monday. In this example, reliable measurement implies consistency over a five-day period. Consistency over time is known as **stability** of measurement and should be high for stable variables.

A variety of methods have been developed for assessing the reliability of measurement procedures. Some of these procedures indicate consistency of measurement at a given time and others indicate stability of measurement across a period of time. Whether a particular method for assessing reliability is appropriate depends on the purposes of the research or evaluation study in which it is used. We will now describe several of the most widely used methods of assessing reliability and discuss the situations in which they are most appropriate.

Suppose you were interested in assessing the reliability of an IQ test in the sense of providing stable measurements across a two-week period. One way you could handle this problem would be to administer the IQ test to a sample selected from the group of individuals you intend to use it with and then administer the test to these same individuals 2 weeks later. You could then calculate the correlation coefficient between the scores resulting from the first administration and the scores resulting from the second. The correlation coefficient you obtained through this procedure would be called a **reliability coefficient**. The higher the correlation coefficient, the higher the reliability of the IQ test. The logic underlying the **test-retest method** of assessing reliability is as follows: if the measurement procedure is reliable, the person who scored highest on the first administration of the test should score at or very near the top on the second administration. The person who scored second highest on the first administration should score high on the second administration, and so on. In short with reliable measurement, people ought to retain their relative scores from one administration of the test to the next. The greater the extent to which this occurs, the higher the reliability coefficient.

The test-retest method of assessing reliability has at least two potential flaws. If it is used with a measurement instrument like an IQ test, there is always the chance that examinees will either remember the items and do better on them the second time around or will remember the answers they gave to specific test items on the first administration and give exactly the same answers on the second administration of the test. This kind of *memory effect* has nothing to do with the *true* reliability of the test, but will tend to increase the correlation between scores obtained on the two administrations. Therefore, the *indicated* reliability of the test when using the correlation between scores at its measurement will be spuriously increased. To handle this problem, publishers of commercially available tests often create several forms of their tests so that one form can be administered initially and a second form, containing totally different items, can be used for the second administration. Test forms that are designed to be interchangeable are known as parallel forms and the use of parallel test forms in a test-retest reliability study is known as the **parallel forms method** of assessing reliability.

The parallel forms method of assessing reliability has advantages beyond its elimination of potential memory effects. In most testing situations, a researcher or evaluator is interested in determining an examinee's performance on an entire domain of skill, rather than the few specific items on one test. A single test merely provides an estimate of how the examinee would perform by sampling from that domain. By using two parallel test forms in assessing reliability, the researcher can get a better estimate of the examinee's performance on the entire domain of items simply because more items are used. In other words, the parallel forms method of reliability estimation includes an important error component, namely the error due to selection of a specific relatively small sample of test items. The test-retest method using a single test form ignores this error component and might therefore provide an unrealistically high estimate of a test's reliability.

Both the test-retest and parallel forms methods of assessing reliability require that a group of individuals be measured twice. This is costly, time-consuming, and inconvenient. It is often difficult, if not impossible, to assemble the same group more than once. A number of reliability assessment methods that require data to be collected on only one occasion have been developed in response to these problems. We will describe the methods that are used most often.

Reliability assessment methods that require data to be collected on only one occasion are called **internal consistency methods**. Although these techniques differ in methodological detail, each of them indicates the degree to which the components of a measurement procedure tend to assess the same underlying variable or construct. When applied to a test, these methods show the extent to which the items on the test tend to measure the same thing.

An easy way to increase reliability is by using a longer test such as giving both parallel forms at one sitting. If you wanted to measure a baseball player's hitting ability, you would get a more reliable measurement of this variable by calculating the average percent of times at bat that the player got a hit in both games of a double-header rather than by calculating the percentage for a single game. In this example, the two baseball games are analogous to two sets of test items such as two parallel forms. You will obtain far more reliable information about an examiner's performance from twice as many test items administered at one time as you will from half as many. In general, the greater the number of measurements involved in a measurement procedure, the more reliable the procedure. So the longer a test—up to the point where examinee fatigue sets in—the more reliable it will be, provided the added measurements assess the same variable as the initial ones. If you were trying to assess a ballplayer's hitting ability, you would not increase measurement reliability by counting the number of times he or she caught a fly ball. Fielding skill doesn't correlate highly with batting skill. The added measurements would have to assess hitting ability methods and indicate whether, and to what degree, all of the measurements used in a procedure assess the same variable.

One of the simplest internal consistency methods for assessing reliability is called the **split-half method** or the **odd-even method**. In this procedure, we make two tests out of one. Thus, a group of examinees is given a test only once, but two scores are calculated for each examinee: one based on the odd-numbered items in the test and the other based on the even-numbered items. These two sets of scores are then correlated to determine whether examinees whose score highest on the odd-numbered items also tend to score highest on the even-numbered items. This correlation coefficient is called the **split-half reliability coefficient**. The higher the split-half reliability coefficient, the higher the consistency is for the items on the test (in the sense that the odd-numbered items measure the same variable as the even-numbered items).

Because the reliability of a measurement procedure increases as the number of measurements is increased, the split-half reliability coefficient will tend to underestimate reliability. For example, if it is applied to a 50-item test then the split-half method really indicates the reliability of two shorter 25-item tests (composed of the odd items and the even items respectively). To estimate the reliability of the full 50-item test it is necessary to adjust the split-half reliability coefficient using a formula known as the **Spearman-Brown prophecy formula**. When you see a reliability coefficient reported as a Spearman-Brown coefficient you can be sure that the author has used the split-half method for assessing reliability and then applied the adjustment formula.

The split-half method of assessing reliability, like all internal consistency methods, will not show the stability of a measurement procedure over time. Nor will it reflect errors that arise from peculiarities in the administration of a measurement procedure on a particular occasion. A split-half reliability coefficient, when adjusted using the Spearman-Brown prophecy formula, might therefore overestimate the true reliability of a measurement procedure.

Two widely used alternatives to the split-half method of estimating reliability are named after the authors of the research paper in which they were first proposed. Both methods can be used only with procedures composed of a series of measurements that are scored 1 for a correct response and 0 for an incorrect response. These procedures are known as the **Kuder-Richadson Formula 20 method** and the **Kuder-Richadson Formula 21 method**. You will often find them abbreviated as KR-20 and KR-21. The KR-21 formula is sometimes used instead of KR-20 because it is simpler and easier to compute. However, it makes the assumption that each of the measurements that compose a procedure (or each of the items that compose a test) is equally difficult. This assumption is often violated in practice resulting in an overestimation of true reliability.

The Kuder-Richardson methods of assessing reliability have all of the advantages and all of the pitfalls of the split-half method. They are sensitive to the same kinds of measurement errors and ignore the same error components as well. They should never be used to assess stability of measurement, situations where the conditions of measurement are difficult to control, and where speed of response is a major factor in performance. In any of these applications they are likely to give a falsely inflated estimate of true measurement reliability.

It is not necessary to use the Spearman-Brown prophecy formula to adjust the reliability estimates obtained through use of the Kuder-Richardson procedures. Unlike the split-half method which requires a correction to be performed with the Spearman-Brown prophecy formula. The correction for reduced test length is built into the KR formulas.

A final method for assessing internal consistency reliability was named after its developer, Lee J. Cronbach. It is known as Cronbach's alpha. Cronbach's procedure like the other internal consistency methods uses measurement data collected on a single occasion. Cronbach's alpha is a generalization of Kuder-Richardson Formula 20, in that the test items (or other components of the measurement procedure) don't have to be scored either 0 or 1. This improvement makes it possible to apply Cronbach's method to many measurement procedures other than tests including attitude instruments in which each item requires a response on a five-point scale that might range "Strongly Disagree" to "Strongly Agree."

In recent years, variants of Cronbach's alpha have been used in more sophisticated studies of reliability known as **generalizability studies**. In these studies, the components of measurement error—such as those due to instability and those due to peculiarities of measurement conditions—are analyzed separately. For example, in a study which has five judges that were used to estimate the work attitudes of a group of employees on three different occasions in two different work settings, generalizability methods would make it possible to determine how much the reliability was affected by disagreement between judges. Furthermore, generalizability methods would determine how much it was affected by changes across time or by inconsistency across work situations. This kind of information could then be used to design a more reliable measurement procedure.

Assessing Validity

When we collect data on a person for the purpose of research in education or the social sciences, we usually measure or observe a *sample* of that person's behavior. We might provide the person with a task to be completed or some set of stimulus materials that require a response. A traditional pencil-and-paper test is a good example. Alternatively, we might just observe the person in some naturally occurring situation and count the incidences of some behavior or action. In recent years, a significant amount of educational research has included observations of students' or teachers' behaviors in classrooms. For example, a researcher might count the number of times in a 15-minute period that a teacher asks his or her students a direct question.

If research merely involved reporting the number of test items a person answered correctly or the number of times a teacher asked his or her students a question, there would be few validity issues. Validity concerns arise because researchers interpret these observable behaviors as an indicator of a person's status on a more general construct such as "ability" or "interest". By generalizing beyond observed behaviors that occur at a particular time and in a particular setting, researchers/evaluators create a multitude of validity issues. The most fundamental concern is the appropriateness of the interpretations of the data. Measurement procedures are not *inherently* valid nor inherently invalid. It is the *interpretations* of measurement results that must be examined for validity.

The nature of measurement validity has been debated for decades and many issues are still far from settled. Entire texts have been devoted to the theory of measurement with regards to the nature and assessment of measurement validity. Our discussion here will be relatively brief and concerned more with practical interpretation than with theoretical underpinnings.

Classical measurement books usually identify four types of evidence pertaining to measurement validity: **construct validity** evidence, **concurrent validity** evidence, **predictive validity** evidence, and **content validity** evidence. Recent contributors to measurement literature, such as Messick (1989), have suggested that content validity evidence is really not concerned with measurement validation and that the other three distinctions are about as confusing as they are helpful. Nevertheless, we think that the four types of evidence form a useful framework for critical reading of many research and evaluation reports. We will therefore discuss each one.

Construct validity evidence is simultaneously the most fundamental and the most complex of the four types of evidence we have mentioned. Messick suggests that it encompasses all of validity and subsumes the other three types. Construct validity evidence is concerned with the total relationship between the results of a particular measurement or set of measurements and the underlying construct that the researcher/evaluator is attempting to measure. In short, it addresses the question "Does this measurement procedure really measure the construct it is claimed to measure?". For example, the question "Does this IQ test really measure intelligence?" is asking about construct validity. As you might imagine, the question is very difficult to answer and a conclusive answer is rarely found. All a researcher can do is to try and amass a balance of evidence in support of the claim that the IQ test measures intelligence. Construct validity is never "proved"; it is merely accepted as long as supporting evidence strongly outweighs contrary evidence. A construct validation study is designed to determine whether a measurement procedure yields results that are consistent with theoretical expectations. For example, the theory of intelligence suggests that older children are on average more intelligent than younger children. If an IQ test is truly a measurement of intelligence, older children should therefore earn a higher average score on that test than younger children—and indeed, on most IQ tests they do. Another theoretical expectation is that more intelligent children on average achieve better grades in school than less intelligent children. The correlation between children's scores on the IQ test and their grade point averages in school should therefore be positive. If the correlation is positive, the construct validity claim that the IQ test measures intelligence is somewhat supported; if the correlation is close to zero or is negative, the claim to construct validity would have suffered a near-fatal blow.

No single piece of evidence is sufficient to establish the construct validity of a measurement procedure. A series of studies that examines every reasonable alternative interpretation of the generated data is necessary before a construct validity claim is adequately supported.

When you read research and evaluation reports that contain discussions involving constructs it is essential that you question the validity of the author's claims. If authors state that one group had more positive attitudes than another, you must ask how these constructs were measured and critically question the author's evidence that their procedures actually measured attitude and motivation.

At this point, you're probably wondering how you can possibly know whether the construct validity of a measurement procedure has been adequately demonstrated. Unfortunately, there is no simple answer. In any study you can quickly note whether the author has even addressed construct validity issues by the presence of any explicit discussion of the problem. When a researcher has used a commercially available measurement instrument, such as a test or an attitude scale, you can confirm the validity of that tool by checking such sources as the Buros Institute's *Mental Measurement Yearbooks*, the *Journal of Educational Measurement*, or *Educational and Psychological Measurement* for professional reviews. As in other aspects of research and evaluation, healthy skepticism on the part of the reader is fully warranted.

When a researcher has developed a new measurement procedure, they might provide evidence of its validity by amassing **concurrent validity** evidence. For example, the Torrance Test of Creativity is one of the most widely accepted measures of what psychologists call "creativity" or "divergent thinking". Suppose that in an evaluation of an educational program designed to foster creativity in children you develop your own procedure for measuring "creativity". In order to provide evidence for this claim you administer your measurement procedure *and* the Torrance Test of Creativity to a large sample of students who have participated in the new educational program. You then compute that the correlation coefficient equals 0.90, suggesting that students who earn relatively high scores on your procedure also earn relatively high scores on the Torrance Test. Furthermore, the students who earn relatively low scores on your measurement also earn relatively low scores on the Torrance Test. Since the Torrance Test is an accepted measurement of creativity you can legitimately reason that your procedure also measures that construct.

The essential features of this example are common to all concurrent validation studies. Such studies involve an accepted measurement of some construct and a new as of yet invalidated measurement. The concurrent validity of the new measurement is estimated by administering it together with the accepted measurement to an appropriate group of subjects. The correlation coefficient between scores on the new measurement and scores on the accepted measurement is known as a **validity coefficient**. The closer the validity coefficient is to 1.00, the higher the concurrent validity of the new measurement. In fact the time interval between the two administrations need only be short enough that no systematic changes in the examinee's behavior are likely to take place. Concurrent validation is often part of a more general validation construct for a new measurement procedure. It should seem reasonable that a new measurement of a given construct should give results that covary positively with the older accepted measurements of that construct. High concurrent validity coefficients thus provide supporting evidence of construct validity.

The primary objective of many applied studies is the *prediction* of subjects' future performance or behavior. For example, a professional polling organization, such as the Gallup Poll, might survey registered voters several months prior to an election in an attempt to predict the election outcome. Such predictions are effective only to the degree that voters' responses to the survey questions indicate their future voting behavior. In an educational context, the most selective colleges require that their applicants for admission complete an aptitude test such as the *Scholastic Aptitude Tests* (SAT) or the *American College Testing Program Tests* (ACT). Colleges then use the applicant's test scores to predict their future college grade point average (GPA) and admit only those applicants who are likely to exceed some minimum GPA.

In studies such as these, the correlation coefficient between the measurement used for prediction (such as the score on the voters' survey questionnaire) and the measurement used to indicate future performance (such as actual voting behavior at the time of the election) is called a **predictive validity coefficient**. A high predictive validity coefficent means that the measurement is a more valid predictor and in this instance predicts actual voting behavior.

Predictive validity evidence is of particular importance in situations involving selection of applicants, placement of applicants into appropriate positions (such as occupational specialties within the military or in industry), and marketing studies that attempt to predict consumer behavior. As is true for all types of validation, a measurement procedure does not have any *inherent* predictive validity. Its predictive validity might be high when predicting some criterion variables and low when predicting others. It might also be high for some groups of individuals and low for others. For example, it might be the case that people who say they intend to vote Democratic really do so a month later whereas people who say they intend to vote Independent often change their minds. In that case, the predictive validity of a survey questionnaire would be high for Democrats, but low for Independents.

When interpreting a research study that involves predictive evidence of validity, make sure that the criterion variable and the group used to collect such evidence are relevant to your own interests and applications. Otherwise, you should regard evidence of predictive validity as being tentative for your own purposes.

The need for **content validity evidence** arises frequently in educational research and evaluation studies, but also in psychiatry and market/political research. In such studies, researchers or evaluators often want to generalize from an individual's (often a student's) performance on a particular test to that individual's ability to solve test problems in some larger **content domain**. For example, suppose that a test contains 20 items (a problem on a test is called an **item**) in which a student is shown four words. In each item, three of the words are spelled correctly and one is spelled incorrectly. The student must identify the word that is spelled incorrectly. In most uses of such a test, the researcher or evaluator would not be interested in the student's ability to discriminate between the *particular* 60 correctly spelled words and the *particular* 20 incorrectly spelled words that make up that test. What is really of interest is the student's ability to discriminate between *all* correctly or incorrectly spelled words that are like the words used to make up the test. Researchers and evaluators routinely generalize students' hypothetical performances on all possible similar test items from their performances on a few specific test items. The content validity issue focuses on exactly these generalizations. To be content valid, the items on the test must constitute a **representative sample** of the domain of items used to generalize the students' performances. For example, if all the words used in the 20 spelling items consist of two syllables and begin with a vowel, it would clearly be unreasonable to generalize the students' performances on all the words in *Webster's New World Dictionary*. The test items would not constitute a representative sample of the words in the dictionary.

Unlike construct validation evidence, concurrent validation evidence, and predictive validation evidence (all of which are substantially statistical [e.g., correlation coefficients]), content validation evidence is solely judgmental. The fundamental issue in content validation is whether the content of the test is consistent with the content of the domain of items/tasks for which the generalization is claimed to hold. In pursuing this fundamental question, one might ask whether each type of item for which representativeness is claimed is actually represented in the test. One might also ask whether each type of item is present in the test in *proportion* to its presence in the domain of content used for generalization. For example, if a mathematics test is claimed to measure students' abilities to solve computational problems involving the four basic arithmetic operations (addition, subtraction, multiplication, and division) then that test should not consist of 80% addition items, 10% subtraction items, 5% division items. The four types of items would normally make up approximately equal proportions of the test.

A good way to begin analyzing the content validity of a commercially available test is to obtain the test publisher's technical manual for the test. More reputable test publishers provide a **test blueprint** or a **table of specifications** that shows how many items on the test were designed to measure each of the different types of behavior that the test is intended assess. Comparison of the test blueprint with a desired domain of content is a useful beginning in assessing a test's content validity for particular application.

Significance testing requires you to clearly define what you are trying to prove and then test if it is in fact real (statistically significant) versus just due to chance. For example, suppose you are trying to prove employee training is effective by comparing pre- and post-training test scores using a sample of employees. If you are using the population data rather than a sample, your observed results are obvious. However, when you use sample data you must test if your results are real or simply due to chance. In other words, if you took another sample would you get similar results?

In this process you must define the level of risk you are willing to take in, assuming your results are real. The only way to be 100% sure is to use the entire population of data. We typically define the risk using a **Cronbach's alpha (α) value**. The most common α-value is 0.05, which corresponds to a 95% confidence level. This means that if you selected 100 different samples from this population you would get similar results in 95 of them. Graphically a 95% confidence level ($\alpha = 0.05$) is depicted in Fig. 11.5.

But let's say that is too much risk and you want to be 99% sure so that in 99 out of 100 samples from the population you get similar results. You would then use an α-value of 0.01.

One way to increase the confidence level and thereby decrease the risk is to use larger and larger samples. Remember to have no risk and thus a 100% confidence level, your sample size would need to be the entire population of data. As sample

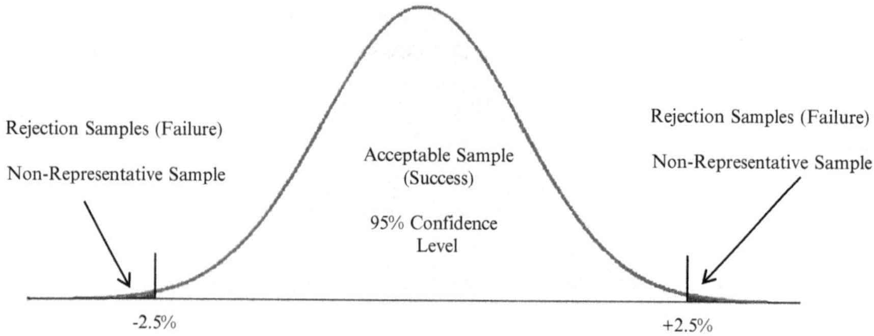

Fig. 11.5 95% confidence level

size increases costs (time and money) may also increase and access to the entire population may be difficult with large sample sizes.

The next step is to establish the relationship between $\boldsymbol{\alpha}$ and the **hypothesis**. We always test for the **null hypothesis**. The significance level is a critical probability when choosing between the **null hypothesis** and the **alternative hypothesis**. The level of significance determines if the probability level is considered too low to warrant support of the **null hypothesis**. Here is an example:

$\boldsymbol{H_0}$**:** The training has no effect. The average test score before training ($\overline{x}_1$) is really the same as the average score after training ($\overline{x}_2$).

$\boldsymbol{H_a}$**:** The test scores before ($\overline{x}_1$) and after training ($\overline{x}_2$) are statistically significantly different within our defined confidence level.

The alpha (α) value defines the tail of the probability distribution. Remember that the model is set up around the null hypothesis. Let's relate the ***p*-value** Excel calculates from the sample to the model we set up in Fig. 11.5. After calculating the ***p*-value** based on the sample data, we compare that with the pre-selected α-value (risk level). If the calculated ***p*-value** is less than our selected α we accept our results are real and not due to chance. Thus, we reject the null hypothesis.

There are many ways to phrase rejecting the null hypothesis as summarized in Fig. 11.6. Likewise Fig. 11.7 summarizes the results for the terminology associated with a 95% confidence level.

The ***p*-value** is the fractional area for the right tail of the ***F*-distribution** above the $\boldsymbol{F}_{\textbf{calculated}}$. $\boldsymbol{F}_{\textbf{calculated}}$ is indicated as simply an $\boldsymbol{F}$ in the Excel output. When the ***p*-value** $< \alpha$ you are fundamentally decreasing the area in the tail and therefore increasing the acceptance region of H_a. On the other hand, when the ***p*-value** $> \alpha$ you are increasing the area in the tail, therefore decreasing the acceptance area of H_a. To prove our results are real we want to minimize the rejection (tail) regions of the probability distribution.

The other test for statistical significance uses the same concepts as with the ***p*-value** comparison, but compares a **test statistic *critical*** with a **test statistic *calculated***. These test statistics are selected based on the type of problem you are trying

- We reject the null hypothesis (H_o)
- Our p-value falls in the H_o rejection zone
- Results we obtained are real
- Results are not due to chance
- We accept the alternate hypothesis (H_a)
- Our p-value falls in the H_a acceptance zone
- Results are statistically significant within our selected confidence interval

Fig. 11.6 Summary of the terminology for **statistically significant** results

- Alpha (α) value of 0.05
- 2 tail values of 2.5% each
- 95 out of 100 samples from the same population will give you similar results
- Range of ±1.96 standard deviations from the mean
- 5/100 samples from the same population will not give you similar results
- H_a acceptance zone of 95%; rejection zone of 5%
- H_o acceptance zone of 5%; rejection zone of 95%
- Significance level is 95%
- p-value of 0.05
- Region of success 95%
- Region of failure 5%

Fig. 11.7 Summary of terminology for **95% confidence level**

to solve. Since our significance tests using the test statistics are based on relative positions between the critical value and the calculated value, it is recommended you work with the absolute values to avoid confusion. The rule to reject H_0 and establish statistical significance is given when the Excel calculated statistic is greater than the critical value of the statistic. So for example with the *F*-test

$$|F\text{calculated}| > |F\text{critical}|$$

Many reports include both types of tests since Excel often provides data to calculate both in the same output. However, the completion of only one of the analyses is sufficient to establish statistical significance.

- **Caution:** If you do run both types of tests and your results are not consistent, go back and check your input is correct and that you have run the Excel tools properly. Mathematically one should always get similar results with the same input data.

The central area in Fig. 11.8 indicates the acceptable range of data for a 95% confidence interval. The acceptable region corresponds to a *z*-score of ±1.96.

Remember if 95% is in the central area then 5% lies in the tails, but you have two tails so each tail must only have 2.5% of the data. The corresponding confidence level is 95% (0.95) with an alpha (α) level of 5% (0.05). This means that in the long run the probability of making an erroneous decision, when H_0 is true, will be fewer than five times in 100. You can think of the alpha areas (or the tails) as the failure region (or the rejection region). The values that lie on the boundary of this rejection region are referred to as **critical values**.

Choosing the Appropriate Significance Test

Just because something is statistically significant doesn't mean that action should be taken in the direction indicated by the test. Other factors such as the need of the application, economics, safety, etc. should be taken into account. Remember significance tests only say that it's actually true.

- Never allow statistics to overrule common sense.

The choice of the best technique depends on the type of questions to be answered, the number of variables, the measurement scale, the sample size, and how much is known about the population. It is important to determine if the solution requires a one-tailed or two-tailed analysis.

One-Tailed Tests

In this situation we test whether one mean is higher than the other. We are only interested in one side of the probability distribution which is illustrated in Fig. 11.9.

In this distribution the shaded region shows the "success region" or where one mean is statistically significantly greater than the other (where the alternative hypothesis is true). The unshaded area represents the "failure region" where the first

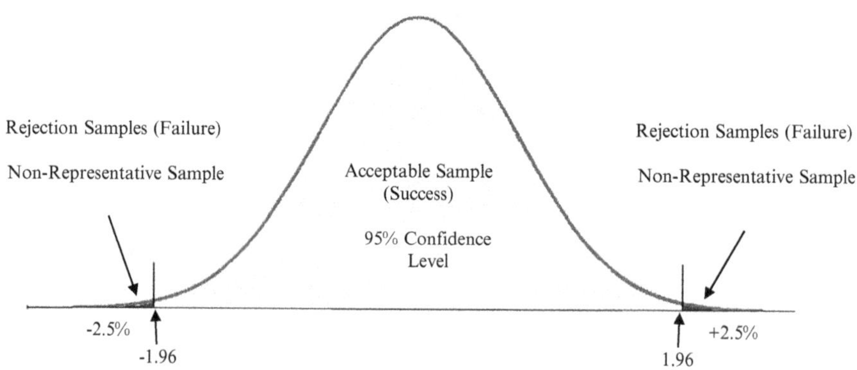

Fig. 11.8 Acceptable/rejection areas as a % and as a *z*-score in a normal distribution

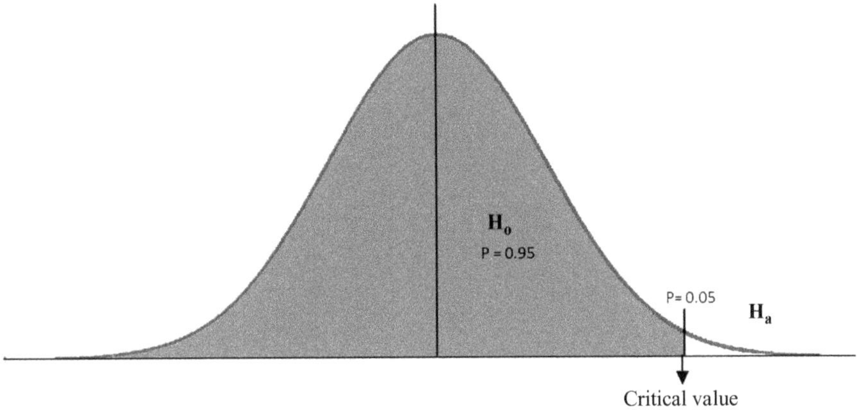

Fig. 11.9 One-tailed significance test

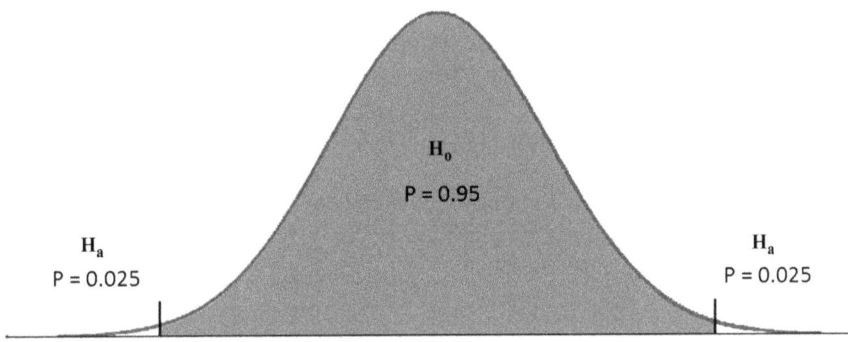

Fig. 11.10 Two-tailed test for 95% confidence level

mean is not greater than the second mean. In other words, where the alternative hypothesis holds true. Because we are only interested in one side of the distribution, or one "tail", this type of test is called a **one-sided** or a **one-tailed test**.

Two-Tailed Tests

A two-tailed test is used to determine if two means are different from one another. As an example, let's assume that we want to check if the pH of a stream has changed significantly in the past year. A useful (although not technically accurate) way to think about *p*-values is to think of them as the probability that the null hypothesis is true. Using this perspective we generally believe the null hypothesis is true unless there is compelling evidence to think otherwise. When there is a low probability that the null hypothesis is true (the *p*-value is less than 0.05), we conclude that the null hypothesis is probably not valid and therefore the alternative hypothesis is probably true. Figure 11.10 illustrates a region of acceptance (or success as 95%) for a two-tailed test.

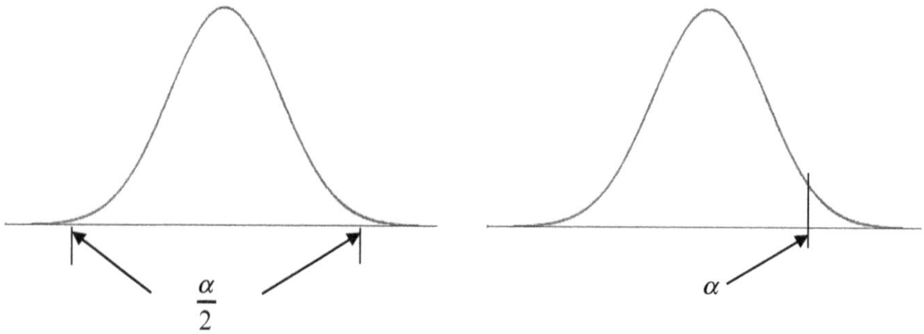

Fig. 11.11 One-tailed versus two-tailed test significance tests

The rejection or failure region is 0.025 × 2 = 0.05. In other words, this could represent a 95% confidence interval. When we run our two-tailed significance test we need to have a calculated p-value that is less than 0.025 to be able to achieve statistical significance. Because we are interested in both sides of the distribution, or two "tails", this type of test is called a **two-sided** or a **two-tailed test**.

In summary, the problem will dictate how many tails are appropriate to include in the significance testing. Figure 11.11 compares the key difference in the design of the two- versus one-tailed tests.

Figure 11.12 summarizes the four basic significance tests that are explained in this book. The F-test and t-test are presented in this chapter while the next two tests are discussed in Chap. 12.

Significance Tests

The next two sections will define the ***F*-test** and the ***t*-test** with examples. The Excel tools will then be discussed for each of these significance tests.

F-*Test*

The F-test can be used effectively under several conditions:

- **Example 1 (One-Way Repeated Measures Using ANOVA):** In this case, each subject is measured several times under different conditions and the results are measured. The repeated measurement ANOVA (Analysis of Variance) ***F*-test** can be used to assess whether any of the conditions are on average different versus the null hypothesis (all the conditions yield the same mean response).

Test	When to Use	One tailed	Two tailed
F-test (Chapter 11)	Repeated measure using ANOVA. Same respondents under different conditions	x	
	Overall regression model significance	NA	NA
	Equality of 2 group means	x	x
	Between Group ANOVA. Test several variables within your group of data	x	x
t-test (Chapter 11)	Equality of means when population std dev unknown. Can use with small samples	x	x
	Paired samples. Before and after analysis with same respondents		x
	Regression. Tests each regression co-efficient	NA	NA
Chi-squared (Chapter 12)	Compares the tallies or counts of categorical responses between two (or more) independent groups	x	x
	Goodness of Fit for model	NA	NA
	Independence of variables	NA	NA
z-test (for normally distributed data) (Chapter 12)	testing the mean of a population versus a standard	x	x
	Comparing the means of two populations, with large ($n \geq 30$) samples	x	x
	Testing the proportion of some characteristic versus a standard proportion	x	x
	Comparing the proportions of two populations	x	x
	Matched pair test. Comparing the means before and after something is done to the samples. Sample should be large ($n \geq 30$)	NA	NA

Fig. 11.12 Selecting the correct significance test

- **Example 2 (Regression Problems):** This is an important test when we run regression models. It is most often used when fitting statistical models to a dataset in order to identify the model that best fits the population from which the data was sampled.
- **Example 3 (*F*-Test for Equality of Two Variances):** The ***F*-test** can also be used to test whether two population variances are equal. It does this by comparing the ratio of the two variances.

- **Example 4 (Between Group ANOVA):** ***F*-tests** can be used to test for the significance of group means and is found within the ANOVA tool in excel.

Excel provides *F*-test analysis within the **Regression** and **ANOVA** tools. Additional Excel tools include the ***F*-test** and the ***F*-test Two-Sample for Variance**.

Rejection Rule

$$\text{Reject } H_0 \text{ if } F_{\text{calculated}} > F_{\text{critical}} \text{ or } p - \text{value}_{\text{calculated}} > p - \text{value}_{\alpha}$$

The **critical value of *F* (F_{critical})** can be thought of as the largest value to occur by chance for the given degrees of freedom. If the $F_{\text{calculated}}$ is larger than F_{critical}, then the null hypothesis of equivalent group means is rejected. In other words, your results are not simply due to chance and are statistically significant. These statistics are depicted in Fig. 11.13.

The ***p*-value** is the probability that a value of ***F*** greater than or equal to the $F_{\text{calculated}}$ could have occurred by chance if there were no difference in the means. The ***p*-value** is the fractional area of the right tail for the ***F*-distribution** above the $F_{\text{calculated}}$. $F_{\text{calculated}}$ is indicated as simply a ***F*** in the Excel output. When the ***p*-value** is less than your chosen confidence level you have statistically significant results so keep going!

Basic Descriptions of F-Test Applications

This first section will describe the general application of each problem type by way of an example and the second section will walk through the step-by-step Excel applications.

Example 1: One-Way Repeated Measures Using ANOVA

The ***F*-test**, in one-way analysis, is used to assess whether the expected values of a quantitative variable within a sample differs from the other sample. In this case, each subject is measured several times under different conditions. For example,

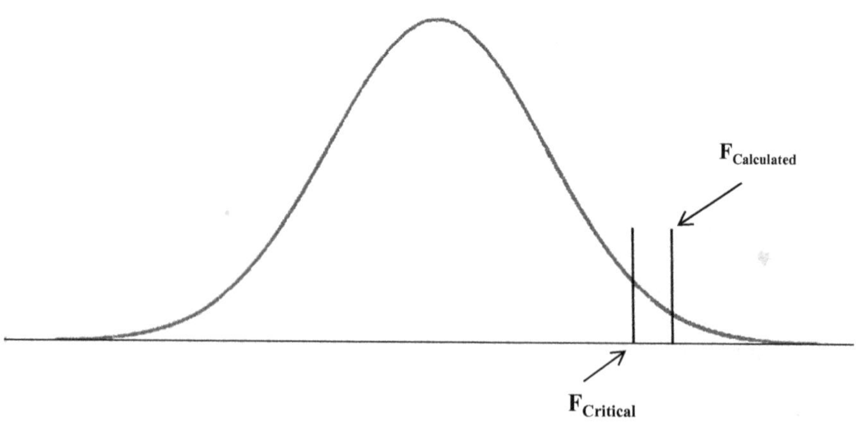

Fig. 11.13 Acceptance of H_a when $F_{\text{calculated}}$ is greater F_{critical}

suppose that a medical trial compares four treatments. The subjects are given each of the four treatments and the results are measured. The ANOVA ***F*-test** can be used to assess whether any of the treatments is on average superior or inferior to the others versus the null hypothesis that all four treatments yield the same mean response. This is an example of an "omnibus" test, meaning that a single test is performed to detect if there are any possible differences that are real differences, not just due to chance. Alternatively, we could carry out pairwise tests among the treatments. For instance in the medical trial example with four treatments, we could carry out six tests among pairs of treatments.

The advantage of the ANOVA ***F*-test** is that we do not need to pre-specify which treatments are to be compared and we do not need to adjust for making multiple comparisons. The disadvantage of the ANOVA ***F*-test** is that if we reject the null hypothesis, we do not know which treatments can be said to be significantly different from the others. Furthermore, if the ***F*-test** is performed at level α we cannot state that the treatment pair with the greatest mean difference is significantly different at level α. We just know there is some real difference in the samples.

Example 2: Regression Problems

This tests the significance of the overall regression model. The overall model is defined by

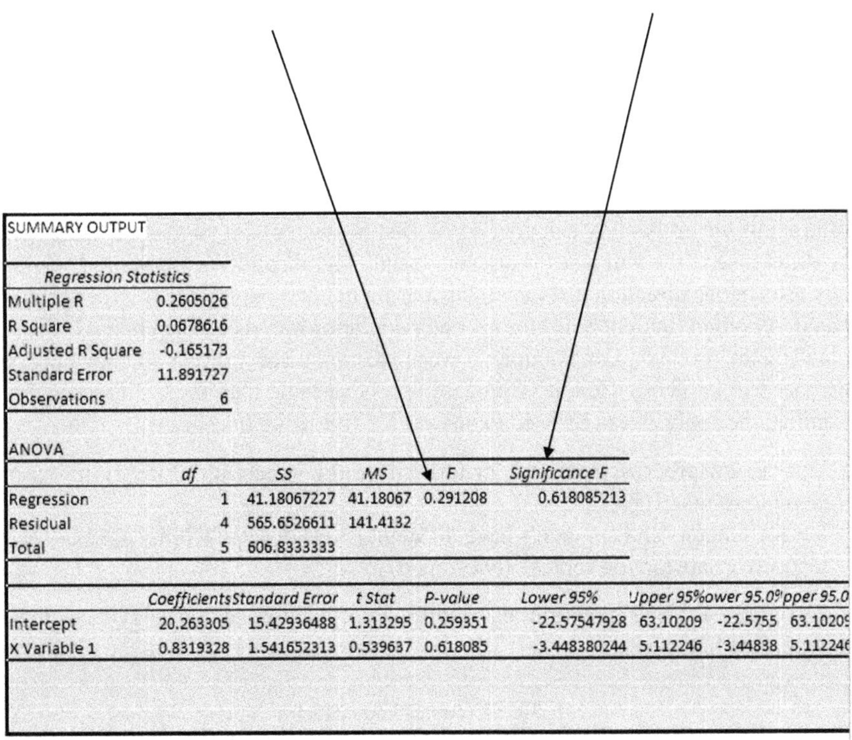

SUMMARY OUTPUT

Regression Statistics	
Multiple R	0.2605026
R Square	0.0678616
Adjusted R Square	-0.165173
Standard Error	11.891727
Observations	6

ANOVA

	df	*SS*	*MS*	*F*	*Significance F*
Regression	1	41.18067227	41.18067	0.291208	0.618085213
Residual	4	565.6526611	141.4132		
Total	5	606.8333333			

	Coefficients	*Standard Error*	*t Stat*	*P-value*	*Lower 95%*	*Upper 95%*	*ower 95.0*'	*pper 95.0*
Intercept	20.263305	15.42936488	1.313295	0.259351	-22.57547928	63.10209	-22.5755	63.1020
X Variable 1	0.8319328	1.541652313	0.539637	0.618085	-3.448380244	5.112246	-3.44838	5.11224

Fig. 11.14 Summary output

$$y = b_1x_1 + b_2x_2 + b_3x_3 + \ldots + b_0$$

In Fig. 11.14, we have included only one independent variable for simplicity in the explanation of significance. Under the ANOVA heading (Analysis of Variance), Excel does not provide the $\boldsymbol{F}_{\textbf{critical}}$ so we can only use the **Significance *F*-value**. The **Significance *F*-value** is really just the ***p*-value**, the probability associated with the $\boldsymbol{F}_{\textbf{calculated}}$ of 0.291208. In other words, there is a 0.618 probability of getting an *F*-value greater than 0.29. This is where the confidence level or the alpha value comes in. Select the desired level for risk or alpha value. Let's select a 95% confidence level, so we have an alpha of 0.05. Now, compare the *p*-value with the selected alpha value. Since a ***p*-value** (**Significance *F***) of 0.618 is greater than 0.05, we do not have statistically significant results at the 95% confidence level. Our regression model will not yield useful results the way it is currently set up. So back to the drawing board and rethink the model.

Example 3: F-Test for Equality of Two Variances

The associated Excel tool is the ***F*-test Two-Sample for Variances** located in the **Data Analysis** tab. For example, you can use this ***F*-test** tool on samples of productivity for two corporate departments. The tool provides the results for a test of the null hypothesis that these two samples come from distributions with equal variances versus the alternative hypothesis that the variances are not equal in the underlying distributions. In simple terms, we confirm or reject the idea that the difference in the results for productivity of these two departments is indeed real and not just due to chance. A value of *F* close to 1 provides evidence that the underlying population variances are equal which is usually not what we want to hear.

This test can be a two-tailed or a one-tailed *F*-test. The two-tailed version returns the two-tailed probability that the variances in the two samples are not significantly different. If the probability calculated in **F.Test** is greater than your alpha value then the difference in the variances is not statistically significant. The one-tailed version only tests in one direction that the variance from the first population is either greater than or less than (but not both) the second population variance. The choice is determined by the problem. For example, if we are testing a new process we may only be interested in knowing if the new process is less variable than the old process. For example, the analysis can be used to answer the following questions:

- Does a new process, treatment, or test reduce the variability of the current process? (one-tailed)
- Do accountants and engineers have different levels of satisfaction based on their respective satisfaction scores? (two-tailed)

The Excel **F.TEST** function also returns two-tailed results. The syntax of the Excel **F.Test** function is:

$$\textbf{F.TEST}(\textbf{array1,array2})$$

where the supplied **array1** and **array2** are two samples of data values.

Example 4: Between Group ANOVA

You may find the need to analyze more than one independent variable or independent variables that have more than two levels. The ANOVA tools which use the ***F*-statistic** allow for analysis in the differences between the group means.

- **One-Way Between Group ANOVA**—the subject is only measured once. For example, you might have data on tire performance for many tire types. You have an overall performance rating and you have data on performance in foul weather conditions. You are interested in seeing if foul weather performance is related to the overall performance rating. ANOVA allows you to break up the group according to the overall performance and then see if foul weather performance is indeed significantly different across overall performance numbers.
- **Two-Way Between Group ANOVA**—the subject is only measured once but you can analyze two factors simultaneously. For example, when you are interested in the role of gender and age on consumer perception rating of truck commercials during the Super Bowl.

Excel

Excel examples will be provided for each of the four *F*-tests problem types:

- Example 1: *F*-Test One-Way Repeated Measures Using ANOVA
- Example 2: Regression
- Example 3: *F*-Test for Equality of Two Variances
- Example 4: Between Group ANOVA

Example 1: One-Way Repeated Measures Using ANOVA

Six office assistants are selected to help choose the best replacement computers for the support staff. The computers have various screen sizes and different styles of keyboards. We need to select the computer that allows the subjects the highest level of accuracy in word documents, so their proficiency scores are recorded for each of the three computers. We need to find the highest scoring computer but we also need to make sure these results are significant at the 0.05 confidence level. Enter the data from Fig. 11.15.

	A	B	C	D	E
1	Subject	computer 1	computer2	computer3	
2	1	19	8	21	
3	2	18	10	31	
4	3	25	10	26	
5	4	20	18	28	
6	5	17	7	14	
7	6	21	16	24	

Fig. 11.15 Input

- Click on the **Data** tab

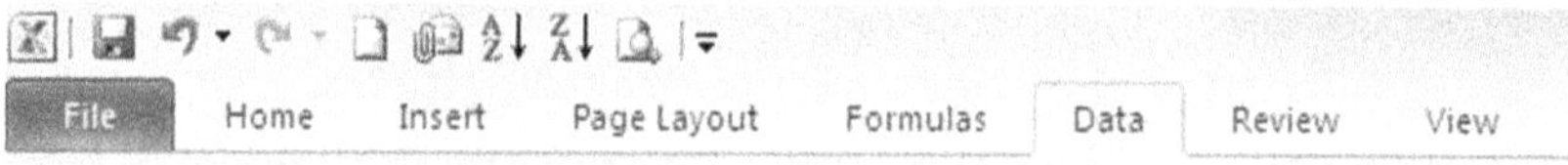

- Click on the **Data Analysis** function

- From the menu select **ANOVA: Two-Factor Without Replication**

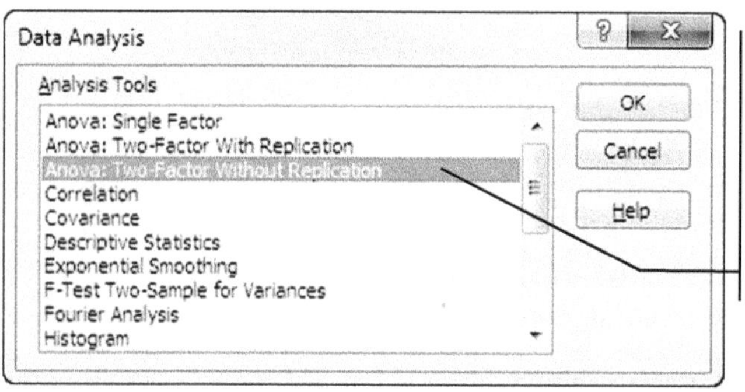

Note: "Without Replication" refers to n=1 subject per cell. "Two-Factor" refers to the subjects and computers.

- Click **OK**
- Click inside the **Input Range** box and highlight the data input area
  - Make sure you include the labels on the rows and the columns to keep track of the results

- Check the **Labels** box

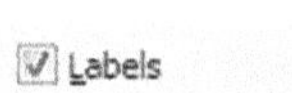

- Input an alpha value of **0.05**

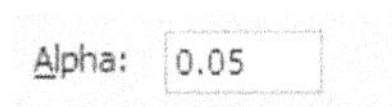

- Select the **Output Range**, click inside box and highlight the cell where you want the output printed

- Click **OK**

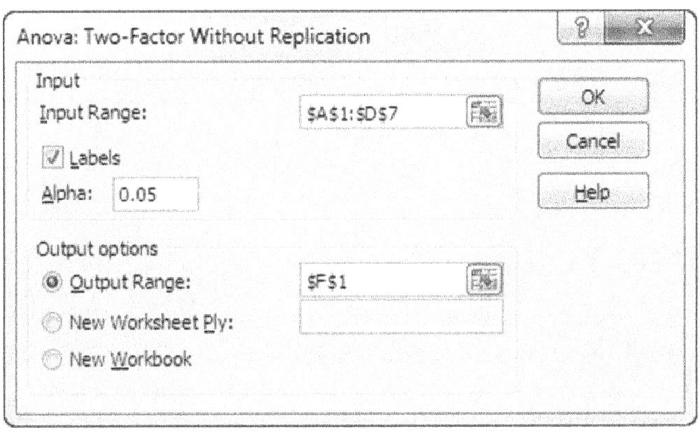

The output as shown in Fig. 11.16 provides summary information about the subjects and the proficiency scores.

The **Count** should be the same as the number of conditions the subject was tested under. In this case each subject tried three computers. In the bottom of this column the number of times each computer was used (6) is provided.

The **Sum** represents the sum of each person's score on all three computers. At the bottom of this column is the sum of all three different people's scores on this same computer. The total score is by computer not person here.

The **Average** represents the mean scores by office assistant and by computer.

The **Variance** represents the variance in the scores by office assistant and by computer.

In the **ANOVA** table we are interested in two sets of results in the **Columns** row for significance testing. The **Columns** row refers to the different computer outcomes which are what we are interested in understanding.

Anova: Two-Factor Without Replication

SUMMARY	*Count*	*Sum*	*Average*	*Variance*
1	3	48	16	49
2	3	59	19.6667	112.3333
3	3	61	20.3333	80.3333
4	3	66	22	28
5	3	38	12.6667	26.3333
6	3	61	20.3333	16.3333
Computer 1	6	120	20	8
Computer 2	6	69	11.5	19.9
Computer 3	6	144	24	35.6

ANOVA

Source of Variation	*SS*	*df*	*MS*	*F*	*P-value*	*F crit*
Rows	181.8333	5	36.3667	2.6806	0.0866	3.3258
Columns	489	2	244.5	18.0221	0.0005	4.1028
Error	135.6667	10	13.5667			
Total	806.5	17				

Fig. 11.16 Output

We can test for significance in two ways:

1. We can compare the ***p*-value** (.0005) for **Columns** with the alpha value we selected (0.05). Since the ***p*-value** is less than alpha, we can confirm that the difference in the computers scores is real at the 95% confidence level.
2. We can compare the ***F*** (18.0221) with the $\boldsymbol{F}_{\mathbf{crit}}$ (4.1028). Since the ***F*** ($\boldsymbol{F}_{\mathbf{calculated}}$ is shown as ***F*** in the ANOVA table) is greater than the $\boldsymbol{F}_{\mathbf{crit}}$, we can conclude our results are in fact statistically significant at the 95% confidence level and not just due to chance.

- If you do run both types of tests and your results are not consistent, go back and check your input is correct and you have run the Excel tools properly. Mathematically speaking one should always get similar results with the same input data.

Our department should select Computer 1 as it had the high proficiency scores with low variance. If we had selected another 100 samples of six office assistants from the population we would expect similar results in 95 of those samples.

Example 2: Regression

A sample of office assistants across the San Francisco-Bay Area was created. It was suggested that the data should be analyzed with a regression model to determine if salary was related to years of experience and aptitude test scores. Use a 99% confidence level. This data is shown in Fig. 11.17.

- There should only be one label line in Excel otherwise Excel will read the extra label lines as numbers and give you an error.
- Input the data from Fig. 11.17
- Click the **Data** tab

- Choose the **Data Analysis** function

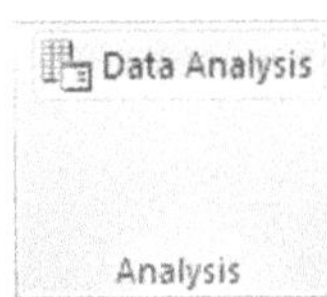

Fig. 11.17 Survey data

Experience(yrs) X_1	Score (out of 100) X_2	Salary ($000) Y
4	78	24
7	100	43
1	86	23.7
5	82	34.3
8	86	35.8
10	84	38
0	75	22.2
1	80	23.1
6	83	30
6	91	33
9	88	38
2	73	26.6
10	75	36.2
5	81	31.6
6	74	29
8	87	34
4	79	30.1
6	94	33.9
3	70	28.9
3	89	30

- Choose **Regression from the menu**

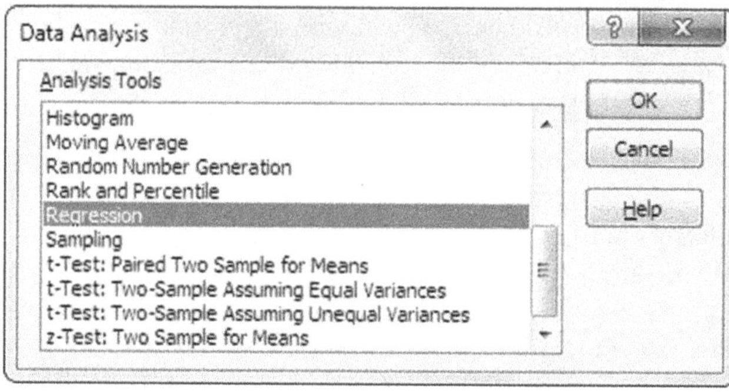

- **Input y Range**

- **Input x Range**

- Select **Labels box**

- Select **Confidence Level** and enter 99 in **Confidence Level** box

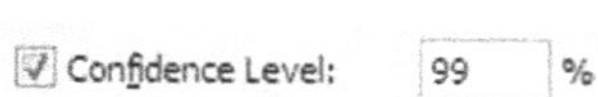

- Select **Output Range** and click inside the **Output Range** box in order to highlight the cell where you want the output printed

- Click **OK**

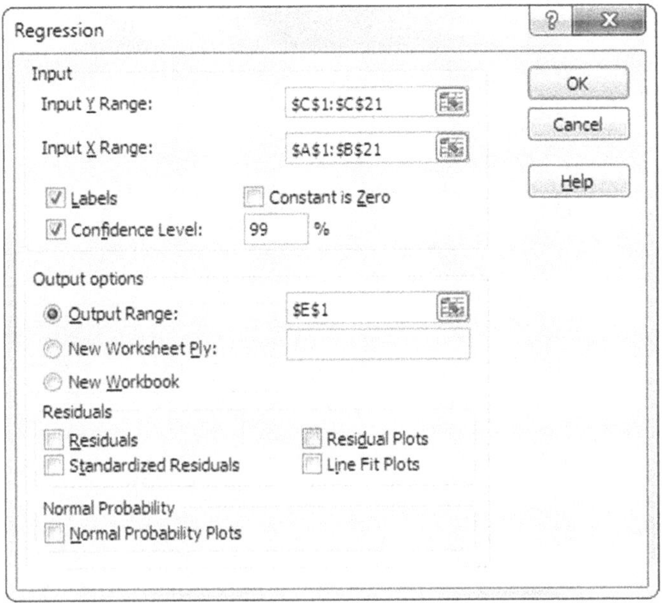

The Excel output is provided in Fig. 11.18. Remember the form of the equation will be

$$\hat{y} = b_0 + b_1 x_1 + b_2 x_2$$

where:

y = annual salary ($000)
x_1 = years of experience
x_2 = score on aptitude test

The resulting equation for this model based on the ANOVA results is

$$y = 3.8651 + 1.4014\ x_1 + 0.2431\ x_2$$

To run significance analysis we compare the **Significance *F*** (p-value) with the alpha value we have selected (0.01). Remember that a confidence level of 99% is the same as an $\boldsymbol{\alpha}$-value of 0.01. The **Significance *F*** is less than 0.01 so the results are significant at the 99% confidence level. In other words, if we selected 100 samples from this population 99 of those analyses would get similar results. We are now

SUMMARY OUTPUT

Regression Statistics	
Multiple R	0.9082
R Square	0.8248
Adjusted R Square	0.8042
Standard Error	2.4782
Observations	20

ANOVA

	df	SS	MS	F	Significance F
Regression	2	491.5994	245.7997	40.0239	0.0000
Residual	17	104.4026	6.1413		
Total	19	596.0020			

	Coefficients	Standard Error	t Stat	P-value	Lower 95%	Upper 95%	Lower 99.0%	Upper 99.0%
Intercept	3.8651	6.3073	0.6128	0.5481	-9.4420	17.1723	-14.4148	22.1451
Experience (yrs)	1.4014	0.2034	6.8883	0.0000	0.9722	1.8306	0.8118	1.9910
Score (out of 100)	0.2431	0.0793	3.0675	0.0070	0.0759	0.4103	0.0134	0.4728

Fig. 11.18 Excel output

confident that the sample data has allowed us to build a regression model that we can use for predictive purposes. Remember this does not tell us anything about the individual x-values, but tells us about the significance of the overall model.

Example 3: Two-Sample for Variances

In this application we can use the one-tailed or two-tailed versions in Excel depending on the model we are trying to test. Let's use the same example but change the confidence level to 95%.

One-Tailed F-Test for Two-Sample for Variances

The one-tailed version only tests in one direction that the mean from the first population is either greater than or less than (but not both) the second population mean. The two-tailed version tests against the alternative that the means are not equal. The choice is determined by the problem. For example, suppose that the two groups of employees report on the time they spend handling customer complaints. The first group has less than 5 years of experience and the second group has at least 15 years of experience. To keep things simple, we will assume that based on the original dataset we have calculated that the two standard deviations are different and now we want to test their significance (group 1 has SD = 1.47 and group 2 has SD = 4.15). However, there is always the possibility that this difference in the standard deviations could have occurred by chance. We need to test if these results are significant at a 95% confidence level. A one-sided ***F*-test** is appropriate in this scenario because

Fig. 11.19 Input

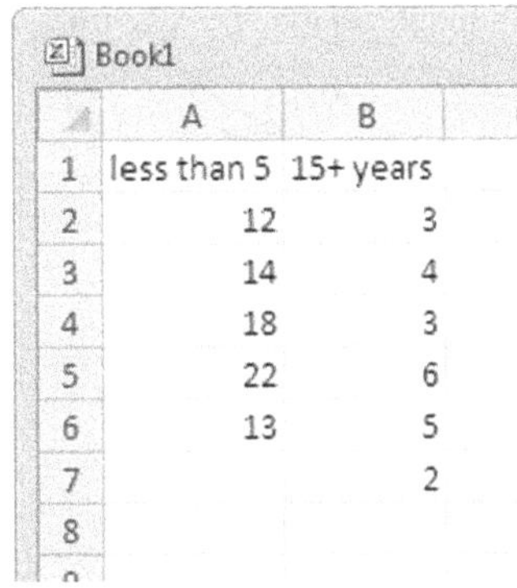

	A	B
1	less than 5	15+ years
2	12	3
3	14	4
4	18	3
5	22	6
6	13	5
7		2
8		

of the claim that one sample standard deviation is "smaller than" another sample standard deviation. Input the data from Fig. 11.19.

- You must put the group with the larger variance in the left-hand column, i.e. variable 1.
- Input the data from Fig. 11.19
- Click on the **Data** tab and select **Data Analysis**
- Select **F-test Two-Sample for Variances** from the drop down menu

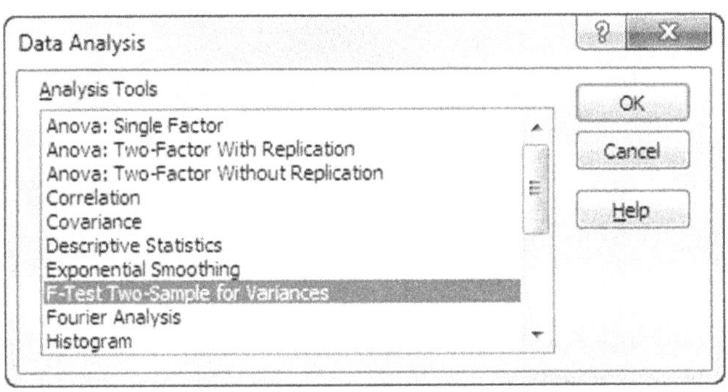

- Click **OK**
- Click inside the **Variable 1 Range box**. Highlight the first column of data including the label

- Note the actual range is A1:A6 but Excel will assume a zero value if no data appears. By including A7 we keep the input field symmetric with the second dataset input field. You will get the same results whichever approach you use.

- Click inside the **Variable 2 Range box**. Highlight the second column of data including the label

- Check the **Labels** box

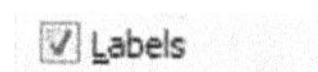

- Select an alpha level of 0.05

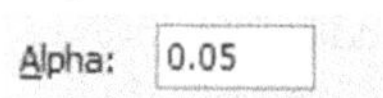

- Click inside the **Output Range box**. Highlight the cell where you want the output

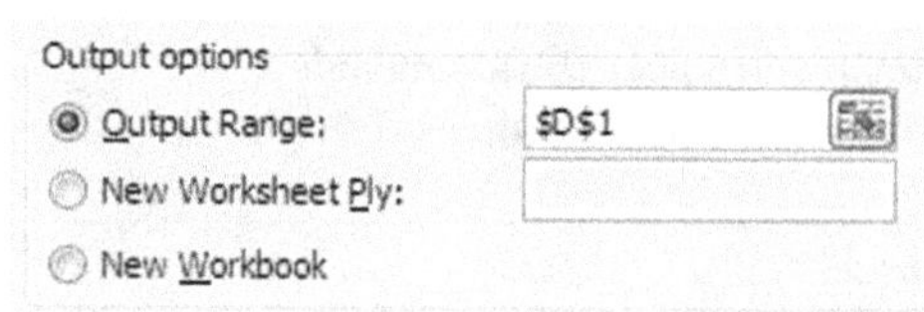

- Click **OK**

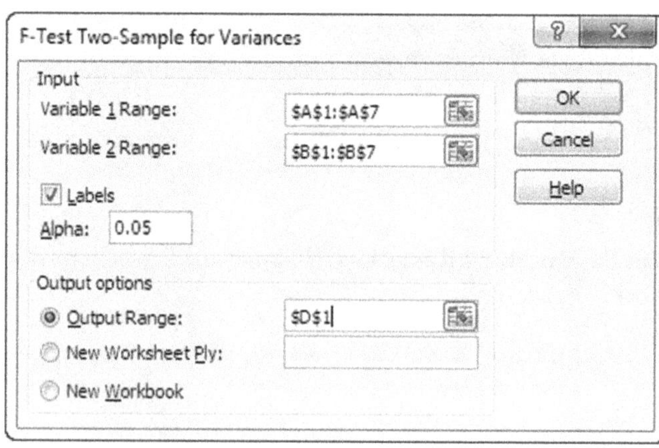

The output data is provided in Fig. 11.20.

Make sure you check all of the data has been input by looking at the **Observations**. The **Observations** indicate we have included all of our data (5,6).

Book1

	D	E	F
1	F-Test Two-Sample for Variances		
2			
3		Less than 5	15+ years
4	Mean	15.8	3.8333
5	Variance	17.2	2.1667
6	Observations	5	6
7	df	4	5
8	F	7.9385	
9	P(F<=f) one-tail	0.0216	
10	F Critical one-tail	5.1922	
11			

Fig. 11.20 Output

The **Mean** and **Variance** are given for each group.

We can use two sets of results for significance testing:

1. We can compare the ***P(F <=f)* one-tail** value (0.0216) with the alpha value we selected (0.05). Since the *p*-value is less than alpha, we can confirm that the difference in the variance is statistically significant at the 95% confidence level.
2. We can compare the ***F*** (7.9385) with the ***F* Critical one-tail** (5.1922). Since the ***F*** is greater than the ***F* Critical one-tail**, we can conclude our results are statistically significant at the 95% confidence level. If we selected 100 other samples from this population, we would get similar results in 95 of those samples.

The variance in satisfaction scores is statistically different so we can assume that the less experienced employees really do have a greater standard deviation than the more experienced employees. In other words, there is a greater variability amongst the members of the less experienced group in terms of how many complaints they handle. However, there is less variability in the results of the more experienced group.

Two-Tailed F-Test for Equality of Two Variances

The **two-tailed *F*-test** determines if one population variance (or standard deviation) is statistically significantly different than another population variance. We have no knowledge nor do we care if one is greater or less than the other, just that they are significantly different.

Two groups of marketing professionals during the past 12 months provided the client satisfaction scores out of 100. Is the variance in these scores real or simply due to chance at the 95% confidence level?

- Because the Excel **F-test** function was written as a one-sided test, three adjustments are required to use it as a two-sided test:

1. *You must put the group with the larger variance in the first column,* i.e. *variable 1. If you forget to do this* ***Fcalc (F)*** *will be less than one and you need to re-enter the data.*
2. *You divide the alpha value in half when you input into Excel.*
3. *The p-value reported by Excel should be doubled to give the correct p-value for the two-sided hypothesis test.*

- **Note:** It is not necessary that both sample sizes be the same.

The two sets of data whose variances are to be compared should be entered in consecutive rows of two columns.

- Input the data from Fig. 11.21

- Click on the **Data** tab and select **Data Analysis**
- Select **F-Test Two-Sample for Variances** from the drop down menu

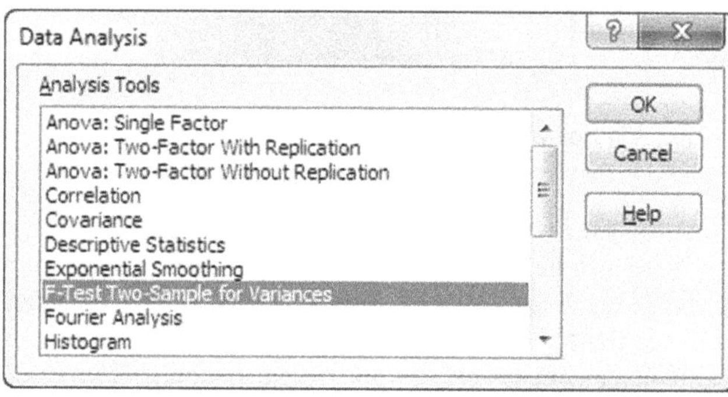

- Click **OK**
- Click inside the **Variable 1 Range box**. Highlight the first column of data including the label

Fig. 11.21 Input

grp 1	grp 2
90.47	90.95
92.02	91.30
93.15	91.48
90.98	92.04
91.73	90.70
	91.33

- Click inside the **Variable 2 Range box**. Highlight the second column of data including the label

- Check the **Labels** box

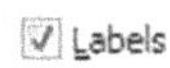

- Select an alpha level of 0.025

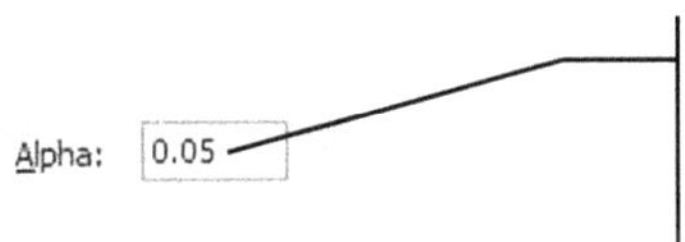

<u>**Note**</u>**:** Because you want a 2-tailed test you need to divide this value (0.05) by 2 and **input 0.025**

- Click inside the **Output Range box**. Highlight the cell where you want the output

- Click **OK**

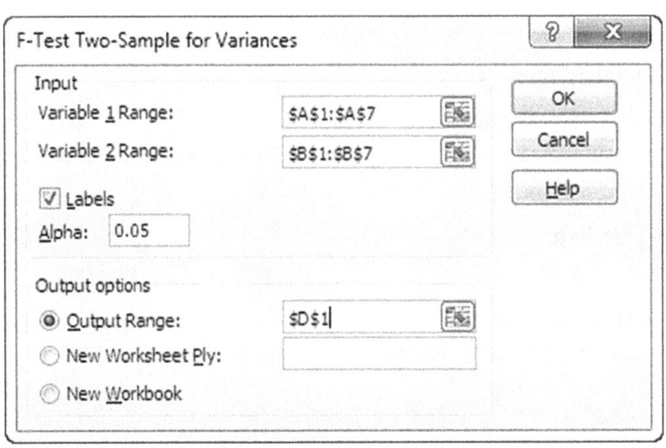

Fig. 11.22 Output

F-Test Two-Sample for Variances

	grp 1	*grp 2*
Mean	91.67	91.3
Variance	1.0582	0.2127
Observations	5	6
df	4	5
F	4.9753	
P(F<=f) one-tail	0.0542	
F Critical one-tail	7.3879	

The output data is presented in Fig. 11.22.

Make sure you check that all of the data has been input by looking at the **Observations**.

The **Mean** and **Variance** are given for each group.

We can use two sets of results for significance testing:

1. Before we can do a comparison, we need to double the ***P*(*F* <=*f*) one-tail** to get 2 × 0.0542 = 0.1084. We can now compare this value with the alpha value we selected (0.05). Since the ***p*-value** is greater than alpha, we confirm that the difference in the variance is not real at the 95% confidence level.
2. We can compare the ***F*** (4.9753) with the ***F* Critical one-tail** (7.3879). Since the ***F* is** less than the ***F* Critical one-tail**, we can conclude our results are not statistically significant at the 95% confidence level. If we selected other samples we would not get similar results.

Example 4: Between Group ANOVA

In this application we can use the one-tailed or two-tailed versions in Excel depending on the model we are trying to test.

One-Tail F-Test Between Group ANOVA

- **Note:** No need to have an equal number of cells for this analysis.

Sixteen customers were assigned to one of three conditions in which they were given truck performance data. The three conditions were:

Fig. 11.23 Input

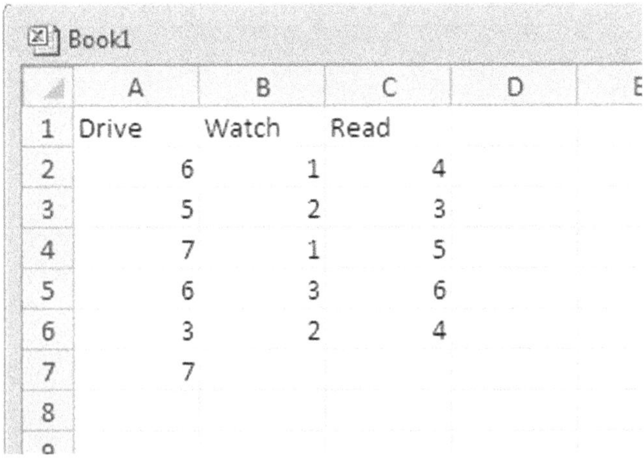

Group 1: They were allowed to test drive the actual truck that was described in the performance data sheet.
Group 2: They watched a video on how these trucks are assembled.
Group 3: They read comparison data on competitor trucks.

The customers were then asked to view a commercial on the truck and rate their perceptions of the truck. They used a 7-point Likert scale with 1 as the lowest score. You are interested in knowing if any of these conditions really influenced the perception rating scores with 95% confidence.

- Input the data from Fig. 11.23

- Click on the **Data** tab and click on **Data Analysis** function
- From the menu select **ANOVA: Single-Factor**

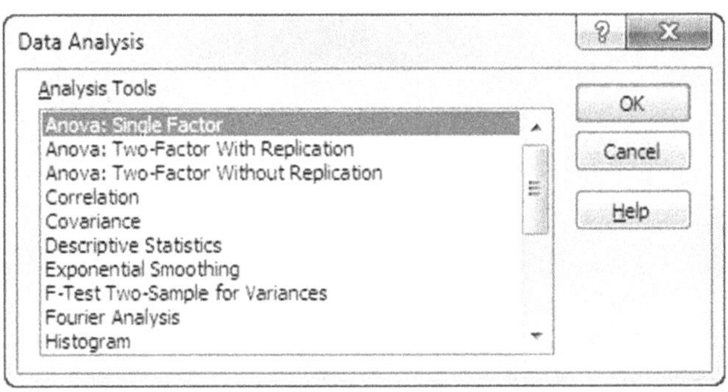

- Click **OK**
- Click inside the **Input Range** box and highlight the data input area. Make sure you include the labels to better keep track of the results

- In the **Grouped by** box select **Columns**

- Check the **Labels in first row** box

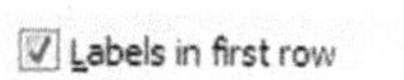

- Input an alpha value of **0.05**. This corresponds to a 95% confidence level

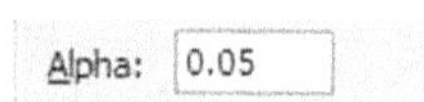

- Select **Output Range** and click inside the **Output Range** box in order to highlight the cell where you want the output printed

- Click **OK**

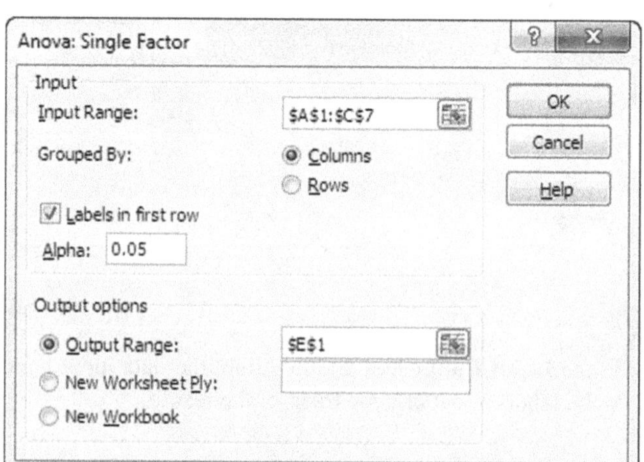

Anova: Single Factor

SUMMARY

Groups	Count	Sum	Average	Variance
Drive	6	34	5.6667	2.2667
Watch	5	9	1.8	0.7
Read	5	22	4.4	1.3

ANOVA

Source of Variation	SS	df	MS	F	P-value	F crit
Between Groups	41.6042	2	20.8021	13.9876	0.0006	3.8056
Within Groups	19.3333	13	1.4872			
Total	60.9375	15				

Fig. 11.24 Output

The results are provided in Fig. 11.24. Make sure the **Count** adds up to the number of participants (16). The Sum, Average, and Variance columns provide the statistics for each condition (category).

The ANOVA table **Between Groups** refers to how differently the three groups ranked the truck.

While the **Within Groups** refers to how differently the customers within each separate group differed from one another.

To determine the significance of these results we can look in two places (Fig. 11.24):

1. We can compare the ***p*-value** (0.0006) with the alpha value we selected (0.05). Since the ***p*-value** is less than alpha, we can confirm that the difference in average scores for each group is real at the 95% confidence level.
2. We can compare the $\boldsymbol{F}$ (13.98761) with the $\boldsymbol{F}_{\textbf{crit}}$ (3.8056). Since the $\boldsymbol{F}$ is greater than the $\boldsymbol{F}_{\textbf{crit}}$, we can conclude our results are in fact statistically significant at the 95% confidence level and not just due to chance.

It seems that the group of customers that got to drive the truck had the highest mean perception score about the truck. So if the company really wanted to do a great job marketing perhaps they need to offer more test drives with their vehicles. An important message for the dealerships!

Two-Tail F-Test Between Group ANOVA

- **Note:** You must have an equal number of cells for this analysis.

Thirty customers were assigned to one of three conditions in which they were given a sheet of truck performance data. The three conditions were:

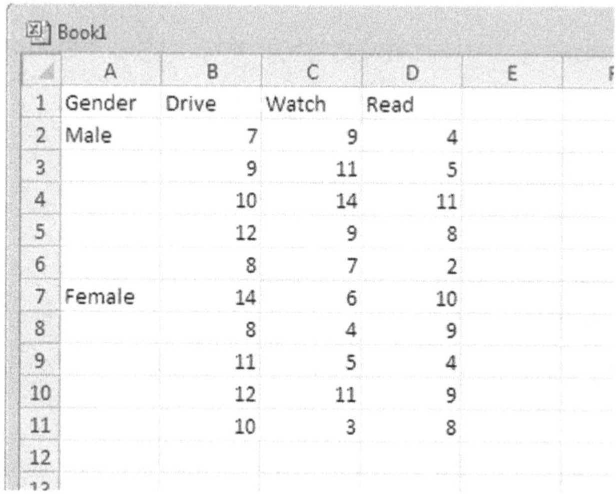

	A	B	C	D	E
1	Gender	Drive	Watch	Read	
2	Male	7	9	4	
3		9	11	5	
4		10	14	11	
5		12	9	8	
6		8	7	2	
7	Female	14	6	10	
8		8	4	9	
9		11	5	4	
10		12	11	9	
11		10	3	8	
12					

Fig. 11.25 Input

Group 1: They were allowed to test drive the actual truck that was described in the performance data sheet.
Group 2: They watched a video on how these trucks are assembled.
Group 3: They read comparison data on competitor trucks.

The customers were then asked to view a commercial on the truck and rate their perceptions of the truck. They used a 7-point Likert scale with 1 as the lowest score. You are interested in knowing if any of these conditions, *as well as their gender*, influenced the perception rating scores with 95% confidence. Now there are two factors so we need to use the **Two-way between-group ANOVA**.

- Input the data from Fig. 11.25

- Click on the **Data** tab and click on **Data Analysis** function
- From the drop down menu select **ANOVA: Two-Factor With Replication**

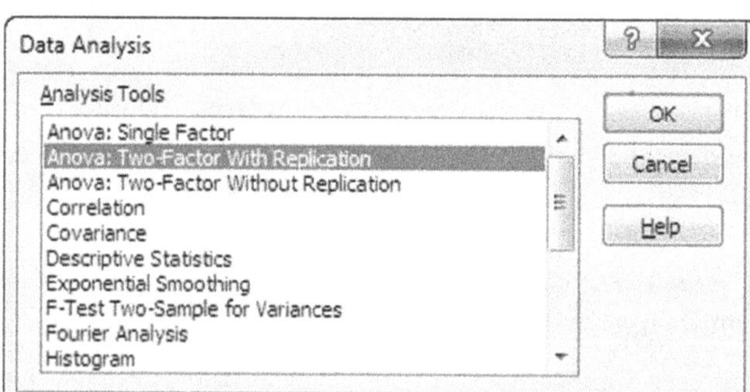

- Click **OK**
- Click inside the **Input Range** box and highlight the data input area. Make sure you include the labels to better keep track of the results

- In the **Rows per sample** box input **5**. Remember for this Excel function you must have an equal number of rows for each group

Note: There is no **Labels box** to check because Excel assumes you have included the labels for both factors in your input range. If you have forgotten to include the labels Excel will treat the first row of data as labels. There are 5 rows of input data.

- Input an alpha value of **0.05** (95% confidence level)

- Select **Output Range** and click inside the **Output Range** box in order to highlight the cell where you want the output printed

- Click **OK**

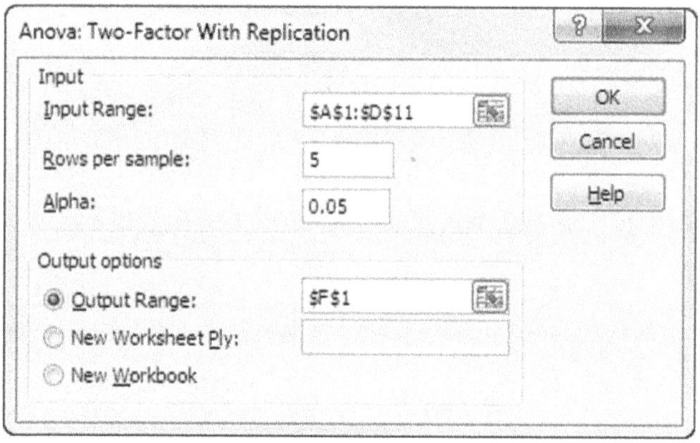

The output data is presented in Fig. 11.26.

Make sure the **Count** adds up to your number of participants (10).

The **Sum, Average,** and **Variance** columns provide the statistics for each condition (category) within each gender, as well as for an overall total.

To determine the significance of these results we can look in two places within the ANOVA output of Fig. 11.25:

1. If any of the ***p*-values** are less than the alpha value we selected (0.05) then the results are statistically significant at the 95% confidence level.
2. If any of the three obtained ***F*-test** values are greater than the **critical *F*-test** then the results are statistically significant at the 95% confidence level.

The **Sample** refers to the first column where we defined the gender categories. The **Columns** refers to the three conditions we had in this study (drive, watch, and read). The **Interaction** refers to the interaction effect between the gender and the condition.

In this case the **Columns *p*-values** for the **condition** (drive, watch, and read) of 0.0458 and the **interaction** (gender *x* condition) of 0.0255 are significant since they are both less than our alpha value of 0.05.

The F_{crit} values provide the same significance information. F_{crit} is less than the ***F*-test** values for the conditions (**Columns: 0.0458**, **Interaction: 0.0255**) so both of these are statistically significant at 95% confidence.

The sample achieved significance. To put it another way, when we divided the groups into gender groups the differences amongst the conditions are statistically important.

Anova: Two-Factor With Replication

SUMMARY	Drive	Watch	Read	Total
Male				
Count	5	5	5	15
Sum	46	50	30	126
Average	9.2	10	6	8.4
Variance	3.7	7	12.5	9.8286
Female				
Count	5	5	5	15
Sum	55	29	40	124
Average	11	5.8	8	8.2667
Variance	5	9.7	5.5	10.6381
Total				
Count	10	10	10	
Sum	101	79	70	
Average	10.1	7.9	7	
Variance	4.7667	12.3222	9.1111	

ANOVA

Source of Variation	*SS*	*df*	*MS*	*F*	*P-value*	*F crit*
Sample	0.1333	1	0.1333	0.0184	0.8931	4.2597
Columns	50.8667	2	25.4333	3.5161	0.0458	3.4028
Interaction	62.0667	2	31.0333	4.2903	0.0255	3.4028
Within	173.6	24	7.2333			
Total	286.6667	29				

Fig. 11.26 Output

In summary, this means that the conditions of driving, watching, and reading have a real effect on the customers' perception ratings of the truck. However, gender also has an effect. The females gave higher scores when they test drove the trucks whereas the males gave higher scores when they got to watch how the trucks were being built. These results are statistically significant.

t-*Test*

The *t*-test can be used effectively in a variety of situations:

- **Example 1 (Regression Problems):** Unlike the ***F*-test**, which tests the statistical significance of the overall regression model, the ***t*-test** looks for statistical significance for each independent variable in the model. This analysis is completed using the Regression tool in Excel. Results are located in the associated ANOVA table.

- **Example 2 (Equality of Two Means):** The ***t*-test** can also be used to test if two population means are equal. It does this by comparing the ratio of the two means and if the means are equal, the ratio of variances will be 1.
- **Example 3 (Before-After Models):** The ***t*-test** may be used in before-after models where the same respondents are measured before and after an intervention. It can be used when you match pairs of respondents on a variable related to the dependent variable. This type of testing is referred to as paired-sample *t*-test, correlated *t*-test, or dependent-sample *t*-test.

Excel provides *t*-test analysis using individual tests located in the **Data Analysis** tab: ***t*-test: Paired Two-Sample for Means, *t*-test: Two-Sample Assuming Equal Variance,** and ***t*-test: Two-Sample Assuming Unequal Variances**. However, all of these tests can also be run using the **T.TEST** function in the formula tab.

Rejection Rule When do we know that our results are real and not just due to chance? In other words when can we reject the null hypothesis:

$$\text{Reject } H_0 \text{ if } t_{\text{calculated}} > t_{\text{critical}}$$

The **critical value of *t*** (**t_{critical}**) can be thought of as the largest value to occur by chance for the given degrees of freedom. If the $t_{\text{calculated}}$ is larger than t_{critical}, then the null hypothesis of equivalent group means is rejected. In other words, your results are simply due to chance and are statistically significant.

The ***p*-value** is the probability that a value of ***t*** greater than or equal to the $t_{\text{calculated}}$ could have occurred by chance if there were no difference in the variances. The ***p*-value** is the fractional area of the right tail of the ***t*-distribution**. $t_{\text{calculated}}$ is indicated as ***F*** in the Excel output. When the ***p*-value** is less than your chosen confidence level you have statistically significant results so keep going!

Basic Descriptions of t-Test Applications

This first section will describe the general application of each problem type by way of an example and the second section will walk through the step-by-step Excel applications for all three problem types.

Example 1: Regression Problems

This tests the statistical significance of each independent variable in the input model for this sample output for the regression model. The overall model is defined by:

$$y = b_1x_1 + b_2x_2 + b_3x_3 + \ldots + b_0$$

In Fig. 11.26 we have included only one independent variable in the input model for simplicity when explaining significance. The test results are shown in the third output table. The results of interest are indicating whether the independent variable

(*X* **Variable 1**) is contributing in a statistically significant way to this regression model. Excel does not provide the $t_{crtical}$ value so we can only use the ***p*-value**. The ***p*-value** (0.618085) is the probability associated with the $t_{calculated}$ (shown as ***t* Stat**) of 0.539637. In other words, there is a 0.618 probability of getting a *t*-value greater than 0.54. This is where the confidence level or alpha value becomes necessary. Select the desired level for risk or alpha value. Let's select a 95% confidence level so we have an alpha of 0.05. Now compare the ***p*-value** with the selected alpha value, because the ***p*-value** of 0.618 is greater than 0.05 we do not have statistically significant results. Since we do not have statistically significant results at the 95% confidence level for this independent variable, our regression model will not yield useful results the way it is currently set up. So we must rethink the model.

Example 2: t-Test for Equality of Means
This test can be a two-tailed test or a one-tailed test. The two-tailed version tests against the alternative that the means are not equal. The one-tailed version only tests in one direction that the mean from the first population is either greater than or less than (but not both) the second population mean. The choice is determined by the problem. For example, if we are testing a new process we may only be interested in knowing if the new mean is less than the mean of the old process. The ***t*-test** can be used to answer the following questions:

1. Do two samples come from populations with equal means? (two-tail)
2. Does a new process, treatment, or test reduce the mean of the current process? (one-tail)

A ***t*-test** returns the two-tailed probability that the means in sample 1 and sample 2 are not significantly different. We can use this function to determine whether two samples have different means. For example, given satisfaction scores from accountants and from engineers you can test whether these professions have different average satisfaction scores.

Example 3: t-Test Paired Samples
This ***t*-test** is used in before-after models where the same respondents are measured before and after an intervention. It can also be used when you match pairs of respondents on a variable related to the dependent variable. Because this ***t*-test** is always a comparison of two means it is a two-sample test.

The two sets of data whose means are to be compared should be entered in consecutive rows of two columns. It is not necessary that both sample sizes be the same.

Excel
Excel examples are provided for each of the problems types:

- Example 1: Regression Problems
- Example 2: Equality of Two Means
- Example 3: Before-After Models

SUMMARY OUTPUT

Regression Statistics	
Multiple R	0.2605026
R Square	0.0678616
Adjusted R Square	-0.165173
Standard Error	11.891727
Observations	6

ANOVA

	df	SS	MS	F	Significance F
Regression	1	41.18067227	41.18067	0.291208	0.618085213
Residual	4	565.6526611	141.4132		
Total	5	606.8333333			

	Coefficients	Standard Error	t Stat	P-value	Lower 95%	Upper 95%	ower 95.0%	pper 95.0
Intercept	20.263305	15.42936488	1.313295	0.259351	-22.57547928	63.10209	-22.5755	63.1020
X Variable 1	0.8319328	1.541652313	0.539637	0.618085	-3.448380244	5.112246	-3.44838	5.11224

Fig. 11.27 Output

Example 1: Regression Problems

A sample of office assistants across the San Francisco-Bay Area was created. It was suggested that the data should be analyzed with a regression model to determine if salary was related to years of experience and aptitude test scores. We want a 99% confidence level. This data is shown in Fig. 11.27.

As a reminder the form of the regression model will be

$$y = b_o + b_1 x_1 + b_2 x_2$$

where:

y = annual salary ($000)
x_1 = years of experience
x_2 = score on aptitude test

- Input the data from Fig. 11.28

- Select **Data** from the menu bar

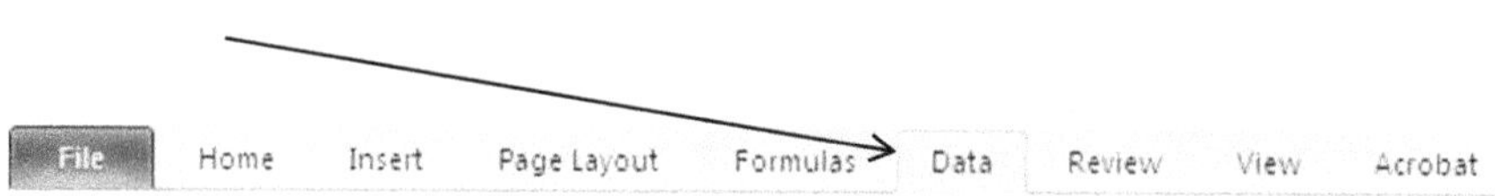

Fig. 11.28 Input

Experience (yrs) X_1	Score (out of 100) X_2	Salary ($000) Y
4	78	24
7	100	43
1	86	23.7
5	82	34.3
8	86	35.8
10	84	38
0	75	22.2
1	80	23.1
6	83	30
6	91	33
9	88	38
2	73	26.6
10	75	36.2
5	81	31.6
6	74	29
8	87	34
4	79	30.1
6	94	33.9
3	70	28.9
3	89	30

- Select the **Data Analysis** function

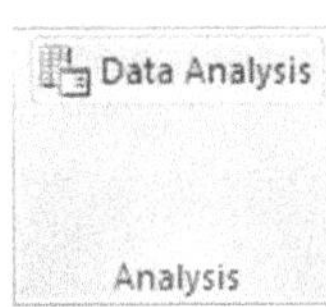

- Select **Regression**

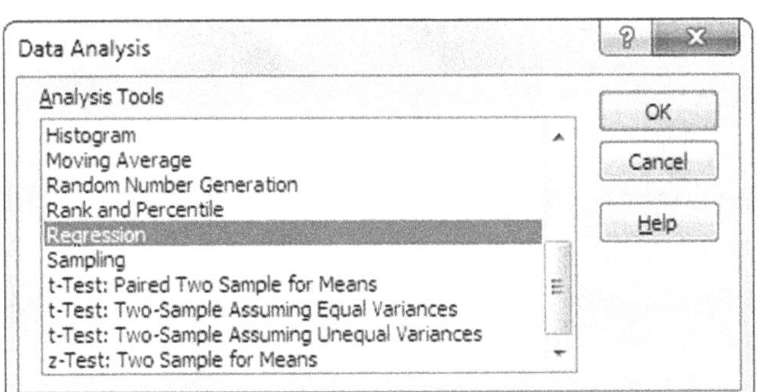

- Click inside the **Input y Range** box to select the correct data

- Click inside the **Input x Range** box to select the correct data

- Check the **Labels** box

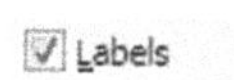

- Select **Confidence Level.** Enter 99 in **Confidence Level** box

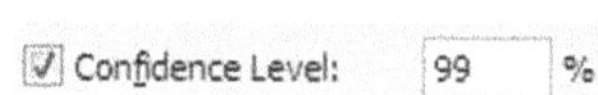

- Select **Output Range** and click inside the **Output Range** box in order to highlight the cell where you want the output printed

- Click **OK**

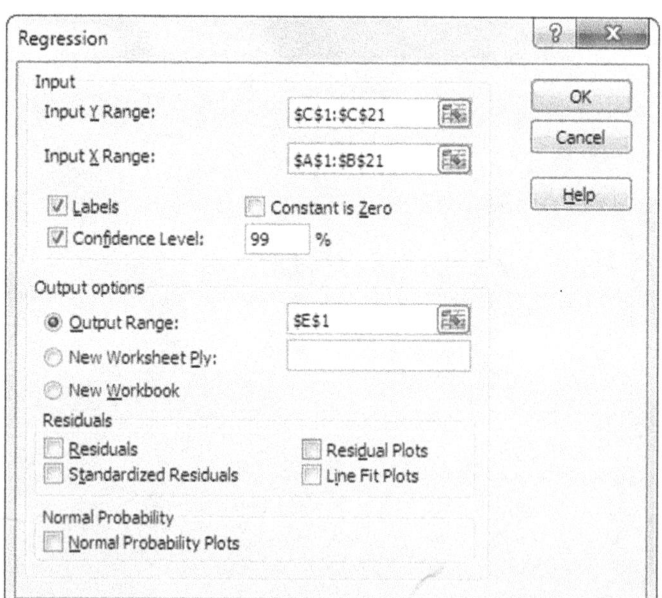

SUMMARY OUTPUT

Regression Statistics	
Multiple R	0.9082
R Square	0.8248
Adjusted R Square	0.8042
Standard Error	2.4782
Observations	20

ANOVA

	df	*SS*	*MS*	*F*	*Significance F*
Regression	2	491.5994	245.7997	40.0239	0.0000
Residual	17	104.4026	6.1413		
Total	19	596.0020			

	Coefficients	*Standard Error*	*t Stat*	*P-value*	*Lower 95%*	*Upper 95%*	*Lower 99.0%*	*Upper 99.0%*
Intercept	3.8651	6.3073	0.6128	0.5481	-9.4420	17.1723	-14.4148	22.1451
Experience (yrs)	1.4014	0.2034	6.8883	0.0000	0.9722	1.8306	0.8118	1.9910
Score (out of 100)	0.2431	0.0793	3.0675	0.0070	0.0759	0.4103	0.0134	0.4728

Fig. 11.29 Output

The Excel output is provided in Fig. 11.29.

The resulting equation is given in the ANOVA table in Fig. 11.29

$$y = 3.8651 + 1.4014x_1 + 0.2431x_2$$

We compared the ***p*-values** with the selected alpha value of 0.01 (99% confidence level) to perform the *t*-test significance analysis. The results in the third table for the ANOVA section of Fig. 11.28 ***p*-value** column indicate that both Experience (x_1) and Score (x_2) are statistically significant. Both ***p*-values** are less than 0.01. In other words, if we selected 100 samples from this population we would get similar results in 99 of those analyses. This suggests that we can use these two independent variables with 99% confidence when predicting our independent variable which in this case is sales. Good model!

Example 2: t-Test for Equality of Means

This test can be run as a two-tailed test or a one-tailed test depending on the model you are trying to solve.

- The one-tailed ***t***-test for equality of means:

A one-sided (or one-tailed) test is appropriate if researchers want to know if the population mean of one of the datasets is larger than that of another set of data or equivalently, if the population mean of one set of data is smaller than that of another set of data. For example, suppose that two groups of employees report on the time (hours) they spend handling customer complaints. The first group has less than 5 years of experience and the second group has at least 15 years of experience. We have already calculated that the two means are different as group 1 spends 19.03 h handling complaints on average and while group 2 spends 4.62 h. However, there is always the possibility that this difference in the means could have occurred by chance. We need to test if these results are significant. A one-sided ***t*-test** is appropriate in the scenario that one sample's mean is "smaller than" the other sample's mean. In this case the more experienced employees on average spend less time handling customer complaints. Assume a 95% confidence level.

- Input the data from Fig. 11.30

Before you select the correct Excel analysis tool you need to determine if you know the population variances. If you do, are they equal? Here are the recommended *t*-test tools based on this information:

less than 5 yrs	15+ yrs
10	1
14	3
16	3
16	3
14	3
24	7
21	5
24	7
27	8
17	4
13	2
27	8
11	1
20	5
23	6
11	2
25	7
20	5
15	3
22	6
24	6
18	4
11	1
19	4
28	8
11	1
27	8
22	6
14	3
27	8

Note: You must put the group with the larger mean in the first column, i.e. **variable 1.**

Fig. 11.30 Input

Problem type	Excel tool
Equal variances	***t*-Test: Two-Sample Assuming Equal Variances**
Unequal variances	***t*-Test: Two-Sample Assuming Unequal Variances**
Unknown variances with sample size >30	***t*-Test: Two-Sample Assuming Equal Variances**
Unknown variances with sample size <30	***t*-Test: Two-Sample Assuming Unequal Variances**
Unknown variances but assumed to be unequal based on other information	***t*-Test: Two-Sample Assuming Unequal Variances**

In this case let's assume we know that the population variances are equal. However, you would follow the same instructions if you had selected the case for unequal variances.

- Click on the **Data** tab and select **Data Analysis** function
- Select **t-Test: Two-Sample Assuming Equal Variances** from the menu

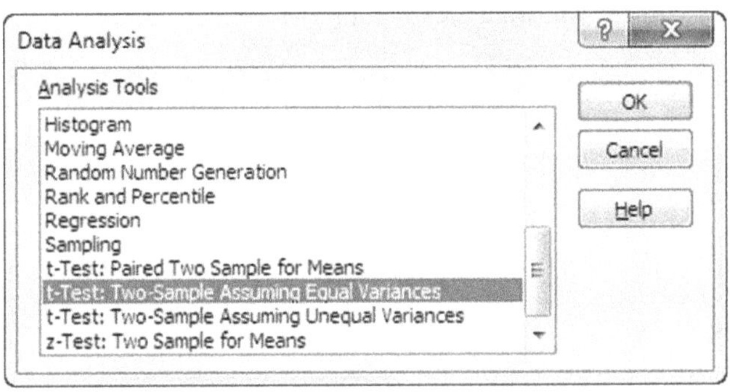

- Click **OK**
- Click inside the **Variable 1 Range box**. Highlight the first column of data including the label

- Click inside the **Variable 2 Range box**. Highlight the second column of data including the label

- Enter **0** in the **Hypothesized Mean Difference** box as we are testing if these two means are equal

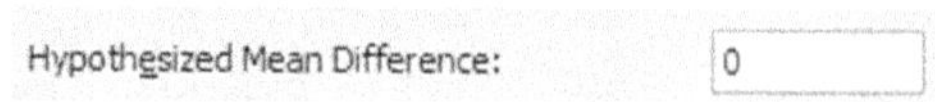

- Check the **Labels** box

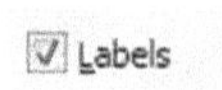

- Select an alpha level of **0.05**

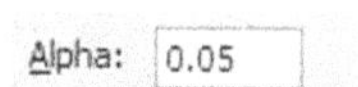

- Select **Output Range** and click inside the **Output Range** box in order to highlight the cell where you want the output printed

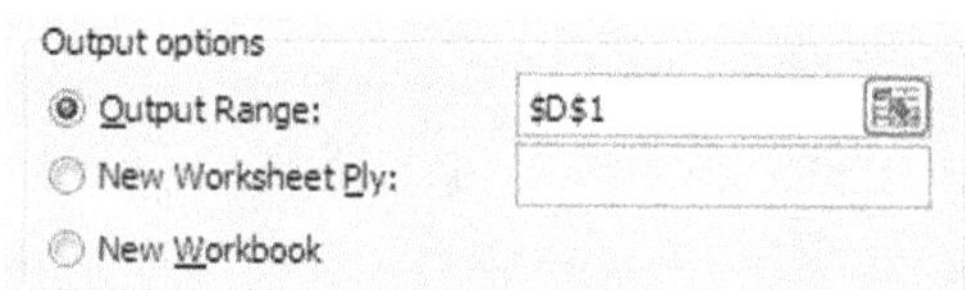

- Click **OK**

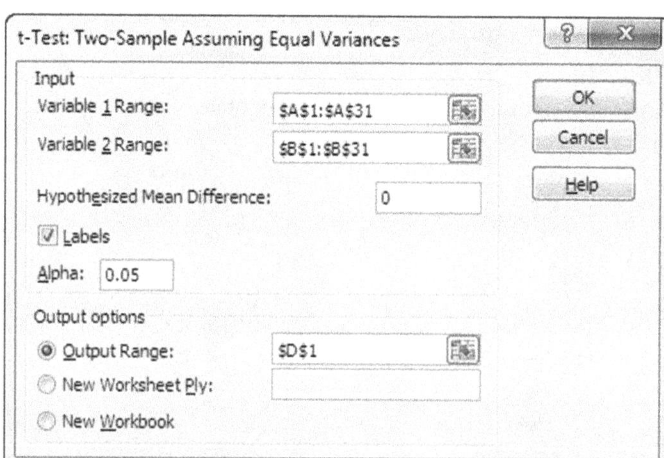

The results will appear as the following in the next output table (Fig. 11.31).

t-Test: Two-Sample Assuming Equal Variances

	less than 5	*15+ yrs*
Mean	19.0333	4.6
Variance	33.6195	5.6966
Observations	30	30
Pooled Variance	19.6580	
Hypothesized Mean Difference	0	
df	58	
t Stat	12.6079	
P(T<=t) one-tail	1.46807E-18	
t Critical one-tail	1.6716	
P(T<=t) two-tail	2.93614E-18	
t Critical two-tail	2.0017	

Fig. 11.31 Output

Make sure that you check all of the data has been input by looking at the **Observations**. The **Observations** will indicate if we have included all of our data (30).

The **Mean** and **Variance** are given for each group.

Pooled Variance is the weighted (using degrees of freedom) average of the two-sample variances. This value is probably not of much use to you.

Hypothesized Mean Difference we set to zero to test if the two means are in fact the same.

df is just the degrees of freedom for the test and is calculated by $(n_1 + n_2 - 2)$.

tStat is the t-value calculated based on the input data (12.6079).

The next four outputs are used for our significance testing. This problem is only testing using one-tail.

One-Tailed Test

1. We can compare the ***P*(*T*<=*t*) one-tail** value (1.46807E−18) with the alpha value we selected (0.05). Since this p-value is less than alpha we can confirm that the difference in the means is statistically significant at the 95% confidence level.
2. We can compare the **tStat** (12.6079) with the ***t* Critical one-tail** (1.6716). Since the **tStat** is greater than the **tcrit**, we can conclude our results are statistically significant at the 95% confidence level. If we selected 100 other samples from this population we would get similar results in 95 of those samples.

The variance in satisfaction scores is statistically different. Thus, we can assume the less experienced employees really do have a greater mean than those more experienced employees. In other words, the less experienced group spends more time on average handling complaints than their more experienced counterparts. This may be worth looking into in terms of understanding best practices and managing time efficiently.

Two-Tailed Test

The **two-tailed *t*-test** determines if one population mean is statistically significantly different than another population mean. We have no knowledge nor do we care if one is greater or less than the other. We only want to determine if they are significantly different. The previous problem would have required a two-tailed analysis if it had been worded "*Is the difference in average time spent handling complaints statistically significant between the two groups of employees with different experience levels?*".

To analyze these results we would follow the same logic as before but word the conclusion somewhat differently.

1. We can compare the ***P(T<=t)* two-tail** value (2.93614E−18) with the alpha value we selected (0.05). Since the *p*-value is less than alpha, we can confirm that the difference in means is statistically significant at the 95% confidence level.
2. We can compare the **tStat** (12.6079) with the ***t* Critical two-tail** (2.0017). Since the **tStat** is greater than the **tcrit**, we can conclude our results are statistically significant at the 95% confidence level. If we selected 100 other samples from this population we would get similar results in 95 of those samples.

In this case we can only conclude the means are statistically significantly different at the 95% confidence level. We cannot infer that one mean is significantly larger than the other. Training may in fact increase or decrease the scores.

Example 3: Before-After Models

A group of marketing professionals provided client satisfaction scores from the first quarter of this year which were rated out of 100. The range of possible scores is 0 to 100 with 100 as the highest possible score. The same marketing professionals then attended a company training program on effective communications and submitted their client satisfaction scores in the following quarter of the year.

- Is the difference in these scores real or simply due to chance at the 95% confidence level? This would be a **two-tailed *t*-test** as we don't care if one is greater than the other just that they are different.
- Are the average employee satisfaction scores higher after training? In other words was training effective? This is a **one-tailed *t*-test** since it determines if one population mean is greater or less than the other in a statistically significantly way.

- Input the following data from Fig. 11.32

Fig. 11.32 Input

Pre-training scores	Post training scores
90.47	90.95
92.02	91.30
93.15	91.48
90.98	92.04
91.73	90.70
94.00	91.33

- Click on the **Data** tab and select **Data Analysis** function
- Select **t-test: Paired Two-Sample for Means**

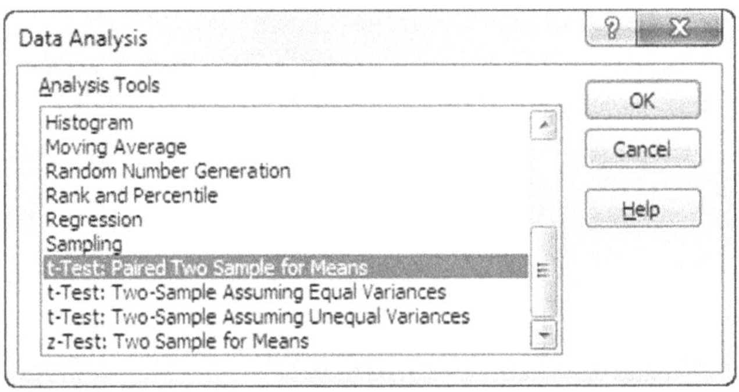

- Click **OK**
- Click inside the **Variable 1 Range** box to select the correct data

- Click inside the **Variable 2 Range** box to select the correct data

- Enter **0** in the **Hypothesized Mean Difference** box, as we are testing if these two means are equal

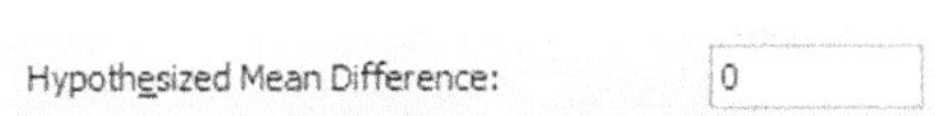

- If you included labels in your input range check the **Labels** box

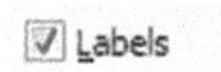

- Enter the level of significance or "**Alpha**" value. The default that should appear is 0.05 which of course is a 95% confidence level. But you can choose whatever alpha value you want

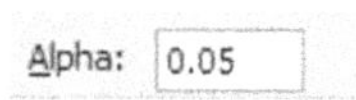

- Select **Output Range** and click inside the **Output Range** box in order to highlight the cell where you want the output printed

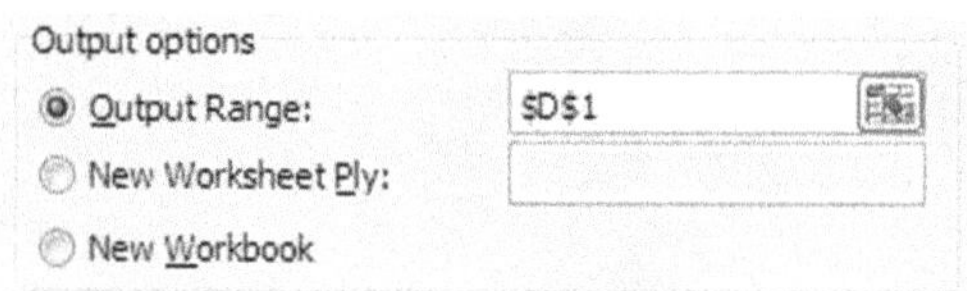

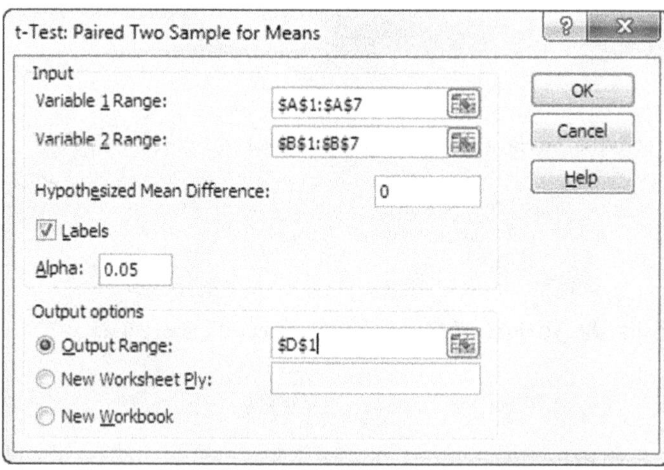

- Click **OK** to see the output of the function (Fig. 11.33)

Output data is provided in Fig. 11.33.

Make sure you check that all of the data has been input by looking at the **Observations**. The **Observations** will indicate if we have included all of our data (6).

t-Test: Paired Two Sample for Means

	Pre-training scores	*Post training scores*
Mean	92.0583	91.3
Variance	1.7513	0.2127
Observations	6	6
Pearson Correlation	0.0687	
Hypothesized Mean Difference	0	
df	5	
t Stat	1.3547	
P(T<=t) one-tail	0.1168	
t Critical one-tail	2.0150	
P(T<=t) two-tail	0.2335	
t Critical two-tail	2.5706	

Fig. 11.33 Output

The **Mean** and **Variance** are given for each group.

Pearson Correlation is the *r*-value (**correlation coefficient**). This number tells us that there is a positive relationship between the two sets of results, but it is weak as the value is only 0.07.

Hypothesized Mean Difference we set to zero to test if the two means are in fact the same.

df is just the degrees of freedom for the test and is calculated $(n - 1)$.

tStat is the *t*-value calculated based on the input data.

The following sections will provide instructions on completing the significance analysis.

One-Tailed Test

We can compare the ***P(T<=t)*** **one-tail** value (0.1168) with the alpha value we selected (0.05). Since the ***p*-value** is greater than alpha, we can confirm that the difference in the variance is not statistically significant at the 95% confidence level.

We can compare the **tStat** (1.3547) with the ***t*** **Critical one-tail** (2.0150). Since the **tStat** is less than the ***t*** **Critical one-tail**, we can conclude our results are not statistically significant at the 95% confidence level. If we selected other samples we would not get similar results.

We can assume the client satisfaction scores are not higher after training since the mean client satisfaction scores are not statistically different. If we are striving for higher scores than we should actually test for that. It seems that training may not have been effective.

Two-Tailed Test

To analyze these results with a two-tailed test we would follow the same logic as before but word the conclusion somewhat differently.

Fig. 11.34 Input

Pre-training scores	Post training scores
90.47	90.95
92.02	91.30
93.15	91.48
90.98	92.04
91.73	90.70
94.00	91.33

We can compare the ***P*(*T*<=*t*) two-tail** value (0.2335) with the alpha value we selected (0.05). Since the *p*-value is greater than alpha, we can confirm that the difference in the means is not statistically significant at the 95% confidence level.

We can compare the **tStat** (1.35467) with the ***t* Critical two-tail** (2.5706). Since the **tStat** is less than the ***t* Critical two-tail**, we can conclude our results are not statistically significant at the 95% confidence level. If we selected other samples we would not get similar results.

In conclusion, the communications training did not have a statistically significant effect on the client satisfaction scores of the employees. There must be another more important factor that is accounting for the client satisfaction results. Perhaps the type of training that was given was not appropriate for this context.

T.TEST

The **T.TEST** tool can also provide the *p*-values for each of the previous three examples. It can be employed by using the Excel tools in **Data Analysis**. The difference with the **T.TEST** results is that you only get information, about the *p*-value, no critical value information, and only one result at a time. In the previously used **Data Analysis** tools you can get multiple results within each analysis.

- Input data from Fig. 11.34

To access the **T.TEST** tool:

- Select a cell and click the Function button

- In the select a category section choose **Statistical**

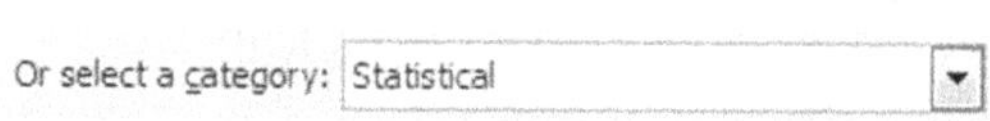

- Select **T.TEST** for Function

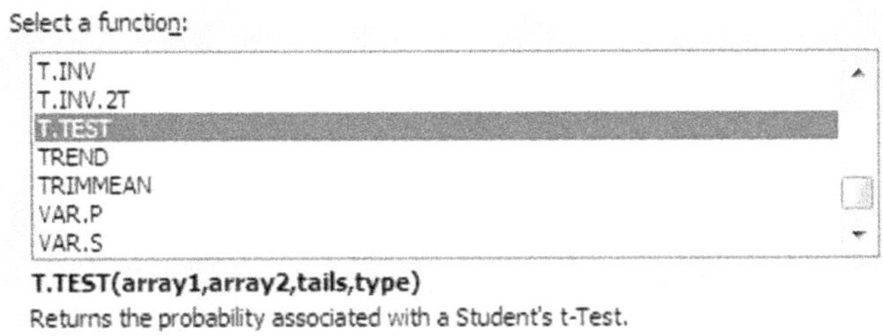

- **Caution:** Do not include labels in your array data with this tool. Since it is operating as a formula you just want the actual data values to be included.

The test type is inputted in the last box labeled **Type**. Each type is listed at the bottom of the input page.

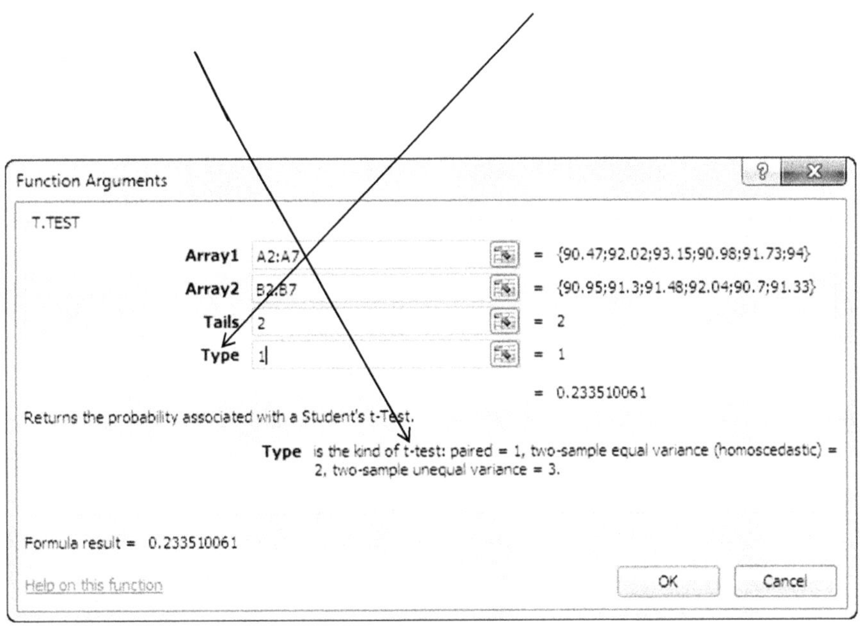

- **Paired two-sample two-tailed** provides the p-value of 0.23351 as shown on the input screen.

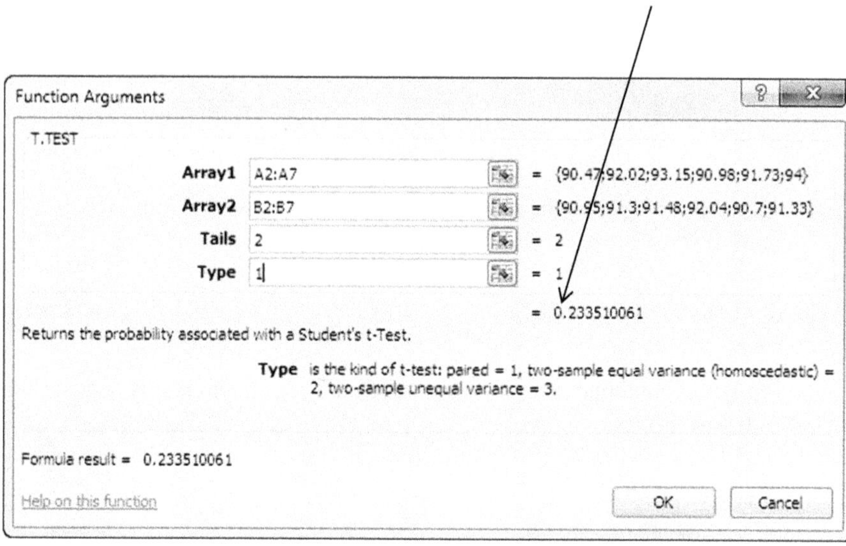

- **Paired two-sample one-tailed** provides the p-value of 0.11675503 as shown on the input screen.

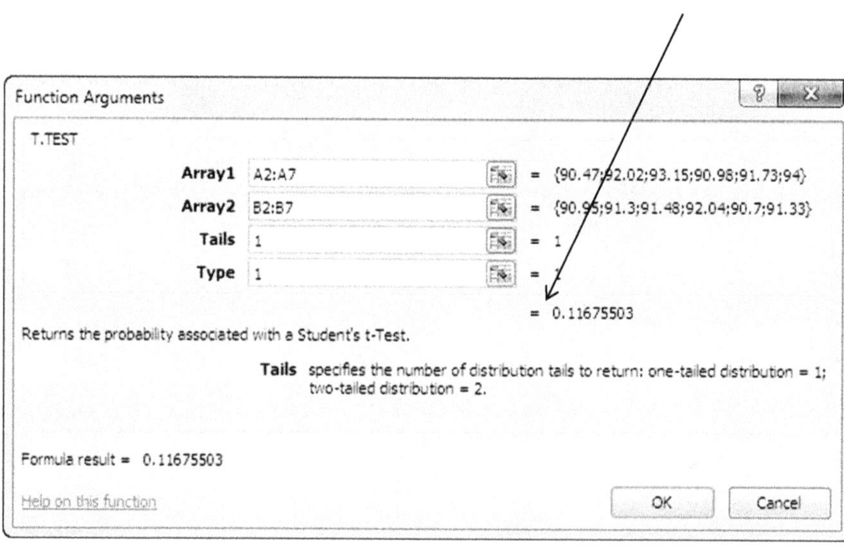

- **Two-sample equal variance (homoscedastic)** provides the ***p*-value** of 0.214504933 as shown on the input screen. Note this can be ran as a one-tailed test by changing the **Tails** to the value of "2".

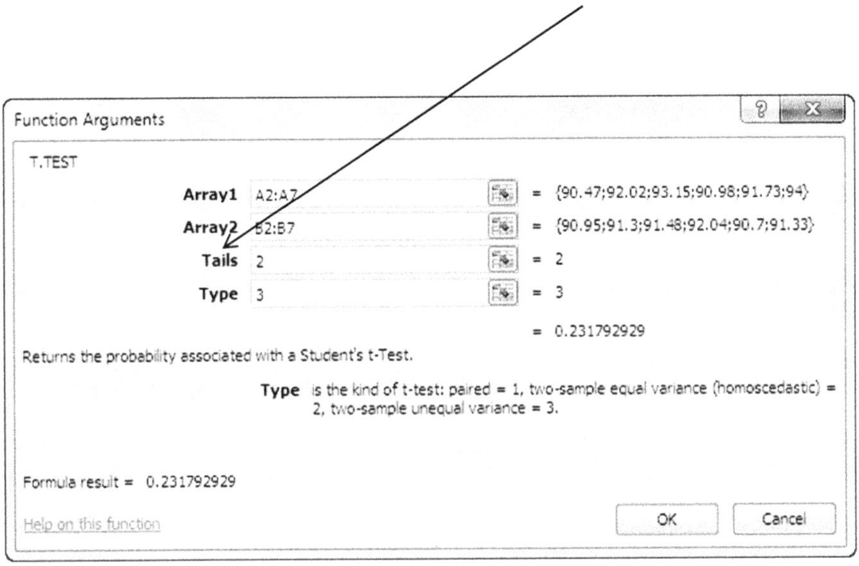

- **Two-sample unequal variance** provides the ***p*-value** of 0.23179 as shown on the input screen. Note this can be ran as a one-tailed test by changing the **Tails** to the value of "2".

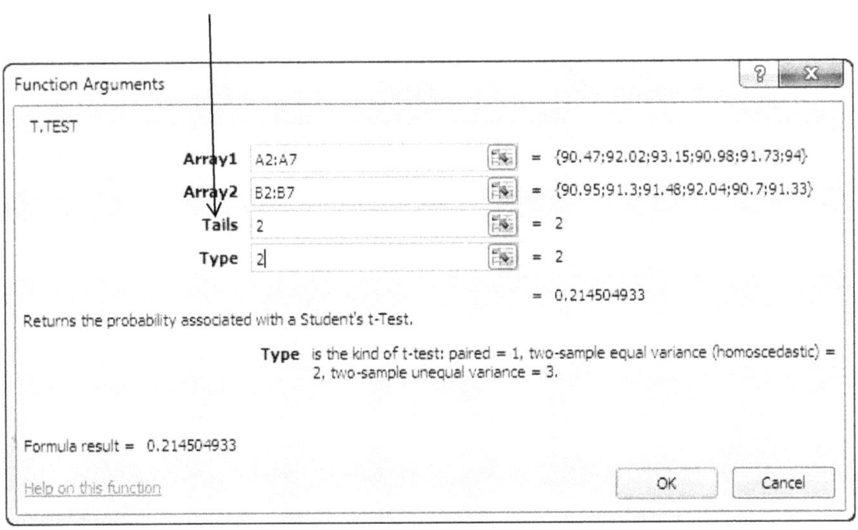

Common Excel Pitfalls

- Make sure you work with absolute values when working with critical values.
- **Two-way between-group ANOVA** must include labels in input arrays.
- Don't include labels in the array in **T.TEST** since this tool operates as a formula.
- **One-tailed *t*-test for equality of means** requires the group with the larger mean to be in the first column.

Final Thoughts and Activities

Practice Problems and Case Studies

1. A professional golf ball manufacturing company wanted to compare the distance traveled by golf balls produced using each of their four different designs. Ten balls were manufactured with each design and were brought to the local golf course for the club professional to test. The order in which the balls were hit with the same club from the first tee was randomized so that the pro did not know which type of ball was being hit. All 40 balls were hit in a short period of time during which the environmental conditions were essentially the same.
   (a) Is there evidence of a difference in the mean distances traveled by the golf balls with different designs at the 0.05 level of significance?
   (b) What assumptions are necessary in (a)?
2. You are planning on stocking your grocery counter with chewing gum but want to know if customers really want this product. The other branches of this store (population) sell on average ten packs of gum per customer every 2 months (population average). You have limited time and limited money so you survey a very small sample of your customers, $n = 15$. You ask them how often they have purchased gum in the last 60 days. You want a 95% confidence level ($\alpha = 0.05$) that all of your customers (population) will also purchase on average 10 packs every 2 months. Test to make sure the average result you calculate for your customers is in fact statistically significantly different from the population average of 10 packs per customer every 60 days. The data is listed below:
   Number of packages of gum purchased in the last 60 days

Packs of gum purchased in 60 days
10
2
4
6
0
15

Packs of gum purchased in 60 days
22
3
8
8
7
10
7
7

3. Comment on the validity and reliability of the following:

   (a) A respondent's reporting of an intention to subscribe to *America Reports* is highly reliable. A researcher believes that this constitutes a valid measurement of dissatisfaction with the economic system and alienation from big business.
   (b) A general interest magazine advertised that the magazine was a better advertising medium than television programs with similar content. Research had indicated that recall scores were higher for the magazine ads than with 30-second commercials for soft drinks and other test products.
   (c) A respondent's report for frequency of magazine reading consistently indicates she reads *Nice Housekeeping*. The respondent also reports that she never reads *Cosmo*.

4. Suppose that shopping times for customers at a local mall are normally distributed with a known population standard deviation of 20 minutes. A random sample of 64 shoppers in the local grocery store had a mean time of 75 minutes. Find the standard error. Also determine the upper and lower confidence limits of a 95% confidence interval.
5. Compare the use of the F-test versus t-test in the following problem types:

   (a) Comparing means
   (b) Comparing variance
   (c) Regression analysis

6. The following data includes the salaries of people who have a degree in economics, medicine, or history. Run a single-factor ANOVA analysis and discuss the F-test results.

	A	B	C	D
1	economics	medicine	history	
2	42	69	35	
3	53	54	40	
4	49	58	53	
5	53	64	42	
6	43	64	50	
7	44	55	39	
8	45	56	55	
9	52		39	
10	54		40	
11				

7. When do you use a one-tailed *t*-test versus a two-tailed *t*-test?
8. Discuss the following results in terms of the statistical significance for the findings while assuming an alpha value of 0.05.

	A	B	C
1	t-test Paired Two Samples for Means		
2			
3	Mean	87.56	
4	Variance	1.8	
5	Observations	30	
6	t-Stat	1.567	
7	p (T<=t) one-tail	0.301	
8	t Critical one -tail	2.11	
9	p(T<=t) two-tail	0.0116	
10	t Critical two -tail	2.45	

Discussion Boards

1. All criminals should take lie detector tests regardless of false positives.
2. The death penalty discourages others from committing murder in the United States.
3. What are the ethical issues involved in hypothesis testing?

Group Activity

1. Choose a company with data available online. Discuss the company's reliability claims. What type of data would this company collect as part of its market research? Discuss the issue of valid research results.
2. There is more than one way to measure reliability but often the actual statistical source is not provided. Complete a web search to see if you can produce any reliability results using a statistic other than a Cronbach's alpha.
3. Explore the following hypotheses using the web:

   H_0: Employees are not more productive working in cubicles.
   H_a: Employees are more productive working in cubicles.

Parting Thought

A statistician is a mathematician broken down by age and sex …

Problem Solutions

1. A professional golf ball manufacturing company wanted to compare the distance traveled by golf balls produced using each of their four different designs. Ten balls were manufactured with each design and were brought to the local golf course for the club professional to test. The order in which the balls were hit with the same club from the first tee was randomized so that the pro did not know which type of ball was being hit. All 40 balls were hit in a short period of time during which the environmental conditions were essentially the same.
   Answer:

   (a) Is there evidence of a difference in the mean distances traveled by the golf balls with different designs at the 0.05 level of significance?
   Because $F = 53.03 > F_{\text{crit}} = 2.8$, there is a significant difference in the mean distances traveled by the golf balls with different designs
   (b) What assumptions are necessary in (a)?
   The assumptions are that the golf ball samples are randomly and independently selected. An additional assumption is that the variances are constant.

ANOVA: Single factor						
Summary						
Groups	*Count*	*Sum*	*Average*	*Variance*		
Design1	10	2069.64	206.964	5.246427		

(continued)

ANOVA: Single factor						
Design2	10	2188.66	218.866	30.66718		
Design3	10	2269.38	226.938	23.18213		
Design4	10	2289.72	228.972	16.10697		
ANOVA						
Source of variation	*SS*	*df*	*MS*	*F*	*P-value*	*F crit*
Between groups	2990.99	3	997.00	53.03	0.00	2.87
Within groups	676.82	36	18.80			
Total	3667.814	39				

2. You are planning on stocking your grocery counter with chewing gum but want to know if customers really want this product. The other branches of this store (population) sell on average ten packs of gum per customer every 2 months (population average). You have limited time and limited money so you survey a very small sample of your customers, $n = 15$. You ask them how often they have purchased gum in the last 60 days. You want a 95% confidence level ($\alpha = 0.05$) that all of your customers (population) will also purchase on average 10 packs every 2 months. Test to make sure the average result you calculate for your customers is in fact statistically significantly different from the population average of 10 packs per customer every 60 days. The data is listed below:

   Number of packages of gum purchased in the last 60 days

Packs of gum purchased in 60 days
10
2
4
6
0
15
22
3
8
8
7
10
7
7

Answer:

In this problem we use a **one-tailed *t*-test** to see if the two averages really are different (alternative hypothesis) or if in fact the sample should be assumed to have the same average as the population (null hypothesis).

	A	B	C	D	E
1	Buying Gum		Variable	Value	How to calculate
2	10		Average	7.785714	use =Average() function
3	2		Hypothesized mean	10	given
4	4		sd	5.549478	use =STDEV.S() function for sample std dev
5	6				
6	0		Count	14	use =COUNT() funtion; this is just n
7	15				
8	22		Standard Error	1.48316	=D4/SQRT(D6)
9	3				
10	8		Alpha	0.05	Given as 95% confidence level
11	8				
12	7		t-stat	-1.49295	=(D2-D3)/D8
13	10				
14	7		Degrees of Fredom	13	=D6-1
15	7				
16			One-Tailed probability	0.079659	use T.DIST function
17					

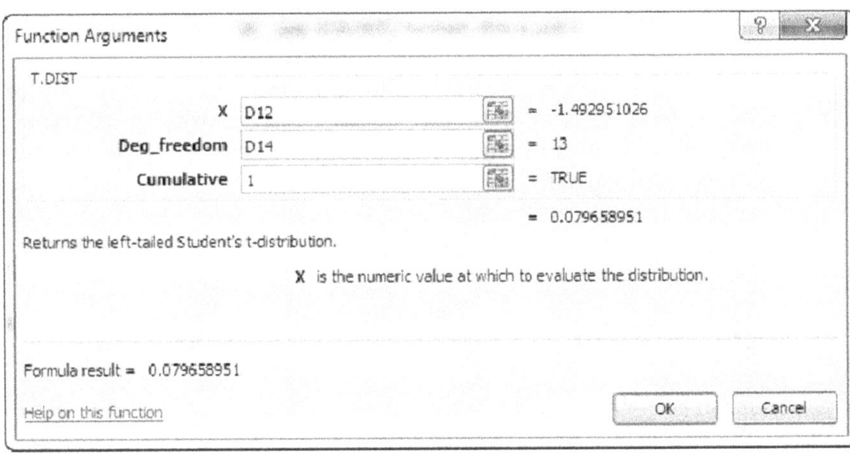

where:

X is the *t*-value you just calculated.
Deg_Freedom is the degrees of freedom ($n - 1$).
Cumulative is where we let Excel know we want the left-hand tail only. You type in the word **TRUE** or the number **1**.

In this case we have a left-hand tail of 0.08 (one-tailed probability) which exceeds our alpha value of 0.05. This means our results are not significantly different. The average gum purchases per customer that our sample indicated (approximately 8 packs in 2 months) is not significantly different from the population average of 10 packs. Therefore, we can assume that in fact the population average of 10 packs can be used to describe these customers as well. If our results had been statistically significant we could safely assume our sample average was real and feel safe to use it to describe our population of customers. Our store owner should go ahead and stock gum in his store as he can expect to sell about as much as the other stores.

Another way of proving significance is to use the t_c calculation where we compare the calculated t with the critical t (based on degrees of freedom). Remember when the $t_{\text{calculated}} < t_{\text{critical}}$ the results are not statistically significant.

3. Comment on the validity and reliability of the following:

   Answer:

   (a) A respondent's reporting of an intention to subscribe to *America Reports* is highly reliable. A researcher believes that this constitutes a valid measurement of dissatisfaction with the economic system and alienation from big business.

   High reliability does not guarantee high validity. You can reliably (consistently) get the wrong answer. More information is required to determine the validity of these suggested factors.

   (b) A general interest magazine advertised that the magazine was a better advertising medium than television programs with similar content. Research had indicated that recall scores were higher for the magazine ads than with 30-second commercials for soft drinks and other test products.

   We don't have enough information to determine the reliability or consistency of the recall scores. Is recall a valid measurement for advertising effectiveness or could recall have more to do with the type of the product etc.? We need more information.

   (c) A respondent's report for frequency of magazine reading consistently indicates she reads *Nice Housekeeping*. The respondent also reports that she never reads *Cosmo*.

   This is a reliable or consistent measurement for magazine preference. Reading the magazine is a valid measurement for consumer interest in the magazine.

4. Suppose that shopping times for customers at a local mall are normally distributed with a known population standard deviation of 20 minutes. A random sample of 64 shoppers in the local grocery store had a mean time of 75 minutes. Find the standard error. Also determine the upper and lower confidence limits of a 95% confidence interval.

   Answer:

   The standard error and the margin of error are as follows:

   $$\text{Standard error} = \frac{\sigma}{\sqrt{n}} = \frac{20}{\sqrt{64}} = 2.5$$

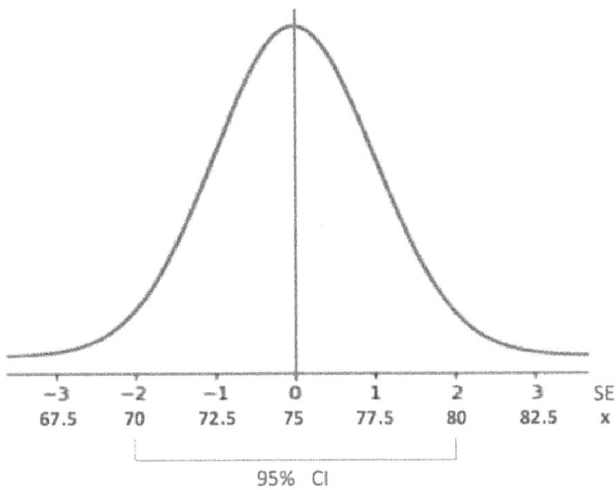

5. Compare the use of the F-test versus t-test in the following problem types:

   (a) Comparing means
   (b) Comparing variance
   (c) Regression analysis

   Answer

   1. Comparing two means:

      - t-test: Consider comparing the mean heights of two groups (Group A and Group B) each consisting of 30 individuals. Since we have sample means and standard deviations for each group we can use a two-sample independent t-test to determine if there is a significant difference between the means.
      - F-test: We have three or more groups (Group A, Group B, and Group C) in which we want to compare means. We have sample means and standard deviations for each group. In this case we can use an analysis of variance (ANOVA) which employs the F-test to determine if there are significant differences among the means for the groups.

   2. Comparing variances:

      - t-test: Compare the sample variances for two groups (Group X and Group Y). We can use a two-sample F-test which is essentially a ratio of the two-sample variances to test if there is a significant difference between the variances of the two groups.
      - F-test: If we have three or more groups (Group X, Group Y, and Group Z) in which we want to compare variances. We can do so by using the F-test within an ANOVA framework. This test will determine if there are significant differences in the variances among the groups.

3. Regression analysis:

   - *t*-test: In linear regression the *t*-test is commonly used to assess the significance of individual regression coefficients. For example, if we have a regression model with several predictor variables we can perform *t*-tests to determine if each coefficient is significantly different from zero.
   - *F*-test: In regression analysis the *F*-test is often used to assess the overall significance of the regression model. It tests whether the model with all its predictor variables provides a better fit than an intercept-only model. The *F*-test can also be used to compare nested models such as when testing if adding additional predictors significantly improves the model fit.

   In summary, the *t*-test is typically used when comparing means/variances between two groups whereas the *F*-test is more appropriate for comparing means/variances among three or more groups. The *F*-test is also more appropriate when assessing the overall significance of regression models.

6. The following data includes the salaries of people who have a degree in economics, medicine, or history. Run a single-factor ANOVA analysis and discuss the *F*-test results.

	A	B	C	D
1	economics	medicine	history	
2	42	69	35	
3	53	54	40	
4	49	58	53	
5	53	64	42	
6	43	64	50	
7	44	55	39	
8	45	56	55	
9	52		39	
10	54		40	
11				

E	F	G	H	I	J	K
Anova: Single Factor						
SUMMARY						
Groups	Count	Sum	Average	Variance		
Column 1	9	435	48.33333	23.5		
Column 2	7	420	60	32.33333		
Column 3	9	393	43.66667	50.5		
ANOVA						
Source of Variation	SS	df	MS	F	P-value	F crit
Between Groups	1085.84	2	542.92	15.19623	7.16E-05	3.443357
Within Groups	786	22	35.72727			
Total	1871.84	24				

Answer:

In this case since the F-value (15.196) > F_{crit} (3.443) this indicates that not all the means are equal. At least one of the means is statistically different. However, ANOVA does not tell you where the difference lies. We need to run a t-test to test each pair of means.

7. When do you use a one-tailed t-test versus a two-tailed t-test?

Answer:

One-tailed t-test: A one-tailed t-test is used when you are specifically predicting the direction of the effect. In other words, you have a prior expectation that the effect will occur in one direction only (either positive or negative). You are focusing the statistical power on detecting that specific directional effect when using a one-tailed test.

Example: Suppose you want to investigate whether a new teaching method improves test scores. You think that the new method will lead to higher scores. In this case you would use a one-tailed t-test because you are only interested in detecting a positive effect (higher scores) and not considering the possibility of a negative effect.

Two-tailed t-test: A two-tailed t-test is used when you do not have a specific directional prediction or when there is a possibility of an effect in either direction. You are testing for the possibility of a significant effect in both positive and negative directions with a two-tailed test.

Example: Let's say you want to examine whether a new medication affects reaction time. Your hypothesis is that the medication will have an effect, but you do not have a specific expectation regarding the direction of the effect. In this case, you would use a two-tailed t-test to assess whether the medication has any significant effect on reaction time regardless of whether it speeds up or slows down the reaction.

Keep in mind that using a one-tailed test requires a stronger justification as you are focusing on a specific direction. While a two-tailed test is more conservative and allows for the detection of effects in both directions.

	A	B	C
1	t-test Paired Two Samples for Means		
2			
3	Mean	87.56	
4	Variance	1.8	
5	Observations	30	
6	t-Stat	1.567	
7	p (T<=t) one-tail	0.301	
8	t Critical one -tail	2.11	
9	p(T<=t) two-tail	0.0116	
10	t Critical two -tail	2.45	

8. Discuss the following results in terms of the statistical significance for the findings while assuming an alpha value of 0.05.

Answer:

One-Tailed Test

We can compare the ***P(T<=t)* one-tail** value (0.301) with the alpha value we selected (0.05). Since the ***p*-value** is greater than alpha, we can confirm that the difference in variance is not statistically significant at the 95% confidence level.

We can compare the **tStat** (1.567) with the ***t* Critical one-tail** (2.11). Since the **tStat** is less than the ***t* Critical one-tail**, we can conclude our results are not statistically significant at the 95% confidence level. If we selected other samples we would not get similar results.

These results are not statistically different and should not be used in decision making.

Two-Tailed Test

We can compare the ***P(T<=t)* two-tail** value (0.0116) with the alpha value we selected (0.05). Since the *p*-value is less than alpha, we can confirm that the difference in means is statistically significant at the 95% confidence level.

We can compare the **tStat** (1.567) with the ***t* Critical two-tail** (1.45). Since the **tStat** is greater than the ***t* Critical two-tail**, we can conclude our results are statistically significant at the 95% confidence level. If we selected other samples we would get similar results.

These results are statistically different and can be used in decision making.

Chapter 12
Significance Tests Part 2

Linda Herkenhoff and John Fogli

Key Concepts

Chi-square test and *z*-test.

Discussion

The significance tests covered in this chapter include the χ^2-test and the *z*-test. The conditions for selecting either of these significance tests are included in this chapter as well as in the previous chapter. See Table 12.1.

Remember, significance only says that the difference is actually true. Never allow statistics to overrule common sense.

The choice of the best technique depends on the type of questions to be answered, the number of variables and the measurement scale, sample size and how much is known about the population and whether you need a one-tailed or two-tailed test. Remember none of these tests are necessary if you have access to the entire population. These tests simply advise you whether you should trust your sample results … are these statistically significant results?

Supplementary Information The online version contains supplementary material available at https://doi.org/10.1007/978-3-031-46371-6_12.

L. Herkenhoff (✉)
Saint Mary's College, Moraga, CA, USA

J. Fogli
Diablo Valley College, Pleasant Hill, CA, USA

L. Herkenhoff, J. Fogli (eds.), *Applied Statistics for Business and Management using Microsoft Excel*, https://doi.org/10.1007/978-3-031-46371-6_12

Table 12.1 Selecting the correct significance test

Test	When to use	One tailed	Two tailed
F-test (Chap. 11)	Repeated measure using ANOVA. Same respondents under different conditions	×	
	Overall regression model significance	NA	NA
	Equality of 2 group means	×	×
	Between Group ANOVA. Test several variables within your group of data	×	×
t-test (Chap. 11)	Equality of means when population std dev unknown. Can use with small samples	×	×
	Matched Pair test. Before and after analysis with same respondents	×	×
	Regression. Tests each regression co-efficient	NA	NA
Chi-squared (this chapter)	Goodness of Fit for model	NA	NA
	Independence of variables	NA	NA
z-test (for normally distributed data) (this chapter)	Testing the mean (or proportion) of a population versus a standard	×	×
	Comparing the means (proportions) of two populations, with large sample size ($n \geq 30$)	×	×
	Matched Pair test. Before and after analysis with same respondents Samples should large ($n \geq 30$)	×	×

Significance Tests

The next two sections will define the χ^2-test and the z-test with examples. The Excel tools will be discussed for each of these significance tests.

χ^2-Test

The χ^2-test can be used effectively under the several conditions:

- *Example 1: Goodness-of-Fit* for estimating how closely an observed distribution of one variable matches its expected distribution.
- *Example 2: Independence of Two Variables* for estimating whether two random variables are independent.

The *Chi-Square statistic* compares the tallies or counts of categorical responses between two (or more) independent groups. There are a number of features of the social world we characterize through categorical variables: religion, political preferences, etc. It can be used to compare the observed frequency distribution with the expected distribution.

If we are concerned about whether the data distribution is normal the simple Excel normality test is simply to enter the data into an Excel histogram and eyeball how close the shape of the graph matches with a normal distribution. If there is still a question, the next (and easiest) normality test is the Chi-Square Goodness-of-Fit test.

The Chi-Square Goodness-of-Fit test is less well known than some other normality tests such as the Kolmogorov-Smirnov test, the Anderson-Darling test, or the Shapiro-Wilk test. The Chi-Square Goodness-of-Fit test is a lot less complicated, and a whole lot easier to implement in Excel than any of the more well-known normality tests.

Rejection Rule

When the computed χ^2 statistic exceeds the critical value for a given probability level, then we can reject the null hypothesis. If you select an $\alpha = 0.05$ (CI = 95%), then you are accepting a 5% chance that your significance is wrong. If the rejection rule provided statistically significant results, there would still be a 5% chance the results are only due to chance.

Example 1: Goodness-of-Fit Test
Consider the case of a gambler playing craps (dice game) in a local casino. The casino and the gamblers are always interested in fairness and cheating in games of chance. Since such games usually involve money, there is an incentive for both sides to try to rig the games in their favor.

The goodness-of-fit test can be used to examine cheating in gambling. In the game of craps, two dice are thrown. Most dice used in wagering have six sides, with each side having a value of one, two, three, four, five, or six. If the dice being used are fair, the chance of any particular number coming up is the same: 1 in 6. However, if the dice have been tampered with, often referred to as "loaded," then certain numbers will have a greater likelihood of appearing, while others will have a lower likelihood. In other words we compare the observed distribution of data with the expected distribution.

Consider the example in which the casino's resident statistician collects data from one of the craps tables (Fig. 12.1). The two dice were thrown 60 times and he recorded the outcomes for one of those dice. The expected value for each possible outcome is 10 in 60 or more simply 1 in 6. These values are given in the last column.

In the Observed Frequency column there are more 1's and 6's than expected, and fewer 2's, 3's, 4's, and 5's than expected. The chi-square function can be used to estimate the likelihood that the values observed on die #1 occurred by pure chance, and therefore prove that there is no relationship between the observed frequency data.

The first step is to compare the observed and expected values. This information is input to Excel to calculate the significance level (p-value) using the chi-square statistic (CHISQ.TEST).

In this example the expected data follows a uniform distribution but often the chi-square test is used when the expected data follows a normal distribution or other distributions.

Value on Die #1	Observed Frequency	Expected Frequency
1	16	10
2	5	10
3	9	10
4	7	10
5	6	10
6	17	10
Total	60	60

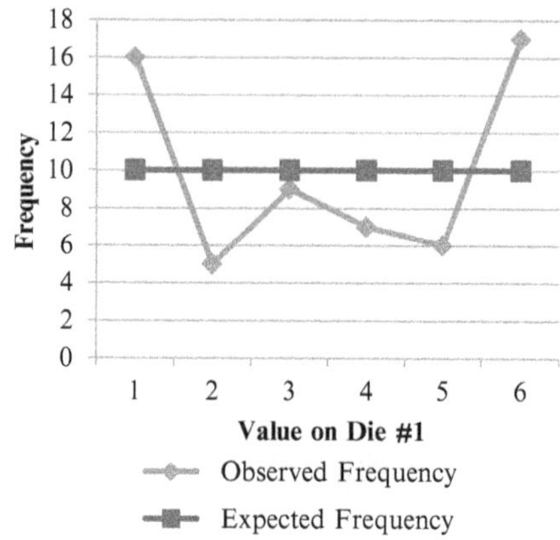

Fig. 12.1 Frequency table for die #1

You may also need the actual chi-square statistic for reporting purposes. So the second step calculates χ^2 by using the CHIINV function in Excel (the inverse of the chi-square) using the p-value and the degrees of freedom.

The p-value is different from the chi-square statistic itself; it is the probability of observing a chi-square at least as large as the one you determine for the dataset, if the null hypothesis is true. The null hypothesis is that the two variables (observed frequency and expected frequency in this case) are unrelated. The alternate hypothesis is that the two variables are related in some way. Generally, if the p-value is lower than some predetermined level, typically 0.05 (α level), then we reject the null hypothesis. By rejecting the null hypothesis means there is some relationship between the expected and observed variables and you can continue with your analysis. In this particular case that would mean the observed outcomes are not due to chance … probably loaded dice!

Example 2: Independence of Two Variables

The other primary use of the chi-square test is to examine whether two variables are independent or not. In this case there are datasets associated with each of the variables, unlike the goodness-of-fit application where you have only observed data for one variable. In this significance test independence means that the two factors (variables) are not related. Typically in social science research, we're interested in finding factors that are related—education and unemployment, stock price and CEO salary, age and consumer behavior. In this case, the chi-square can be used to assess whether two variables are independent or not.

More generally, we say that the y variable is "not correlated with" the x variable if more of one is not associated with more of another. If two categorical variables are correlated their values tend to move together, either in the same direction ($+r$) or in the opposite ($-r$). In most cases we hope for a relationship between the two factors.

In this example the two variables are gender and promotions. We have recorded how many promotions were handed out and how many people did not get promotions. Within those two datasets we kept track of gender. We want to know whether

Fig. 12.2 Input data

	Promotion	No Promotion	Total
Females	46	71	117
Males	37	83	120
Total	83	154	237

male or female nurses at a local hospital are more likely to gain promotions within the first 2 years of employment with a confidence level of 95% ($\alpha = 0.05$). In other words are the promotions independent of gender? Is there a statistically significant relationship between gender and promotions (H_a) or are these differences simply due to chance (H_0)? The data is provided in Fig. 12.2.

In Fig. 12.2 there are 237 nurses at this hospital, 117 of whom are female while 120 are male. At first review of the data it appears that female nurses have received more promotions (46) than male nurses (37). The null hypothesis is that the two variables are independent, that the likelihood of getting promoted is the same for males and females.

Rejection Rule The goodness-of-fit problems, when the calculated χ^2 statistic (χ^2_{calc}) exceeds the critical χ^2-value (χ^2_{α}) for your selected probability level, then we can reject the null hypothesis of equal distributions. If you select an $\alpha = 0.05$ (CI = 95%) then you are accepting a 5% chance that your significance finding is wrong. If the rejection rule provided significant results, there will still be a 5% chance the results are not significant. If the results were indicated as nonsignificant, there would still be a 5% chance the results are significant.

Excel

Excel examples will be provided for each of the two χ^2 problem types:

- Example 1: Goodness-of-Fit
- Example 2: Independence of Two Variables

Example 1: Goodness-of-Fit

You are planning on stocking your grocery counter with chewing gum so you need to know the preferred brand. You have limited time and limited money so you survey a very small sample of your customers, $n = 15$. You ask them which is their preferred brand of gum {Brand 1, Brand 2, Brand 3}. You want a 90% confidence level ($\alpha = 0.10$). Test to make sure the results are in fact statistically significant. The data is listed below:

Brand 1: 10
Brand 2: 3
Brand 3: 2

This is a problem involving goodness-of-fit with the χ^2 test.

- Enter the data and the labels as indicated in Fig. 12.2

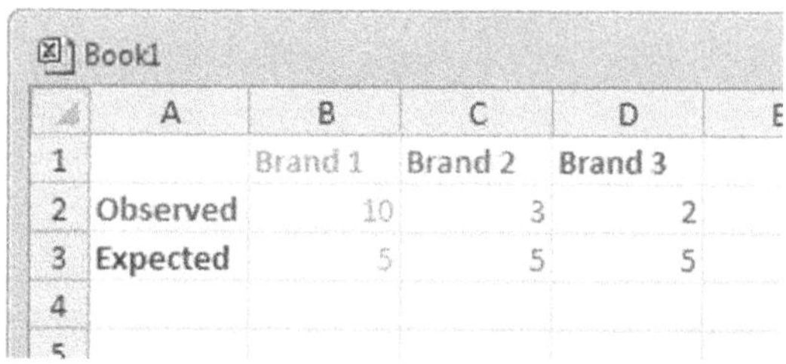

To calculate the expected values, we know that the brands all have an equal chance of being selected and there are 15 possible outcomes, so each gum has a 1 in 5 chance of being selected.

- Click on cell B5
- Click on white input ribbon and type = CHISQ.TEST(B2:D2, B3:D3)

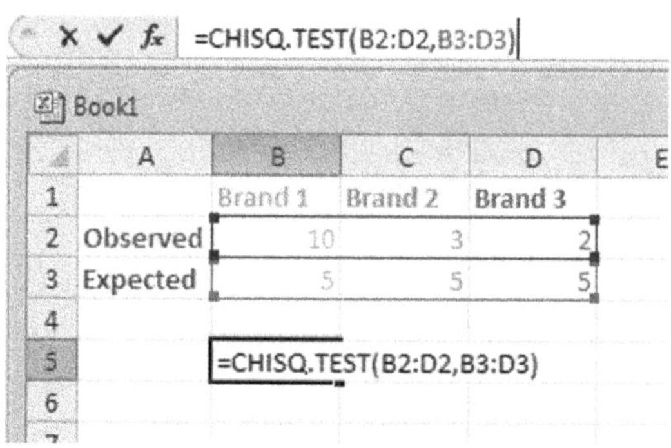

- Click on the checkmark

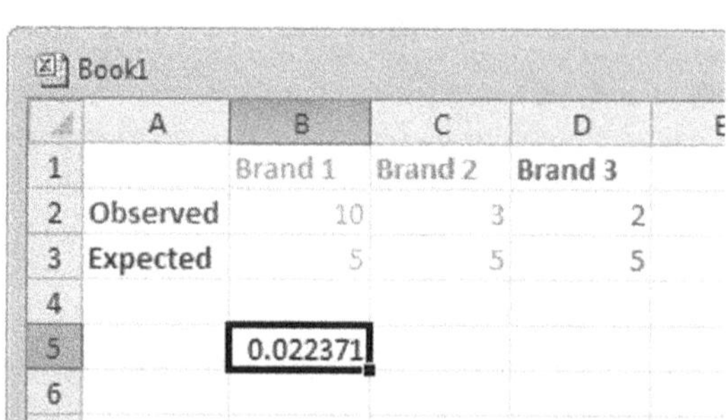

This gives the result **0.0224**. This *p*-value indicates that there is a significant difference between the observed frequencies and the expected frequencies, which is unlikely to be simply due to chance error since the calculated *p*-value is less than the alpha value of 0.05. These results are statistically significant; in other words, you should stock more Brand 1 because your customers really are more likely to purchase this type of gum.

You could also run the significance test using the X^2 statistic test where you compare the χ^2_{calc} with the $\chi^2_{critical}$. If $\chi^2_{calc} > \chi^2_{critical}$ you can reject H_0 and acknowledge that the observed distribution is real and not just due to chance.

Example 2: Testing Independence

Your company conducted an insecticide trial on a group of plants. The data represents two datasets in which one group of plants received insecticide and the other group did not. Within these two datasets you also recorded whether the plant lived or died. You hypothesized that the plants being treated with the insecticide would survive better than those that did not receive the insecticide (H_a). The null hypothesis (H_o), that you hope is not true, would be that there is no statistically significant relationship between plant survival and insecticide treatment.

H_o: The survival of the plants is independent of insecticide treatment
H_a: The survival of the plants is associated with insecticide treatment
Note: $H_a = H_\alpha$ in terms of terminology.

You conduct the study and collect the data as shown in Fig. 12.3.

Before you build your marketing claims you need to be sure that there is a real relationship between the insecticide treatment and the number of surviving plants at the 95% confidence level. Your company wants to avoid being sued for false advertising.

- Input **Dead** into B1
- Input **Alive** into C1
- Input **Total** into D1
- Input **Treated** into A2
- Input **Not Treated** into A3
- Input **Total** into A4
- Fill in cells B2 through D4 with the data as given in Fig. 12.3

Fig. 12.3 Number of plants that survived a treatment

	Dead	Alive	Total
Treated	36	14	**50**
Not treated	30	25	**55**
Total	**66**	**39**	**105**

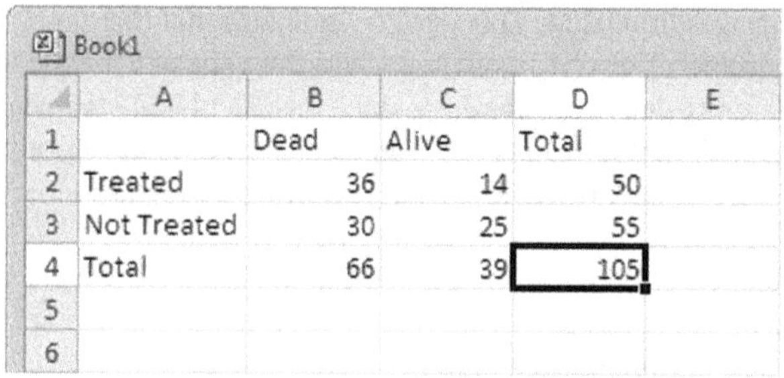

Calculate expected values $E(x) = (\text{row total} \times \text{column total})/N$

- In cell B6 type **= D2 * B4/D4** into the input ribbon

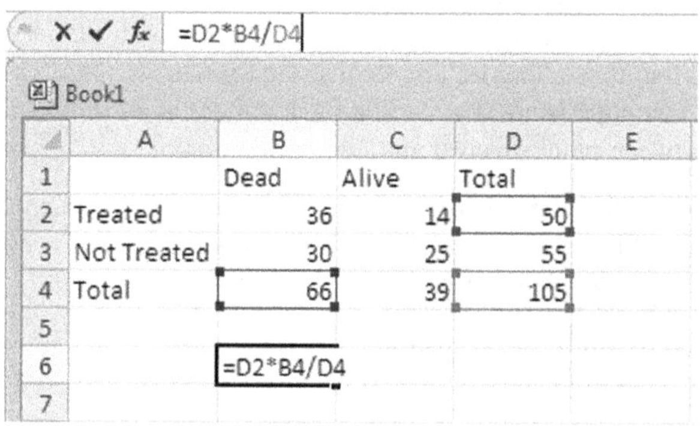

- Click green checkmark
- In cell B7 type **= D3 * B4/D4** into the input ribbon

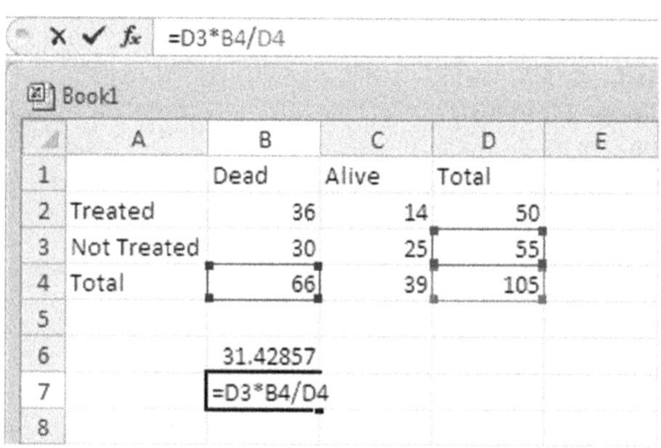

- Click the green checkmark
- In cell C6 type **= D2 * C4/D4** into the input ribbon

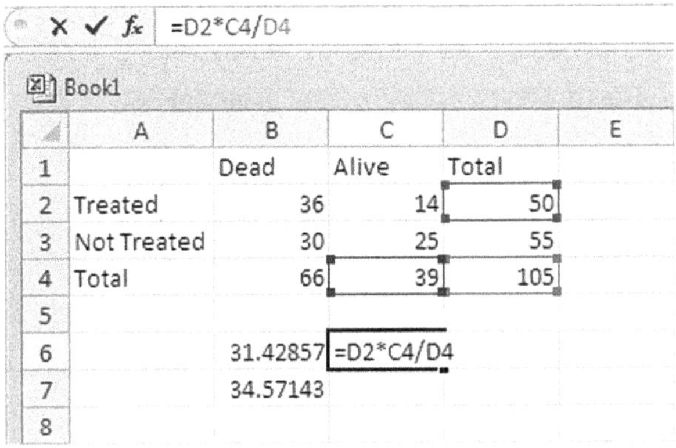

- Click the green checkmark
- In cell C7 type **= D3 * C4/D4** into the input ribbon

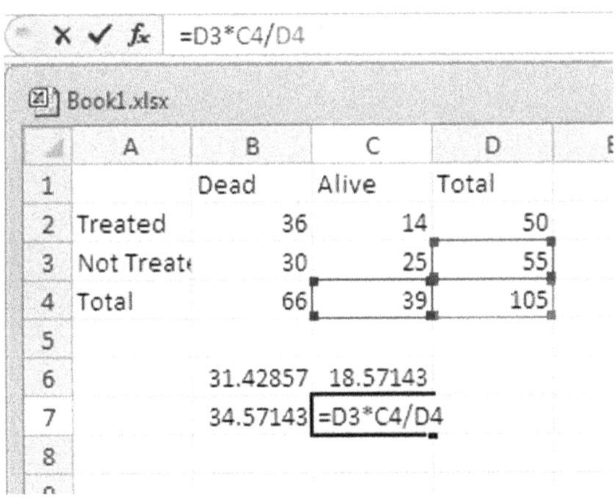

- Click the green checkmark

Calculate **degrees of freedom**: (number of columns minus one) * (number of rows minus one). No real easy way to do this except to actually count the rows and columns in your pivot table and use Excel as a calculator.

- In cell E2 type = **(2 – 1) * (2 – 1)** into the input ribbon

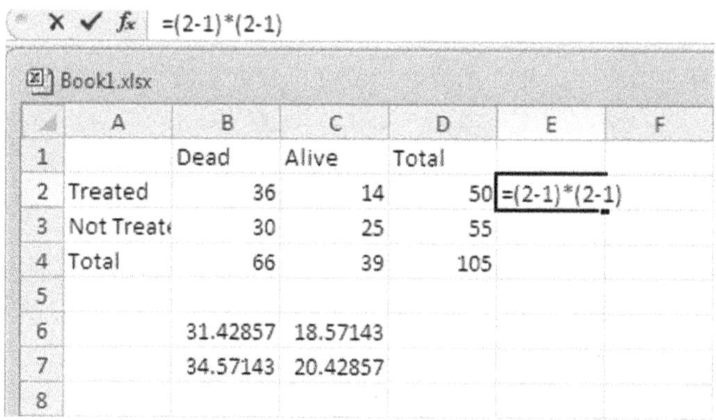

- Click the green checkmark

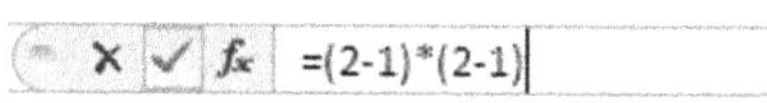

Calculate the *p*-value

- Click on cell F2
- Click on the ***fx*** button

- A dialog box will appear. Select **Statistical** from drop-down menu next to "or select a category"

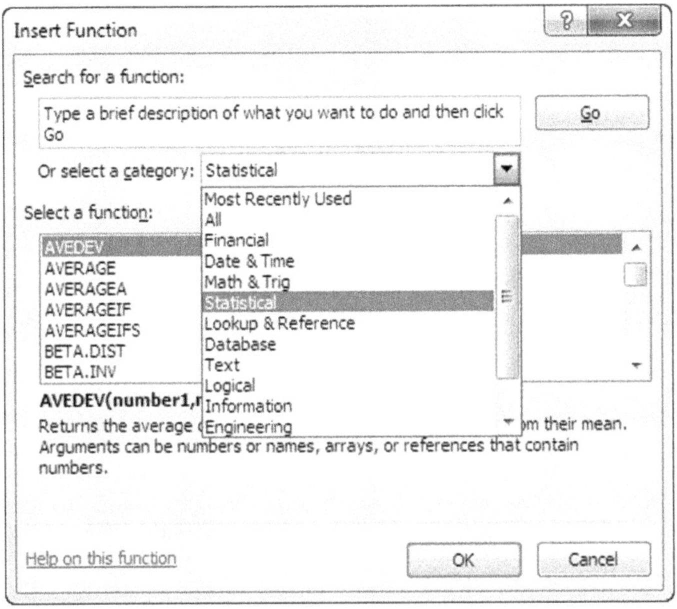

- Select **CHISQ.TEST** under "Select a function":

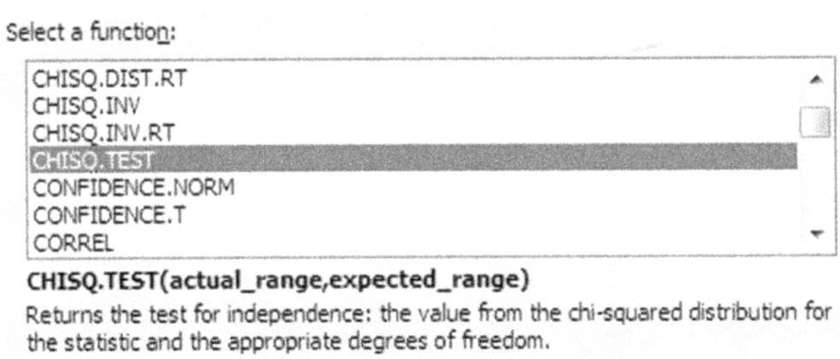

- Click **OK**

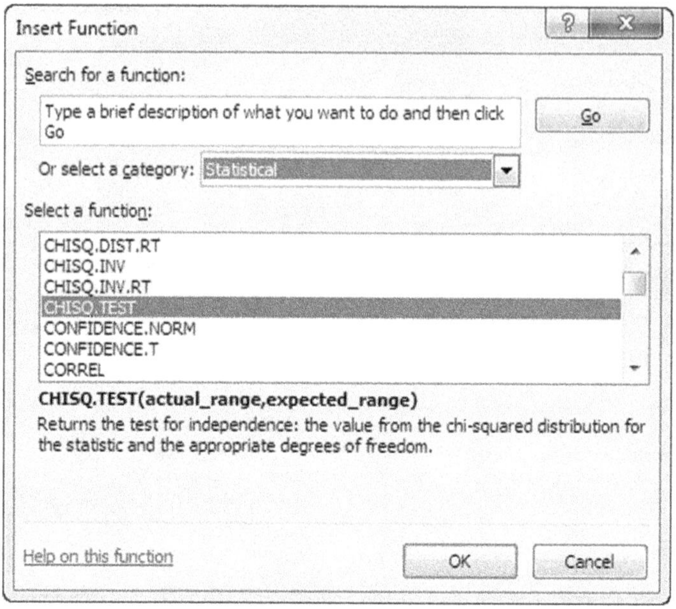

- Select the **Actual Range** (cells B2 through C3)

- Place cursor on **Expected Range** (cells B5 through C7)

- Click **OK**

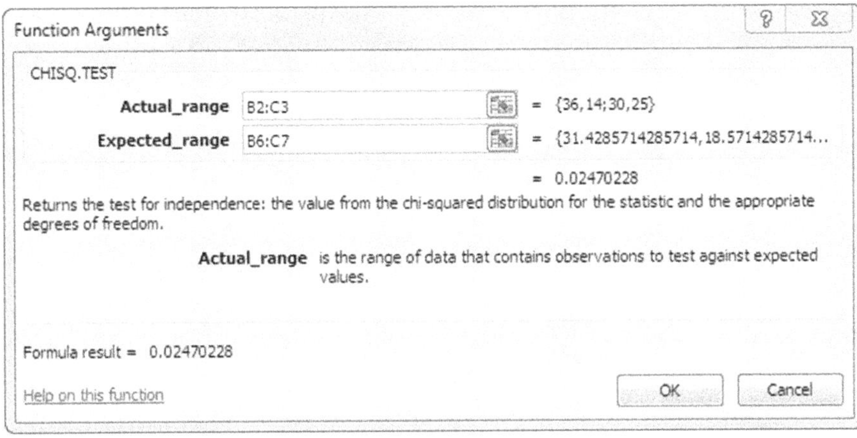

Calculate the actual test statistic χ^2 if you need it for reporting purposes.

- **Note:** In most cases knowing the *p*-value is sufficient for basic analysis.
- In cell G2, type =(((B2 – B6)^2)/B6) + (((B3 – B7)^2)/B7) + (((C2 – C6)^2)/C6) + (((C3 – C7)^2)/C7)

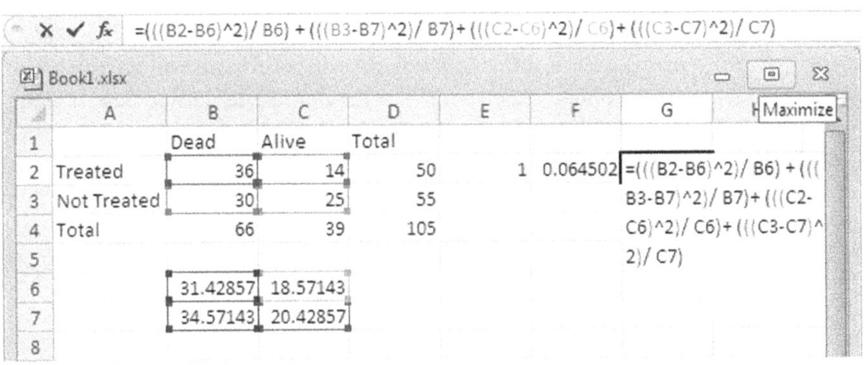

- Click the green checkmark

The final worksheet will look like Fig. 12.4. The reported outcome statistics using just the *p*-value data would read χ^2 (1) = 3.42, *p* = 0.065. These are not statistically significant results for our required level of confidence since the calculated *p*-value (0.065) is more than the *p*-value of our selected confidence level (0.05). Likewise, if we had calculated the $\chi^2_{critical}$ we would have found that the $\chi^2_{calc} < \chi^2_{critical}$. We can state that H_0 is true; the survival of the plants is not dependent on the insecticide treatment. We cannot claim our insecticide has a statistically significant relationship with the health of plants.

Book1.xlsx

	A	B	C	D	E	F	G	H
1		Dead	Alive	Total				
2	Treated	36	14	50	1	0.064502	3.417673	
3	Not Treated	30	25	55				
4	Total	66	39	105				
5								
6		31.42857	18.57143					
7		34.57143	20.42857					
8								

Fig. 12.4 Output

z-Test

The ***z*-test** is used when you **know the population mean and standard deviation.** Thus, it is obviously more accurate than the *t*-test where you must approximate these input statistics.

- The *z*-test works best with datasets that have at least 30 data points.

  The *z*-test can be used in the following situations:

- **Example 1 (Mean or proportion vs. a standard)**: The recent CPA exam results which are normally distributed in California averaged 78% but the US standard is 82%. We could use a *z*-test to determine if these results are real or simply due to chance in a 90% confidence interval. The data follows a normal distribution.
- **Example 2 (Testing the means or proportions of two populations/Matched pair test)**: To use this test to compare the means of two populations you need large ($n \geq 30$) samples. The average employee ethics score for the Plano office was 66%. The average score for the Detroit office was 71%. The *z*-test can help determine if these results are statistically significant or just due to chance with 90% confidence. The data follows a normal distribution.

The second part of this test includes the "Matched pair" which is used to compare the means before and after something is done to the samples. The *z*-test is used when the samples are large ($n \geq 30$). The variable is the difference between the before and after measurements. The employees of Aaron Building Supplies completed a workshop on effective communication. The company is wondering if the workshop is effective so they measure communication patterns before and after the workshop.

This test establishes the critical *z*-values in a normal distribution for a given confidence level. It determines if your sample statistic falls within the confidence level. In the ***z*-test** the critical value is determined solely by your chosen significance level, whereas in a ***t*-test** the critical values are determined by the significance level and the sample size.

Rejection Rule

Note $z_{\text{calculated}}$ is calculated from the problem you are solving.

$$\text{Reject } H_0 \text{ if } \text{ABS}\left[z_{\text{calculated}}\right] > \text{ABS}\left[z_{\text{critical}}\right]$$

The α indicates the confidence level you choose for the analysis. In other words the $z_{\alpha/2}$ is the critical or boundary value defining the rejection areas in the tail(s). If you select an $\alpha = 0.05$ (CI = 95%) then you are accepting a 5% chance that your significance finding is wrong. On the flipside if the results were indicated as nonsignificant, there would still be a 5% chance the results are significant. Remember the only way you can be 100% sure with your results is to use the entire population of data available, otherwise there is always a probability associated with your results.

Excel

Excel examples will be provided for each of the z-test problem types.

Example 1: Mean or proportion vs. a standard
Example 2: Testing the means or proportions of two populations/Matched pair test

The **Z.TEST** Excel tool will also be described with an example as an alternate means of comparing mean or proportion data with a standard.

Example 1: Z-Test One-Sample Mean Versus a Standard
- **Note: This can also be used in the same way for Proportion vs. a Standard.**

Of all of the individuals (population) who develop sunburn, suppose the mean recovery time for individuals is 30 days with standard deviation equal to 6 days. A pharmaceutical company manufacturing a certain type of aloe lotion wishes to determine whether the lotion shortens, extends, or has no effect on the recovery time. The company chooses a random sample of 100 individuals who have used the lotion, and determines that the mean recovery time for these individuals was 28.5 days. They want the analysis to incorporate a 95% confidence level (alpha = 0.05).

(a) Is the lotion effective in reducing the recovery time? (One-tailed z-test)
(b) Does the lotion have any effect at all? (Two-tailed z-test)

- Prepare an output table by typing in labels and data from Fig. 12.5

- To calculate **Standard Error of the Mean** input the formula in cell B6 (across from its label) **= B4/SQRT (B5)**

Fig. 12.5 Input

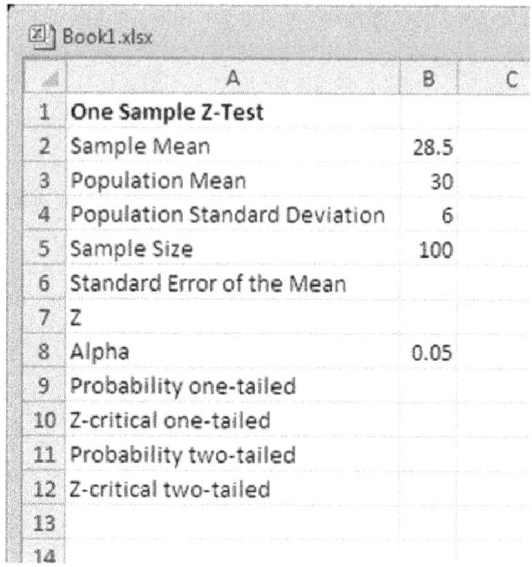

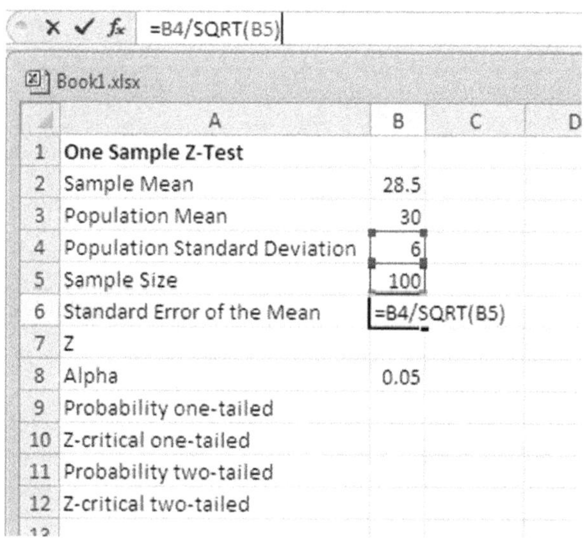

- Remember the standard error of the mean provides an approximation of the average amount by which the sample mean deviates from the population mean.
- Click the **checkmark**

- Click on cell B7. To calculate the z-statistic click on the white formula ribbon and insert = **(B2 – B3)/(B6)**

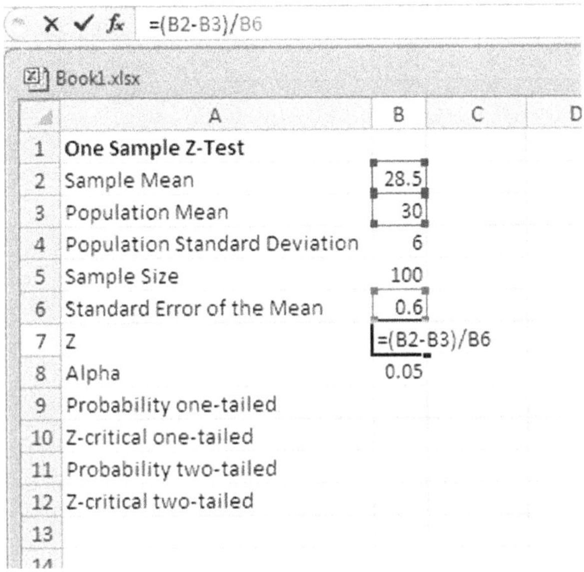

- Click the checkmark
- Click on cell B9. To calculate the probability for a one-tailed test click on the white formula ribbon and insert = **1-**. Now click on the *fx* symbol to the left of the input ribbon

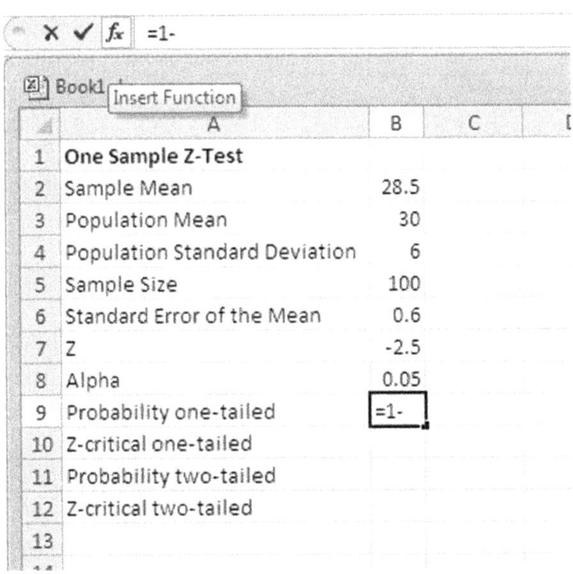

- In the pop-up window, select the category **Statistical**

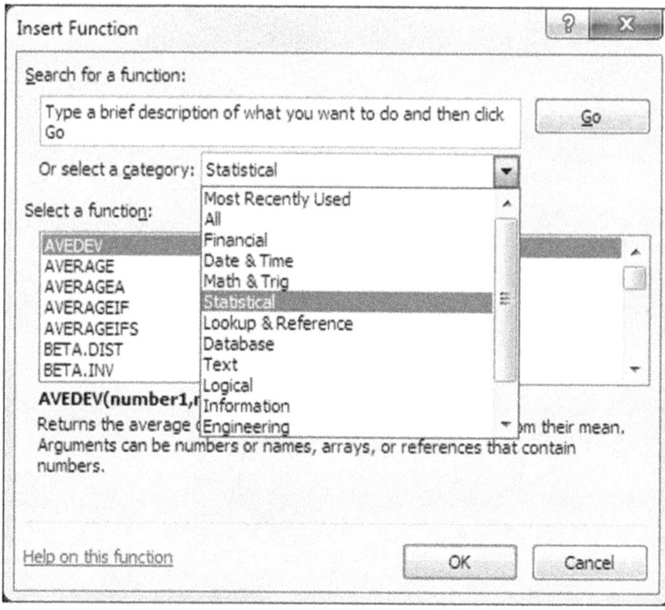

- Select **NORM.S.DIST** (Remember from Chap. 4 **NORM.S.DIST** tells you the distribution to the left of this z-statistic for a mean of zero and a SD of one)

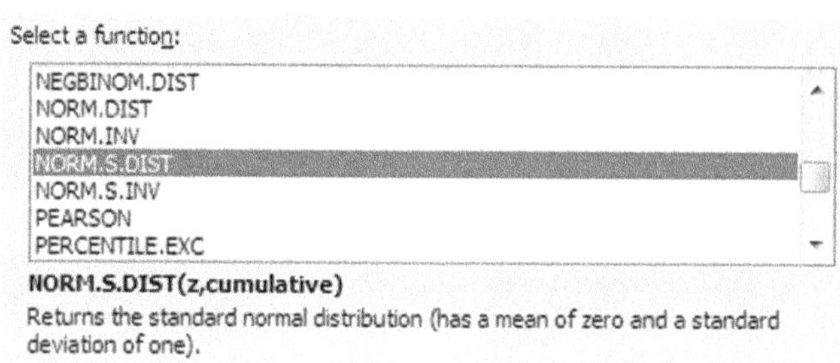

- Click **OK**
- Type **ABS(B7)** for **Z** which provides the absolute value which makes the task of interpreting the statistical significance of z easier

- Type **TRUE** in the cumulative box to ensure the cumulative (everything to the left) distribution is calculated

- Click **OK**

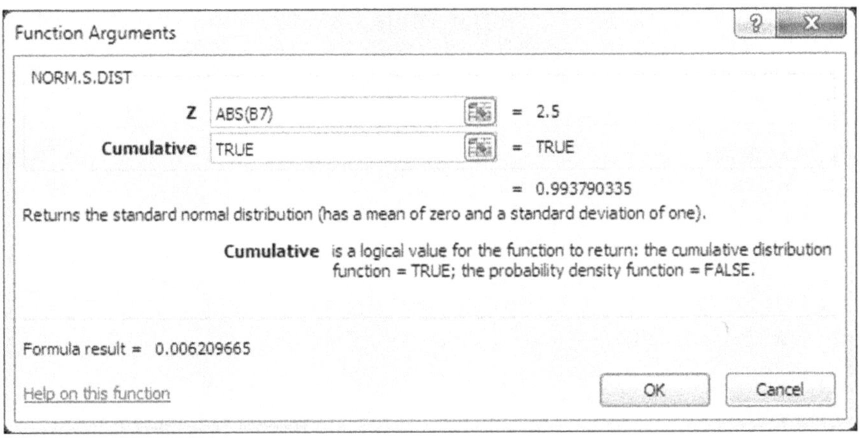

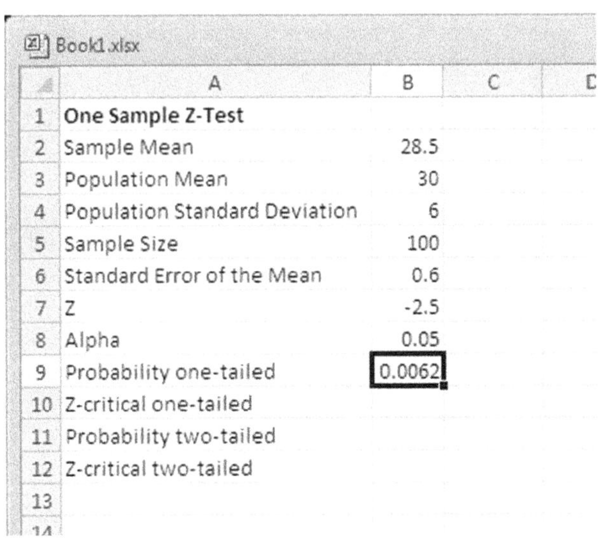

- Click on cell B10. To calculate the *z*-Critical for a one-tailed test click on the white formula ribbon and insert = **ABS(NORM.S.INV(0.05))**

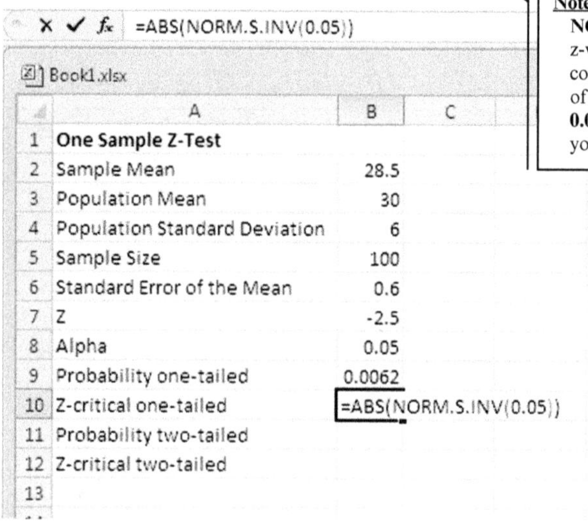

Note: Remember from Chapter 4 **NORM.S.INV** tells you the z-value associated with your confidence interval usinga mean of zero and a SD of one. Use **0.05** for the **Probability** which is your alpha value for a CI of 95%.

- Click the **Checkmark**
- **Note:** You will get an error if you do not include closing brackets at the end of the formula. So remember to add the brackets.

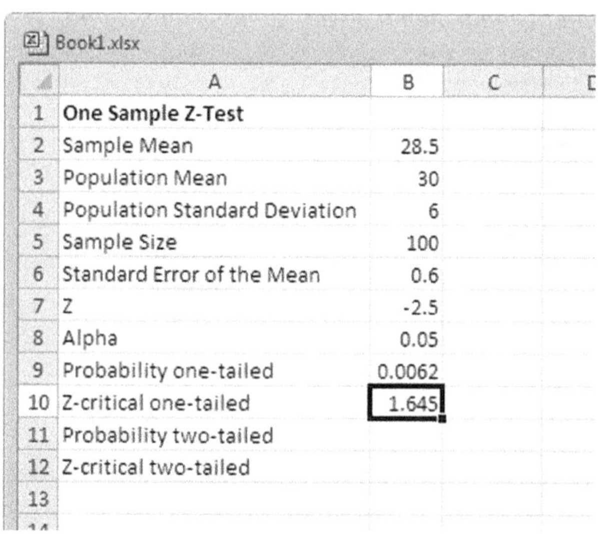

- **Note:** You will always get this same answer for z-critical in all problems whenever you use a confidence level of 95% in a one-tailed test. You could also look this up in the standard z-table.
- Click on cell **B11**. Because the two-tailed probability is just twice the one-tailed probability type in the formula ribbon = **2*B9**

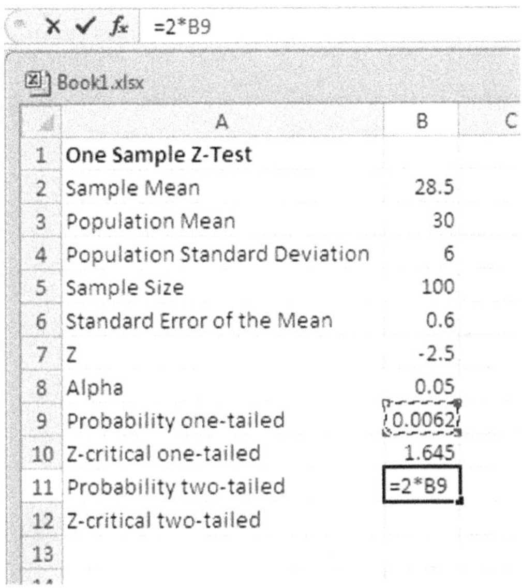

=2*B9

Book1.xlsx

	A	B	C
1	**One Sample Z-Test**		
2	Sample Mean	28.5	
3	Population Mean	30	
4	Population Standard Deviation	6	
5	Sample Size	100	
6	Standard Error of the Mean	0.6	
7	Z	-2.5	
8	Alpha	0.05	
9	Probability one-tailed	0.0062	
10	Z-critical one-tailed	1.645	
11	Probability two-tailed	=2*B9	
12	Z-critical two-tailed		
13			

- Click on the **Checkmark**

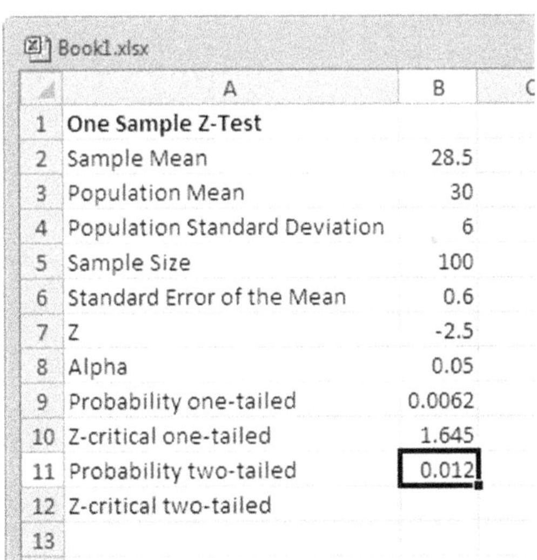

Book1.xlsx

	A	B	C
1	**One Sample Z-Test**		
2	Sample Mean	28.5	
3	Population Mean	30	
4	Population Standard Deviation	6	
5	Sample Size	100	
6	Standard Error of the Mean	0.6	
7	Z	-2.5	
8	Alpha	0.05	
9	Probability one-tailed	0.0062	
10	Z-critical one-tailed	1.645	
11	Probability two-tailed	0.012	
12	Z-critical two-tailed		
13			

- Click on cell B12. To calculate the *z*-Critical for a two-tailed test click on the white formula ribbon and insert = **ABS(NORM.S.INV(0.025))**

Note: Use **0.025** for **Probability**, which is your alpha value for a CI of 95% divided into two tails not just one. As such we need to divide the alpha value in half.

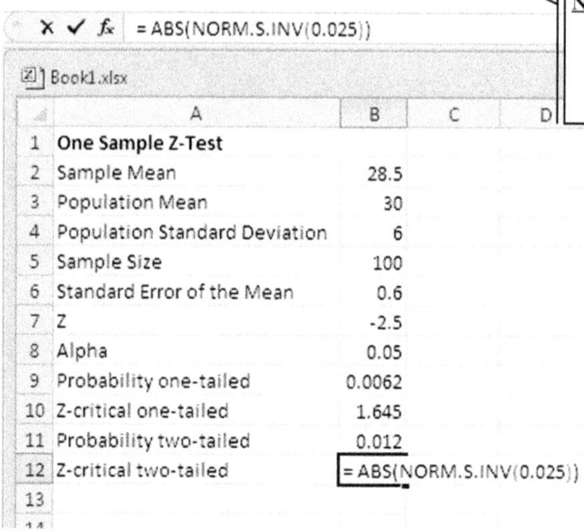

- Click on the **Checkmark**

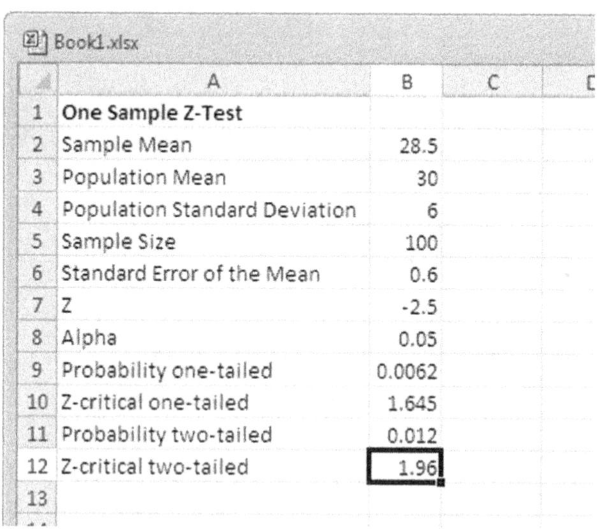

One-Tailed Results

The calculated *z*-**statistic** (**Z**) for this problem is −2.5. The probability of obtaining this $z_{\text{calculated}}$ in the one-tailed test is 0.006 (Probability one-tailed). We check to see if the probability of getting this *z*-**statistic** is less than our alpha value. Since 0.006 is less than 0.05 (Alpha) we can conclude that our results of obtaining a *z*-**statistic** of −2.5 are statistically significant.

Based on our chosen alpha value of 0.05, we obtain a critical z_{critical} that we need to compare to the absolute value of the $z_{\text{calculated}}$ of 2.5. Since the z_{critical} of 1.64 (Z-critical one-tail) is less that the obtained *z*-**value** (absolute value) of 2.5 (Z), the mean value of the sample is real and not just due to chance. Therefore the shorter recovery time of 28.5 days versus 30 days indicates the lotion is effective.

Two-Tailed Results

For the two-tailed test the calculated *z*-**statistic** (**Z**) for this problem is still −2.5. The probability of obtaining this *z*-**value** in the two-tailed test is 0.012 (Probability two-tailed). We check to see if the probability of getting this *z*-**statistic** is less than our alpha value. Since 0.012 is less than 0.05 (Alpha) we can conclude that our results of obtaining a *z*-**statistic** of −2.5 are statistically significant.

Based on our chosen alpha value of 0.05, we obtain a z_{critical} that we need to compare to the obtained absolute value of the *z*-**statistic** of 2.5 (Z). Since the z_{critical} of 1.96 (Z-critical two-tailed) is less that the obtained *z*-**value** (absolute value) of 2.5, the mean value of the sample is real and not just due to chance. Therefore the mean recovery time associated with the use of this lotion is significantly different for the mean of the population. With the two-tailed test we don't know if the sample mean is greater than or less than the population mean, just that the difference is real and not simply due to chance.

Example 2: Testing the Means of Two Populations

- **Note:** This can also be used in the same way for Two-Sample Proportions. This can also be used in a Matched Pair application where you need to test whether before and after results are in fact significant.

In this example we have data from the two samples, and want to test that the differences between the two are real and not just due to chance. The two samples are collected from the same sales organization; the first column included only male respondents and the second column represents female respondents. These employees recorded how many complaints they received about the company from customers over a 12-month window. Is there a real difference between the average complaints received by male employees vs. female employees?

First enter in the data for your two samples, one in Column A and the other in Column B (Fig. 12.6). Note to use this test the number of entries in each column

Fig. 12.6 Input data

Males	Females
43.9781	39.5925
59.4639	74.1280
76.7843	65.4352
67.4777	88.8866
62.8863	73.2028
74.6958	99.2154
81.0606	90.8101
74.1874	89.8319
70.3135	79.9945
76.4352	96.5206
78.9252	88.5916
84.6146	102.2015
63.5628	54.4017
64.2910	71.3893
49.7899	87.9951
68.6028	88.4589
69.5705	84.9577
58.1655	86.8707
39.6463	77.3022
73.3056	92.7214
48.6142	75.3770
50.9503	86.0096
74.3111	67.9071
75.5680	95.6901
75.0190	63.1654
72.2472	48.6237
65.5014	94.5908
61.6964	66.7961
71.6934	82.5949
68.8099	85.2731

does not have to be the same, but the rule of thumb is to have at least 30 entries in each column for the test to work properly.

- Open the **Data** tab and select the **Data Analysis** tool, then select **Descriptive Statistics**

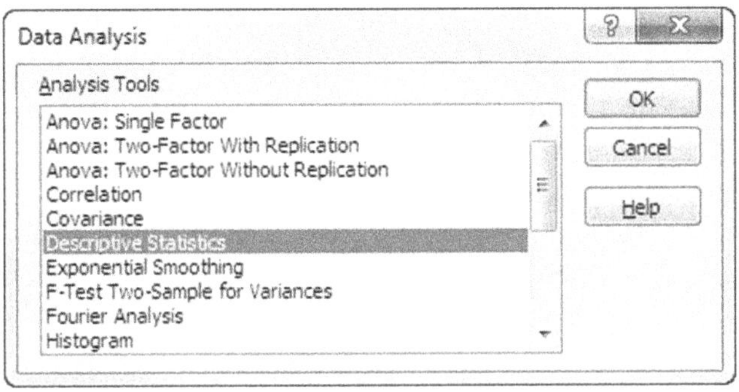

- Highlight the two columns of data as your input field. Check the **Labels in first row box**. Select **Output Range**, click inside the box, and highlight a cell for you output location. Check the **Summary statistics** box

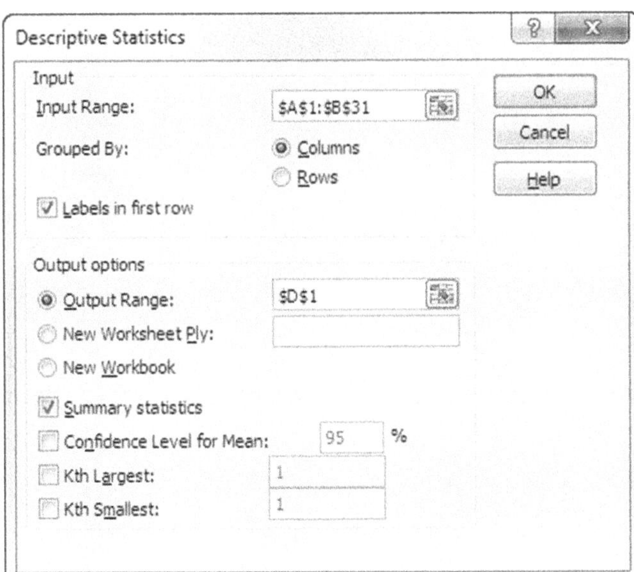

- Click OK

Males		Females	
Mean	66.73892	Mean	79.95118
Standard Error	2.039626	Standard Error	2.769339
Median	69.19016	Median	85.11539
Mode	#N/A	Mode	#N/A
Standard Deviation	11.17149	Standard Deviation	15.1683
Sample Variance	124.8022	Sample Variance	230.0772
Kurtosis	0.145944	Kurtosis	0.563807
Skewness	-0.84898	Skewness	-0.94007
Range	44.96824	Range	62.60904
Minimum	39.64632	Minimum	39.59246
Maximum	84.61456	Maximum	102.2015
Sum	2002.168	Sum	2398.535
Count	30	Count	30

Note: The "**#N/A**" indicates that no number in the data set is replicated; thus there is no mode.

- Select the **Data Analysis** tool and select *z*-**Test: Two Sample for Means**

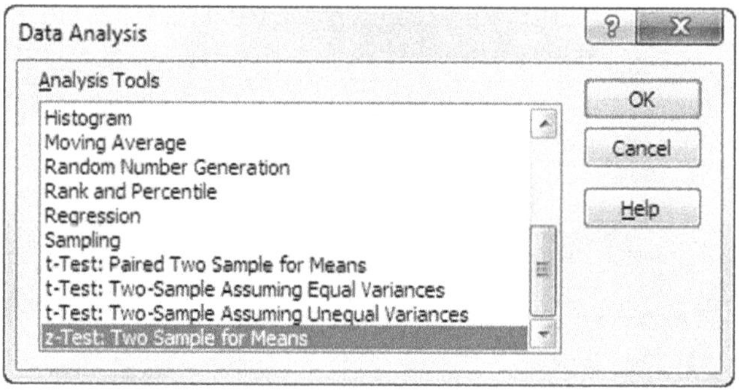

- In the **Variable 1 Range** box, indicate the range of the first sample in Column A (**$A$1:$A$31**)

- In **Variable 2 Range**, indicate the range of the second sample in Column B (**$B$1:$B$31**)

- In the **Hypothesized Mean Difference** box, enter the figure **0**. Since the null hypothesis, which we are attempting to refute, says that both samples come from the same population, we are hypothesizing that the difference between the means for the two populations is 0

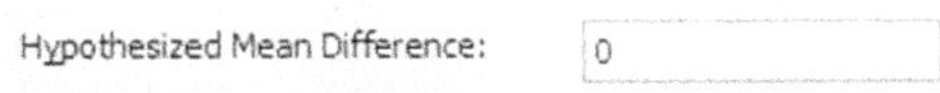

- In the **Variable 1 Variance (known)** box enter the value for the variance for the sample in the A Column from the Descriptive Statistics output. For this sample enter **124.8**. Remember that the variance is just the square of the standard deviation

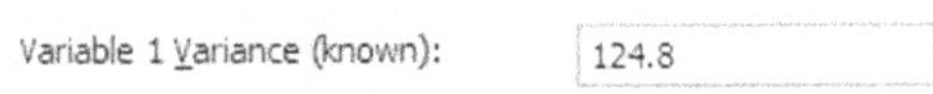

- In the **Variable 2 Variance (known)** box enter the corresponding figure for the variance of the second sample. For this sample enter **230.1**

- Check **Labels**

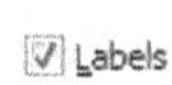

- In the **Alpha**: box the number 0.05 should already appear. If it is empty or shows a number different from 0.05, then enter the number 0.05. The Alpha figure indicates the confidence level for this test. A figure of 0.05 states that you want to be 95% certain of the result or, in other words, that you want the probability of being wrong to be 0.05 or lower

- Select **Output Range**: then in the **Output Range**: box, click on the cell where you want the output to appear. The Output Range table will take up 3 columns and 12 horizontal rows

- Click **OK**

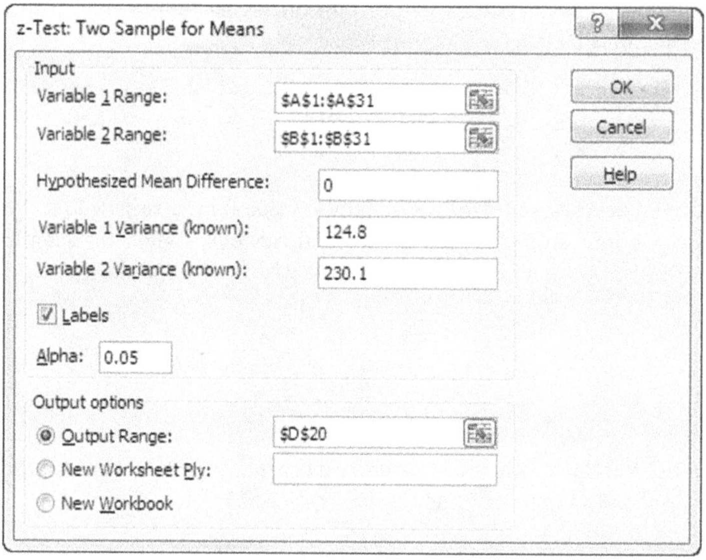

The output table in Fig. 12.7 has the heading: ***z*-Test: Two Samples for Means**. You will need to widen the left-hand column of the table in order to read the names of the items. In the table there are results for the following items: **Mean, Known Variance, Observations, Hypothesized Mean Difference, *z*, *P*(*Z*<=*z*) one-tail, *Z*-critical One-tail, *P*(*Z*<=*z*) two-tail, *Z*-critical two-tail**.

The **Mean** values give the arithmetical average for each sample. It should be the same as the value for the **Mean** in your Descriptive Statistics chart you generated earlier.

The **Known Variance** similarly gives the variance for each sample and corresponds to the variance results generated earlier (these are the values you entered into the ***z*-Test** dialogue box).

The **Observations** is the number of items in each sample.

The **Hypothesized Mean Difference** should be 0 (the value you entered in the *z*-Test dialogue box earlier).

The ***z*-value** indicates in standard deviation units how far from the mean the value for the difference between your two samples is located. Remember that the normal curve for all the differences between all the possible pairs of samples from the population has a mean of 0. Your two samples did not have the same mean; thus they fall away from the mean in the normal distribution. The ***z*-value** tells you how far away the difference falls.

Following the ***z*-value** there are four lines of results, two concerning one-tail and two concerning two-tail testing. You will use one or the other of these pairs of figures, not both. The one you use will depend upon on how you defined your alternative hypothesis, or in other words, how you define failure.

z-Test: Two Sample for Means

	Males	Females
Mean	66.7389	79.9512
Known Variance	124.8	230.1
Observations	30	30
Hypothesized Mean Difference	0	
z	-3.8414	
P(Z<=z) one-tail	0.0001	
z Critical one-tail	1.6449	
P(Z<=z) two-tail	0.0001	
z Critical two-tail	1.9600	

Fig. 12.7 Output

If you have set up your alternative hypothesis such that it argues that one of the populations will have a higher value than the other on your variable of interest, then you are interested only in one end of the distribution curve. This means you should use the one-tail output. In this case an example might be whether the women's higher mean number of customer complaints (80) is real when compared to the mean number of customer complaints that the men receive (67).

However, if your alternative hypothesis simply asserts that there will be a significant difference between the populations, without saying which will be higher or lower, then you need the two-tail output. For example, "There is a difference in the number of customer complaints about the company that male employees receive versus the number female employees receive."

In Fig. 12.7 the *P(Z<=z)* values indicate the probability that the two populations are the same, so you want these values to be very small. In Fig. 12.7 the one-tail *p*-test of 0.0001 suggests that the difference in the mean number of customer complaints received by women is in fact higher than those received by the men. The results are real and not just due to chance. We can be more specific by choosing our level of accepted risk with this outcome by choosing a confidence level. If we choose a 95% confidence level, then the α-value is 0.05. The one-tail *p*-test of 0.0001 is smaller than the value of 0.05. This means that we can be 95% confident that the mean number of customer complaints received by women is in fact higher than those received by the men. This finding is statistically significant.

The two-tail p-test of 0.0001 is also very small indicating that the difference in the means of the two samples is real and not due to chance. If we choose a 95% confidence level we accept that 95 out of 100 times we will get means that are different, but 5 out the 100 times we will not get these results. We can say that the means for customer complaints about the company are different for men and women and this difference is mathematically significant at the 95% confidence level.

Another method to determine if the difference in means is real is to compare the z-value with the ***z*-Critical** value. If the ***z*-value** is less than **the *z*-Critical** value, you affirm the null hypothesis; if the z-value is greater than the ***z*-Critical** value, you reject it, which says your difference in the means is real. In this example the ***z*-value** of 3.84 is greater than the ***z*-Critical one-tail** (1.6449) and ***z*-Critical two-tail** (1.96). In both of these models the difference in the means is real and not due to chance. Remember that by changing the confidence level in a ***z*-test**, the ***z*-Critical** values will also change.

Z.TEST Tool for Comparing a Mean or Proportion with a Standard

Excel also provides a **Z.TEST** tool in the formula function area. This is another method for comparing a mean or proportion with a standard. The **Z.TEST** function returns the one-tailed p-value of a z-test. It can be modified to also return two-tailed results. The input format to use this **Z.TEST** function in Excel is:

$$Z.\mathbf{TEST}\left(\text{array of data, the population mean, the population std. deviation}\right)$$

However we could get the same results if we used our NORMDIST formula from Chap. 4:

$$= 1 - \text{NORMSDIST}\left(\left(\text{sample mean} - \text{population mean}\right) / \left(\text{population SD} / \text{SQRT}\left(\text{sample size}\right)\right)\right)$$

Example Problem

In this section we will use the **Z.TEST** as the Excel tool. This example is based on collecting engineering proficiency quotient scores which happen to follow a normal distribution. The population standard deviation is 15 points, and the mean is 60 points. A sample of 30 engineers is collected and you need to know if based on this sample, that the sample mean is significant or a chance occurrence, at a 95% confidence level.

- Input the data in Fig. 12.8
- Select cell D1 in your blank Excel worksheet, and then paste the entries so that the array data fills cells D1:D31 in your worksheet

Scores
43.98
59.46
76.78
67.48
62.89
74.70
81.06
74.19
70.31
76.44
78.93
84.61
63.56
64.29
49.79
68.60
69.57
58.17
39.65
73.31
48.61
50.95
74.31
75.57
75.02
72.25
65.50
61.70
71.69
68.81

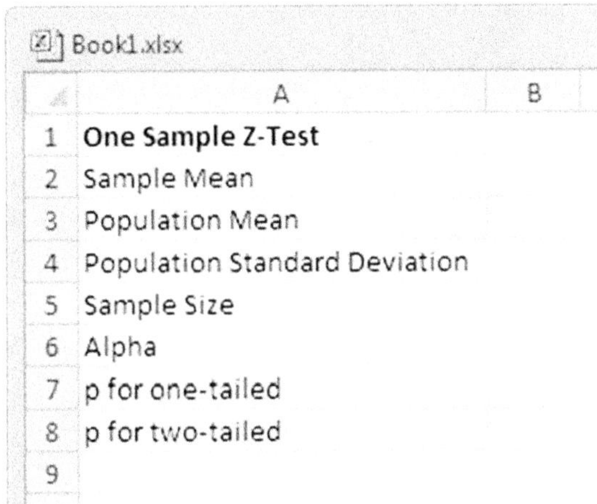
Book1.xlsx

	A	B
1	**One Sample Z-Test**	
2	Sample Mean	
3	Population Mean	
4	Population Standard Deviation	
5	Sample Size	
6	Alpha	
7	p for one-tailed	
8	p for two-tailed	
9		

Fig. 12.8 Input data and template

- Input the population mean of **60** in B3

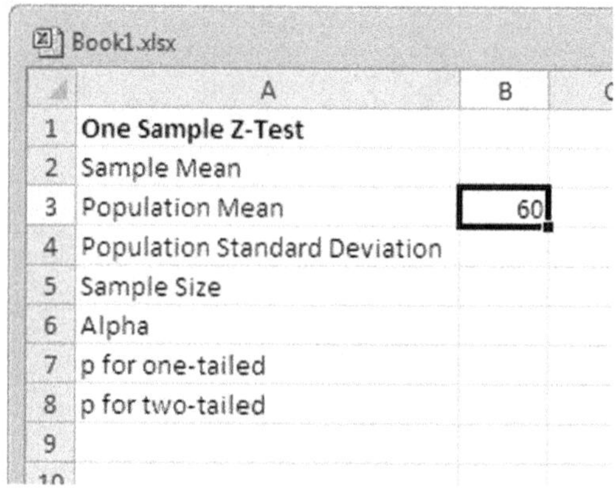

- In cell B4 input the population std. deviation of **15**

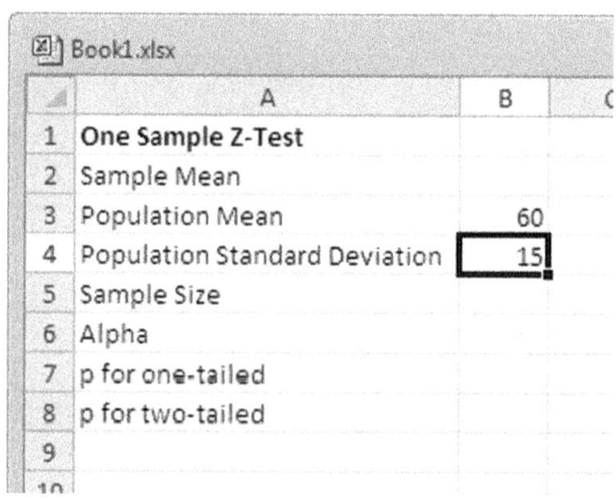

- In cell B5 input the sample size of **30**

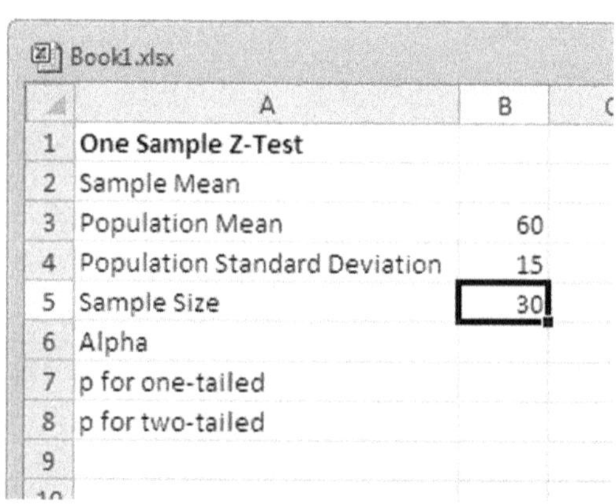

- Click on cell B2. Calculate the mean for the input data by clicking on the formula ribbon and typing = **AVERAGE(D2:D31)**

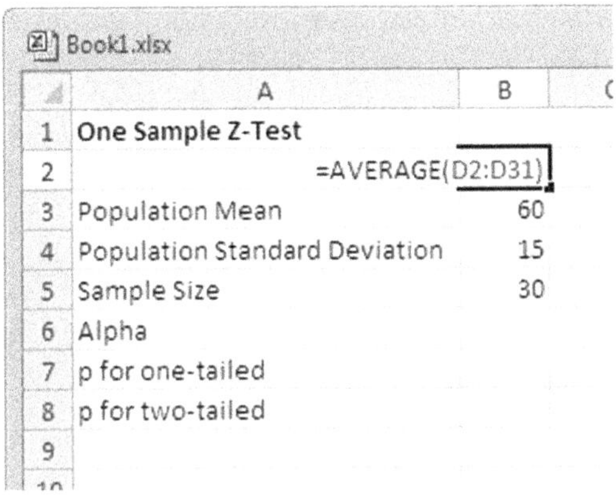
Book1.xlsx

	A	B
1	**One Sample Z-Test**	
2	=AVERAGE(D2:D31)	
3	Population Mean	60
4	Population Standard Deviation	15
5	Sample Size	30
6	Alpha	
7	p for one-tailed	
8	p for two-tailed	
9		

The average engineering proficiency score for this sample is **66.74** (Cell B2).

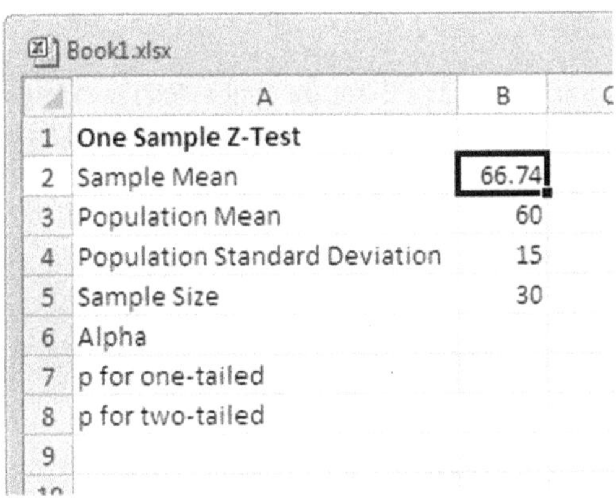
Book1.xlsx

	A	B
1	**One Sample Z-Test**	
2	Sample Mean	66.74
3	Population Mean	60
4	Population Standard Deviation	15
5	Sample Size	30
6	Alpha	
7	p for one-tailed	
8	p for two-tailed	
9		

- Click on cell B6 and input the chosen alpha value **0.05**

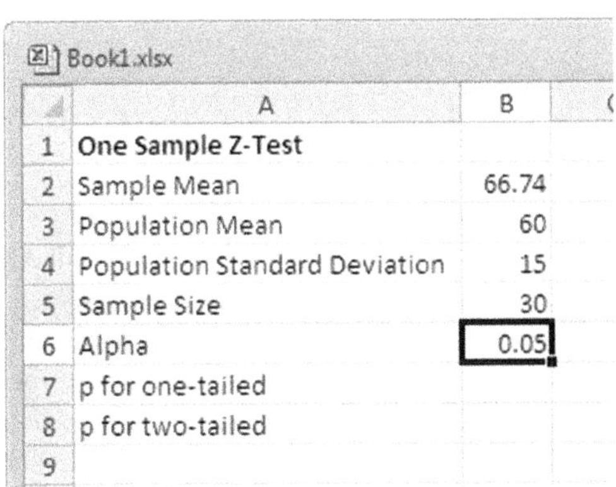
Book1.xlsx

	A	B
1	**One Sample Z-Test**	
2	Sample Mean	66.74
3	Population Mean	60
4	Population Standard Deviation	15
5	Sample Size	30
6	Alpha	0.05
7	p for one-tailed	
8	p for two-tailed	
9		

- Click on cell B7. Calculate the **z-statistic** by clicking on the formula ribbon and typing = **Z.TEST(D2:D31,B3,B4)**
- **Note:** When the population SD is not known, the optional third argument in Z. TEST can be left blank. In this case, Excel calculates the SD from the sample data.

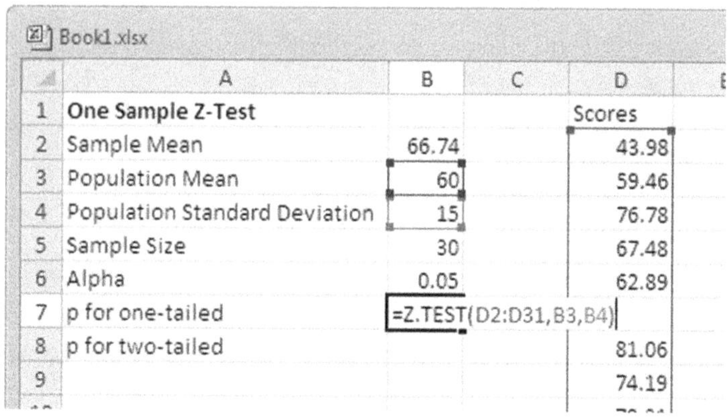

Book1.xlsx

	A	B	C	D
1	**One Sample Z-Test**			Scores
2	Sample Mean	66.74		43.98
3	Population Mean	60		59.46
4	Population Standard Deviation	15		76.78
5	Sample Size	30		67.48
6	Alpha	0.05		62.89
7	p for one-tailed	=Z.TEST(D2:D31,B3,B4)		
8	p for two-tailed			81.06
9				74.19

The *p*-value returned by Z.TEST for the one-sided (or one-tailed) test is in cell B7 (0.0069). Z.TEST confirms that if the true mean of the underlying normal distribution from which all 30 engineering scores were drawn is in fact 60 points, and the true standard deviation is 15, a sample mean that is higher than 66.74 would occur with probability 0.0069.

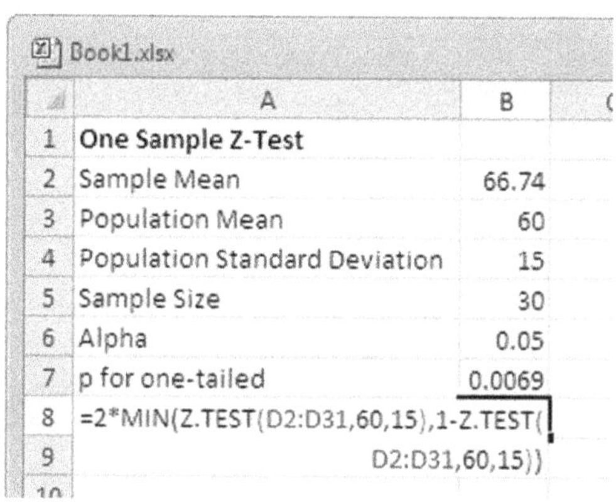

Book1.xlsx

	A	B
1	**One Sample Z-Test**	
2	Sample Mean	66.74
3	Population Mean	60
4	Population Standard Deviation	15
5	Sample Size	30
6	Alpha	0.05
7	p for one-tailed	0.0069
8	=2*MIN(Z.TEST(D2:D31,60,15),1-Z.TEST(	
9	D2:D31,60,15))	

Since we chose a significance level of 0.05 for this one-tailed test and the *p*-value of 0.0069 is less than this significance level, we have established significant results. These results are not due to chance.

Z.TEST is designed to give a one-tailed result, but it can be adapted to give a two-tailed result. For this example the question might be whether we expect the sample mean of the proficiency scores to be either larger or smaller than 66.74 points.

The formula for the two-tailed z-test is:

$$= \mathbf{2 * MIN}\left(\begin{matrix}\mathbf{Z.TEST(array\ ofsample\ data,\ population\ mean,\ population\ SD),} \\ \mathbf{1 - Z.TEST(array,\ population\ mean,\ population\ SD)}\end{matrix}\right)$$

- Click on cell B8. Input into the formula ribbon

$$= \mathbf{2 * MIN\ (Z.TEST(D2:D31,60,15),1 - Z.TEST(D2:D31,60,15))}$$

Book1.xlsx

	A	B
1	**One Sample Z-Test**	
2	Sample Mean	66.74
3	Population Mean	60
4	Population Standard Deviation	15
5	Sample Size	30
6	Alpha	0.05
7	p for one-tailed	0.0069
8	p for two-tailed	0.0139
9		

The p-value returned by **Z.TEST** for the two-sided test in cell B8 (0.0139) confirms that if the true mean of the underlying normal distribution from which all 30 engineering scores were drawn is in fact 60 points, and the true standard deviation is 15, a sample mean that is higher or lower than 66.74 would occur with probability 0.0139.

Since we choose a significance level of 0.05 for this two-tailed test, and the p-value of 0.0139 is less than this significance level, we have established significant results. These results are not due to chance. We are 95% confident that the average engineering proficiency quotient score is 60.

Common Excel Pitfalls

- z-test works best with $n > 30$.
- Make sure you have an equal amount of "(" as you do ")". If you do not have enough closed parentheses, Excel will give you an error.
- When doing the z-test make sure that you are using the correct formulas and templates for your problem. Using the wrong formula or template will produce incorrect answers.
- By default answers and calculations will display as many decimals as it can to fill a cell. To reduce the number of digits displayed after the decimal, format the cell to display no more than four decimal places.

Final Thoughts and Activities

Practice Problems and Case Studies

1. A teacher wanted to know how well the gifted students in her class perform relative to her other classes. She administers a standardized test with a mean of 50 points and standard deviation of 10 points. Her class of 31 students has an average score of 55. Test if this new sample mean has statistical significance at the 95% confidence level.
2. A rental car company claims the mean time to rent a car on their website is 60 s with a standard deviation of 30 s. A random sample of 36 customers attempted to rent a car on the website. The mean time to rent was 75 s. Is this enough evidence to contradict the company's claim?
3. Suppose that in a particular company, the overall mean and standard deviation of scores on a safety awareness test are 100 points, and 12 points, respectively. Our interest is in the scores of 55 employees in a new office that just opened in Plano, Texas. They achieved a mean score of 96. Is their mean score significantly lower than the overall company mean: that is, would the scores of the employees in this office be comparable to another simple random sample of 55 employees from the company, or are their scores surprisingly low? In other words, if we took another sample would we get similar results?
4. Many companies are finding that customers are using various types of online content before purchasing products. The following results are the percentages of adults and youths who use various sources of online content. Suppose the survey is based on 100 adults and 100 youths.

	Use online content	
Type of online content	Adult	Youth
Customer product ratings/reviews	74	84
For sale listings with seller ratings	72	80
For sale listings without seller ratings	61	68
Online classified ads	60	69
Message-board posts	60	74
Web blogs	58	70
Dating side profiles/personals	52	62
Peer-generated and peer-reference information	52	71
Peer-posted event listings	49	74

For "Customer product ratings/reviews" and "For sale listings with seller ratings" determine whether there is a difference between adults and youths in the proportion who use the type of online content at the 0.05 level of confidence. Use the template provided in **Chi-Square Test** file.

5. A company is considering an organizational change by adopting the use of self-managed work teams. To assess the attitudes of employees of the company toward this change, a sample of 400 employees is selected and asked whether they favor the institution of self-managed work teams in the organization. Three responses are permitted: favor, neutral, or oppose. The results of the survey, cross-classified by type of job and attitude toward self-managed work teams are summarized as follow:

	Self-managed work teams			
Type of job	Favor	Neutral	Oppose	Total
Hourly worker	110	49	70	229
Supervisor	20	15	29	64
Middle management	37	17	25	79
Upper management	26	10	8	44
Total	193	91	132	416

At the 0.05 level of significance, is there evidence of a relationship between attitude toward self-managed work and type of job? Use the template provided in **Chi-Square Test** file.

6. We want to know whether male or female nurses at a local hospital are more likely to gain promotions within the first 2 years of employment within a confidence level of 95% ($\alpha = 0.05$). In other words are the promotions independent of gender? Is there a statistically significant relationship between gender and promotions (H_a) or are these differences simply due to chance (H_0)? The data is provided below.

	Promotion	No promotion	Total
Females	46	71	117
Males	37	83	120
Total	83	154	237

7. This data was collected in an attempt to better understand the management distribution in a partnership between Norwegian and US companies. Given the following Excel output calculate the χ^2 statistic and the degrees of freedom for this dataset. The χ^2 critical is given as Interpret these statistics.

	A	B	C	D	E
1	**Observed**	**Mgr**	**Non-mgt**	**Total**	
2	American	56	81	137	
3	Norwegian	47	93	140	
4	Total	103	174	277	
5					
6	**Expected**				
7	American	50.97	86.03		
8	Norwegian	52.03	87.97		
9					
10	**Calculations**				
11	*p* value given		0.073287		
12					
13	X2	0.496388071	0.294094		
14		0.486275226	0.287608		

8. From the following list of conditions which ones are important for using a *z*-test?

Discussion Boards

1. If you want to inspire confidence, give plenty of statistics. It does not matter that they should be accurate, or even intelligible, as long as there is enough of them.
2. Economic data is not strong enough to bear the weight of elaborate mathematics and statistics.
3. The formal and mathematical theory of statistics was largely invented in the 1880s by eugenicists. Eugenics lost favor because of misunderstood statistics.
4. No one really knows how many people are malnourished.

Group Activity

1. What's the weak global economy to do when one of its strongest players, China, is growing at its slowest pace in 3 years? What organizations are responsible for reporting global economic statistics? How do they collect this data and how do they define economic growth?

2. The Bureau of Labor Statistics issues the "U.S. Employment Report" monthly. However, it focuses on the gain or decline from the previous month. This does not take into account seasonal variations, such as teachers who go on unpaid leave in the summer. For this reason, a better way to look at employment statistics is year-over-year. Compare and contrast the two methods.
3. Explore the World Hunger Organization's most recent report (www.worldhunger.org). Discuss the statistical pros and cons of this report.

Parting Thought

An unsophisticated forecaster uses statistics as a drunken man uses lamp posts … for support rather than for illumination.

Problem Solutions

1. A teacher wanted to know how well the gifted students in her class perform relative to her other classes. She administers a standardized test with a mean of 55 points and standard deviation of 10 points. Her class of 31 students has an average score of 50. Test if this new sample mean has statistical significance at the 95% confidence level.

Book1.xlsx

	A	B	C
1	**One Sample Z-Test**		
2	Sample Mean	50	
3	Population Mean	55	
4	Population Standard Deviation	10	
5	Sample Size	31	
6	Standard Error of the Mean	1.796	=B4/SQRT(B5)
7	Z	-2.784	=(B2-B3)/B6
8	Alpha	0.05	
9	Probability one-tailed	0.0027	=1-NORM.S.DIST(ABS(B7),TRUE)
10	Z-critical one-tailed	1.645	=ABS(NORM.S.INV(0.05))
11	Probability two-tailed	0.005	=2*B9
12	Z-critical two-tailed	1.96	= ABS(NORM.S.INV(0.025))
13			

Answer:

Testing for significance we use the two-tailed results since we are only interested if the means are different, not if one is larger or smaller than the other.

- The p-value of 0.005(B11) is smaller than the alpha value of 0.05 (B8). Therefore these results are statistically significant.
- The z-critical value of 1.96 (B12) is less than the z-calculated ABS (2.78) (B7) which also reinforces that these results are statistically significant. Remember when doing this test you need to work with absolute values.

The teacher can use this statistically significant data to show that the gifted students in her class perform differently from her other classes. The difference in the standardized test means is real at the 95% confidence level. If you wanted to know if they performed better or worse you would need to redefine the problem that way and use the one-tailed test.

2. A rental car company claims the mean time to rent a car on their website is 60 s with a standard deviation of 30 s. A random sample 36 customers attempted to rent a car on the website. The mean time to rent was 75 s. Is this enough evidence to contradict the company's claim with 95% confidence?

Book1.xlsx

	A	B	C	D
1	**One Sample Z-Test**			
2	Sample Mean	75		
3	Population Mean	60		
4	Population Standard Deviation	30		
5	Sample Size	36		
6	Standard Error of the Mean	5	=B4/SQRT(B5)	
7	Z	3	=(B2-B3)/B6	
8	Alpha	0.05		
9	Probability one-tailed	0.0013	=1-NORM.S.DIST(ABS(B7),TRUE)	
10	Z-critical one-tailed	1.645	=ABS(NORM.S.INV(0.05))	
11	Probability two-tailed	0.003	=2*B9	
12	Z-critical two-tailed	1.96	= ABS(NORM.S.INV(0.025))	
13				

Answer:

Testing for significance we use the two-tailed results since we are only interested if the means are different, not if one is larger or smaller than the other.

- The p-value of 0.003 (B11) is smaller than the alpha value of 0.05 (B8). Therefore these results are statistically significant.
- The z-critical value of 1.96 (B12) is less than the z-calculated 3.00 (B7) which also reinforces that these results are statistically significant. Remember when doing this test you need to work with absolute values.

We can use these statistically significant results as evidence to contradict the company's claim with 95% confidence. The difference in the mean time to rent is real at the 95% confidence level. Best to change your claim of 60 s to avoid a lawsuit.

3. Suppose that in a particular company, the overall mean and standard deviation of scores on a safety awareness test are 100 points, and 12 points, respectively. Our interest is in the scores of 55 employees in a new office that just opened in Plano, Texas. They achieved a mean score of 96. Is their mean score significantly lower than the overall company mean—that is, would the scores of the employees in this office be comparable to another simple random sample of 55 employees from the company, or are their scores surprisingly low? In other words, if we took another sample would we get similar results?

Answer:

Since the company is only interested in the sample mean score being unusually low, we are only interested in one tail.

Book1.xlsx

	A	B	C	D
1	**One Sample Z-Test**			
2	Sample Mean	96		
3	Population Mean	100		
4	Population Standard Deviation	12		
5	Sample Size	55		
6	Standard Error of the Mean	1.618	=B4/SQRT(B5)	
7	Z	-2.472	=(B2-B3)/B6	
8	Alpha	0.05		
9	Probability one-tailed	0.0067	=1-NORM.S.DIST(ABS(B7),TRUE)	
10	Z-critical one-tailed	1.645	=ABS(NORM.S.INV(0.05))	
11	Probability two-tailed	0.013	=2*B9	
12	Z-critical two-tailed	1.96	= ABS(NORM.S.INV(0.025))	
13				

- The one-sided p-value of 0.007 (B9) tells us that the z-value is statistically significant. Since the p-value for the one-tailed test is less than the chosen alpha value of 0.05 (B8), the results of the Plano office are real at the 0.05 level. The z-test tells us that the 55 employees of interest have an unusually low mean test score. We could also say that there is a 95% confidence that any other random sample from this same population would generate similar results. These results are not just due to chance.
- We can also consider the z-critical values. In this case the z-critical value of 1.645 (B10) is less than the ABS z-calculated of 2.47 (B7); since we are working with absolute values, we drop the negative sign which also reinforces that these results are statistically significant.

Their mean score is statistically significantly lower than the overall company mean: that is, the scores of the employees in this office are not comparable to another simple random sample of 55 employees from the company; their scores are surprisingly low.

4. Many companies are finding that customers are using various types of online content before purchasing products. The following results are the percentages of adults and youths who use various sources of online content. Suppose the survey is based on 100 adults and 100 youths.

	Use online content	
Type of online content	Adult	Youth
Customer product ratings/reviews	74	84
For sale listings with seller ratings	72	80
For sail listings without seller ratings	61	68
Online classified ads	60	69
Message-board posts	60	74
Web blogs	58	70
Dating side profiles/personals	52	62
Peer-generated and peer-reference information	52	71
Peer-posted event listings	49	74

For "Customer product ratings/reviews" and "For sale listings with seller ratings" determine whether there is a difference between adults and youths in the proportion that use the type of online content at the 0.05 level of confidence.

Answer:

Adult		Youth	
Mean	59.77778	Mean	72.44444
Standard error	2.871207	Standard error	2.186519
Median	60	Median	71
Mode	60	Mode	74
Standard deviation	8.61362	Standard deviation	6.559556
Sample variance	74.19444	Sample variance	43.02778
Kurtosis	−0.4815	Kurtosis	0.285721
Skewness	0.63015	Skewness	0.396455
Range	25	Range	22
Minimum	49	Minimum	62
Maximum	74	Maximum	84
Sum	538	Sum	652
Count	9	Count	9

z-test: two sample for means		
Customer product ratings/reviews		
	Variable 1	Variable 2
Mean	74	84
Known variance	74.19444	43.02778
Observations	1	1
Hypothesized mean difference	0	
Z	−0.92362	
P(Z<=z) one-tail	0.177841	
z-critical one-tail	1.644854	
P(Z<=z) two-tail	0.355682	
z-critical two-tail	1.959964	
Mean	72	80
Known variance	74.19444	43.02778
Observations	1	1
Hypothesized mean difference	0	
Z	−0.7389	
P(Z<=z) one-tail	0.229984	
z-critical one-tail	1.644854	
P(Z<=z) two-tail	0.459968	
z-critical two-tail	1.959964	

Based upon the online content that they used, there is no difference between what the youth and the adults used when looking at customer product ratings/ reviews or sales listings with seller ratings.

5. A company is considering an organizational change by adopting the use of self-managed work teams. To assess the attitudes of employees of the company toward this change, a sample of 400 employees is selected and asked whether they favor the institution of self-managed work teams in the organization. Three responses are permitted: favor, neutral, or oppose. The results of the survey, cross-classified by type of job and attitude toward self-managed work teams are summarized as follow:

	Self-managed work teams			
Type of job	Favor	Neutral	Oppose	Total
Hourly worker	110	49	70	229
Supervisor	20	15	29	64
Middle management	37	17	25	79
Upper management	26	10	8	44
Total	193	91	132	416

At the 0.05 level of significance, is there evidence of a relationship between attitude toward self-managed work and type of job? Use the template provided in **Chi-Square Test** file.

Answer:
Chi-Square Test

	Observed frequencies			
	Column variable			
Row variable	Favor	Neutral	Oppose	Total
Hourly worker	110	49	70	**229**
Supervisor	20	15	29	**64**
Middle management	37	17	25	**79**
Upper management	26	10	8	**44**
Total	**193**	**91**	**132**	**416**

	Expected frequencies			
	Column variable			
Row variable	Favor	Neutral	Oppose	Total
Hourly worker	106.2428	50.09375	72.66346	229
Supervisor	29.69231	14	20.30769	64
Middle management	36.65144	17.28125	25.06731	79
Upper management	20.41346	9.625	13.96154	44
Total	193	91	132	416

Calculations

(Freq observed) – (Freq expected)		
3.757212	−1.09375	−2.66346
−9.69231	1	8.692308
0.348558	−0.28125	−0.06731
5.586538	0.375	−5.96154

((Freq observed) – (Freq expected))^2/(Freq expected)		
0.132871	0.023881	0.097629
3.16381	0.071429	3.720571
0.003315	0.004577	0.000181
1.528864	0.01461	2.545561

Data	
Level of significance	0.05
Number of rows	4
Number of columns	3
Degrees of freedom	1

Results	
Critical value	3.841459
Chi-square test statistic	11.89531
p-value	0.000563
Reject the null hypothesis	

Reject the null hypothesis, which means that we found a significant difference between attitude toward self-managed work and an employee's type of job classification.

6. We want to know whether male or female nurses at a local hospital are more likely to gain promotions within the first 2 years of employment within a confidence level of 95% ($\alpha = 0.05$). In other words are the promotions independent of gender? Is there a statistically significant relationship between gender and promotions (H_a) or are these differences simply due to chance (H_0)? The data is provided below.

	Promotion	No promotion	Total
Females	46	71	117
Males	37	83	120
Total	83	154	237

Answer:

The data shows that there are 237 nurses at this hospital, 117 of whom are female while 120 are male. At first review of the data it appears that male nurses have received fewer promotions (37) than female nurses (46). The null hypothesis is that the two variables are independent, that the likelihood of getting promoted is the same for males and females.

Table of observed and expected values

Book1.xlsx

	A	B	C	D	E
2	females	46	71	117	
3	males	37	83	120	
4	Total	83	154	237	
5					
6	EXPECTED				
7	females	40.97	76.03		
8	males	42.03	77.97		
9					

As with all χ^2 testing we need the observed and expected values. Expected values are calculated based on the row and column totals from the table.

With two variables the expected value for each cell of the table can be calculated using the following formula:

$$(\text{Row Total} \times \text{Column Total}) / \text{Total } N \text{ count for overall table}$$

There is no Excel function that will do these calculations automatically but you can input them as formula for each cell.

By using CHISQ.TEST in Excel we generate a p-value of 0.171 (B10), but we must refer to the number of degrees of freedom to be totally correct. There is no Excel function that will automatically calculate this so you can input the formula

$$\text{df} = (\text{the number of columns} - 1) \times (\text{number of rows} - 1)$$

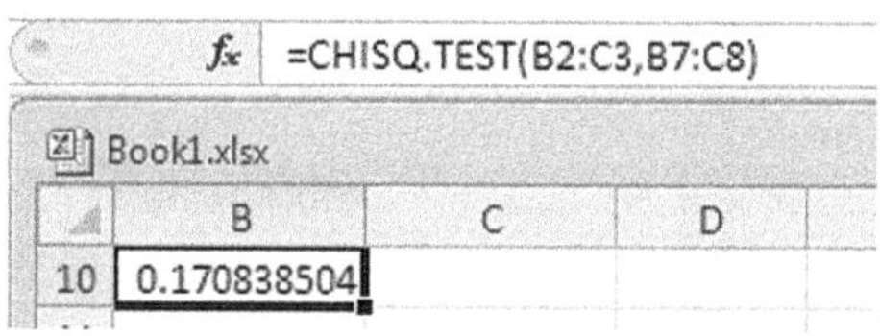

In this table, there were two rows and two columns. Therefore, the number of degrees of freedom is one. We can now state that the p-value was calculated with one degree of freedom as 0.17 (B10); this value exceeds our p_{critical} of 0.05. Our results are not statistically significant; we cannot reject the null hypothesis. The relationship between gender and promotions is not significant, and so we can assume is these results are simply due to chance.

For reporting we may also want to report the actual χ^2 test statistic. Unfortunately Excel does not provide this to the user in any simple way. Once again you need to input a formula:

$$X^2 = \Sigma(\text{observed value}_i - \text{expected value}_i)^2 / \text{expected value}_i$$

To formally report our results, we would say "no statistically significant relationship was found between gender and promotions, $\chi^2(1) = 1.88$, $p = 0.17$."

fx =(B2-B7)^2/B7

Book1.xlsx

	A	B	C	D	E
1	OBSERVED	promotion	no promo	Total	
2	females	46	71	117	
3	males	37	83	120	
4	Total	83	154	237	
5					
6	EXPECTED				
7	females	40.97	76.03		
8	males	42.03	77.97		
9					
10		0.170838504			
11					
12	Xsq test stat	0.61632704	0.33277522		1.8756
13		0.60197240	0.32449532		

7. This data was collected in an attempt to better understand the management distribution in a partnership between Norwegian and US companies. Given the following Excel output calculate the χ^2 statistic and the degrees of freedom for this dataset. The χ^2 critical is given as Interpret these statistics.

	A	B	C	D	E
1	**Observed**	**Mgr**	**Non-mgt**	**Total**	
2	American	56	81	137	
3	Norwegian	47	93	140	
4	Total	103	174	277	
5					
6	**Expected**				
7	American	50.97	86.03		
8	Norwegian	52.03	87.97		
9					
10	**Calculations**				
11	*p* value given		0.073287		
12					
13	X2	0.496388071	0.294094		
14		0.486275226	0.287608		

Answer:

$$\text{df} = (\text{the number of columns} - 1) \times (\text{number of rows} - 1)$$
$$= (2-1) * (2-1) = 1 \text{there is one degree of freedom}$$

This statistic is taken into consideration when calculating the p-value (0.673287). In this dataset since $p > 0.05$ we consider this as not statistically significant.

$$X2 = \Sigma\left(\text{observed value}_i - \text{expected value}_i\right)^2 / \text{expected value}_i$$

In this case you need to calculate all four χ^2 values and sum them together

0.496388071	0.294094
0.486275226	0.287608
SUM χ^2 =	1.564365494

$$\Sigma \text{calculated} = 1.564365494$$

$$\chi^2 \text{ critical is given as } 1.66324288$$

Since $\chi^2_{\text{calc}} < \chi^2_{\text{critical}}$ we cannot reject H_0 and acknowledge that the observed distribution is not real and just due to chance. In other words not statistically significant. This is not data that should be used in decision-making. Back to the drawing board.

8. What are the conditions for using a z-test?

   Answer:

1. The sample size must be more than 30.
2. The sample data should always be random. Otherwise, the test statistic results may turn out to be inaccurate.
3. The data points must not be similar. Besides that, they must not overlap with each other.
4. The data must reflect a standard normal distribution.
5. You need to know the population mean.
6. You need to know the standard deviation.
7. Outliers need to be removed from data. The z-test can be sensitive to outliers in the data, especially when dealing with small sample sizes. Outliers can significantly affect the estimated mean and inflate the standard deviation, potentially leading to incorrect conclusions.
8. Limited to means: The z-test is specifically designed to compare means between two populations. It may not be suitable for analyzing other types of statistical comparisons, such as proportions or variances.

Chapter 13
Multiple Regression

Linda Herkenhoff and John Fogli

Key Concepts

Adjusted coefficient of determination, Law of parsimony, Multicollinearity, and Multiple regression.

Discussion

The chapter discusses **multiple regression** by building on the basic concepts of simple linear regression. Now instead of only one independent variable we have several independent variables to predict the dependent variable. To visualize the difference between simple and multiple regression see Figs. 13.1 and 13.2. In simple linear regression the data is fit to a straight line. In multiple regression the data is fit to a plane.

Figures 13.1 and 13.2 display the basic models. Excel cannot draw a multidimensional graph so if you actually need to see a multiple regression graph more sophisticated software will be required.

Supplementary Information The online version contains supplementary material available at https://doi.org/10.1007/978-3-031-46371-6_13.

L. Herkenhoff (✉)
Saint Mary's College, Moraga, CA, USA

J. Fogli
Diablo Valley College, Pleasant Hill, CA, USA

L. Herkenhoff, J. Fogli (eds.), *Applied Statistics for Business and Management using Microsoft Excel*, https://doi.org/10.1007/978-3-031-46371-6_13

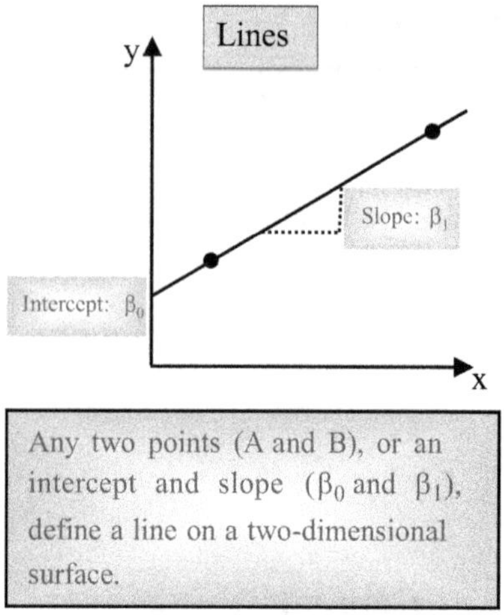

Fig. 13.1 Simple linear regression

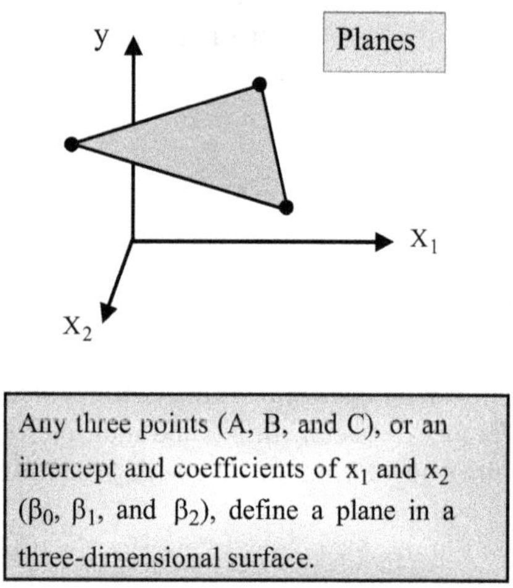

Fig. 13.2 Multiple regression

Figure 13.3 compares how we fit a simple regression model and a multiple regression model by minimizing the residuals or the error terms. The simple linear regression data is fit as a line while the multiple regression data is fit to a plane.

The **multiple regression** equation combines a calculated amount of each of the x variables with a y-intercept. The **regression coefficients** indicate how much of each x variable to include in your model.

Consider the multiple regression equation $\mathbf{y = 2x_1 + 5.2x_2 - .78x_3 + 22}$

In this case the regression coefficients are 2, 5.2, and −.78. There are three independent variables (x_1, x_2, x_3) and the y-intercept is indicated as the "b_0" term (22 in this case).

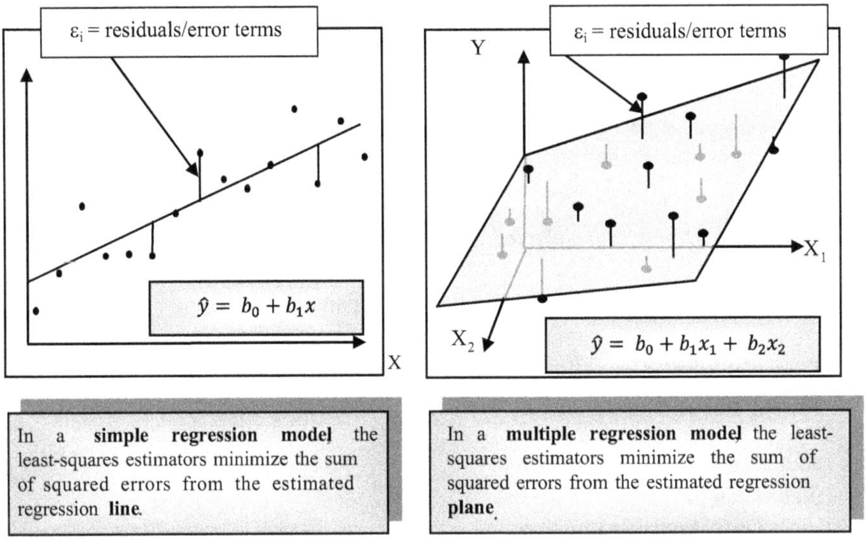

Fig. 13.3 Minimizing the residuals

Adjusted Coefficient of Determination Remember in simple linear regression the coefficient of determination describes how much of the variation in *y* can be explained by one *x*. The **adjusted coefficient of determination** describes how much of the variation in *y* can be explained by all of the *x*'s. The term "adjusted" is included to reflect the fact that there is more than one *x* in the model and we want to statistically "adjust out" (or remove) the effect of simply adding more independent variables to the model when Excel calculates the coefficient of determination.

Law of Parsimony This law states that the optimum model is one with the fewest number of independent variables. The more variables you have the more difficult it becomes to collect good data for all of the variables each time. You can gain insight on which independent variable(s) should be deleted by completing correlation tests and significance tests.

The first step is to run a correlation matrix to identify which of the *x* variables have high correlations with the other *x* variables. When two or more of the independent variables have a strong coefficient of correlation we refer to this as **multicollinearity**. In other words, multicollinear variables are considered redundant and should be eliminated. High *x* to *x* correlations suggest the *x*'s are contributing similar effects to the model and you probably only need one of them.

The second step is to consider the associated significance tests that can indicate which of the independent variables are not worth keeping in the model as robust predictors of the dependent variable.

Excel

In this first example we will test if sales can be predicted by knowing how much money has been spent on radio advertising and newspaper ads. We need to run a multiple regression model rather a simple linear regression model because there are two x's.

Step 1: Fit the Model with Selected Independent Variables

$$y = x_1\left(\text{radio advertising\$}\right) + x_2\left(\text{newspaper advertising\$}\right) + b$$

where $y \rightarrow$ sales, $x_1 \rightarrow$ radio ads, $x_2 \rightarrow$ newspaper ads

Step 2: Does Multicollinearity Exist? Run a Correlation Matrix

- Input data from Fig. 13.4
- Click on **Data** tab
- Select **Data Analysis** function
- Select **Correlation** from the list of Analysis Tools

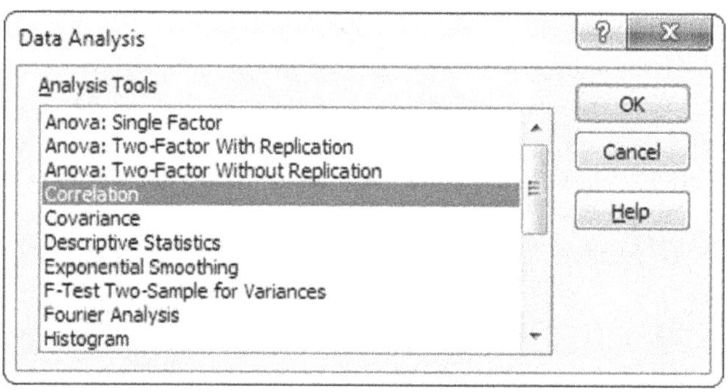

- Click **OK**
- Click inside the **Input Range** text box and then highlight all columns for the x variables and the y variable

- Remember to include labels so you can keep track of which variables correlate to each other
- Check the **Labels in first row** box

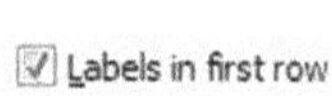

Fig. 13.4 Input data

Sales	Radio	Newspaper
973	0	40
1119	0	40
875	25	25
625	25	25
910	30	30
971	30	30
931	35	35
1177	35	35
882	40	25
982	40	25
1628	45	45
1577	45	45
1044	50	0
914	50	0
1329	55	25
1330	55	25
1405	60	30
1436	60	30
1521	65	35
1741	65	35
1866	70	40
1717	70	40

- Select **Output Range** then click inside the text box to input the output range

- Click **OK**

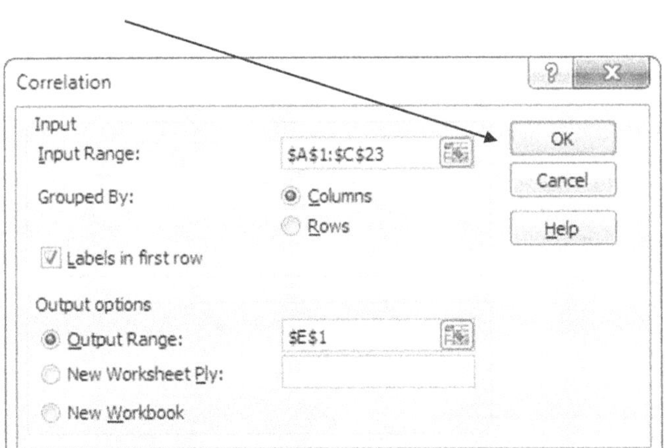

Book1.xlsx

	A	B	C
1	Sales	Radio	Newspaper
2	973	0	40
3	1119	0	40
4	875	25	25
5	625	25	25
6	910	30	30
7	971	30	30
8	931	35	35
9	1177	35	35
10	882	40	25
11	982	40	25
12	1628	45	45
13	1577	45	45
14	1044	50	0
15	914	50	0
16	1329	55	25
17	1330	55	25
18	1405	60	30
19	1436	60	30
20	1521	65	35
21	1741	65	35
22	1866	70	40
23	1717	70	40
24			

E	F	G	H
	Sales	*Radio*	*Newspaper*
Sales	1		
Radio	0.696608	1	
Newspaper	0.502113	-0.09213	1

Correlation matrix

The rule of thumb suggests that r values (correlation coefficients) greater than +0.6 or less than −0.6 between **x variables** should be flagged as possible multicollinear variables. In this example, the relationship between **Radio** (x_1) and **Newspaper** (x_2) is only −0.09 (G4) which is within the acceptable level. Therefore, no multicollinearity is noted so no need to delete any of the x variables in this model.

But we also would like to consider how strongly each x variable correlates with y (sales). Radio is stronger with an r of .7 while newspaper only achieves an r value of .5. We need to flag the newspaper variable for further consideration before deleting it from the model.

Step 3: Run Regression Model

- **Note:** Columns of data must be adjacent to one another
- Click on **Data** tab
- Select **Data Analysis** function

- Select **Regression** from the list of Analysis Tools

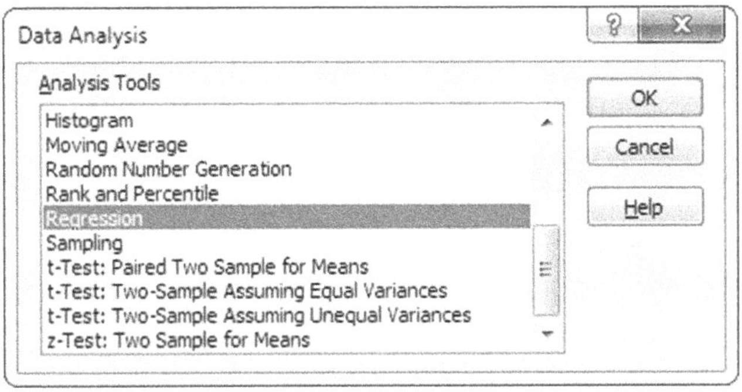

- **Input Y Range:** you need to highlight the column of *y* data

- **Input X Range:** you need to highlight **all** columns of **all x variables**

- Check the **Labels** box if the first row of your data includes labels

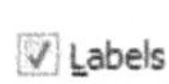

- Open a new work page by clicking on **Sheet 2** at the bottom of the page
- Select the **Output Range** option and click on the cell where you want the output displayed

- Click **OK**

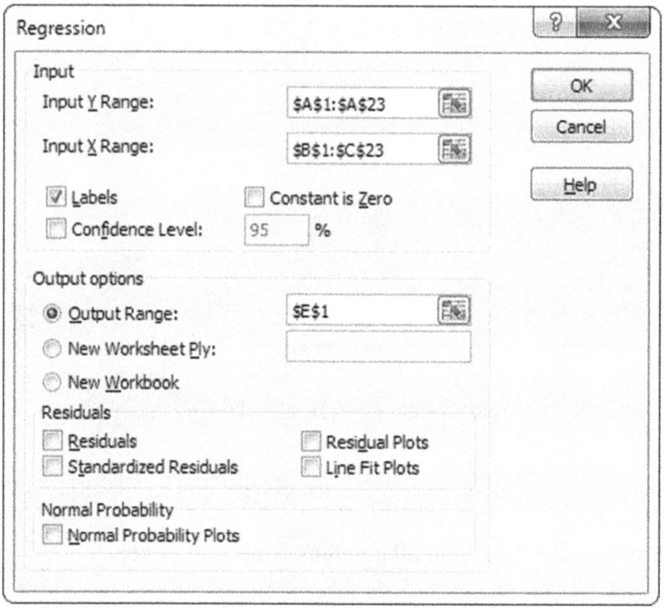

Results

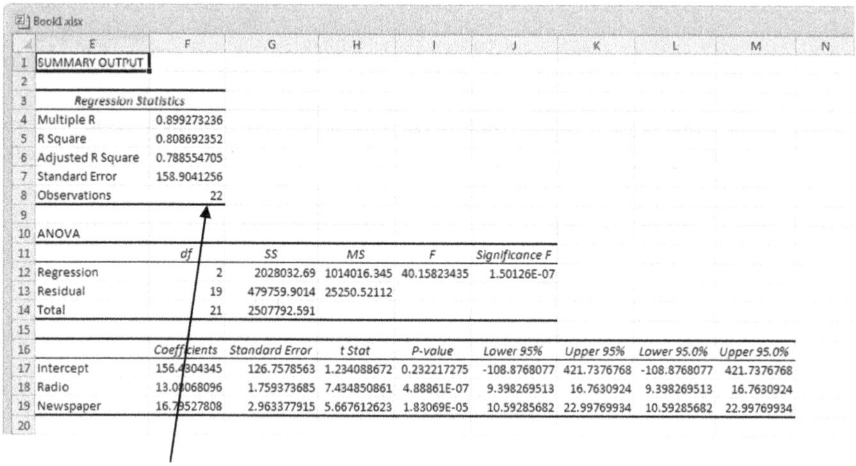

SUMMARY OUTPUT								
Regression Statistics								
Multiple R	0.899273236							
R Square	0.808692352							
Adjusted R Square	0.788554705							
Standard Error	158.9041256							
Observations	22							
ANOVA								
	df	*SS*	*MS*	*F*	*Significance F*			
Regression	2	2028032.69	1014016.345	40.15823435	1.50126E-07			
Residual	19	479759.9014	25250.52112					
Total	21	2507792.591						
	Coefficients	*Standard Error*	*t Stat*	*P-value*	*Lower 95%*	*Upper 95%*	*Lower 95.0%*	*Upper 95.0%*
Intercept	156.4304345	126.7578563	1.234088672	0.232217275	-108.8768077	421.7376768	-108.8768077	421.7376768
Radio	13.08068096	1.759373685	7.434850861	4.88861E-07	9.398269513	16.7630924	9.398269513	16.7630924
Newspaper	16.79527808	2.963377915	5.667612623	1.83069E-05	10.59285682	22.99769934	10.59285682	22.99769934

- Remember to check that all of your data has been included by looking at the **Observation**s count

Step 4: Are the Assumptions of Regression Satisfied?
Check the model in Fig. 13.5 to see if the assumptions are satisfied

Step 5: Test Overall Model Significance (F-Test)

The significance test results are included in the **Regression Output**. Remember to check if the **Significance F** for the overall model significance is less than 0.05 (for 95% confidence level etc.). Also check that the **p-values** are less than 0.05 (for 95% confidence level etc.) for each of the independent variables as discussed in the previous chapters on Significance Testing.

In this example, the **Significance F** is very small (much less than the threshold of 0.05) which is good. If it is greater than or equal to 0.05 you need to go back and review the x variables you have included in the model. Make sure you included all columns of x data when rerunning the model.

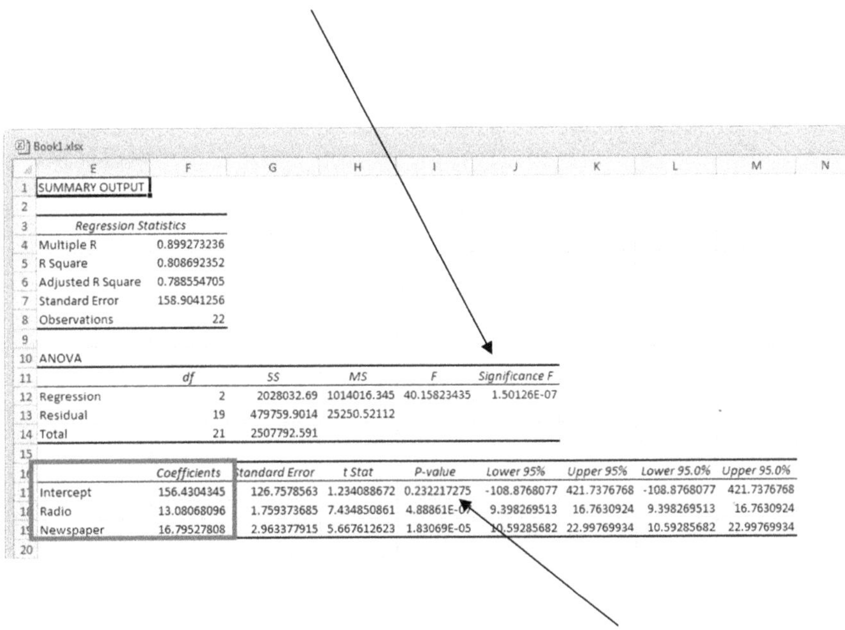

Book1.xlsx

SUMMARY OUTPUT

Regression Statistics	
Multiple R	0.899273236
R Square	0.808692352
Adjusted R Square	0.788554705
Standard Error	158.9041256
Observations	22

ANOVA

	df	SS	MS	F	Significance F
Regression	2	2028032.69	1014016.345	40.15823435	1.50126E-07
Residual	19	479759.9014	25250.52112		
Total	21	2507792.591			

	Coefficients	Standard Error	t Stat	P-value	Lower 95%	Upper 95%	Lower 95.0%	Upper 95.0%
Intercept	156.4304345	126.7578563	1.234088672	0.232217275	-108.8768077	421.7376768	-108.8768077	421.7376768
Radio	13.08068096	1.759373685	7.434850861	4.88861E-07	9.398269513	16.7630924	9.398269513	16.7630924
Newspaper	16.79527808	2.963377915	5.667612623	1.83069E-05	10.59285682	22.99769934	10.59285682	22.99769934

Step 6: Check p-Values for Independent Variables Meet Significance Criteria (t-Test)

Identify the weak x's by considering the p-values greater than 0.05. Sometimes the **p-values** can be greater than 0.05 because there is a low sample size for the variable relative to the other variable sample sizes. If you know the variable is important to the model you may decide to include it despite the fact that it does not pass the significance test. It is important to remove one x at a time when cleaning up the model. Rerun your model if you need to remove nonsignificant x variables. In this example the **p-values** are all smaller than 0.05 so there is no need to delete any x data at this point. No need to rerun the model.

- No need to look at the p-value for the y-intercept

Step 7: Run Model for Prediction and Estimation

The resulting equation can now be used to predict new values of *y* by inputting data for each of the significant *x* variables. Consider the resulting model:

$$y = 13.08\ x_1 + 16.80\ x_2 + 156.43$$

The adjusted R^2 of this model suggests that about 79% of sales can be explained by knowing the amount spent on radio and newspaper advertising. As you gain more data it is always recommended to rerun the regression analysis and recalibrate the model.

Figure 13.5 summarizes the step by step road map for developing the best fitting regression model. Remember if there is only one *x* variable a simple linear regression can be used. Multiple regression models are used when there is more than one *x* variable.

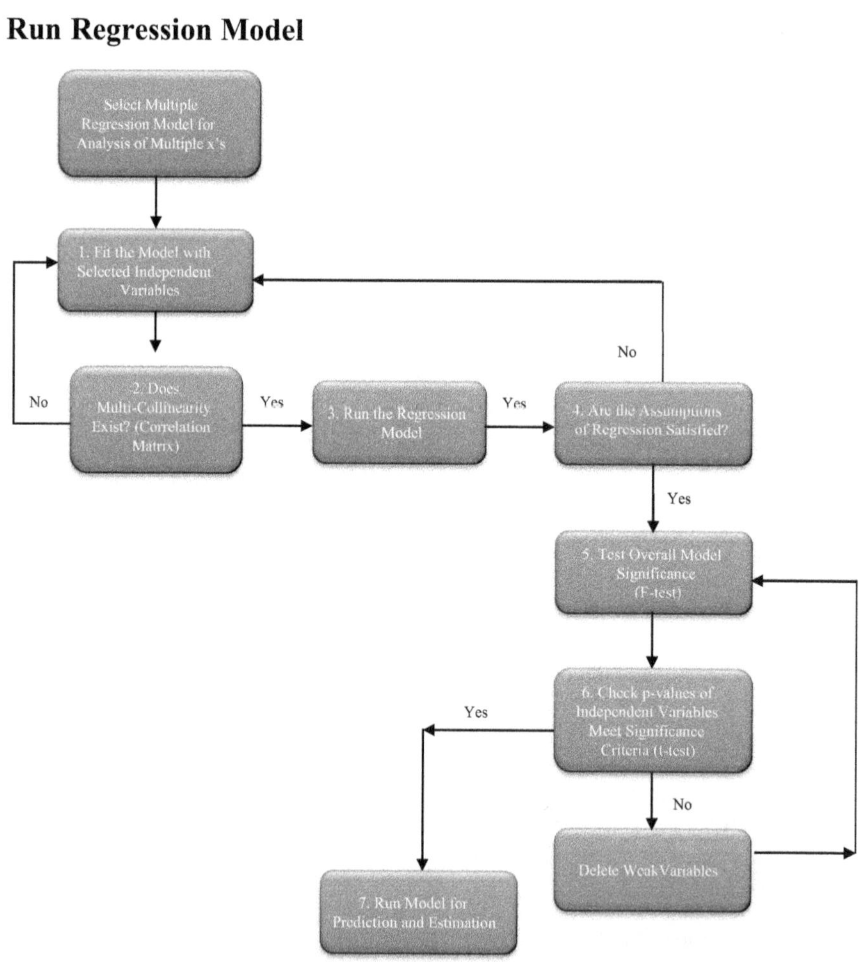

Fig. 13.5 Road map for selecting the best fitting regression model

Run Regression Model

Final Thoughts and Activities

Practice Problems

1. Generate a multiple regression equation for the following data where y is sales ($000), x_1 is money spent on decorating the store ($000), and x_2 is area of the store (hundreds of square feet). The owner is trying to determine how much he can expect in sales from a store that is 4000 ft^2 and will require $4000 worth of decorating. Can you help him out?

Data		
Costs of decorating ($000) ($x_1$)	Area (00 ft^2) (x_2)	Sales ($000) (y)
0.4	19.7	19.7
2.8	19.1	19.3
4.0	18.2	18.6
6.0	5.2	7.9
1.1	4.3	4.4
2.6	9.3	9.6
7.1	3.6	8.0
5.3	14.8	15.7
9.7	11.9	15.4
3.1	9.3	9.8
9.9	2.8	10.3
5.3	9.9	11.2
6.7	15.4	16.8
4.3	2.7	5.1
6.1	10.6	12.2
9.0	16.6	18.9
4.2	11.4	12.2
4.5	18.8	19.3
5.2	15.6	16.5
4.3	17.9	18.4

2. A survey of recently sold single-family houses in a small city is selected. The houses in the city were reassessed 6-months prior to the study. Develop a model to predict the selling price (in thousands of dollars) by using the assessed value (in thousands of dollars) as well as the time period when sold (in months since reassessment). Use the results in the file **Housing** for your analysis.

   (a) Is there multicollinearity?
   (b) Determine whether there is a significant relationship between selling price and the two dependent variables (assessed value and time period) at the 0.05 level of confidence.
   (c) Determine the adjusted R^2.

3. A consumer products company is trying to measure the effectiveness of different types of advertising media in the promotion of its products. Specifically, the company is interested in the effectiveness of digital advertising and television advertising (including the cost of discount coupons). During a one-month test period data was collected from a sample of 22 cities with approximately equal populations. Each city is allocated a specific expenditure level for digital advertising and for television advertising. The sales of the product (in thousands of dollars) and also the levels of media expenditure (in thousands of dollars) during the test month are recorded with the following results shown below.

City	Sales ($Thousands)	Radio advertising ($Thousands)	Newspaper advertising ($Thousands)
1	973	0	40
2	1119	0	40
3	875	25	25
4	625	25	25
5	910	30	30
6	971	30	30
7	931	35	35
8	1177	35	35
9	882	40	25
10	982	40	25
11	1628	45	45
12	1577	45	45
13	1044	50	0
14	914	50	0
15	1329	55	25
16	1330	55	25
17	1405	60	30
18	1446	60	30
19	1521	65	35
20	1741	65	35
21	1866	70	40
22	1717	70	40

(a) State the multiple regression equation.
(b) Interpret the meaning of the regression coefficients for x_1 and x_2 in this problem.
(c) Interpret the meaning of the y-intercept.
(d) Are the two advertising variables significant?

4. Develop a model to predict the assessed value (in thousands of dollars) using the size of the houses (in thousands of square feet) and the age of the houses (in years) from the following table.

House	Assessed value ($Thousands)	Size of house (thousands of square feet)	Age (years)
1	184.4	2	3.42
2	177.4	1.71	11.5
3	175.7	1.45	8.33
4	185.9	1.76	0
5	179.1	1.93	7.42
6	170.4	1.2	32
7	175.8	1.55	16
8	185.9	1.93	2
9	178.5	1.59	1.75
10	179.2	1.5	2.75
11	186.7	1.9	0
12	179.3	1.39	0
13	174.5	1.54	12.58
14	183.8	1.89	2.75
15	176.8	1.59	7.17

(a) State the multiple regression equation.
(b) Interpret the meaning of the slopes in this equation.
(c) Predict the assessed value for a house that has a size of 1750 ft^2 and is 10 years old.
(d) Perform a residual analysis on the results and determine whether the regression assumptions are valid.
(e) Determine whether there is a significant relationship between the assessed value and the two independent variables (size and age) at the 0.05 level of significance.
(f) Interpret the meaning of the adjusted r^2.
(g) At the 0.05 level of significance determine whether each independent variable makes a significant contribution to the regression model. Indicate the most appropriate regression model for this set of data.
(h) The real estate assessor's office has been publicly quoted as saying that the age of a house has no bearing on its assessed value. Based on your answers above do you agree with this statement? Explain.

5. The following data produces the AVOVA shown below where CARS is y whereas HH SIZE and Cubed HH Size are the x variables. Interpret the key statistics to determine if this is a statistically viable model.

	A	B	C
1	CARS	HH SIZE	CUBED HH SIZE
2	1	1	1
3	2	2	8
4	2	3	27
5	2	4	64
6	3	5	125

8	SUMMARY OUTPUT						
9							
10	*Regression Statistics*						
11	Multiple R	0.895828018					
12	R Square	0.802507837					
13	Adjusted R Square	0.605015674					
14	Standard Error	0.444400903					
15	Observations	5					
16							
17	ANOVA						
18		*df*	*SS*	*MS*	*F*	*Significance F*	
19	Regression	2	1.605015674	0.802507837	4.063492063	0.197492163	
20	Residual	2	0.394984326	0.197492163			
21	Total	4	2				
22							
23		*Coefficients*	*Standard Error*	*t Stat*	*P-value*	*Lower 95%*	*Upper 95%*
24	Intercept	0.896551724	0.764398064	1.172885918	0.361624318	-2.39238769	4.185491139
25	HH SIZE	0.33646813	0.422703764	0.79599038	0.509506953	-1.482279374	2.155215634
26	CUBED HH SIZE	0.002089864	0.013113794	0.159363815	0.888021498	-0.054334235	0.058513964

6. Discuss the assumptions of multiple regression analysis and explain their importance in interpreting the results.
7. Discuss the issue of multicollinearity when running multiple regressions.
8. This model has a sample size of 20. Based on these results is this model statistically viable? Explain.

Discussion Boards

1. When a university accepts a new student it is basically predicting that the student can be successful in his or her studies. What predictor variables do admissions offices use? Some studies suggest SAT scores are not useful. Discuss.
2. Why is "parsimony" important in multiple regression?

Parting Thought

The 50–50–90 rule states that anytime you have a 50–50 chance of getting something right there is a 90% probability you will get it wrong.

Problem Solutions

1. Generate a multiple regression line for the following data where y is sales ($000), x_1 is money spent on decorating the store ($000), and x_2 is area of the store (hundreds of square feet). The owner is trying to determine how much he can expect in sales from a store that is 4000 ft^2 and will require $4000 worth of decorating. Can you help?

Costs of decorating ($000) ($x_1$)	Area (00 ft^2) (x_2)	Sales ($000) ($y$)
0.4	19.7	19.7
2.8	19.1	19.3
4.0	18.2	18.6
6.0	5.2	7.9
1.1	4.3	4.4
2.6	9.3	9.6
7.1	3.6	8.0
5.3	14.8	15.7
9.7	11.9	15.4
3.1	9.3	9.8
9.9	2.8	10.3
5.3	9.9	11.2
6.7	15.4	16.8
4.3	2.7	5.1
6.1	10.6	12.2
9.0	16.6	18.9
4.2	11.4	12.2
4.5	18.8	19.3
5.2	15.6	16.5
4.3	17.9	18.4

Answer:

Step 1: Fit the Model with Selected Independent Variables
Step 2: Check for Multicollinearity

	Costs of Decorating ($000)	Area (00ft^2)	Sales ($000)
Costs of Decorating ($000)	1		
Area (00ft^2)	-0.21838896	1	
Sales ($000)	0.035218554	0.95487407	1

The r-value of x_1 and x_2 is less than the threshold value of ±0.6 thus indicating there is no multicollinearity in this model.

Step 3: Run Multiple Regression Model

The **Adjusted R-square value** is nice and high at 0.97. This means that 97% of the change in sales can be explained by the cost of decorating (x_1) and the area of the store (x_2). When we check the **Observations** we see that all of our input data has been included (20).

SUMMARY OUTPUT	
Regression Statistics	
Multiple R	
R Square	
Adjusted R Square	
Standard Error	
Observations	

ANOVA					
	df	*SS*	*MS*	*F*	*Significance F*
Regression	2	463.71	231.86	320.64	0.00
Residual	17	12.29	0.72		
Total	19	476.01			

	Coefficients	*Standard Error*	*t Stat*	*P-value*	*Lower 95%*	*Upper 95%*	*Lower 95.0%*	*Upper 95.0%*
Intercept	**0.68**	0.65	1.04	0.31	-0.70	2.06	-0.70	2.06
Costs of Decorating ($)	**0.50**	0.08	6.41	0.00	0.33	0.66	0.33	0.66
Area (ft2)	**0.86**	0.03	25.31	0.00	0.79	0.94	0.79	0.94

The resulting y-intercept is 0.68 with both x variables significantly contributing to the model. The final equation is

$$y = .5x_1 + .86x_2 + .68$$

Step 4: Test Assumptions of Regression

Make sure all the assumptions for regression are met.

Step 5: Review F-Tests

The overall model is significant with a **Significance F**-value of 0.00.

Step 6: Review Significance t-Tests

Both x_1 and x_2 have significant **p-values** of 0.00.

Step 7: Run Model for Prediction and Estimation

This multiple regression model seems to be a reasonable model for predicting sales using the cost of decorating and the area of the store as input data. Now plug in x_1 = \$4000(input as 4) and x_2 = 4000 ft^2 (input as 40) to the equation and the equation is ready to use.

Be very careful with the input units

$$y = .5(4) + .86\ (40) + .68 = \$37080$$

In this example the value for predicted sales (y) is \$37,080.

2. A survey of recently sold single-family houses in a small city is selected. Develop a model to predict the selling price (in thousands of dollars) by using the assessed value (in thousands of dollars) as well as time period when sold (in months since reassessment). The houses in the city have been reassessed at full value a year prior to the study. Use the results in the file **Housing** for your analysis.

(a) Is there an issue with multicollinearity?

	Price(\$ooo)	*Assessed Value*	*Time*
Price(\$000)	1		
Assessed Value	0.962084132	1	
Time	0.252413854	**0.126265739**	1

There is no multicollinearity between the two x variables (assessed value and time). The r-value of x_1 and x_2 is less than the threshold value of ±0.6.

(b) Determine whether there is a significant relationship between selling price and the two dependent variables (assessed value and time period) at the 0.05 level of confidence. Note although number of foreclosures in the area were provided the model did not request that data be included.

Yes the relationship is significant. The **p-values** of the x variables are 0.00 and 0.01 which are both significant. The overall model is also significant with a **Significance F** of 0.00.

(c) Determine the adjusted R^2.

0.94. This means that 94% of the variation in y (selling price) can be explained by considering these the two x variables (assessed value and time period).

Output Table

SUMMARY OUTPUT	
Regression Statistics	
Multiple R	0.97
R Square	0.94
Adjusted R Square	**0.94**
Standard Error	3.10
Observations	**30**

ANOVA

	df	*SS*	*MS*	*F*	***Significance F***
Regression	2	4285.85	2142.92	223.46	**0.00**
Residual	27	258.93	9.59		
Total	29	4544.78			

	Coefficients	*Standard Error*	*t Stat*	***P-value***	*Lower 95%*	*Upper 95%*	*Lower 95.0%*	*Upper 95.0%*
Intercept	-149.83	19.56	-7.66	**0.00**	-189.96	-109.70	-189.96	-109.70
Assessed Value	1.75	0.09	20.41	**0.00**	1.57	1.93	1.57	1.93
Time	0.74	0.26	2.87	**0.01**	0.21	1.26	0.21	1.26

3. A consumer products company is trying to measure the effectiveness of different types of advertising media in the promotion of its products. Specifically, the company is interested in the effectiveness of radio advertising and newspaper advertising (including the cost of discount coupons). During a one-month test period data was collected from a sample of 22 cities with approximately equal populations. Each city is allocated a specific expenditure level for radio advertising and for newspaper advertising. The sales of the product (in thousands of dollars) and also the levels of media expenditure (in thousands of dollars) during the test month are recorded with the following results shown below.

City	Sales ($Thousands)	Radio advertising ($Thousands)	Newspaper advertising ($Thousands)
1	973	0	40
2	1119	0	40
3	875	25	25
4	625	25	25
5	910	30	30
6	971	30	30
7	931	35	35
8	1177	35	35

(continued)

City	Sales ($Thousands)	Radio advertising ($Thousands)	Newspaper advertising ($Thousands)
9	882	40	25
10	982	40	25
11	1628	45	45
12	1577	45	45
13	1044	50	0
14	914	50	0
15	1329	55	25
16	1330	55	25
17	1405	60	30
18	1446	60	30
19	1521	65	35
20	1741	65	35
21	1866	70	40
22	1717	70	40

(a) State the multiple regression equation.

Answer:

$$y = 156.4 + 13.081X_1 + 16.795X_2$$

where X_1 = radio advertising and X_2 = newspaper advertising

(b) Interpret the meaning of the regression coefficients for x_1 and x_2 in this problem.

Each increase by $1000 in radio advertising is estimated to result in an increase in sales of $13,081 for a given number of newspaper advertisements. Each increase by $1000 in newspaper advertising is estimated to result in an increase in sales of $16,795 for a given number of radio advertisements.

(c) Interpret the meaning of the y-intercept.

When there is no money spent on radio advertising and newspaper advertising the estimated mean sales is $156,430.44.

(d) Are the two advertising variables significant?

Yes, according to the ANOVA table both advertising variables are significant as their p-values are less than 0.01.

ANOVA

	df	SS	MS	F	Significance F
Regression	2	2032436.022	1016218.01	40.2530293	1.47431E-07
Residual	19	479669.2958	25245.7524		
Total	21	2512105.318			

	Coefficients	Standard Error	t Stat	P-value	Lower 95%	Upper 95%	Lower 95.0%	Upper 95.0%
Intercept	155.898726	126.7458862	1.23001015	0.23370639	-109.3834625	421.1809147	-109.3834625	421.1809147
Radio Advertising ($Thousands)	13.1012979	1.759207543	7.44727248	4.7727E-07	9.419234224	16.78336163	9.419234224	16.78336163
Newspaper Advertising ($Thousands)	16.7984773	2.963098076	5.66922756	1.8243E-05	10.59664172	23.00031281	10.59664172	23.00031281

4. Develop a model to predict the assessed value (in thousands of dollars) using the size of the houses (in thousands of square feet) and the age of the houses (in years) from the following table.

House	Assessed value ($Thousands)	Size of house (thousands of square feet)	Age (years)
1	184.4	2	3.42
2	177.4	1.71	11.5
3	175.7	1.45	8.33
4	185.9	1.76	0
5	179.1	1.93	7.42
6	170.4	1.2	32
7	175.8	1.55	16
8	185.9	1.93	2
9	178.5	1.59	1.75
10	179.2	1.5	2.75
11	186.7	1.9	0
12	179.3	1.39	0
13	174.5	1.54	12.58
14	183.8	1.89	2.75
15	176.8	1.59	7.17

(a) State the multiple regression equation.

Answer: $y = 163.7751 + 10.7252X_1 - 0.2843X_2$ where X_1 = size and X_2 = age.

(b) Interpret the meaning of the regression coefficients in this equation.

Holding age constant, for each additional thousand square feet the assessed value is estimated to increase by $10.7252 thousand. Holding size constant, for each additional year the assessed value is estimated to decrease by $0.2843 thousand.

(c) Predict the assessed value for a house that has a size of 1750 ft^2 and is 10 years old.

Answer: y = 163.7751 + 10.7252 (1.75) − 0.2843 (10) = 179.7017 thousand dollars.

(d) Perform a residual analysis on the results and determine whether the regression assumptions are valid.

Based on a residual analysis the errors appear to be normally distributed. The equal-variance assumption appears to be valid. There might be a violation of the linearity assumption for age. You might want to include a second order polynomial in the model for age.

(e) Determine whether there is a significant relationship between the assessed value and the two independent variables (size and age) at the 0.05 level of significance.

F_{STAT} = 28.58 and p-value = 0.0000272776. Because p-value 0.0000 < 0.05 reject H_0. There is evidence of a significant relationship between the assessed value and the two independent variables (size and age).

(f) Interpret the meaning of the adjusted r^2.

Answer: $r^2_{\text{adj}} = 0.7976$ thus 79.76% of the variation in the assessed value is explained by variation in size and age in this multiple regression model.

(g) At the 0.05 level of significance determine whether each independent variable makes a significant contribution to the regression model. Indicate the most appropriate regression model for this set of data.

For X_1: $t_{STAT} = 3.5581$ and p-value $= 0.0039 < 0.05$. Reject H_0. The size of a house makes a significant contribution and should be included in the model. For X_2: $t_{STAT} = -3.4002$ and p-value $= 0.0053 < 0.05$. Reject H_0. The age of a house makes a significant contribution and should be included in the model. Based on these results the regression model with the two independent variables should be used.

(h) The real estate assessor's office has been publicly quoted as saying that the age of a house has no bearing on its assessed value. Based on your answers above do you agree with this statement?

Explain. The age of a house does have an effect on its assessed value.

5. The following data produces the AVOVA shown below where CARS is y whereas HH SIZE and Cubed HH Size are the x variables. Interpret the key statistics to determine if this is a statistically viable model.

	A	B	C
1	CARS	HH SIZE	CUBED HH SIZE
2	1	1	1
3	2	2	8
4	2	3	27
5	2	4	64
6	3	5	125

8	SUMMARY OUTPUT						
9							
10	*Regression Statistics*						
11	Multiple R	0.895828018					
12	R Square	0.802507837					
13	Adjusted R Square	0.605015674					
14	Standard Error	0.444400903					
15	Observations	5					
16							
17	ANOVA						
18		*df*	*SS*	*MS*	*F*	*Significance F*	
19	Regression	2	1.605015674	0.802507837	4.063492063	0.197492163	
20	Residual	2	0.394984326	0.197492163			
21	Total	4	2				
22							
23		*Coefficients*	*Standard Error*	*t Stat*	*P-value*	*Lower 95%*	*Upper 95%*
24	Intercept	0.896551724	0.764398064	1.172885918	0.361624318	-2.39238769	4.185491139
25	HH SIZE	0.33646813	0.422703764	0.79599038	0.509506953	-1.482279374	2.155215634
26	CUBED HH SIZE	0.002089864	0.013113794	0.159363815	0.888021498	-0.054334235	0.058513964

Answer:

- All data has been included since $n = 5$
- Adjusted R-Square of .605 is strong
- The F-statistic is >0.05. The overall fit is not good
- Neither of the x variables achieve significance (p-values > 0.05)

Conclusion is that this is not a statistically significant model. Back to the drawing board.

6. Discuss the assumptions of multiple regression analysis and explain their importance in interpreting the results.

Answer:

Multiple regression analysis relies on several assumptions that need to be met for accurate interpretation of the results. The key assumptions include:

- Linearity: The relationship between the dependent variable and independent variables should be linear. This assumption ensures that the estimated regression coefficients reflect the true relationship.
- Independence: The observations should be independent of each other. This assumption prevents issues such as autocorrelation where the residuals may be correlated over time.
- Homoscedasticity: The variance of the residuals should be constant across different levels of the independent variables. Violations of this assumption indicate heteroscedasticity which may affect the accuracy of the regression estimates.
- No multicollinearity: The independent variables should not be highly correlated with each other. Multicollinearity can make it difficult to isolate the individual effects of the independent variables and can lead to unstable regression estimates.
- Normality: The residuals should follow a normal distribution. Departures from normality may affect the validity of statistical tests and confidence intervals.

These assumptions are important because violating them can lead to biased/inefficient estimates, affect the interpretation of regression coefficients, and impact the reliability of the statistical tests. Checking these assumptions and if necessary, addressing any violations is crucial for obtaining accurate/valid results from multiple regression analysis in business decision-making.

7. Discuss the issue of multicollinearity when running multiple regression.

Answer:

Multicollinearity is a common issue that can arise when running multiple regression analysis. It refers to a high correlation between independent variables in the regression model. This high correlation can cause challenges in interpreting the results and undermine the accuracy and stability of the regression estimates. Here are some key points to discuss regarding the issue of multicollinearity:

- Definition: Multicollinearity occurs when there is a strong linear relationship between two or more independent variables in a regression model. It indicates that the independent variables are highly correlated and may share redundant information thus making it difficult to isolate their individual effects on the dependent variable.
- Consequences of Multicollinearity: (a) Inflated Standard Errors: Multicollinearity inflates the standard errors of the regression coefficients thus leading to wider confidence intervals. This makes it more difficult to determine the statistical significance of individual predictors. (b) Unreliable and Unstable Coefficient Estimates: High multicollinearity can cause unstable coefficient estimates as small changes in the data can result in significant changes in the estimated coefficients. This instability makes it challenging to interpret the impact of each variable accurately. (c) Ambiguous Interpretation: Multicollinearity makes it difficult to discern the individual contributions of correlated variables to the dependent variable. It becomes challenging to determine which variable is truly driving the observed changes in the dependent variable. (d) Decreased Predictive Power: Multicollinearity reduces the predictive power of the model since correlated variables may exhibit similar relationships with the dependent variable. This redundancy may lead to less accurate predictions.
- Detection of Multicollinearity: (a) Correlation Matrix: A correlation matrix helps identify highly correlated variables. A correlation coefficient close to +1 or −1 indicates a strong linear relationship between two variables. (b) Variance Inflation Factor (VIF): VIF quantifies the extent of multicollinearity by measuring how much the variance of the estimated regression coefficient is inflated due to correlations with other variables. VIF values above 5 or 10 are often considered indicative of problematic multicollinearity.
- Dealing with Multicollinearity: (a) Feature Selection: If multicollinearity is detected consider removing one or more variables that are highly correlated with others. Prioritize variables based on theoretical relevance or domain knowledge.
- Data Collection and Experimental Design: Collecting additional data or using experimental design techniques to reduce the correlation between variables can help mitigate multicollinearity. Addressing multicollinearity is crucial to ensure the accuracy and interpretability of the regression results. Analysts can obtain reliable and meaningful insights from multiple regression models by identifying/addressing multicollinearity.

8. This model has a sample size of 20. Based on these results is this model statistically viable? Explain.

SUMMARY OUTPUT

Regression Statistics	
Multiple R	0.857
R Square	0.734
Adjusted R Square	0.703
Standard Error	5.366
Observations	20

ANOVA

	df	*SS*	*MS*	*F*	*Significance F*
Regression	2	1350.76	675.38	23.46	0.00
Residual	17	489.44	28.79		
Total	19	1840.20			

	Coefficients	*Standard Error*	*t Stat*	*P-value*	*Lower 95%*	*Upper 95%*
Intercept	67.67	2.82	24.03	0.00	61.73	73.61
hours	5.56	0.90	6.18	0.00	3.66	7.45
prep_exams	-0.60	0.91	-0.66	0.52	-2.53	1.33

- Strong adjusted *R*-square of 0.703
- All data is included as $n = 20$
- Significance $F < 0.05$
- *p*-values: hours is significant (<0.05) and prep exams is not significant (>0.05)

Based on the ANOVA table this model requires removal of the prep exam variable to better fit the data.

Chapter 14
Nonlinear Regression

Linda Herkenhoff and John Fogli

Key Concepts

Exponential function, Logarithmic function, Polynomial function, and Power function.

Discussion

Since not all x, y relationships are linear, this chapter will discuss some of the most common nonlinear regression tools: power, polynomial, exponential, logarithmic. Excel can provide a graphical display with the associated trendline equation and R^2 for single independent variable models. The best way to get started is to insert various forms of trendlines on a scatterplot and observe the resulting R^2-values. Remember, in all trend fitting, we want the model that best fits the data with the least number of variables or terms.

Supplementary Information The online version contains supplementary material available at https://doi.org/10.1007/978-3-031-46371-6_14.

L. Herkenhoff (✉)
Saint Mary's College, Moraga, CA, USA

J. Fogli
Diablo Valley College, Pleasant Hill, CA, USA

L. Herkenhoff, J. Fogli (eds.), *Applied Statistics for Business and Management using Microsoft Excel*, https://doi.org/10.1007/978-3-031-46371-6_14

Power

The formula for this trendline is:

$$y = a^* \ x^b$$

Example: $y = 3.2x^{1.67}$

This means that the independent variable x is raised to a power. The value for the power must be positive, and does not have to be a whole number. There is only one regression coefficient in this case and that is shown in front of the x. As we increase the power, the graph climbs more steeply as shown in Fig. 14.1.

Polynomial

The formula for this trendline requires you to state the highest power you want x to be raised to:

For power (order) of 2 $y = a * x^2 + b * x + c$

Example: $y = 3.2x^2 + 5x + 10$

For power (order) of 3 $y = d * x^3 + a * x^2 + b * x + c$

Example: $y = 6x^3 + 3.2x^2 + 5x + 10$

This means that the independent variable x is raised to several powers. The convention that Excel follows is the power level(s) as the order, so a third-order polynomial has x raised to the highest power of 3. A second-order polynomial would have x raised to the highest power of 2. The value for the power must always be positive, and will always include values as whole numbers in sequential order from

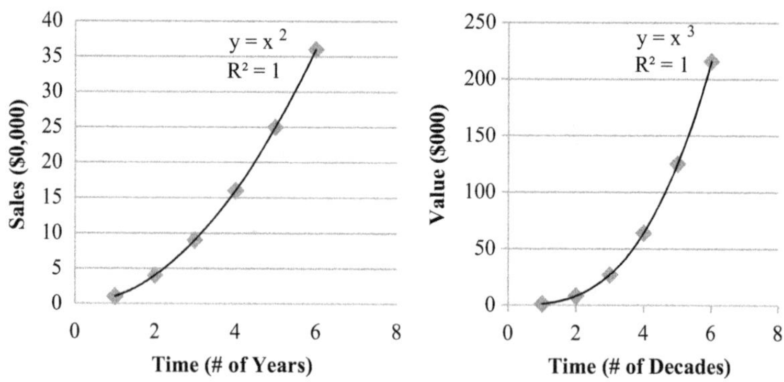

Fig. 14.1 Increasing power levels

the highest power you select. If you selected the highest power of 4, Excel would provide a trendline equation with x^4, x^3, x^2, and x. The values for x can be positive or negative.

When data seems to fit a simple curve or bulge you might want to try a second-order polynomial. But in cases where you have an "s" shape, in other words, a peak and a trough in your data, you should try a third-order polynomial. See Fig. 14.2. As you add more powers, the shape of the model is wavier.

- But remember each time you add a higher power the more complicated your model. Try to get the lowest order model with a reasonable R^2.

Exponential

The formula for this trendline is:

$$y = a * e^{bx}$$

Example: $y = 3.2e^{1.67x}$

This means that the independent variable x is now showing up in the power. The value for "e" is always a numerical constant that is equal to 2.71828. The independent variable x, must have only positive values, and does not have to be a whole number. There are two regression coefficients in this case, one in front of the e (3.2) and one in front of the x (1.67).

Examples of a negative power and a positive power are provided in Figs. 14.3 and 14.4.

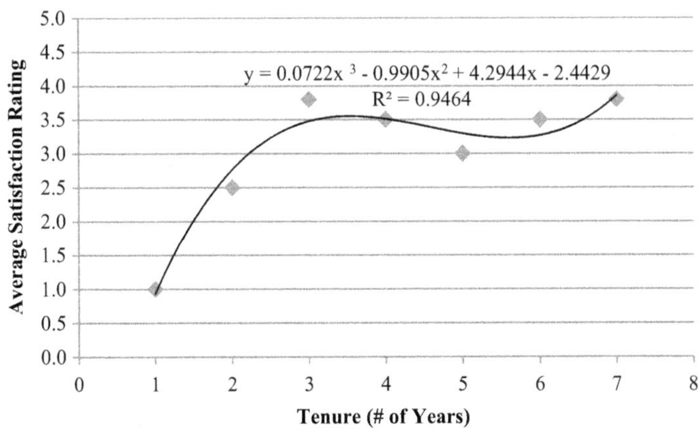

Fig. 14.2 Third-order (cubic) polynomial

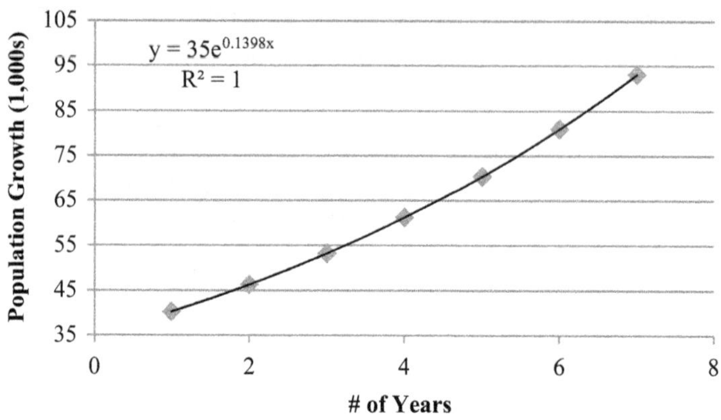

Fig. 14.3 Positive exponential model

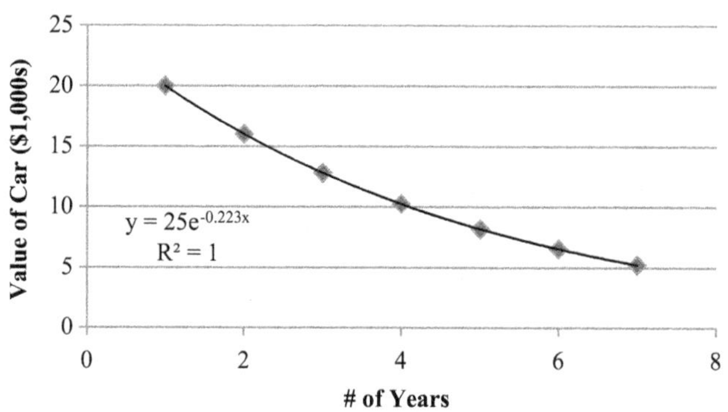

Fig. 14.4 Negative exponential model

Logarithmic

The formula for this trend line is (Fig. 14.5):

$$y = a^* \ln(x) + b$$

Example: $y = 6.7 \ln(1) + 1$

The natural logarithm of a number x, written as $\ln(x)$ is the power to which e would have to be raised to equal x. Here are some important relationships:

- ln (7.389 …) is 2, because $e^2 = 7.389$ ….
- The natural log of e, $\ln(e)$, is 1, since $e^1 = e$.
- The natural logarithm of 1, ln(1) is 0, since $e^0 = 1$.

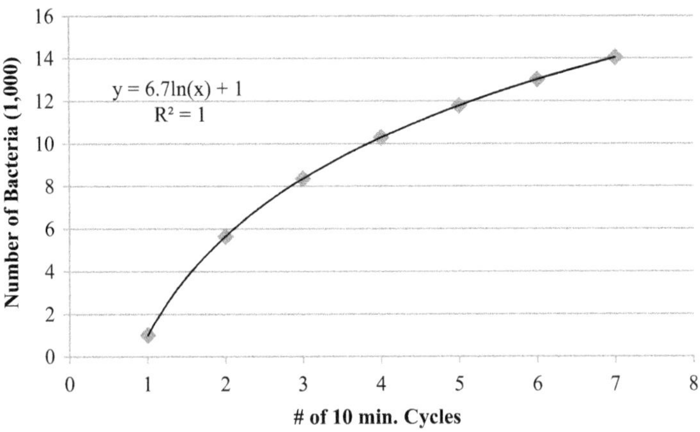

Fig. 14.5 Logarithmic function

Although logarithms are a more complicated function type to understand, there are still many places where they are used. These include:

- Working with the data obtained from magnitude scales used to measure earthquakes
- Compounding interest calculations
- Calculating how long it will take a bank deposit at a set interest rate to reach a specified higher amount
- Understanding global warming by calculating alterations in atmospheric CO_2
- Calculating radioactive decay-dating estimations

Excel

The input dataset below is used for exploring nonlinear regression models in this chapter. We will consider the relationship between the independent variable, *advertising dollars spent*, and the dependent variable, *resulting sales* (Fig. 14.6).

If we want to see the trendline graph output, we can only have one "*x*" and one "*y*" in our models. Excel cannot create multidimensional graphs. If you want to have a visual output of a multiple nonlinear regression model, you need to use more sophisticated modeling software such as SAS or XGobi. In this chapter, we will consider simple (only one *x*) nonlinear models, thereby allowing us to show trendlines on the scatterplots for each of the model types which include

- Power
- Polynomial
- Exponential
- Logarithmic

Fig. 14.6 Input dataset

advertising ($000)	sales ($000)
74	900
27.5	266.4
169	155.2
497	4320
270.5	2707.2
44.5	439.2
63	1209.6
189.5	2966.4

Create the Trendline Graphs

- Highlight *x* and *y* data on your spreadsheet with *x* as the first column followed by the *y* data in the next column
- Select **Insert** from the top ribbon
- Select the **Scatter** icon
- From the drop down menu, highlight the upper left-hand corner display that shows markers without lines

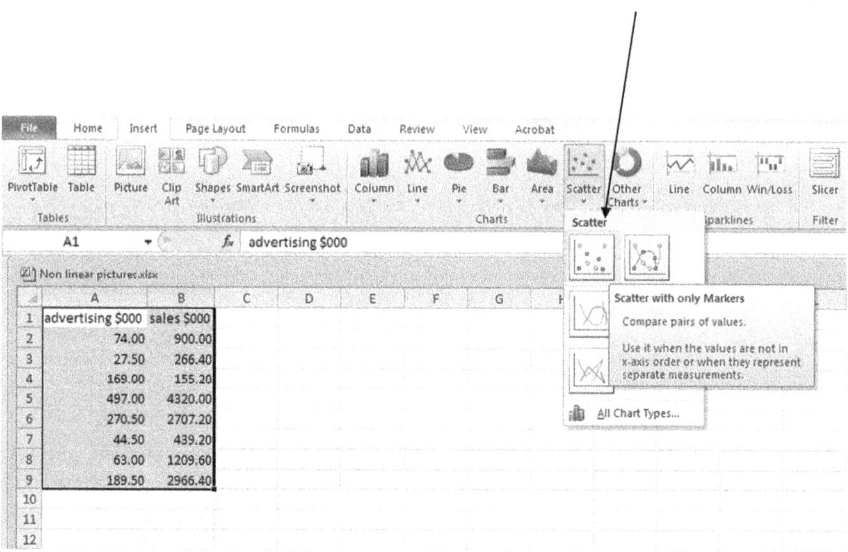

- Right click on any data point and select the **Add Trendline** option

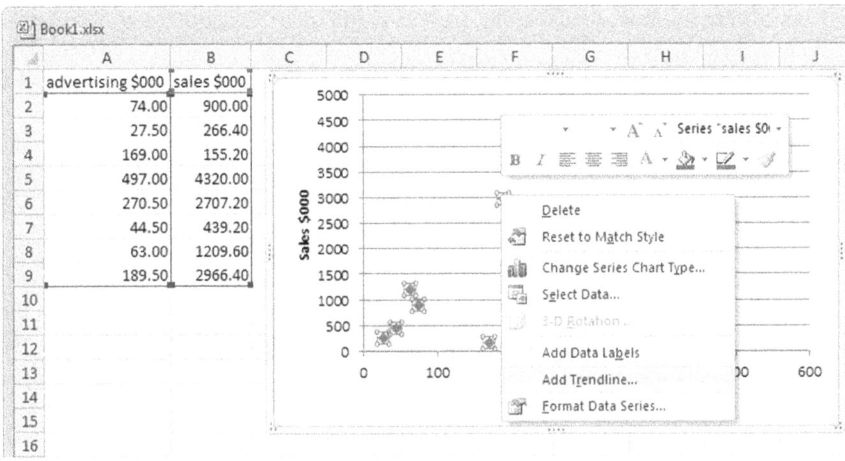

- Select the **Trendline Options** you want by clicking on the corresponding box

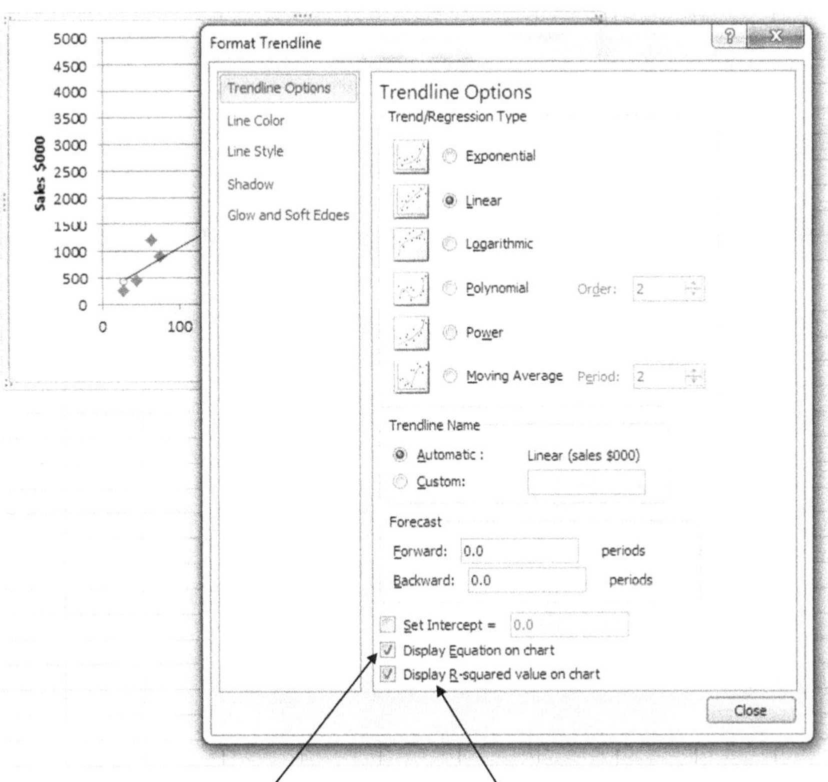

- Also check **Display Equation on chart** and **Display R-squared value on chart**
- You can add several different trendlines on the same graph to see which one looks like it fits best. Just keeping right clicking on a data point to choose a new type of trend line.

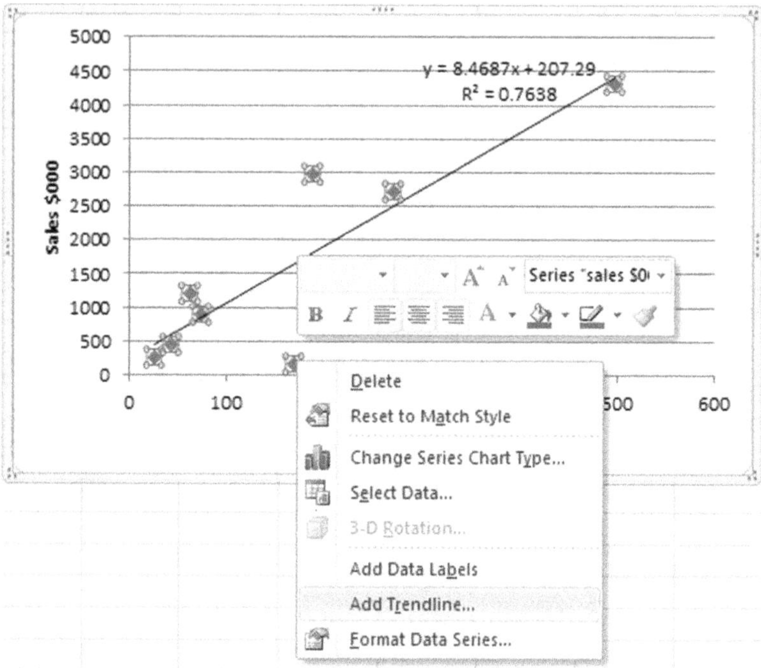

To be able to differentiate the various trendlines you might want to change the color of each of the lines. To do this:

- Right click on **Format Trendline** to change color of trendline, select the "bucket," go down menu to **Color**

- Choose the color desired from the drop down menu on the right

To change the color of the equation follow the same instructions after clicking on the equation.

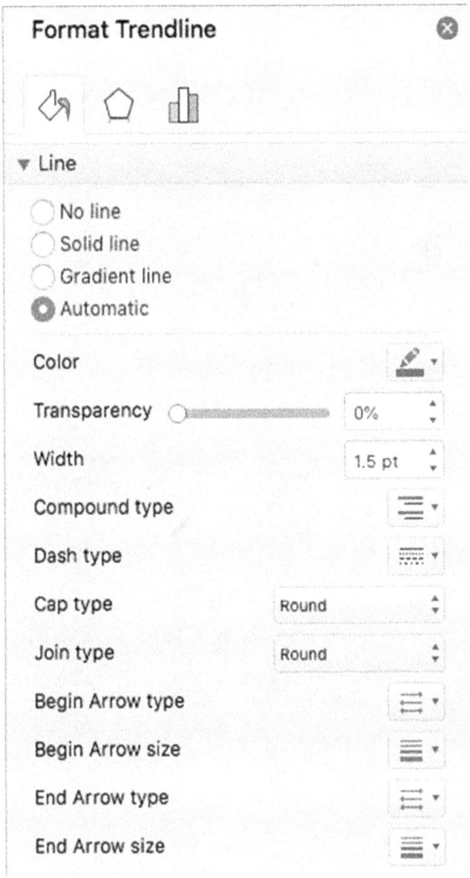

Comparing Trendline Models

Figure 14.7 illustrates the various output trendlines. Note we have used different colors to more easily associate the equations with their corresponding trendlines. We can compare the R^2 values to see which model explains more of the variation in the y variable. In this case it seems that either the simple linear or the third-order polynomial models have R^2 of 77%. After considering the R^2 it is usually best to select the simplest model; in this case we would select the simple linear regression model.

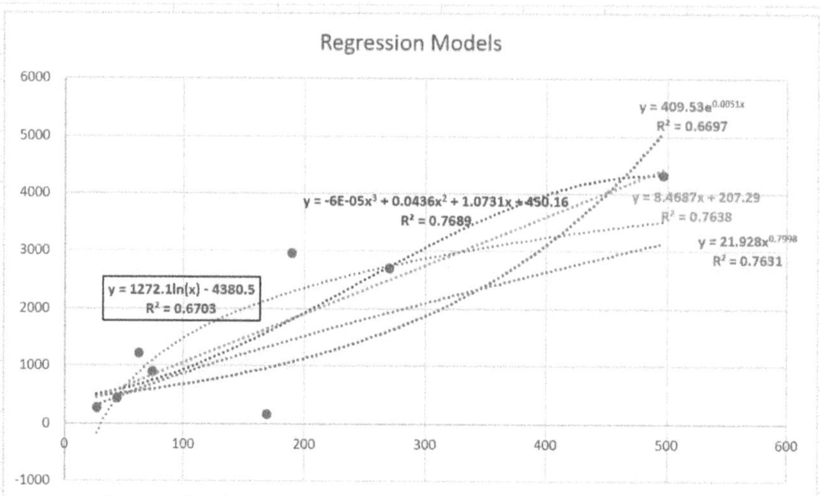

Fig. 14.7 Output data

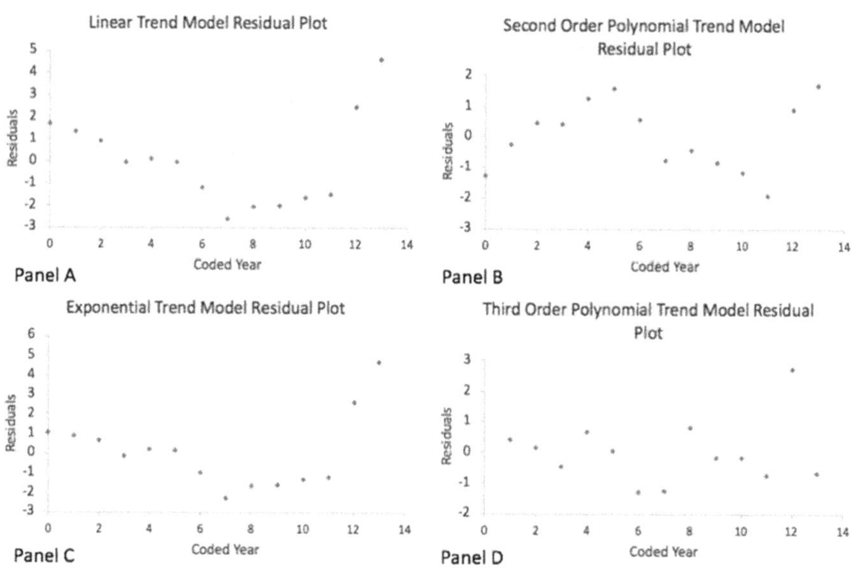

Fig. 14.8 Residual plots for four nonlinear trend models

We can also look at the residual plots for different models. A comparison of four nonlinear trend models is given in Fig. 14.8.

Consider the Coca-Cola Company's revenue data. To illustrate the model selection process, you can compare four of the forecasting models: the linear model, the second-order polynomial model, the exponential model, and the third-order polynomial model.

Figure 14.8 displays the residual plots for the four models. In reaching conclusions from these residual plots, you must use caution because there are only 23 values.

In Fig. 14.8, observe the systematic structure of the residuals in the linear model (Panel A), second-order polynomial model (Panel B), and exponential model (Panel C). For the third-order polynomial model (Panel D), the residuals appear more random.

To summarize, on the basis of the residual analysis of all four forecasting models, it appears that the third-order polynomial model is the most appropriate, and the linear, second-order polynomial, and exponential models are less appropriate.

Using the Nonlinear Regression Trendline for Prediction

To use these models we need to input the resulting trendline equations as formulas for each of these functions.

- Highlight a cell on the page
- In the formula ribbon area, type: =

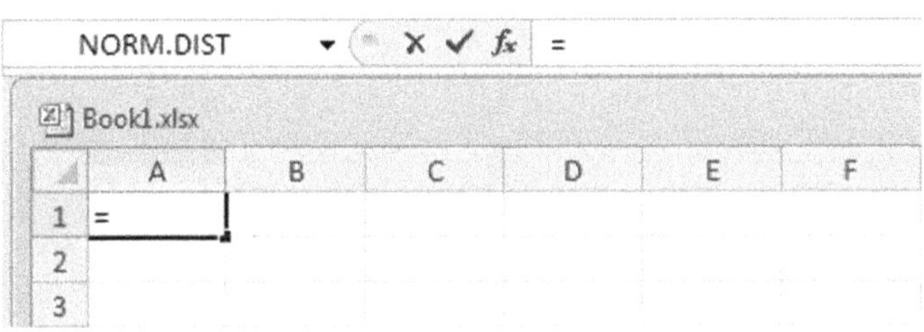

Now you can begin typing your actual equation. Here's how to type in the equations for each trendline model with **x** being located in cell **B2.**

Trendline type	Example of type equation	Input to Excel	Let B2 have a value of 3 for x. So y equals
Power	$y = 3.2x^{1.67}$	=3.2 * **B2** ^ 1.67	20.04
Polynomial	$y = 6x^3 + 3.2x^2 + 5x + 10$	=6 * **B2** ^ 3 + 3.2 * **B2** ^ 2 + 5 * **B2** + 10	215.80
Exponential	$y = 3.2e^{1.67x}$	=3.2 * EXP(1.67 * **B2**)	479.70
Logarithmic	$y = 3.2 \text{ Ln } (x) + 5$	=3.2 * LN(**B2**) + 5	8.52

Check that you know how to input these equations by typing a value of 3 into cell B2. When you input the equations from the third column of the table above, you should get the answers as indicated in the fourth column from the above table.

After you have typed in the formula click the green **checkmark**.

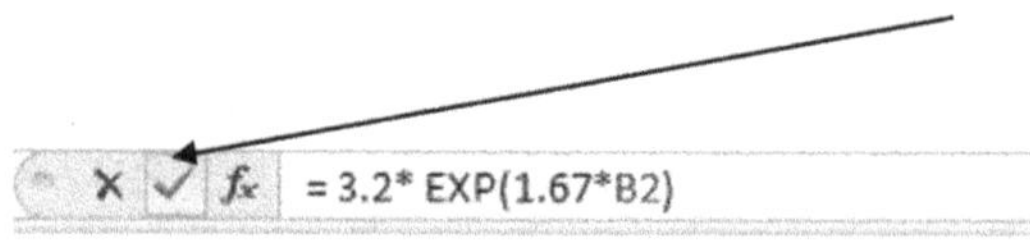

This formula will be used to calculate the y value corresponding to the x value you input from cell B2.

To calculate other y values for given x values:

- Click on the cell where you first used the formula
- Select **Copy**

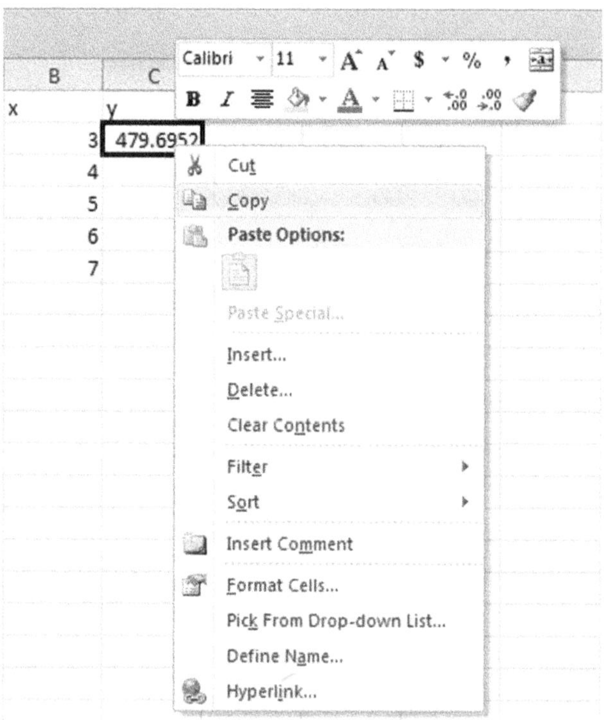

- Highlight the other empty *y* cells and type **Paste**

  Before **Paste**:

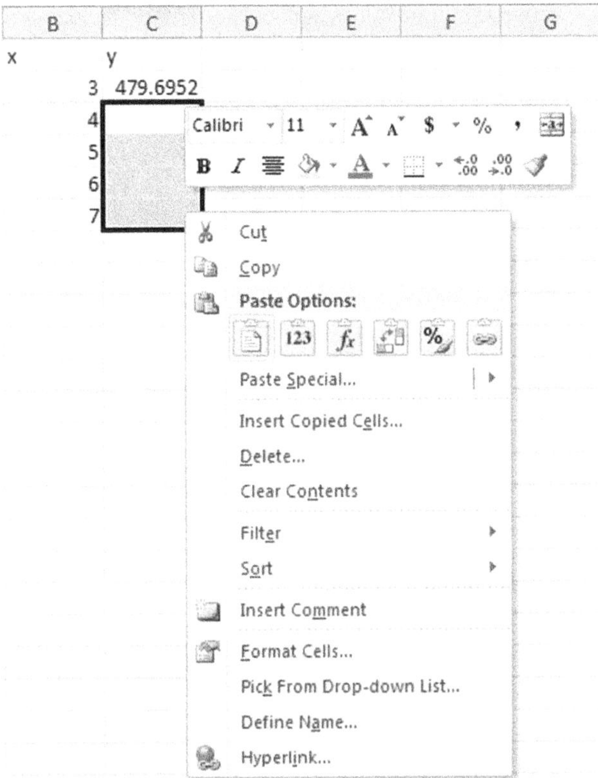

After **Paste**:

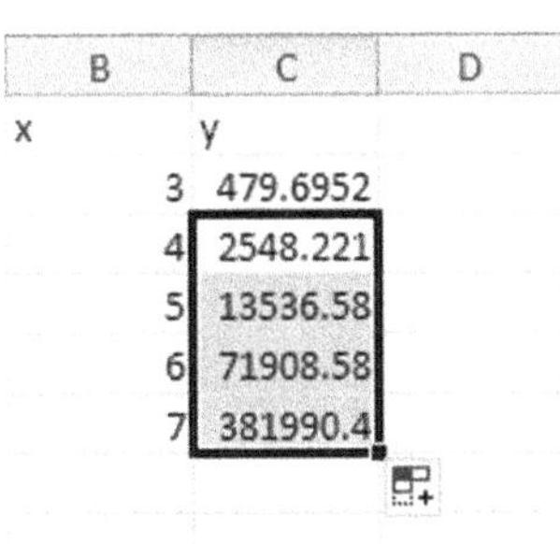

Common Excel Pitfalls

- All x values have to be positive to run a power or exponential model. If you do have some negative values or values equal to zero, you can simply add a constant to all of the x data values. This will not affect the trendline fit.

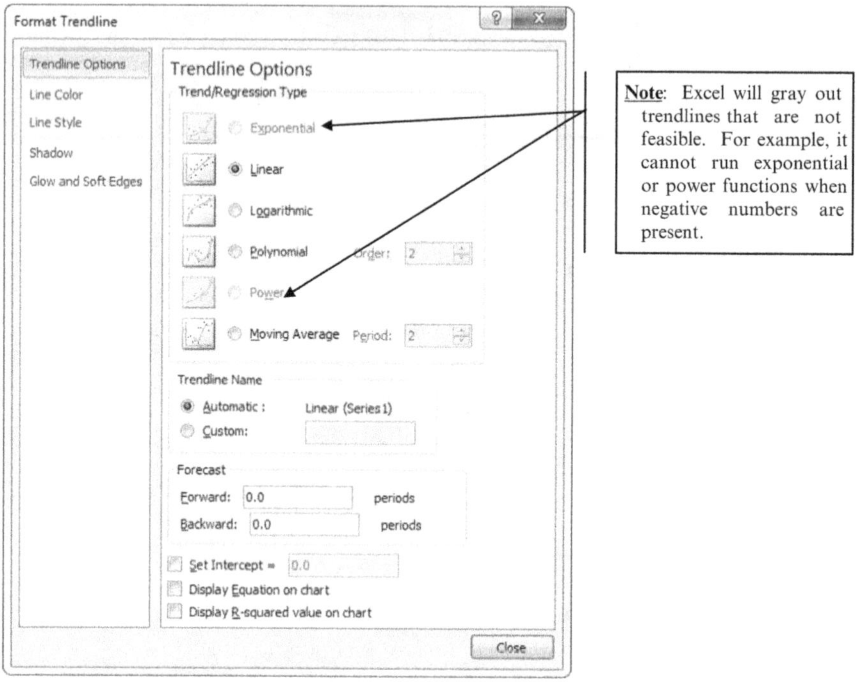

- Make sure you know what form of logarithm you need to work with. We have only used the natural logarithm in this chapter. If you want to use a base other than the constant e, you must use a different form of this function.

y = LN(number)	Natural log
y = LOG (number, base)	Any base
y = LOG10 (number)	Base 10

- When adding in trend lines, all the lines will appear in the default color, unless you format the color of the line.
- Make sure you have the correct number of parentheses (opening and closing). If you do not have enough close parentheses, Excel will give you an error or may not calculate your equation correctly.

Final Thoughts and Activities

Practice Problems and Case Studies

1. A technology manager of a small company would like to better understand the trends in laptop depreciation prior to purchasing a new set of laptops for the employees. She reviews laptop prices over the course of 9 months with the following results:

# of months	Price ($)
1	2750
2	2000
3	1550
4	1250
5	950
6	750
7	625
8	500
9	375

   (a) Construct a scatter plot.
   (b) Determine the appropriate nonlinear function that matches the data.
   (c) What does the model tell you about the relationship between laptop price and time?

2. A travel expense coordinator at a hotel needs to understand the fluctuation of gas prices over the course of a year to better plan the department's travel budget. After doing some research, he acquired the following data:

Month	# of months	Average gas price ($)
January	1	3.07
February	2	3.20
March	3	3.55
April	4	3.77
May	5	3.95
June	6	3.72
July	7	3.61
August	8	3.58
September	9	3.50
October	10	3.40
November	11	3.30
December	12	3.27

   (a) Create a frequency plot.
   (b) Determine the appropriate nonlinear function that matches the data.
   (c) What does the model tell about the relationship between average gas prices over the course of 12 months?

3. An ad agency wanted to understand how the number of ads shown in one day related to the number of people who were aware of the advertisement. The data they gathered is displayed below:

Number of ads/day	# of people aware (0000s)
2	1.00
3	1.25
4	2.25
5	6.25
6	22.25
7	86.25
8	342.25

   (a) Construct a scatter plot.
   (b) Determine the appropriate nonlinear function that best matches the data.
   (c) What does the model tell you about the relationship the number of ads shown in one day related to the number of people who were aware of the advertisement?

4. The data shown in the following table represents the yearly amount of solar power installed (in megawatts) in the United States from 2000 through 2008:

Year	Amount of solar power installed
2000	18
2001	27
2002	44
2003	68
2004	83
2005	100
2006	140
2007	210
2008	250

   (a) Plot the data.
   (b) Compute a linear trend forecasting equation and plot the trend line.
   (c) Compute a second-order polynomial trend forecasting equation and plot the results.
   (d) Compute an exponential trend forecasting equation and plot the results.

5. Bed Shower & Beyond Inc. is a nationwide chain of retail stores that sell a wide assortment of merchandise, principally including domestics merchandise and home furnishings, as well as food, giftware, and health and beauty care items. The following data shows the number of stores open at the end of the fiscal year from 1993 to 2009:

Year	Stores open	Year	Stores open
1993	38	2002	396
1994	45	2003	519
1995	61	2004	629
1996	80	2005	721
1997	108	2006	809
1998	141	2007	888
1999	186	2008	971
2000	241	2009	1037
2001	311		

(a) Plot the data.
(b) Compute a linear trend forecasting equation and plot the trend line.
(c) Compute a second-order polynomial trend forecasting equation and plot the results.
(d) Compute an exponential trend forecasting equation and plot the results.
(e) How can you explain the differences in the three forecasts in (e)? What forecast do you think you should use? Why?

6. Use the following power equation to solve for y when $x = 3$.

$$Y = 3.2x^{1.67}$$

7. Use the following exponential equation to solve for y when $x = 3$.

$$Y = 3.2e^{1.67x}$$

Discussion Boards

1. The pricing of oil is a complex process. Consumers are often confused that the "price at the pump" does not seem to track with corporate profits or the discovery of new oil fields. Collect the price at the pump data and plot it against the corporate net profit figures for one of the major oil companies over the past 24 months. Are you surprised at the shape of the resulting graph?
2. Most statistical factors, relationships and processes involving people have a behavioral component. This means they do not follow a linear model. Why is that?

Group Activity

Change management processes are often evaluated using linear models. In other words as change is implemented, performance improvement is expected to follow. In fact performance may decline, flatten or improve. The punctuated equilibrium model describes periods of inertia followed by periods of positive energy. Choose a recent change initiative in your own organization. Which of the nonlinear models might best fit the process and why?

Parting Thought

Statistically speaking, in China, even if you are a one in a million kind of guy, there are a thousand more just like you.

Problem Solutions

1. A technology manager of a small company would like to better understand the trends in laptop depreciation prior to purchasing a new set of laptops for the employees. She reviews laptop prices over the course of 9 months with the following results:

# of months	Price ($)
1	2750
2	2000
3	1550
4	1250
5	950
6	750
7	625
8	500
9	375

(a) Construct a scatter plot.

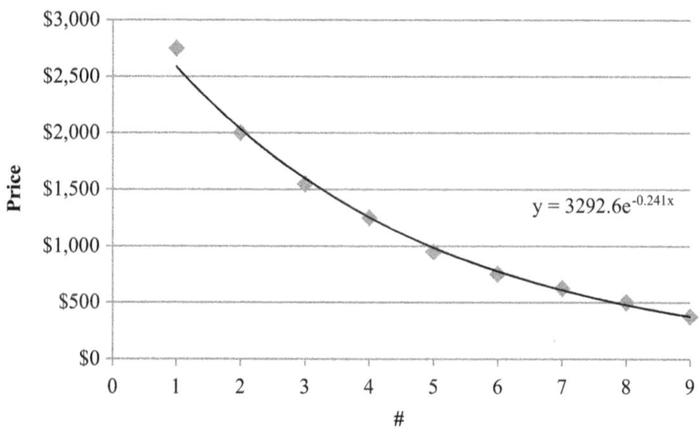

(b) Determine the appropriate nonlinear function that matches the data.

A negative exponential function fits the chart best.

(c) What does the model tell about the relationship between laptop price depreciation over time?

The data shows that as time progresses the depreciation of value for a laptop is exponentially negative, or drops exponentially.

2. A travel expense coordinator at a hotel needs to understand the fluctuation of gas prices over the course of a year to better plan the department's travel budget. After doing some research, he acquired the following data:

Month	# of months	Average gas price ($)
January	1	3.07
February	2	3.20
March	3	3.55
April	4	3.77
May	5	3.95
June	6	3.72
July	7	3.61
August	8	3.58
September	9	3.50
October	10	3.40
November	11	3.30
December	12	3.27

(a) Plot the data.

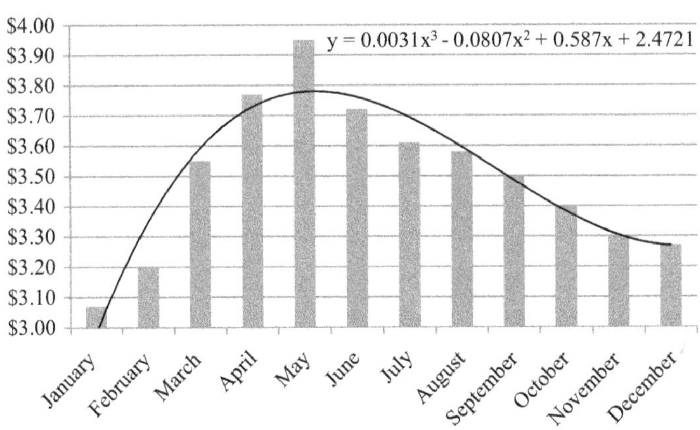

(b) Determine the appropriate nonlinear function that matches the data.

A polynomial trendline best fits the data.

(c) What does the model tell about the relationship between average gas prices over the course of 1 year?

The trendline shows that over the course of 1 year gas prices fluctuate and reach a peak in the summer months but decrease for the fall and winter.

3. An ad agency wanted to understand how the number of ads shown in one day related to the number of people who were aware of the advertisement. The data they gathered is displayed below:

Number of ads/day	# of people aware (0000s)
2	1.00
3	1.25
4	2.25
5	6.25
6	22.25
7	86.25
8	342.25

(a) Construct a scatter plot.

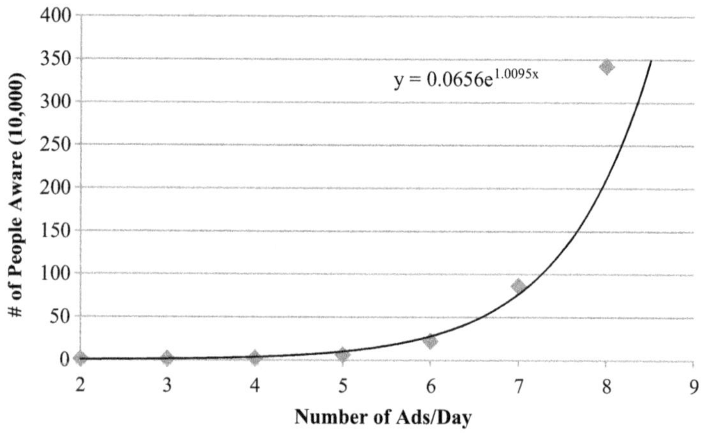

(b) Determine the nonlinear function that matches the data.

$$y = 0.0656e^{1.0095x}$$

(c) What does the model tell about the relationship between the number of ads and the number of people who are aware of the ad?

The chart shows that the number of ads is positively exponentially related to the number of people who are aware of the ad.

4. The data shown in the following table represents the yearly amount of solar power installed (in megawatts) in the United States from 2000 through 2008:

Year	Amount of solar power installed
2000	18
2001	27
2002	44
2003	68
2004	83
2005	100
2006	140
2007	210
2008	250

(a) Plot the data. Highlight your data. You can use the Insert Tab and select the scatter graph

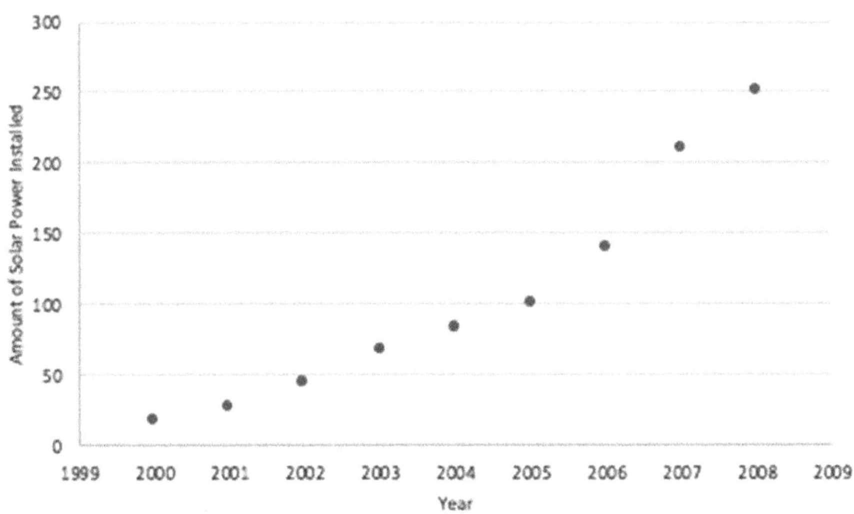

(b) Compute a linear trend forecasting equation and plot the trend line. Now you can add the trendline by right clicking on any data point and selecting which type of trendline you want to try. You can also add the actual equation to the graph by clicking on Display Equation on Chart.

$$y = 28.35x \; - 5609.$$

(c) Compute a second-order polynomial trend forecasting equation and plot the results. Click on any data point on the graph and Add Trendline by selecting Polynomial and order 2. Select Add Equation

$$y = 3.44361x\hat{}2 - 137744x + 1\hat{} + 07$$

(d) Compute an exponential trend forecasting equation and plot the results.

$$y = (6E - 280)(e^\wedge 0.323x)$$

5. Bed Shower & More Inc. is a nationwide chain of retail stores that sell a wide assortment of merchandise, principally including domestics merchandise and home furnishings, as well as food, giftware, and health and beauty care items. The following data shows the number of stores open at the end of the fiscal year from 1993 to 2009:

Year	Stores open	Year	Stores open
1993	38	2002	396
1994	45	2003	519
1995	61	2004	629
1996	80	2005	721
1997	108	2006	809
1998	141	2007	888
1999	186	2008	971
2000	241	2009	1037
2001	311		

(a) Plot the data.

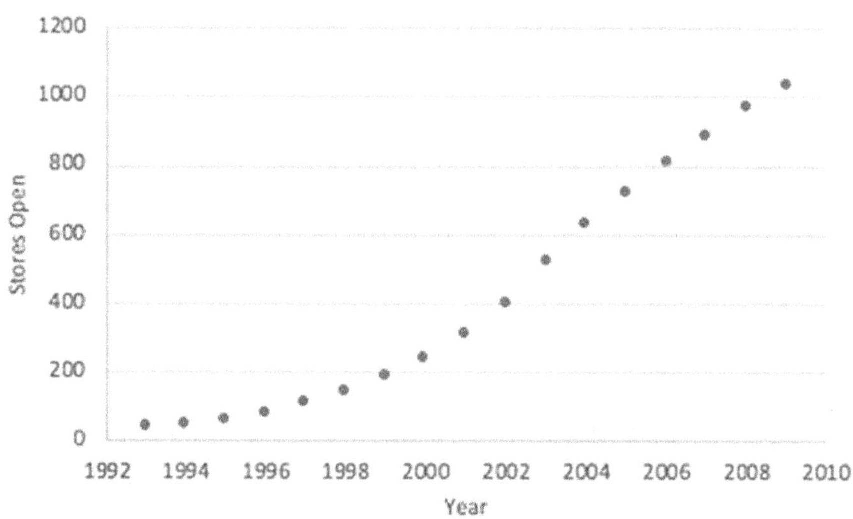

(b) Compute a linear trend forecasting equation and plot the trend line.

$$y = 68.181x - 136009$$

(c) Compute a second-order polynomial trend forecasting equation and plot the results.

$$y = -13499x + 1E07$$

(d) Compute an exponential trend forecasting equation and plot the results.

$$y = 2\hat{} - 190e\hat{}0.2211x$$

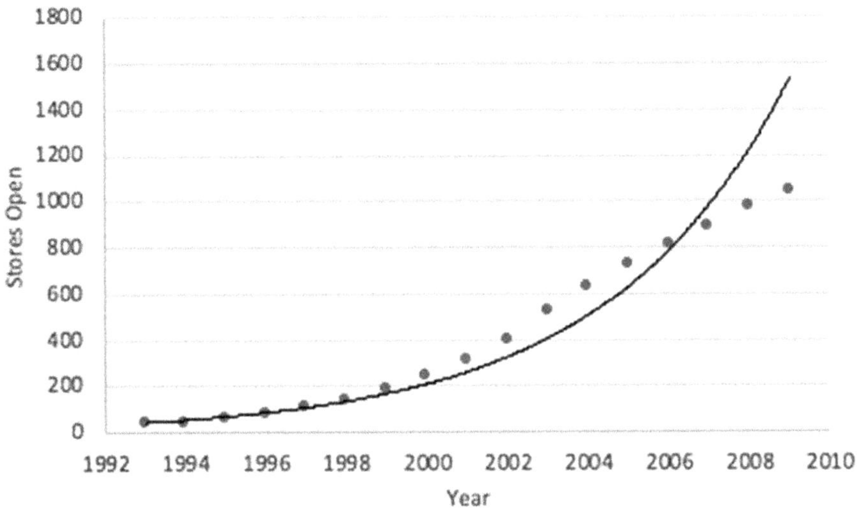

6. Use the following power equation to solve for y when $x = 3$.

$$Y = 3.2x^{1.67}$$

Year	Stores open
1993	38
1994	45
1995	61
1996	80
1997	108
1998	141
1999	186
2000	241
2001	311
2002	396
2003	519
2004	629
2005	721
2006	809
2007	888
2008	971
2009	1037

Answer:
Input into Excel as an equation

	A	B	C
1	x	Equation	Substitute
2	3	3.2*3^1.67	20.04206534

7. Use the following exponential equation to solve for y when $x = 3$.

$$Y = 3.2e^{1.67x}$$

Answer:
Input into Excel as an equation

	A	B	C
1	x	Equation	Substitute
2	3	3.2 *EXP(1.67*x)	479.6951557
3			

Chapter 15
Survey Reports

Linda Herkenhoff and John Fogli

Case Study: Creating an Impactful Report for Infinity Auto Insurance
To better understand the Hispanic market in Houston, Texas, Infinity Auto Insurance sampled 400 respondents in Houston, Texas area. The sample design and specifications of the study are summarized in the table below.

Design	Specification
Methodology	Telephone survey
Population	Residents of the City of Houston, Texas
Sample size	200

(continued)

Supplementary Information The online version contains supplementary material available at https://doi.org/10.1007/978-3-031-46371-6_15.

L. Herkenhoff (✉)
Saint Mary's College, Moraga, CA, USA

J. Fogli
Diablo Valley College, Pleasant Hill, CA, USA

L. Herkenhoff, J. Fogli (eds.), *Applied Statistics for Business and Management using Microsoft Excel*, https://doi.org/10.1007/978-3-031-46371-6_15

Create a report with your new found knowledge of statistics and survey reporting using the data provided below to help Infinity Auto Insurance better understand the gathered data. Write a report that includes this survey's sampling methodology, survey creation, results, conclusions, and recommendations with what you have learned from this book.

	Frequency	Valid %
Own vehicle		
Yes	179	89.5
No	21	10.5
Total	200	100.0
Own insurance		
Yes	150	75
No	50	25
Total	200	100.0
Switch insurance		
Not at all likely	92	61.3
Not likely	9	6.0
Somewhat likely	29	19.3
Likely	13	8.7
Extremely likely	7	4.7
Total	150	100.0

Note: This question is only asked of people who do own vehicle insurance

	Frequency	Valid %
Heard of Fu Auto Insurance		
Yes	150	75.0
No	50	25.0
Total	200	100.0
Heard of ES Auto Insurance		
Yes	50	25.0
No	150	75.0
Total	200	100.0
Heard of Infinity Auto		
Yes	100	50.0
No	100	50.0
Total	200	100.0
Seen Infinity Auto Ad		
Yes	86	86.0
No	14	14.0
Total	100	100.0
Seen Fu Auto Insurance Ad		
Yes	125	83.3
No	25	16.7
Total	150	100.0
Seen ES Auto Insurance Ad		
Yes	30	60.0
No	20	40.0
Total	50	100.0
Fu Auto Insurance Favorability		
Very unfavorable	3	2
Unfavorable	6	4
Somewhat favorable	32	21.3
Favorable	51	34
Very favorable	58	38.7
Total	150	100.0

Fu Auto Insurance Favorability Mean: 4.03

	Frequency	Valid %
ES Auto Insurance Favorability		
Very unfavorable	2	4
Unfavorable	3	6
Somewhat favorable	17	34
Favorable	16	32
Very favorable	12	24
Total	50	100.0

ES Auto Insurance Favorability Mean: 3.66

	Frequency	Valid %
Infinity Auto Favorability		
Very unfavorable	9	9.0
Unfavorable	7	7.0
Somewhat favorable	22	22.0
Favorable	32	32.0
Very favorable	30	30.0
Total	100	100.0

Infinity Auto Favorability Mean: 3.67
Demographics

	Frequency	Valid %
Spoken language		
English	32	16.0
Spanish	168	84.0
Total	200	100.0

	Frequency	Valid %
Ad Language preference		
English-language advertisement	29	14.5
Spanish-language advertisement	171	85.5
Total	200	100.0
Gender		
Male	97	48.5
Female	103	51.5
Total	200	100.0

Key Concepts

Executive summary, Survey report, Large metropolitan area, Staff satisfaction, Cross-tabulation, Conclusions and recommendations, Executive summary, Multi-analysis report, Overall report, and Pilot testing.

Discussion

The earlier chapters introduced the basics of business statistics and this final chapter will integrate those concepts into the discussion of survey reports and presentations.

We often hear that a picture is worth a thousand words. In this chapter we focus our attention on how to use Excel to visually enhance our reports. The audience is often critical about how much information is necessary to display and report. For example, XYZ, Inc. has 30,000 stores worldwide. Operational results can vary from one part of the world to another: a store manager in New York City is probably not concerned about the operational results from a store located in Milan, Italy. Therefore, we must target our results to our audience of interest.

Even after all the data has been collected and the statistical analysis conducted, the research is usually not complete until a final report is written. The report should clearly explain to the intended audience the essential findings of the study/research. There are many ways to approach the survey report process. The following fundamentals of a standard survey report are discussed in this chapter.

I. Title Page
II. Table of Contents
III. Executive Summary
IV. Background and Objectives
V. Research Methodology
VI. Overview of Results
VII. Conclusions and Recommendations
VIII. Bibliography
IX. Appendices

When your survey and analysis has been completed, the final step in the survey process is to present your findings which involves the creation of a research report. This report should include a background of why you conducted the survey, a breakdown of the results, and conclusions/recommendations supported by this material.

Title Page

The title for a report should be simple yet attention-grabbing. Generally the title will give the recipient of the report an overview or highlight of its contents. The title should be long enough to relay the point across but short enough to hold the recipient's interest. An example of an attention-grabbing title would be: "Los Angeles: Extremely Loyal Market Place with Company Reputation."

Include the names of those who prepared the report, to whom it will be presented, and the submission date of the report.

Table of Contents

List the sections in your report. Here is where you give a high-level overview of the topics to be discussed in the order they are presented in the report. Depending on the length of your report you should consider including a list of all charts/graphs so that your audience can quickly locate them.

Executive Summary

The Executive Summary is located at the beginning of the report following the Table of Contents. The Executive Summary provides the recipient of the report with a high-level overview of the study. This section usually encompasses no more than two pages and is preferably only a few paragraphs. It includes a quick summary of the survey's background, objective, results, and sometimes a few key conclusions/recommendations. The outline explains what the recipient will be seeing on the pages to follow.

Example of a Key Finding in the Executive Summary

This study shows that over the past three years, people's impression of the Company has steadily increased. The Company has the highest rate of favorability compared to its competitors in the Los Angeles and Orange County area.

Background and Objectives

The Background section explains what led up to conducting the study and its purpose whereas the Objectives section explains the survey study's goals. The Background section details to the recipient (who may or may not have any prior knowledge of the study) who requested the study, why the study was requested, the purpose of the study, what previous studies may have been completed, and what the study measured.

The Objectives section describes the main purpose (or goal of the study) and the supporting rationale for doing the study. Often times survey studies have multiple objectives or secondary objectives. Each of the objectives should be stated/outlined and each separate objective should be explained as necessary along with any goals (or expected outcomes for the objectives).

Example of an Objective

Objective *Taking the research and coalescing the information together in an annual report will integrate all the avenues of research and create a more meaningful picture regarding the organization's target market areas and population.*

Research Methodology

The Research Methodology details every step taken to prepare, execute, and analyze this study. It explains the process of the survey study and provides the audience with a general timeline of events. Include a copy of the survey in the Appendix section of the report.

The first step detailed in the Research Methodology section is usually the sampling plan. The sampling plan should explain the rationale for the sample size as well as discuss the sample representativeness.

The next section should describe the type of survey and the procedures taken to create, administer, and collect the survey/data. When explaining the survey creation specify who participated in the creation of the survey, what type of questions were included, and who approved the final survey. **Pilot testing** involves sending out the survey to "test the waters"; it is an opportunity to get feedback on the survey to allow for improvements before the final survey goes out. The report should include comments on any pilot testing that was completed. The survey administration and data collection section illuminates the procedures taken before/during the data collection process. It can include pilot test data information, how the survey was distributed or collected (via mail, email, telephone, etc.), and the basic details of the data collection process (including any challenges that were overcome).

A quick summary of the sample methodology at the end of the Research Methodology section can prove helpful. Placing the sample methodology into a table shares the necessary information in a concise manner. Figure 15.1 is an example of a summary table.

DESIGN	SPECIFICATION
Methodology	Telephone survey
Population	Residents of the City of Dallas, Texas
Sampling Method	Random Digit Dialer (RDD)
Calls made	19,474
Respondents	600 respondents who own vehicles
Confidence Level	95% confidence level
Survey language	Approximately 10% English; 90% Spanish
Data collection period	November 19- December 11, 2012

Fig. 15.1 Example summary table

Results

The Results section gives the audience the basic results of the study. In this section, the survey results are explained in detail usually in words rather than in charts and figures. For readability purposes the results usually follow the order in which they appear in the survey. Categorizing the sections of the survey helps us group the results into more meaningful themes and concepts. The rationale for why the results for a specific question occur should be explained in this section. By explaining the results we help the audience understand not just the basic numbers but what the numbers actually mean. This is the section where you can discuss the demographics (measurable characteristics) of the sample.

Example of Results

The sample of 600 respondents is 51% male. Ninety percent of respondents primarily speak Spanish at home and 86% prefer advertisements in Spanish as opposed to in English.

Cross-Tabulate Relevant Pairs of Questions

Most questions don't have much meaning on their own so you need to compare or **cross-tabulate** them with other questions. If the answers to a cross-tabulation are not statistically significant but are still important to the organization you may still choose to report the results for completeness. For example, "Question 3 produced no statistically significant differences at the .05 level for different genders, age groups, or geographical areas."

Report, remember that you can only squeeze about 8 columns across a page and about 50 rows down when putting tables into a report. If there are too many different answers you'll need to recode the question to reduce the number of columns (or sometimes rows).

Make sure that readers cannot misinterpret the percentages. The normal convention is that percentages add to 100 downwards. If they add across (usually done to fit the table onto a single page) you must show this very clearly. See Fig. 15.2.

Note in Fig. 15.2 that the sample sizes are shown as percentages and the overall significance results are provided. For a review of significance results see Chaps. 11 and 12.

Primary Home Ownership	Japan (n = 600,000)	Iceland (n = 512,000)	Tonga (n = 25,000)
	% of respondents in that country		
Single Family Home	58	61	76
Condominium	12	18	0
Multi-family Home	2	0	20
Apartment	28	21	4
Chi-squared = 22.74 (10 degrees of freedom) Significance of Difference = .008**			

Fig. 15.2 Cross-tab example

Conclusions and Recommendations

The Conclusions and Recommendations section is the main emphasis of the report. Often the Conclusions and Recommendations are grouped together. The Conclusions should summarize and explain the results, whereas the Recommendations give these results applications and meanings. This section does not include graphs, charts, or tables. It emphasizes the purpose of the report and what it means. The conclusions should explain what inferences can be drawn from the data. The Recommendations should draw upon the inferences from the Conclusions and provide limitations of the study with possible improvements for future work. When writing recommendations always draw on the information gathered from the survey and restate it in the Conclusions. We need to ensure that a link exists between the actual data and the recommendations.

Example of a Conclusion and Recommendation

Because the Los Angeles/Orange County market area is doing well we suggest that the Company conduct focus groups in that area to see what is working best and what trends have had the most impact on that area. These focus groups would not only help Los Angeles sustain its favorability rating but also give other marketing areas insights on how to improve their own visibility and favorability.

Bibliography

The Bibliography lists the sources you consulted during your survey and any additional sources you may have referred to in the report. The Bibliography section is necessary if we refer to different studies/references in journals or if we've used information from external sources (such as newspapers, the Internet, etc., in our report). For example, we want to explain our rationale for using both landlines and mobile numbers for our telephone survey. From our Internet research we found a journal article stating the percentage people use their mobile phone as their main phone number. If we refer to the percentage in the journal article, we will need to site it in our bibliography.

Example of a Bibliography Entry

Blumberg, S. J., & Luke, J. V. (2009). Reevaluating the need for concern regarding noncoverage bias in landline surveys. American Journal of Public Health, 99(10), 1806–10. doi: https://doi.org/10.2105/AJPH.2008.152835

Appendices

The Appendix section should include any supplementary information that will help your audience understand the survey study better such as any additional graphs, charts, and tables. One should also include a copy of the actual survey, the invitation email, etc., and any other relevant study documents. This is a good place to put transcripts of historic documents and other lengthy bits of documentation that do not fit comfortably in the main body of your report.

Useful Hints and Phrases for the Report

Make sure you clearly note on the cover page of the report the level of privacy associated with the content. The more common labels for privacy include:

- **Anonymous**: Name of respondent remains unknown
- **Confidential**: Private information not for general distribution and that is usually limited to only approved individuals (often only shared to just those in the presentation room)
- **For Management Review Only:** Distribution limited to management levels and which will not be shared to employees below the rank of manager

It is common for certain phrases to get repeated over and over throughout the report. Here are several common phrases with examples of how to say the same thing:

- **To introduce:** *The purpose/aim of this report, As requested, This survey was carried out/conducted by means of…, The questionnaire consisted of, etc.*
- **To generalize:** *In general, Generally, On the whole, etc.*
- **To refer to a fact:** *The fact is that…, In fact, In practice, etc.*
- **To conclude/summarize:** *In conclusion, All things considered, To sum up, all in all, It is not easy to reach any definite conclusions, If any conclusions may be drawn from the data, It is clear that, The survey shows/indicates/demonstrates, etc.*

Present tenses, reported speech, and an impersonal style should be used in survey reports. Use a variety of reporting verbs such as claim, state, report, agree, complain, suggest, etc.

Effective PowerPoint Presentations with Excel

Often the delivery of the report content is also requested as a PowerPoint presentation. Excel slides can be too complex and confusing for the audience to easily follow so here are some useful hints to share the Excel data in a more "audience-friendly" way.

1. Use charts and graphs to illustrate big ideas. Keep it simple and don't give too much detail.
2. If reproduction will be in color then use contrasting colors. However, if any copies may get made in black and white use different black and white patterns. Check if the audience or client has brand colors that are important to utilize.
3. Use graphics for instructional purposes or reasons. Don't add clip-art just to add art that can distract the audience.
4. Don't crowd your reports with too much information. Make sure to use appropriate amounts of white space.
5. Make good use of the space available in the slide. Enlarge the graphs and have the text large enough that it is easy to read from across a room.
6. One graph per slide is ideal. Two graphs maximum per slide. This will make your data easier to understand. If you must have two visuals make sure the accompanying text is simple.
7. Use fonts appropriately. Serif fonts are typically easier to read. However, the audience or client may have a font preference. Reports should not have more than two fonts throughout the report. Experiment with type styles, sizes, and colors. Don't be afraid to bold, underline, or italicize text if you are trying to emphasize a point.
8. Keep titles short. About five to seven words will get your point across.

9. Avoid using busy slide backgrounds. Multiple colors or gradients can make text hard to read.
10. Avoid regurgitating a lot of data without interpretation of that data. The details are appropriate for the Appendix of the report, not for the presentation or main body of the report.
11. Make sure your presentation "stands on its own." Some people may forward the presentation to other interested parties, such as those employees who missed the initial presentation. Make sure they will be able to understand your work and recommendations by reading the slides.

Executive Summary

1. Background

Infinity Auto Insurance is looking to expand their customer base by starting with large metropolitan areas across the United States. Because the Latino/Hispanic population is the largest growing population in the United States, Infinity Auto Insurance wishes to better understand this market segment and expand its advertising campaigns to include this population. They have decided to concentrate their strategic efforts to large metropolitan areas in four different states: California, Arizona, Texas, and Florida. For this initial study they sampled 200 people in Houston, TX.

2. Methodology

Design	Specification
Methodology	Telephone survey
Population	Residents of the City of Houston, Texas
Sample size	200

3. Results

   (a) Over 85% of respondents own a vehicle.
   (b) Roughly 75% of respondents own auto insurance.
   (c) Most respondents who own auto insurance do not plan to switch insurance in the next 12 months.
   (d) About 50% of people have heard of Infinity.

Methodology

Sampling Plan and Survey Creation

Infinity Property and Casualty Corporation (Infinity Auto Insurance) headquartered in Birmingham, Alabama provides personal automobile insurance with a concentration on nonstandard auto insurance. Nonstandard insurance serves individuals unable to obtain coverage through standard insurance companies which can be due to a driving record with accidents and/or tickets, prior DUI, the driver's age, vehicle type, etc. Infinity Auto Insurance's products include personal automobile insurance for individuals, commercial vehicle insurance for businesses, and classic collector insurance for individuals with classic/antique automobiles. Infinity Auto Insurance distributes its products primarily through independent agencies and brokers. Infinity Auto Insurance, a top-performing Infinity brand, provides nonstandard car insurance through more than 12,500 independent agents. Infinity Auto Insurance utilizes Internet-based software applications to provide many of its agents with real-time underwriting, claims, and policy information. The Company is licensed to write insurance in all 50 states and the District of Columbia.

Infinity Auto Insurance is looking to expand their customer base by starting with large metropolitan areas across the United States. Because the Latino/Hispanic population is the largest growing population in the United States, Infinity Auto Insurance wishes to better understand this market segment and expand its advertising campaigns to include this population. They have decided to concentrate their strategic efforts to large metropolitan areas in four different states: California, Arizona, Texas, and Florida. For this initial study they sampled 200 people in Houston, TX.

Roughly 2.1 Million Hispanics live in Houston, Texas. Infinity Auto Insurance has asked you to help them create a market research survey for the Houston, Texas Area. The Infinity marketing group wants to better understand the Hispanic automobile insurance market in this large metropolitan area. Infinity Auto Insurance has decided that a telephone survey would reach their target population (less acculturated Hispanics) better than any other data collection method. To get a representative sample, Infinity Auto Insurance has decided that 200 respondents would be sufficient to draw some conclusions.

Infinity Auto Insurance wanted to better understand some of the following information: what percentage of Hispanics are insured, what companies are Hispanics insured with, how willing are Hispanics to buy or switch insurance companies, etc. Finally, Infinity Auto Insurance would like to know what attributes keep Hispanics buying auto insurance in the Houston area.

After speaking with Subject Matter Experts (SMEs) and doing research on the Houston area we created a survey based on Infinity Auto Insurance's needs.

Results

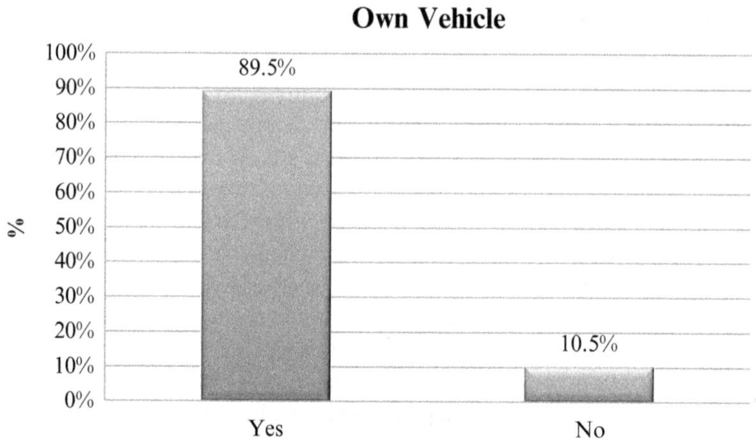

The above figure depicts the responses for those who own a vehicle. Of the sample of 200 respondents, 89.5% of people own a vehicle.

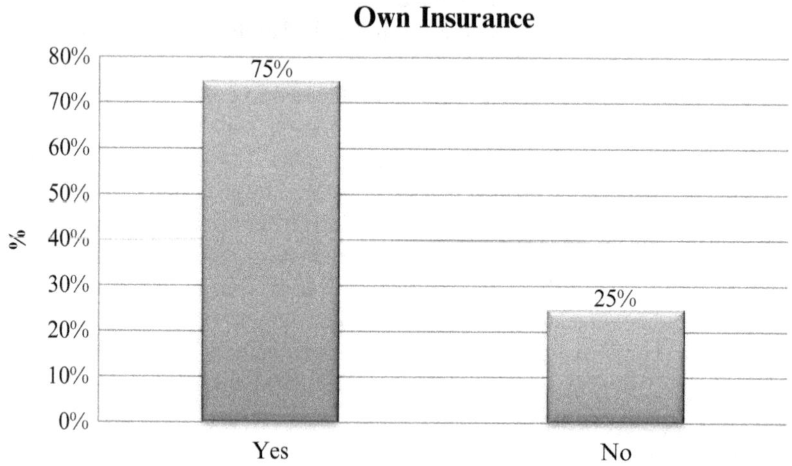

The above figure depicts the responses for those who own a vehicle. Of the sample of 200 respondents, 75% of people own auto insurance.

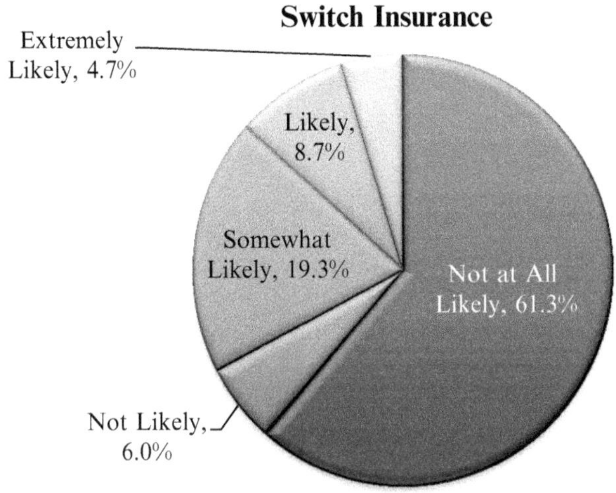

The figure above shows the responses to the question: "How likely or unlikely would you be to switch auto insurance companies in the next 12 months?" When insured respondents were asked whether or not they would switch auto insurance companies, 61.3% of respondents said that they were "Not at All Likely" to switch.

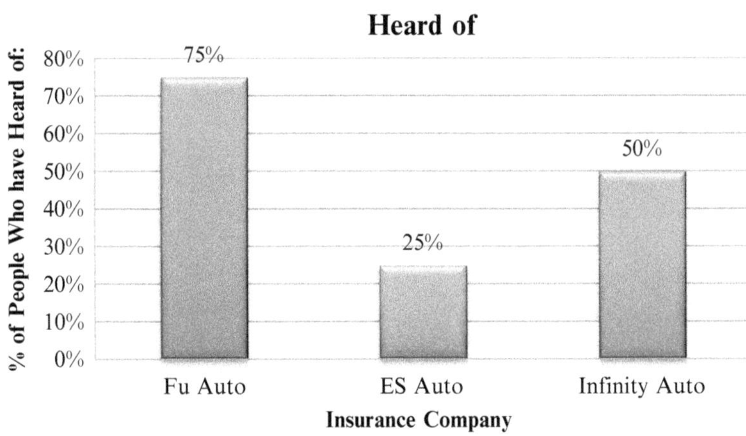

The figure above shows the responses to the question: "Have you heard of [Fu Auto, ES Auto, Infinity Auto] Insurance?" When asked if respondents have heard of Fu Auto Insurance, ES Auto Insurance and Infinity Auto, most respondents said they have heard of Fu Auto Insurance.

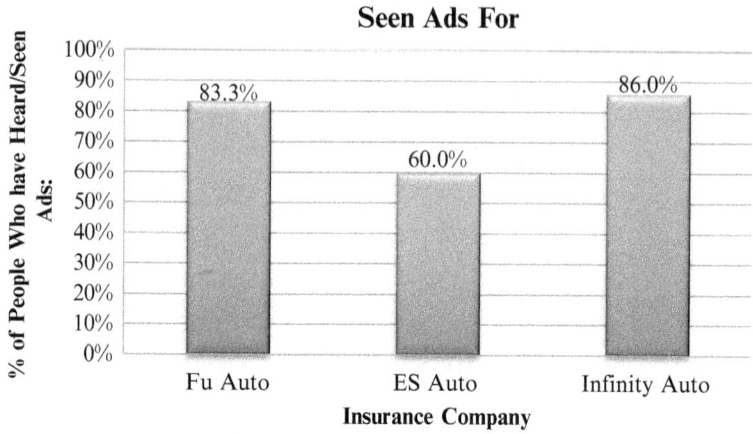

The figure above shows the responses to the question: "Have you seen ads for [Fu Auto, ES Auto, Infinity Auto] Insurance?" When asked if respondents have seen ads for Fu Auto Insurance, ES Auto Insurance, and Infinity Auto, most respondents said they have seen Infinity Auto Insurance advertisements.

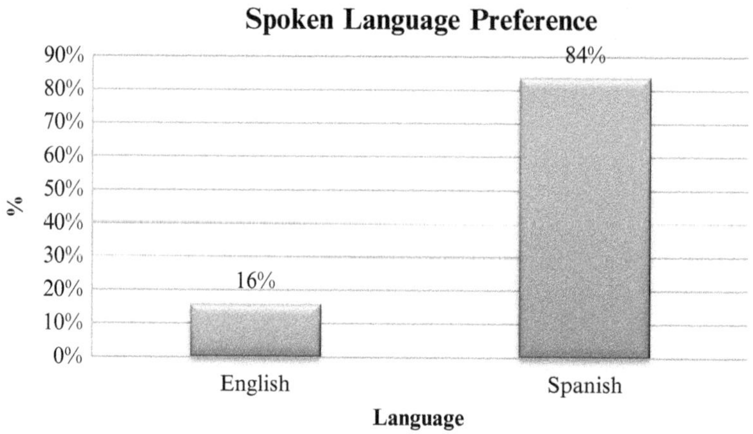

The figure above shows the responses to the question: "What language do you primarily speak in your home?" Most respondents (84%) stated that they spoke Spanish at home.

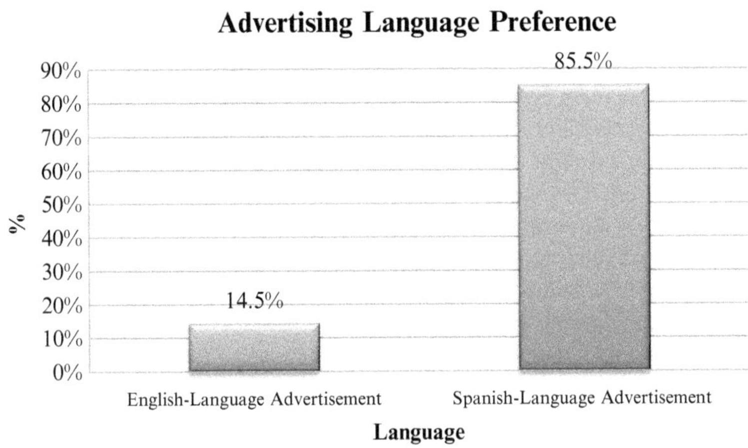

The figure above shows the responses to the question: "Do you prefer English-language advertising or Spanish-language advertising?" Most respondents (85.5%) stated that they prefer Spanish-language advertising.

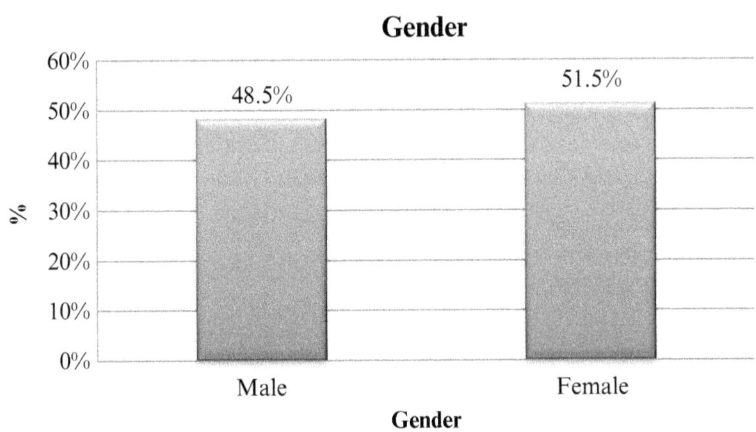

The figure above shows the responses to the question: "What gender are you?" 51.5% of respondents denoted that they are female.

Conclusions and Recommendations

- This is the first study done in the Houston, Texas area. We suggest annually conducting the same random telephone interviews to keep track of Infinity Auto Insurance's brand reputation in the Houston area. Each market area has some individual differences and should be monitored as significant advertising changes are made in the Houston market.
- The Houston market area is relatively brand loyal with only about 33% of respondents willing to switch auto insurance companies in the next 12 months. Infinity Marketing could coordinate a few insight sessions with Hispanic customers in order to reveal interesting auto insurance preferences. These insight sessions can help build on how to get Hispanic customers to switch, attract (advertise), and retain future Infinity Auto Insurance.

Final Thoughts and Activities

Practice Problems and Case Studies

1. Why is the sampling plan included in the survey report?
2. In which section of the report should you discuss the survey demographics?
3. The Bibliography lists the sources you consulted during your survey. Which of the following sources need not be included in the Bibliography?
    - Newspaper articles
    - Internet references
    - Personal interviews
    - Journal articles
    - Youtube clips
4. How would you address potential biases and limitations associated with survey data in your report?
5. Why is it important to include ethical wording in a report based on survey data?
6. Why is pilot testing an important tool in survey work?
7. How would you improve this executive summary? Remember the purpose of this section in the report is to provide a concise overview of a report's key findings, recommendations, and highlights to enable busy executives (and stakeholders) to grasp the main points (and make informed decisions) without reading the entire document.

    Executive Summary: The impact of COVID-19 on overhead costs in our IT business has been significant. The pandemic has caused a decrease in revenue and increased expenses, leading to a rise in overhead costs. This report analyzes the effects of COVID-19 on various overhead expenses and provides recommendations for cost reduction.

8. Review the attached report and identify areas of inaccuracy as well as areas for improvement.

++
++++++++++

Tiga Consulting
Lunch Room Report for EIGC
To: S.M. Riesling, General Manager
From: Paula Simms
Date: June 11
Subject: Lunch Room Vending Machine

1. At the monthly staff meeting on May 21, 20XX, you requested our company provide a report about staff satisfaction with the new high end vending machine. The aim of this report is to present this information with recommendations.
2. Since the move to the new office location in Brownsville employees have had difficulty in finding a nearby place to buy lunch. The new office is in a factory area and restaurants are limited. A new state of the art bistro vending machine was purchased. However, some employees expressed dissatisfaction with this vending machine.
3. All 122 employees were surveyed by an online questionnaire on May 21 and May 22.
4. In general, staff members are not satisfied with the prices and food choices in the vending machine.
5. Some were dissatisfied with the location of the vending machine, but most were dissatisfied with its reliability.
6. On analyzing the data two distinct groups of staff emerged. The first includes 68 staff that had usually eaten in the office when the office was in West Grove. The second group of 40 staff had usually eaten outside the facility in West Grove.
7. The overall response rate was 90% with 108 responses.
8. 25% of Group B staff said that the vending machine had adequate food choices. Less than a third of Group A staff were satisfied, giving a total for both groups of less than 30%.
9. About 65% (100%–35%) of Group B staff said that the prices were too high and were not satisfied with the prices. For example, one member of staff noted that a Caesar salad was $12, which feels like airport prices. However, slightly more than half of Group A staff (55%) indicated satisfaction with the current prices, giving a total satisfaction for both groups of just under 50%.
10. Both groups of staff thought that the location of the vending machine was acceptable. Only 29% of the staff wanted a better position for the vending machine.
11. 95% were satisfied with the reliability, meaning when they put their money in they got the correct product and if appropriate they received the correct change.
12. The findings show that staff members, especially staff in Group B, were not satisfied with prices and food selections. We should therefore consider buying a

larger vending machine or another separate vending machine so there are more selections at broader range of pricing.

13. The company should consider providing a microwave and a blending station in addition to the bistro vending machine. This would allow the staff to prepare their own lunches with food from home.
14. A frequency analysis of pricing within the current machine indicates that most (56%) of the items are priced above $10.00. (See the graph that follows).

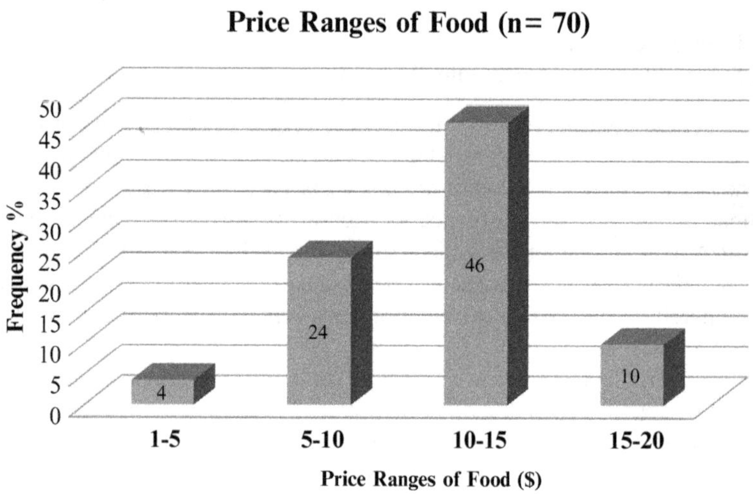

15. A second more detailed survey should be undertaken to better understand the acceptable price points and food preferences for this group of employees.

Discussion Boards

1. Reports need to avoid quantitative terms as most clients don't really understand statistics.
2. Sample size can make or break your final report.
3. Executive reports should not be more than two pages.

Group Activity

Find a report online for the financial investment industry, the wine industry, and the Environmental Protection Agency (EPA). What are some of the commonalities and differences in how these reports are organized? Consider the mission and motivations of these groups. Do these factors influence what is included in the reports and how it is presented? Give specific examples.

Parting Thought

Statistics means never having to say you're certain.

Problem Solutions

1. Why is the sampling plan included in the survey report?

   Answer: It explains the rationale for the sample size and it discusses the sample representativeness.

2. In which section of the report should you discuss the survey demographics?

   Answer: This information is included in the Results section of the report. Explaining the measurable characteristics of the sample allows the audience to formulate a more complete understanding and applicability of the results.

3. The Bibliography lists the sources you consulted during your survey. Which of the following sources need not be included in the Bibliography?

   - Newspaper articles
   - Internet references
   - Personal interviews
   - Journal articles
   - Youtube clips
   - Answer:
   - All of these references should be included in the Bibliography.

4. How would you address potential biases and limitations associated with survey data in your report?

   Answer:
   Biases/limitations can affect the generalizability and accuracy of survey data. There are several strategies you would employ to identify, address, and mitigate potential biases/limitations when interpreting and reporting survey data. These strategies may involve discussing the steps taken to ensure the representativeness of the sample (e.g., random sampling and stratification), acknowledging/accounting for nonresponse bias (or self-selection bias), and addressing any limitations such as a small sample size (or survey response limitations). Furthermore you could consider how you would critically evaluate the findings in light of potential biases/limitations and discuss any alternative explanations (or potential sources of error) to provide a comprehensive analysis in your report.

5. Why is it important to include ethical wording in a report based on survey data?

   Answer:
   Including ethical wording in a report based on survey data is essential for several reasons. Ethical considerations play a crucial role in maintaining the integrity of the research process and ensuring the well-being (and rights) of survey participants. Here are the key reasons why ethical wording should be included in a report:

   1. Participant Protection: Ethical wording demonstrates a commitment to protecting the participants' rights and ensuring their well-being throughout the survey process. It includes aspects such as informed consent, confidentiality, and voluntary participation. By explicitly stating these ethical considerations in the report readers can understand the steps taken to protect the participants and maintain their privacy/anonymity.
   2. Research Transparency: Ethical wording enhances research transparency by providing a clear understanding of how the survey was conducted, any potential risks (or benefits) associated with participation, and how participant confidentiality (and data privacy) were ensured. This transparency fosters trust and allows readers to evaluate the reliability/validity of the research findings.
   3. Compliance with Ethical Guidelines: Including ethical wording in the report demonstrates adherence to ethical guidelines set forth by professional bodies, institutions, or regulatory authorities. These guidelines may require researchers to obtain informed consent, handle participant data confidentially, and ensure the well-being of participants. Researchers demonstrate their commitment to ethical research practices by explicitly stating compliance with ethical guidelines.
   4. Ethical Accountability: Ethical wording in the report holds researchers accountable for their actions and provides a framework for ethical decision making. It showcases the researcher's commitment to conducting research with integrity and ensures that any potential ethical concerns were addressed appropriately throughout the study.
   5. Replicability and Reproducibility: Ethical wording provides important documentation for future researchers who may want to replicate or build upon the study. By including details about the ethical considerations in the report researchers facilitate the replication and reproducibility of the study as other researchers can understand the steps taken to ensure ethical research conduct.

In summary, including ethical wording in a report based on survey data is crucial to protect participant rights, maintain research transparency, adhere to ethical guidelines, foster accountability, and facilitate the replicability/reproducibility of the study. It demonstrates a commitment to ethical research practices and helps maintain the credibility/integrity of the research findings.

6. Why is pilot testing an important tool in survey work?

   Answer:
   Pilot testing is an important tool in survey work for several reasons. It involves conducting a trial run of the survey instrument on a small group of participants who are similar to the target population. Here are the key reasons why pilot testing is essential:

   - Identify and Correct Errors: Pilot testing allows researchers to identify and correct any errors, ambiguities, or problems in the survey instrument before it is administered to the full sample. It helps identify issues related to question clarity, response options, skip patterns, or formatting. By testing the survey on a smaller scale researchers can catch and rectify these errors ensuring the accuracy/reliability of the final survey instrument.
   - Enhance Questionnaire Validity: Pilot testing helps evaluate the validity of the survey questions. Researchers can assess whether the questions effectively capture the intended information and whether respondents can interpret them as intended. Through pilot testing researchers can make necessary revisions to the wording, phrasing, or order of questions to enhance their validity and ensure that they align with the research objectives.
   - Assess Survey Length and Respondent Burden: Pilot testing allows researchers to assess the length of the survey and the burden it may impose on participants. By measuring the time it takes participants to complete the survey during the pilot phase researchers can estimate the average completion time for the full sample. This information helps in managing respondent fatigue and optimizing the survey length to maintain participant engagement (and reduce the likelihood of incomplete responses or dropouts).
   - Test Survey Logistics: Pilot testing provides an opportunity to test the logistical aspects of the survey such as the mode of administration (such as online or paper-based), data collection procedures, and any technical requirements. This ensures that the survey runs smoothly and that data collection processes are efficient (and error-free) when administered to the larger sample.
   - Improve Participant Understanding and Feedback: Pilot testing allows researchers to gather feedback from participants regarding their experience with the survey. This feedback can reveal insights into participants' understanding of the questions, any confusion (or ambiguity) encountered, and suggestions for improvement. Researchers can use this feedback to refine the survey instrument, ultimately making it more user-friendly and enhancing participant comprehension.

In summary, pilot testing is a critical step in survey work as it helps identify/rectify errors, enhance questionnaire validity, assess survey length (and respondent burden), test survey logistics, and gather valuable participant feedback. By conducting a trial run on a smaller scale researchers can improve the quality/effectiveness of the survey instrument, ultimately leading to a more reliable/valid data set from the larger sample.

7. How would you improve this executive summary? Remember the purpose of this section in the report is to provide a concise overview of a report's key findings, recommendations, and highlights to enable busy executives and stakeholders to grasp the main points and make informed decisions without reading the entire document.

   Executive Summary: The impact of COVID-19 on overhead costs in our IT business has been significant. The pandemic has caused a decrease in revenue and increased expenses, leading to a rise in overhead costs. This report analyzes the effects of COVID-19 on various overhead expenses and provides recommendations for cost reduction.
   Answer:
   This executive summary provides an overview of the impact of COVID-19 on overhead costs in our IT business. The COVID-19 pandemic has introduced significant challenges/disruptions to our operations thus leading to both revenue reduction and increased expenses. This report aims to analyze the specific effects of the pandemic on various overhead expenses and presents actionable recommendations for cost reduction strategies.
   The findings of this report indicate that COVID-19 has had a notable impact on our IT business's overhead costs. The decrease in revenue resulted from reduced client demand and project cancellations. The increased expenses were primarily driven by additional health policies, safety measures, remote work infrastructure, and cybersecurity requirements.
   To mitigate the impact of COVID-19 on overhead costs several key recommendations have been identified. First, implementation of stricter cost control measures by scrutinizing nonessential expenses and optimizing procurement processes could lead to immediate cost savings. Additionally, exploring opportunities for remote work optimization and evaluating the necessity of physical office space can significantly reduce facility-related costs. By implementing the recommended cost reduction strategies outlined in this report we can effectively mitigate the impact of COVID-19 on our overhead costs, improve our financial resilience, and position ourselves for future growth.

8. Review the report and indicate errors as well as areas for improvement include:

   Answer:

   - Missing Title Page
   - Missing Table of Contents, but since the report is so short one is not really necessary
   - Complete date missing in header
   - The report needs separate headings for each section
   - Executive Summary missing

- Graph not numbered
- Frequencies in Graph 1 don't add up to 100%
- No sample size provided in Table 1
- Categories overlap in Graph 1
- Page numbers missing
- No copy of the actual survey attached. What are the questions? Did all the questions get answered or were some confusing?
- Was this anonymous/confidential? If so, make sure you mention this
- No discussion of the average data in Table 1
- Survey source and development details missing
- Statistical significance discussion missing
- Missing section on ethical guidelines

The Corrected Report Follows

Tiga Consulting
Lunch Room Report for EIGC
June 11 20–

Consultant: Dr. Paula Simms
888-745-4444
psimms@gazelle.nz

To: S.M. Riesling, General Manager
From: Paula Simms
Date: June 11/20–
Subject: Lunch Room Vending Machine

Executive Summary

Employees are not satisfied with the current lunchroom situation. This survey was conducted to explore the details of the reasons for this dissatisfaction. The lunchroom has been outfitted with a new bistro style vending machine as this is the only place to buy food in the area. Outside the actual office is an industrial area with no available restaurants. Staff are not as dissatisfied with the location of the machine and its reliability. The issues primarily focus on the price of food and the limited food choices.

The recommendations we suggest based on this exploratory survey include buying an additional vending machine with lower priced healthy foods. A follow-up survey is recommended that specifically addresses price points and food choices to better understand the needs of employees.

Background and Objectives

At the monthly staff meeting on May 21, 20–, you requested our company provide a report about staff satisfaction with the new high end vending machine. The aim of this report is to present this information with recommendations.

Since the move to the new office location in Brownsville employees have had difficulty in finding a nearby place to buy lunch. The new office is in a factory area and restaurants are limited. A new state of the art bistro vending machine was purchased. However, some employees expressed dissatisfaction with this vending machine.

Research Methodology

All 122 employees were surveyed by an online questionnaire on May 21, 22, 20—. The resulting response rate was 89%, representing 110 completed surveys. The survey included 28 questions with a Likert scale that went from 1 to 5, with 5 being the most satisfied and 1 being the most dissatisfied. The survey was developed by Tiga Consulting and was based on initial interviews with EIGC employees. The survey is included in Appendix 1.

The data was analyzed in Excel using the frequency analysis tool to show the results graphically. There were no outliers in the data that required deletion.

Overview of Results

In general staff are not satisfied with the prices and food choices in the vending machine. A minority was dissatisfied with the location of the vending machine, but most were satisfied with its reliability.

On analyzing the data two distinct groups of staff emerged. The first includes 68 staff who had usually lunched in the office when the office was in West Grove (Group A). The second group of 40 staff had usually lunched outside in West Grove (Group B). The overall response rate was 90% with 108 responses (Table 15.1).

Twenty-five percent of Group B staff said that the vending machine had adequate food choices. Less than a third of Group A staff were satisfied with the food choices, giving a total for both groups of less than 30%.

Table 15.1 Staff satisfaction with the vending machine

Feature	Staff satisfaction (n = 108) Group A Usually lunched in the office (n = 68)	Group B Usually lunched outside the office (n = 40)	Both groups
Price	55%	35%	48%
Food choices	30%	25%	28%
Location	70%	75%	71%
Reliability	95%	95%	95%
Overall avg	62.5%	57.5%	60.6%

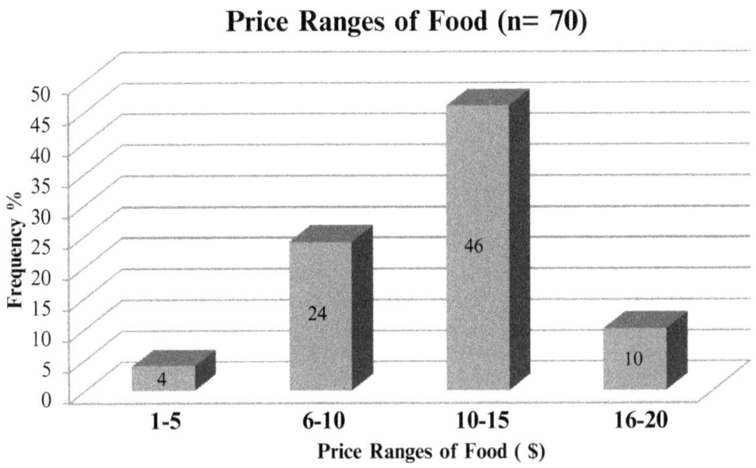

Graph 15.1 Price ranges of food

About 65% of Group B staff said that the prices were too high. For example, one member of staff noted that a Caesar salad was $18, which feels like airport prices. However, slightly more than half of Group A staff indicated satisfaction with the current prices, giving a total satisfaction for both groups of just under 50%.

Both groups of staff thought that the location of the vending machine was acceptable. Only 29% of the staff wanted a better position for the vending machine.

Ninety-five percent were satisfied with the reliability, meaning when they put their money in they got the correct product and if appropriate they received the correct change.

The findings show that staff members, especially staff in Group B, were not satisfied with prices and food selections.

A frequency analysis of pricing within the current machine indicates that most (56%) of the items are priced above $10.00 (Graph 15.1).

Conclusions and Recommendations

We should therefore consider buying a larger vending machine or another separate vending machine so there are more selections with a broader range of pricing.

A second more detailed survey should be undertaken to better understand the acceptable price points and food preferences for this group of employees.

The company should consider providing a microwave and a blending station in addition to the bistro vending machine. This would allow the staff to prepare their own lunches with food from home. Perhaps a microwave and blending machine may be more appealing than the vending machine. Future surveys should address how demand for vending machines may decline if other equipment (such as a microwave and blending machine) was installed. These results may impact the decision to purchase a second vending machine or to even retain the existing vending machine.

Bibliography

Freedman, D., Pisani, R., Purves, R., 1998. Statistics (third edition). Norton and Company, NY, USA.

Nelson, D., Quick, J., 2006. Organizational Behavior (fifth edition). Thomson/Southwestern, Ohio, USA

Appendix 1

Include the actual survey questions here

Chapter 16
Data Visualization

Linda Herkenhoff and John Fogli

Key Concepts

Data visualization, Conditional formatting, Heat maps, Pareto chart, People graph, Power BI Desktop sparklines, and Standard charts.

Discussion

This chapter will discuss and define the importance of data visualization. In Chap. 2, we discuss the most popular charts inserted in Excel. Data visualization is a method used to help explain statistics in a graphical way. Data visualization can often be interchangeably referred as information graphics, information visualization, and statistical graphics.[1] With the advent of online real-time reporting on smartphones and computers, data visualization has grown in popularity among many industries. The goal of this chapter will be to explore Excel's data visualiza-

[1] https://www.techtarget.com/searchbusinessanalytics/definition/data-visualization#:~:text=The%20term%20is%20often%20used,information%20visualization%20and%20statistical%20graphics

Supplementary Information The online version contains supplementary material available at https://doi.org/10.1007/978-3-031-46371-6_16.

L. Herkenhoff (✉)
Saint Mary's College, Moraga, CA, USA

J. Fogli
Diablo Valley College, Pleasant Hill, CA, USA

L. Herkenhoff, J. Fogli (eds.), *Applied Statistics for Business and Management using Microsoft Excel*, https://doi.org/10.1007/978-3-031-46371-6_16

tion capabilities and provide recommendations on how to provide the best way to graphically present your statistical data to your audience.

What Is Data Visualization?

Data visualization is a process where a summary of data is used to make a graphical representation. Indeed, data visualization has evolved from the first statistical graph in 1644 from the Dutch astronomer, Michael Florent van Langren.[2] He was credited for the first representation of statistical data, showing the wide range of estimates of the distance in longitude between Toledo and Rome.

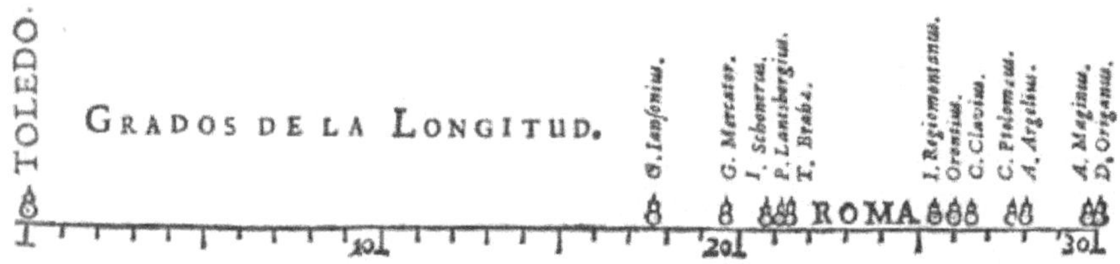

In the late 1990s, the emergence of the personal computer and increased computer processing allow statisticians to analyze large volumes of data using data visualization techniques to present statistical findings. Today, data visualization is becoming extremely popular in many careers of study. In business management, data is constantly being collected, stored, and analyzed for strategic purposes. Businesses are using small and large amounts of data to quickly visualize the business enterprise. Data visualization becomes a user-friendly way to help management quickly display massive amounts of numerical information.

Common Data Visualization (Standard) Charts in Excel

The following are the most common data visualization techniques used in Excel in alphabetical order: (These were discussed in Chaps. 2 and 3.)

1. Area charts
2. Bar charts
3. Column charts
4. Bubble charts
5. Histograms
6. Line charts

[2] https://en.wikipedia.org/wiki/Michael_van_Langren

7. Pie charts
8. Pivot tables
9. Other charts include
   - Stock charts
   - Surface charts
   - Bubble charts
   - Radar charts

Other Data Visualization Used in Excel

The following data visualization techniques require a little more sophistication to be created and can be considered as more advanced techniques to be used with more complex illustrations of datasets:

1. Bullet charts
2. Gantt charts
3. Heat maps
4. Pareto charts
5. Pictographs
6. People graphs
7. Sparklines
8. Power BI desktop

Bullet Charts

The bullet chart is useful to make comparisons of performance metrics. Stephen Few, a data visualization expert, consultant, and author invented this chart.[3] Performance metrics in a variety of industries can use bullet charts to compare sales data to sales targets with a region or a product. The following table can be used to create a bullet chart:

	Region A	Region B	Region C	Region D
Poor	10	15	5	30
Good	20	20	15	15
Excellent	30	35	35	15
Target Revenue	45	75	70	80
2023 Revenue	60	70	73	60

[3] https://www.perceptualedge.com/about.php

- **Highlight** the previous dataset above or create your own dataset
- Select the **Insert** tab

- Select the **stacked column** chart

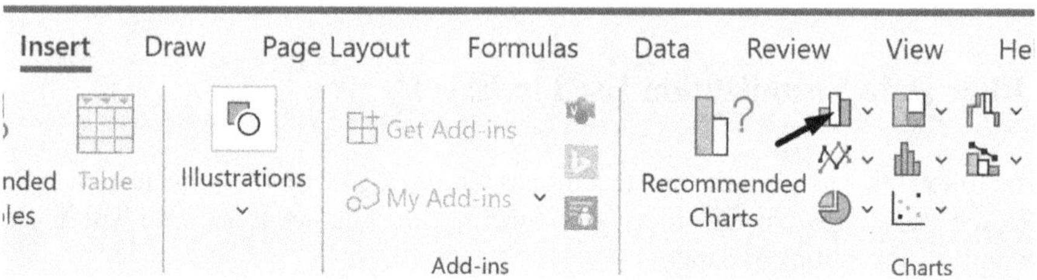

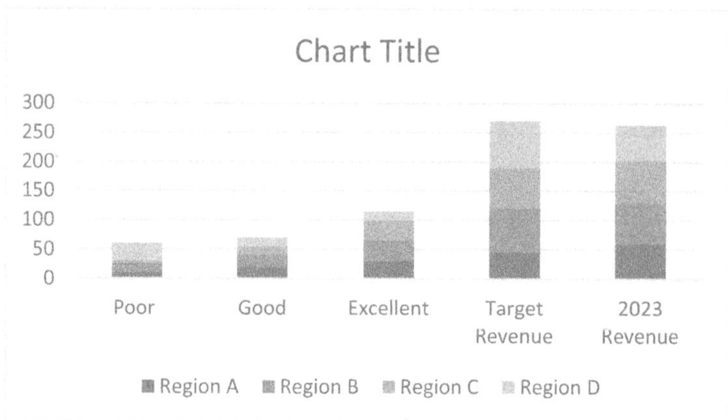

- Click anywhere on the chart
- Click the **Chart Design** tab
- Select the **Switch Row/Column** function

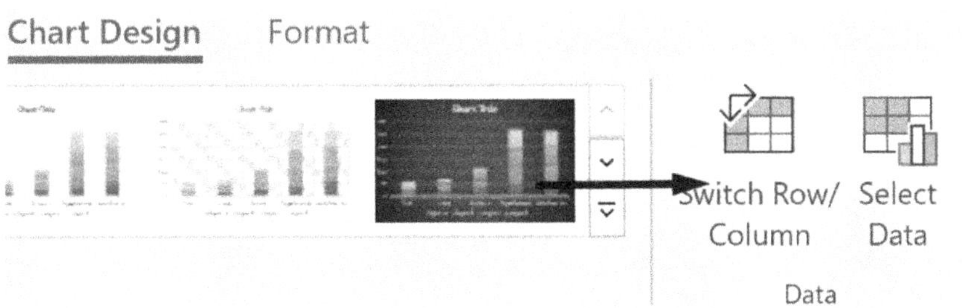

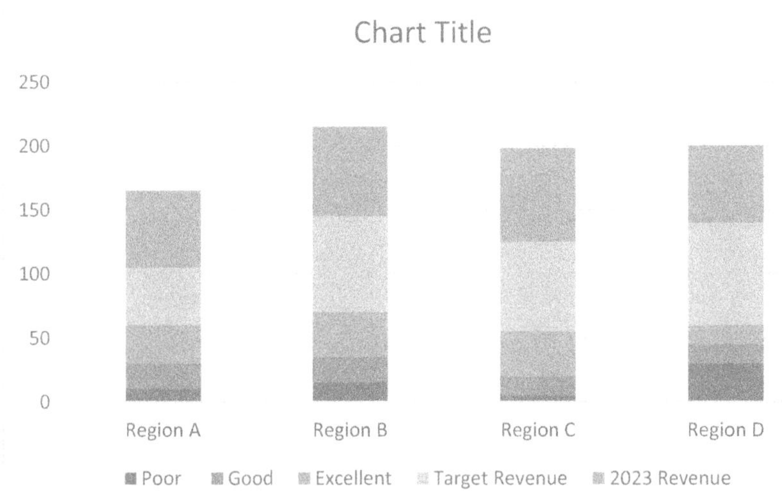

- Click anywhere on the **2023 Revenue** chart area
- Right click to get to the **Format Data Series**

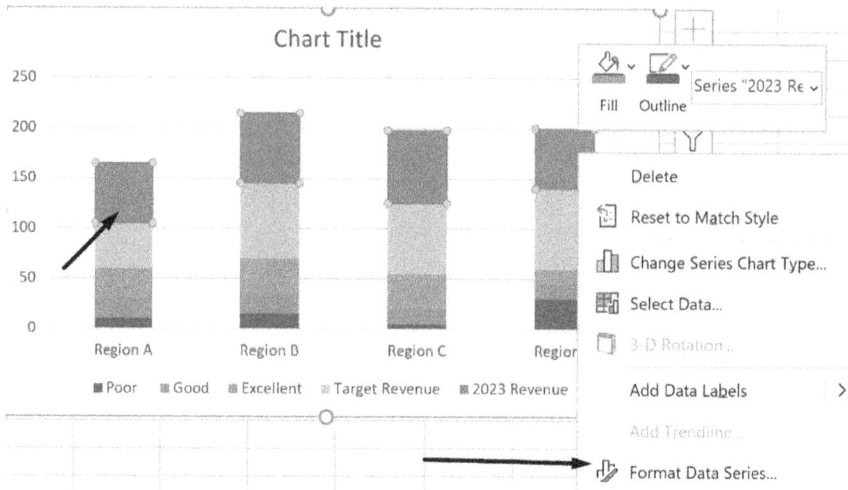

- Click on **Secondary Axis**

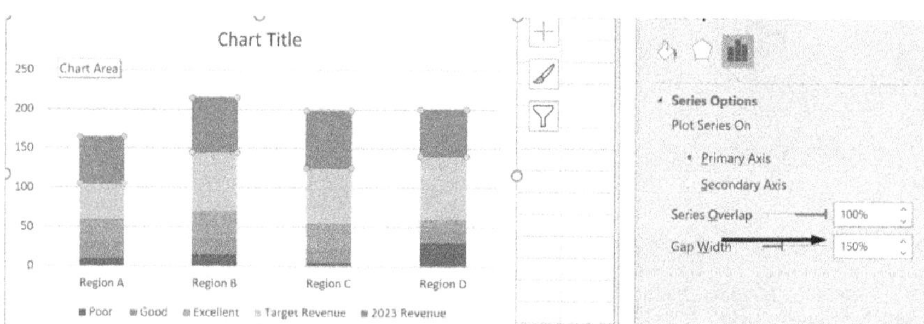

- Click on the **Gap Width** and set it to 500%

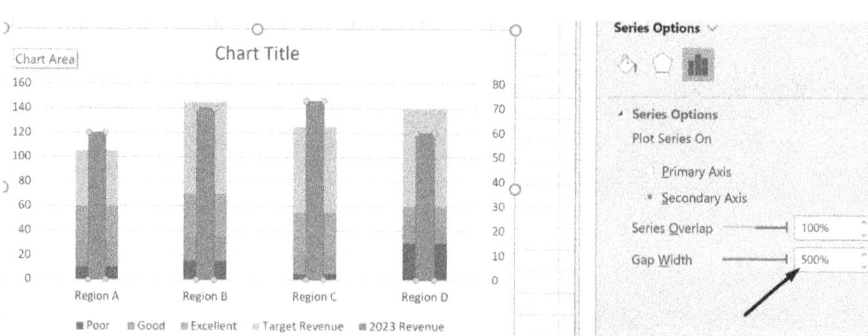

- Click anywhere on the **Target Revenue** area chart

- Right click to get to the **Format Data Series**
- Select **Change Series Chart Type**

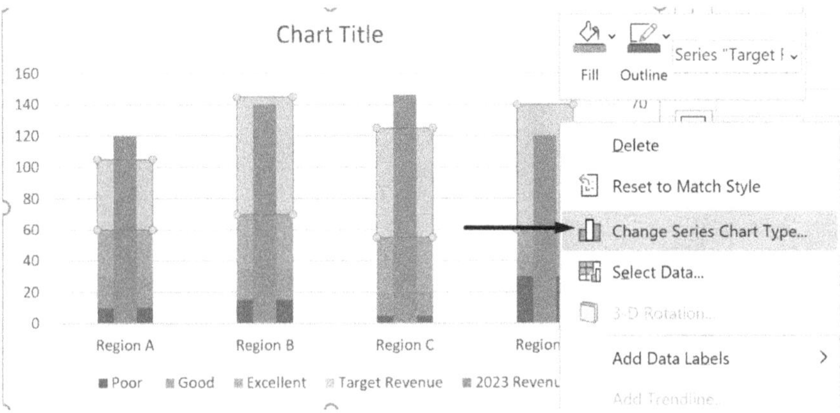

- Select the **Combo** with the chart type as **Stacked Line with Markers**
- Make sure the **Secondary Axis** is checked

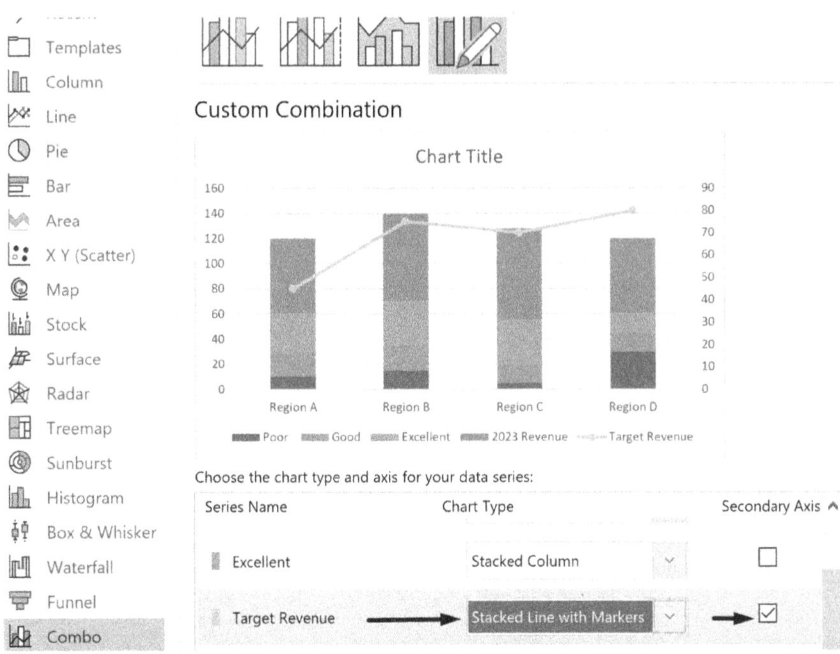

You have completed your bullet chart. However, you should a chart title or even format your data series to fill the color charts areas using red for "Poor" performance, yellow for "Good" performance, and green for "Excellent" performance.

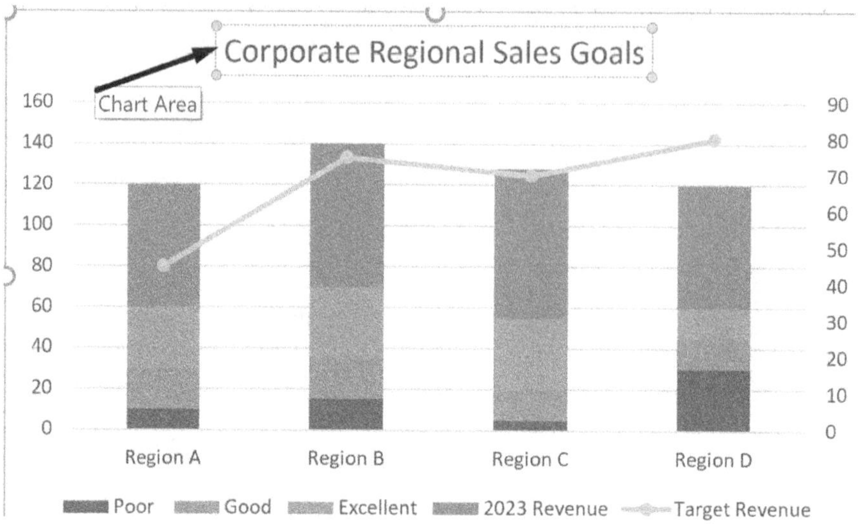

Gantt Charts

The Gantt chart is used as a project management visualization chart that tracks a project with a schedule tracking tasks. The chart allows managers to project plan start and end dates of tasks, milestones, dependencies, assignees, and other management resources needed. Henry Laurence Gantt, a mechanical engineer, invented the popular chart that was widely used in the 1920s.

The following table can be used to create a Gantt chart:

Tasks	Task duration	Start date	End date
1. Initial meeting with client	2	1-May	3-May
2. Focus groups	5	6-May	11-May
3. Draft of customer satisfaction survey	10	16-May	26-May
4. Edits of customer satisfaction survey	2	18-May	20-May
5. Program of online survey	4	22-May	26-May
6. Final survey to client	2	24-May	26-May
7. Send out survey to customers	1	25-May	26-May
8. Days survey open to complete	15	9-June	24-June
9. Final survey results report	8	17-June	25-June

- Select the **stacked 2-D Bar** chart

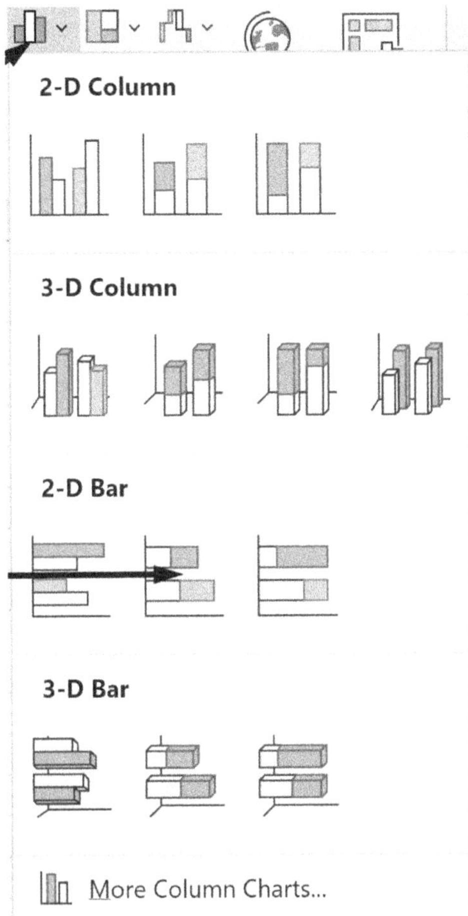

You will see a blank 2-D bar chart inserted

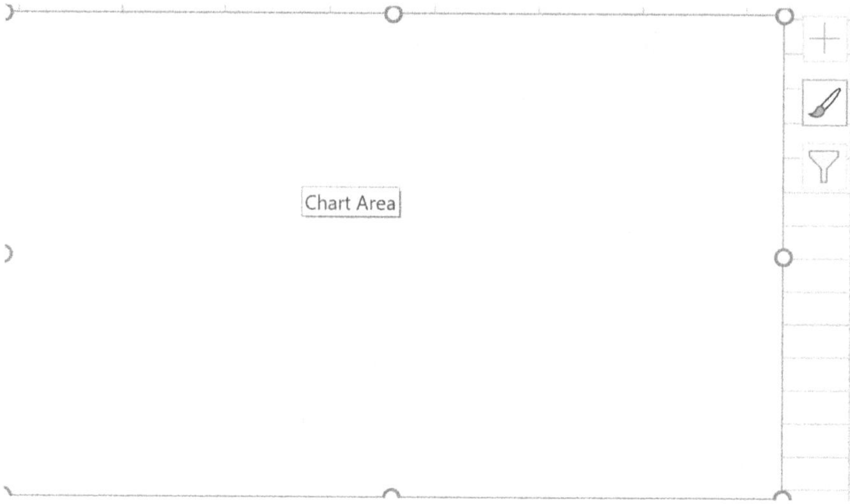

- Select **Chart Design** and click on **Select Data**

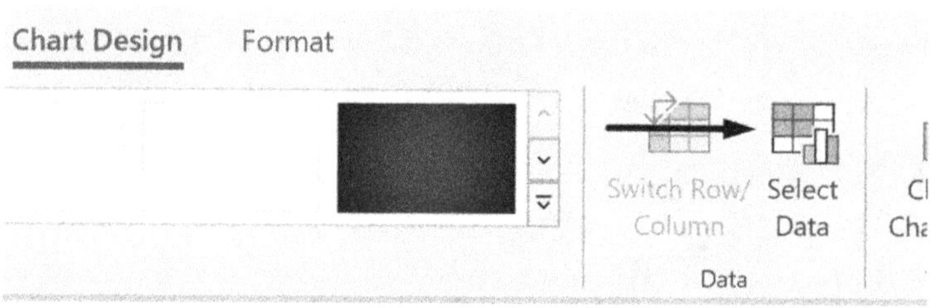

- Click on **Add** data source below Legend Entries (Series)

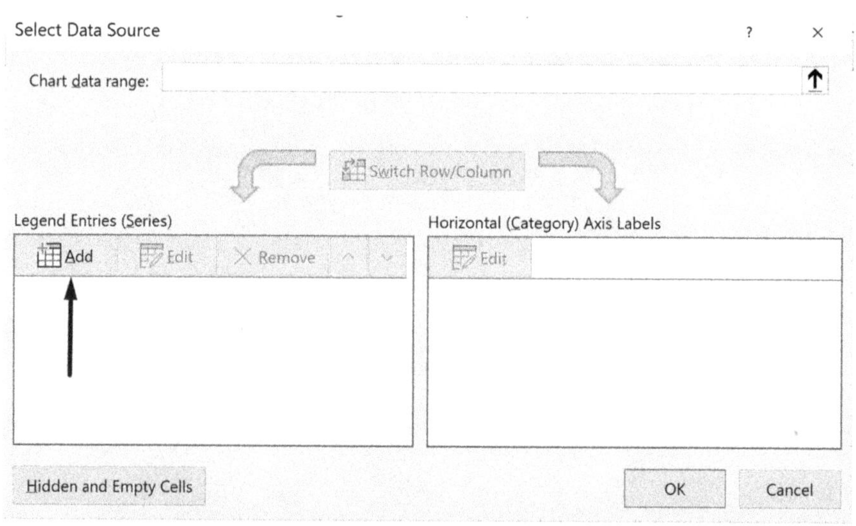

- Add your start date in the **Series name**
- Add the column data for all your start dates in the **Series values**

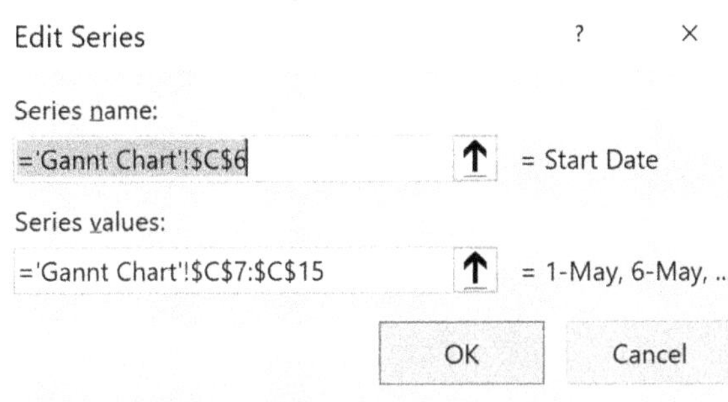

- Click on **Add** data source
- Add your task duration in the **Series name**
- Add the column data for all your task duration days in the **Series values**

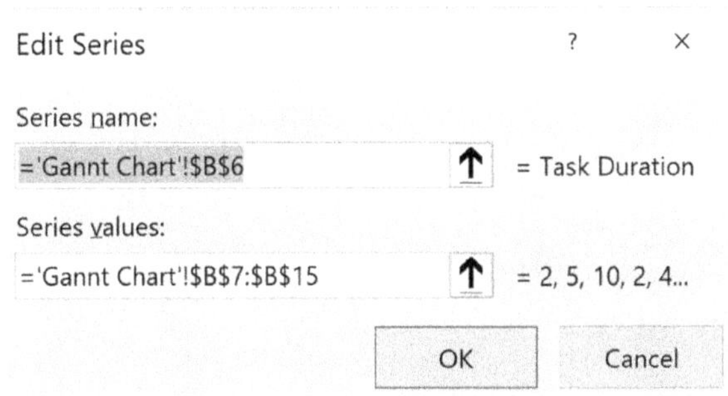

- Select **Chart Design** and click on **Select Data**
- Click on **Edit** data source below Horizontal (Category) Axis Label

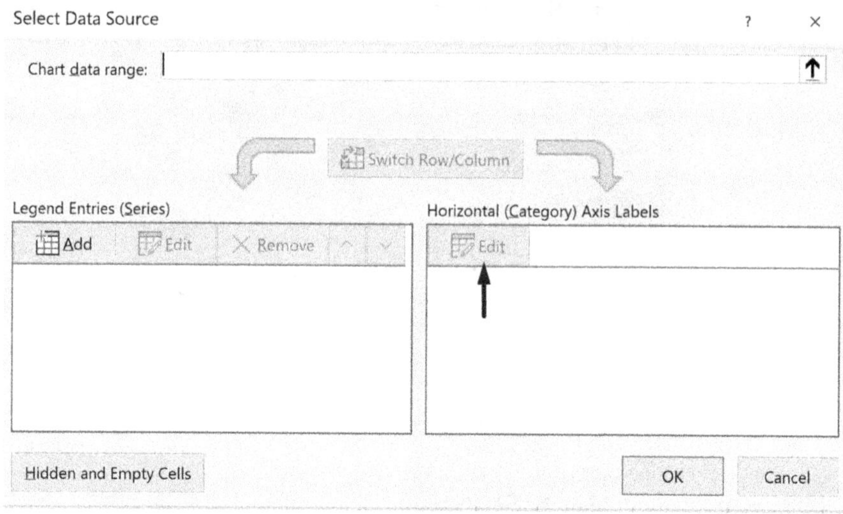

- Click the Axis label range and include all your tasks from 1. Initial Meeting to 9. Final survey results report

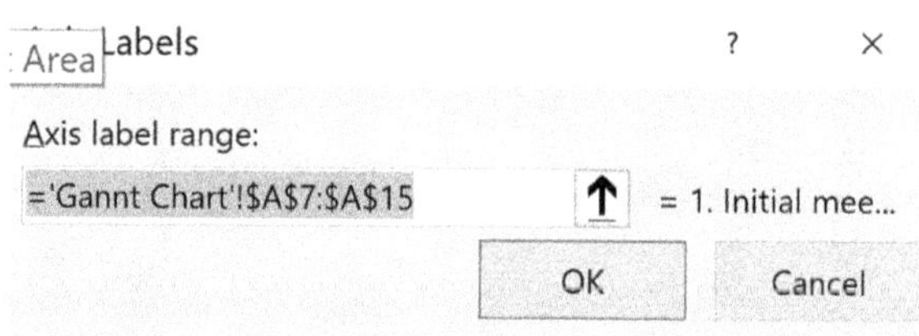

Your data source should be completed to look like the following data source:

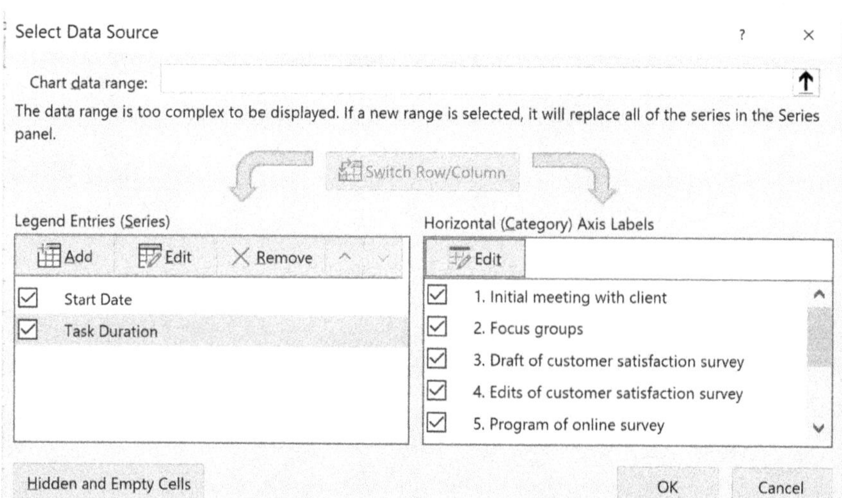

- Click **OK**

You should have the following initial (not final) Gantt chart displayed:

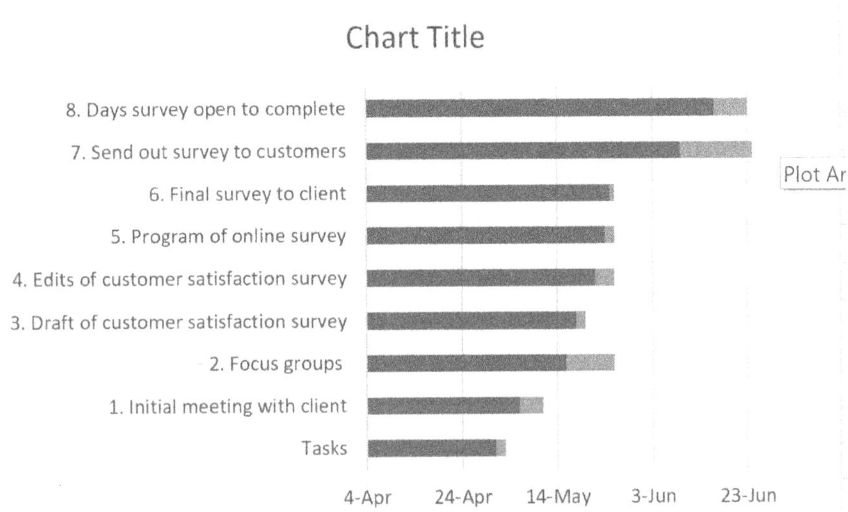

You will want to remove the blue bar chart because the start date is irrelevant.

- Click and select the blue bar chart area
- Right click and scroll down to **Format Data Series**

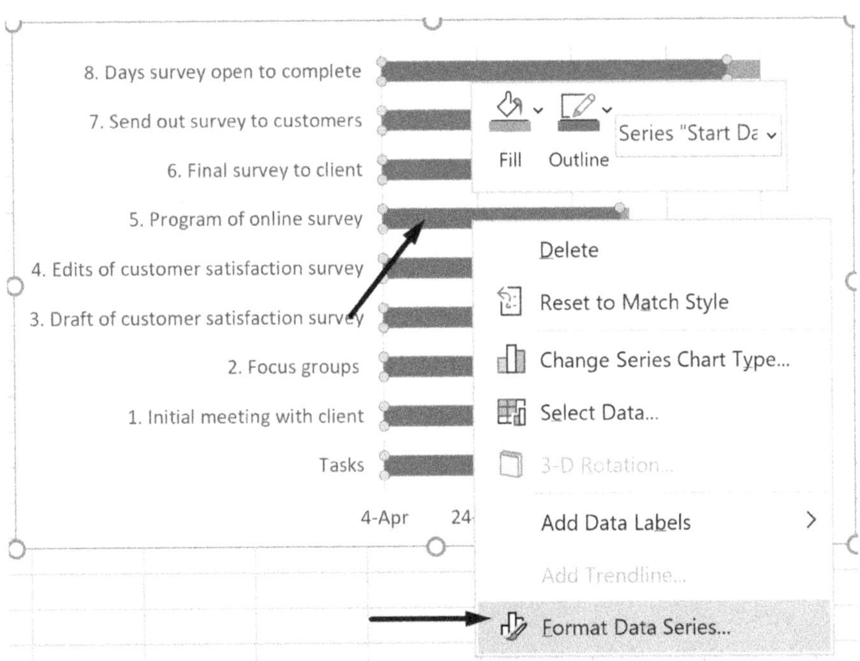

- Select the Series Options that has the paint bucket
- Click on the **No fill** option

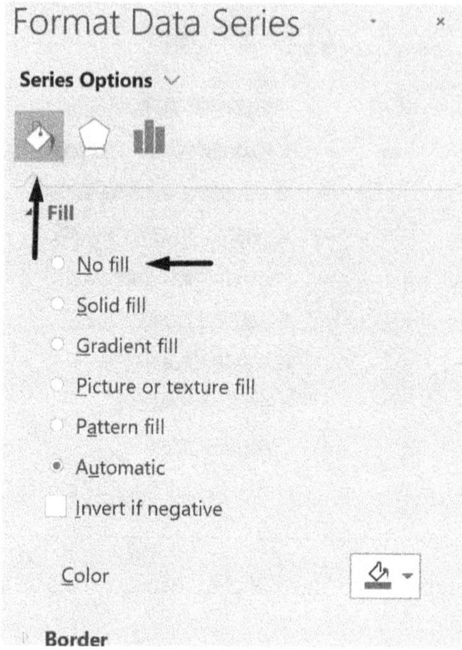

The blue bar chart will have disappeared.

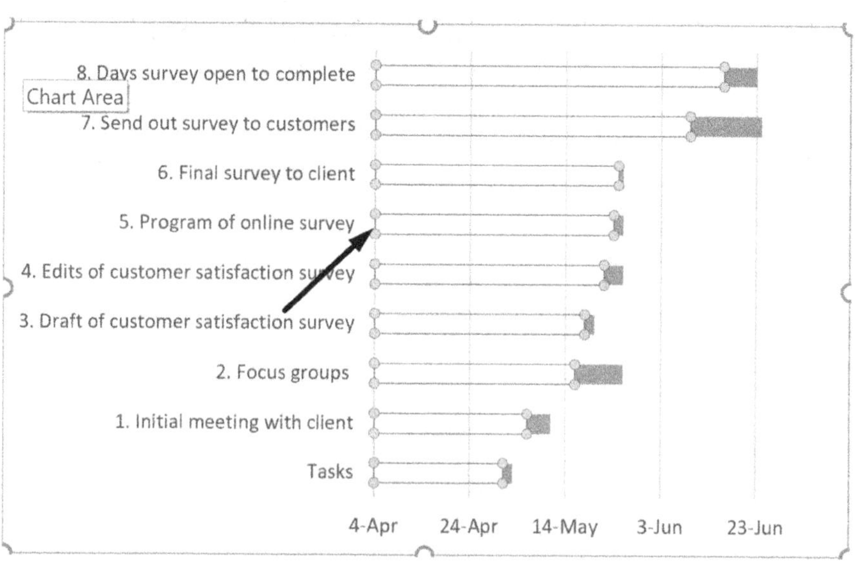

The Gantt chart no longer shows the start date but rather the number of days to complete the tasks.

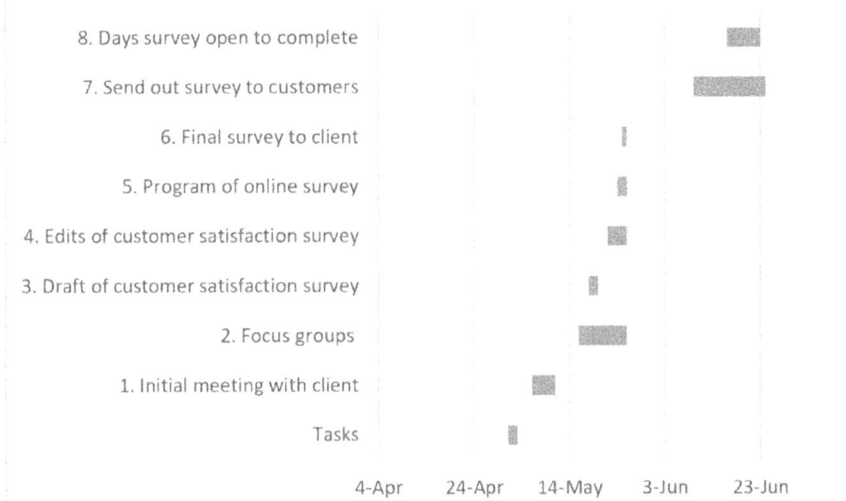

- Click on the list of task items on the chart
- Right click and scroll down to **Format axis**
- Click on the check box **Categories in reverse order**

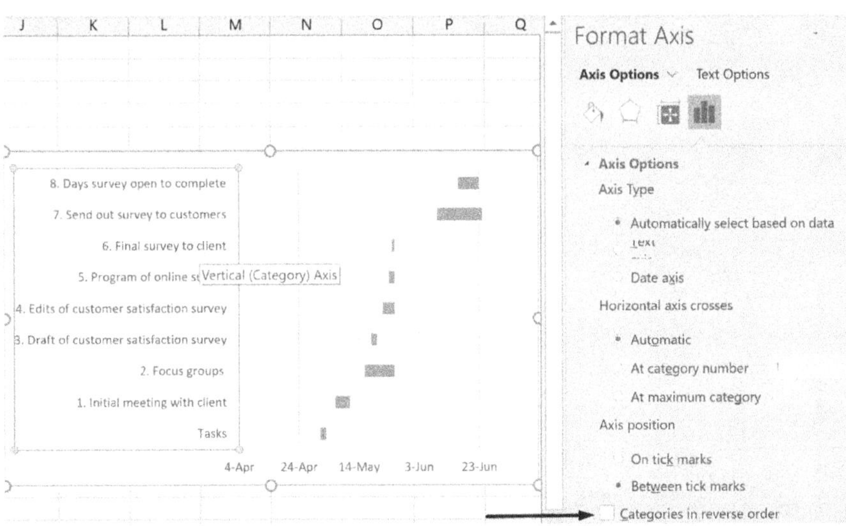

The tasks should now be listed from top to bottom in ascending order.

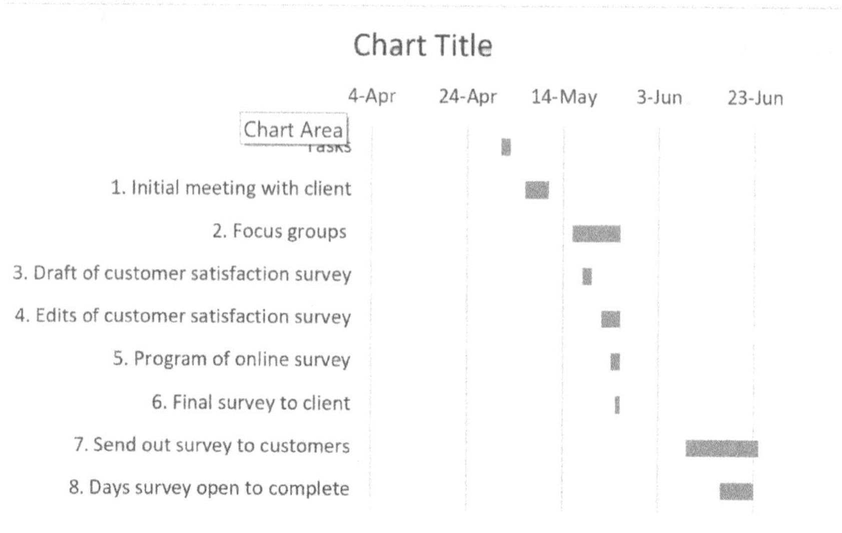

The start dates on the horizontal top axis needs to be formatted.

- Click on the initial start date of 1-May
- Rather than **Custom** format, scroll down to the **Number** format

Microsoft Excel will change your date of 1-May start date into a number that is **45047.00**

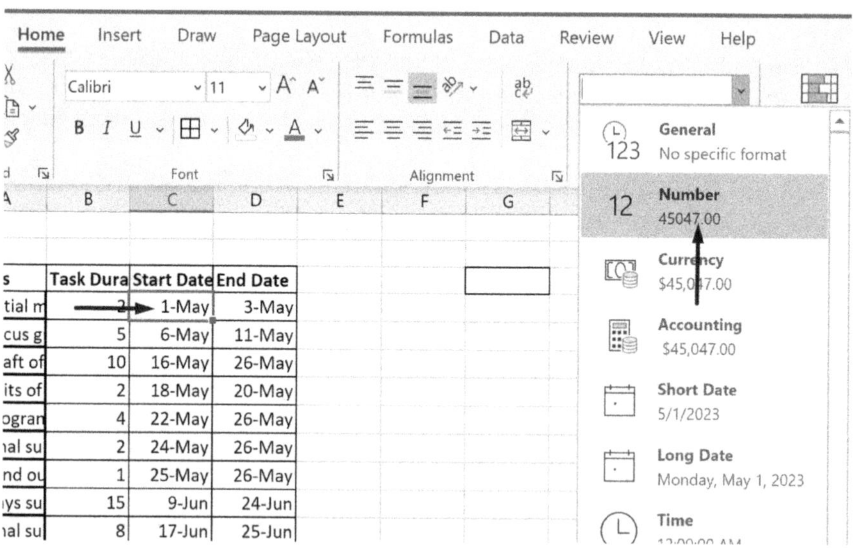

- Click on the dates listed on your chart
- Right click on the Format axis and make sure the column chart icon is selected
- Enter the **Minimum Axis Option** to be 45047

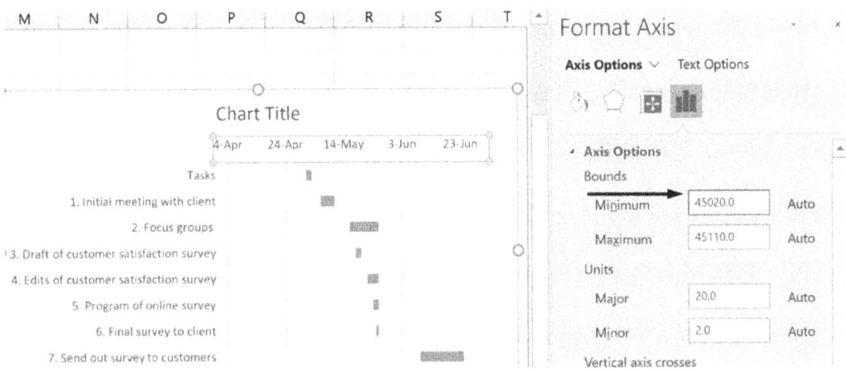

The orange bar charts will now have the correct start dates on the top horizontal axis.

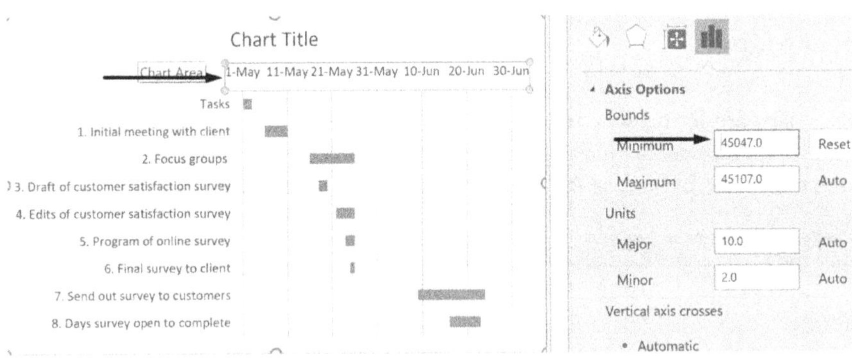

You have completed your Gantt chart. However, you should include a chart title. If I were the project manager, I would be concerned about the number of days after the survey closes to when the final survey results report is due.

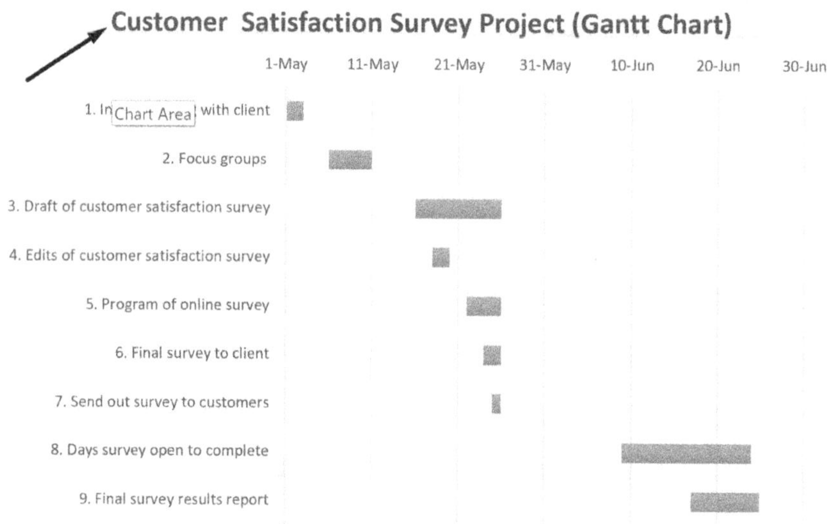

Heat Maps

Heat maps are useful way to look at comparisons of data inside of a spreadsheet. Excel's conditional formatting allows us to quickly shade values based upon numeric values. Let us use the following sales data from a retail store.

Time of day	Day (December) - Customer Sales in Millions ($)																														
	1	2	3	4	5	6	7	8	9	10	11	12	13	14	15	16	17	18	19	20	21	22	23	24	25	26	27	28	29	30	31
8an	296	380	600	172	540	134	548	285	290	460	126	393	552	226	573	123	203	240	194	456	407	572	263	582	129	408	389	543	501	243	444
9an	489	257	250	558	409	284	565	493	285	381	261	307	182	454	375	561	247	510	362	415	420	116	202	253	439	589	345	132	132	485	273
10a	215	282	303	194	235	532	410	175	247	199	315	349	388	368	415	172	377	417	453	259	204	186	122	372	202	307	452	508	460	545	170
11a	406	305	475	256	541	469	529	121	529	528	574	179	384	276	598	205	521	408	204	533	194	144	386	444	481	119	286	195	263	267	297
12p	432	584	105	547	108	328	439	403	233	587	236	571	495	240	396	581	239	429	524	259	587	347	422	327	597	312	556	154	566	376	572
1pn	123	263	249	281	439	344	594	204	319	515	164	197	600	291	473	157	438	548	201	299	177	158	299	177	579	567	483	492	356	412	228
2pn	383	501	116	442	341	310	459	144	332	190	315	138	269	126	152	130	430	599	374	506	154	219	133	194	511	458	522	282	161	262	319
3pn	268	490	307	246	370	214	446	426	218	159	188	417	546	441	270	160	573	200	559	433	224	517	421	188	565	446	165	422	198	525	521
4pn	504	492	373	403	531	459	438	536	566	443	324	399	114	111	397	422	360	277	357	597	105	276	538	350	142	325	103	489	411	510	169
5pn	483	198	596	428	460	479	111	441	488	465	383	240	402	136	316	271	527	228	499	462	562	329	158	194	233	407	554	313	245	296	540
6pn	700	645	850	700	655	800	855	755	500	700	688	565	900	734	680	495	500	555	655	880	775	655	490	478	424	480	780	795	999	455	375
7pn	224	364	168	137	367	235	182	223	235	489	447	139	278	350	534	148	272	587	522	197	334	429	535	463	108	156	410	415	373	166	432
8pn	268	508	536	209	335	170	470	538	470	268	111	303	263	289	310	359	425	193	217	141	100	318	483	538	334	555	594	411	509	592	298
9pn	139	204	310	138	346	328	575	117	521	416	381	468	160	328	164	300	168	137	179	524	268	480	351	562	561	140	533	110	574	381	139

- Highlight the data cells that you wish to create a heat map on all sales data
- Click on **Conditional Formatting** and select the **Color Scales** with Green, Yellow, and Red schemes

You will notice that Excel's has used the green, yellow, and red color schemes to highlight sales. The darker the green the more sales and the darker the red the less sales generated on a given day in December.

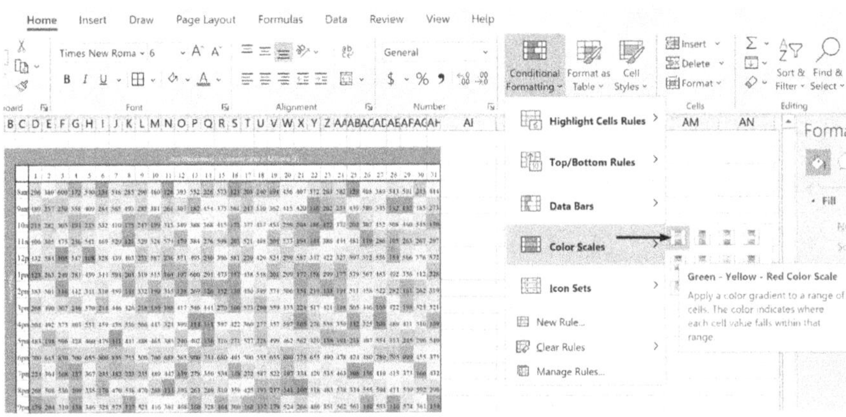

- Highlight the sales data cells and right click on **Format cells**

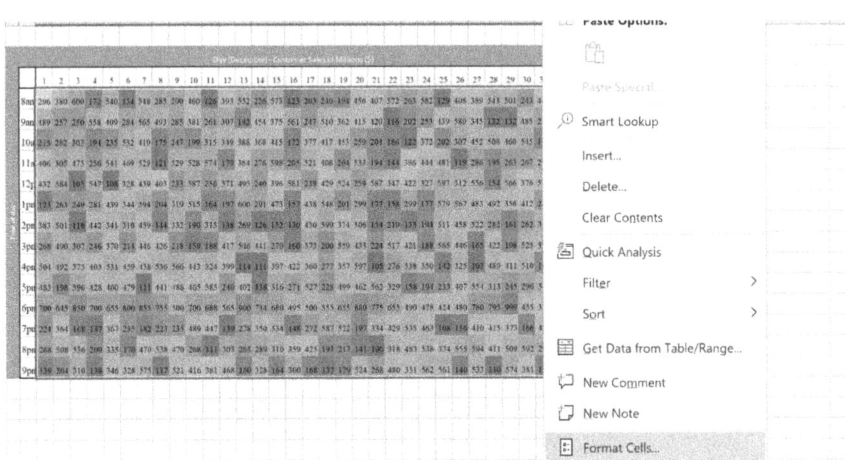

- Click on **Number Category** and select **Custom**
- To remove the numbers from the table, simply add three ":::" semi-colons the **Type** dialog box

- Click on **OK**

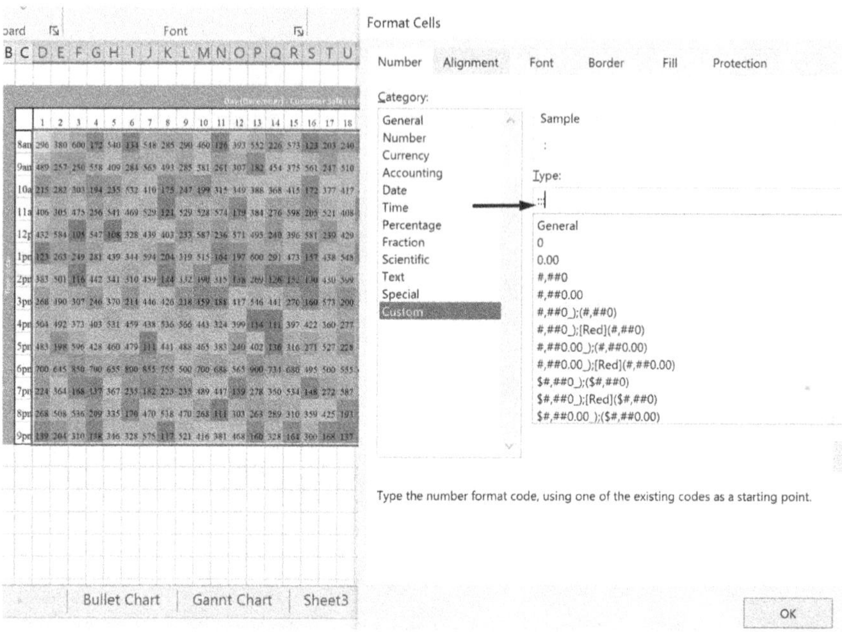

You have completed your heat map using conditional formatting. The heat map below shows that more sales (darker green) are created around 6 pm and typically on the weekends and especially on the Memorial Day holiday weekend.

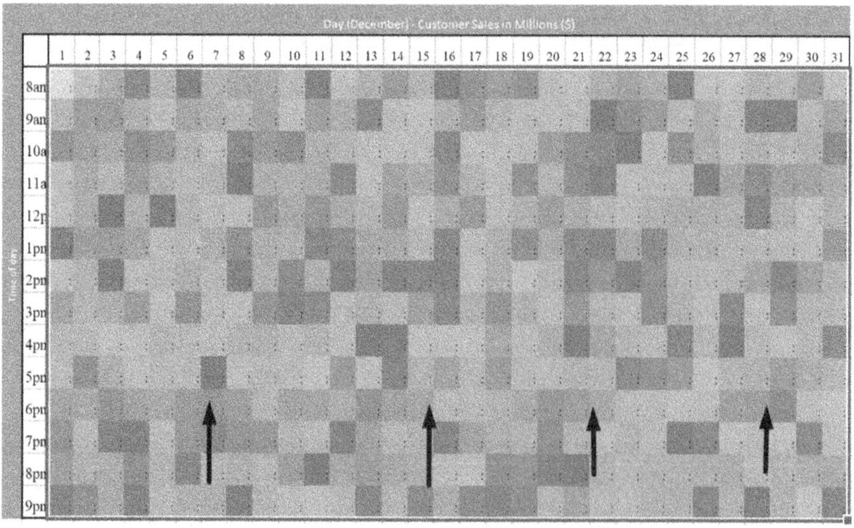

Pareto Charts

Pareto charts are used in management to identify skewed distributions or the most popular frequency units of measurement with a bar chart and a line graph to depict the cumulative total percentage across any unit of measurement. Pareto charts have three axes of measurement. The left vertical axis is typically the frequency of occurrence but can represent another unit of measurement. The right vertical axis is the cumulative percentage measured using a line graph. And the horizontal axis is any unit of measurement of interest. A quality control manager will measure the number of defects in a production line, a human resource manager will measure the top reasons for arriving to work late, or a nursing manager might want to know the most common reasons for a slip and fall in a hospital.

The Italian economist, Vilfredo Pareto, devised the Pareto chart in 1897 when creating a formula representing the uneven distribution of wealth—what later came to be known as the 80–20 rule or the Pareto Principle. He observed that 80% of the wealth in Italy belonged to 20% of the population.

A tennis coach was interested in knowing what the most common errors made during the 2050 Women's Tennis Open tournament. The coach tabulated the following errors committed during the weeklong tournament.

2030 Women's Tennis Open Tournament Errors	
Type of error	Errors made
Backhand groundstroke error	1310
Forehand groundstroke error	837
Forehand volley error	225
Backhand volley error	210
Unreturnable (winner)	185
Overhead error	55
Drop shot error	45
Total	2867

A total of 2867 tennis errors were captured during the tournament. A Pareto chart would be a great way to visualize these errors to help improve practicing among women tennis players.

- Start by creating a new column for **Total %**
- To calculate the backhand groundstroke errors **Total %** you must enter the formula with the correct designated column and row and in this case the formula is = **SUM(B27/B$27:B$33)**. This is the total backhand groundstroke errors percentage of 1310 divided by the total number of errors of 2867. <u>**Note**</u>: The use of the dollar sign in the formula of B$27:B$33 keeps the constant of 2867 errors in the denominator

- To format the other type of errors you must move your mouse cursor to the lower right-hand corner of the Total % for Backhand groundstroke error until the + sign appears
- Make sure to apply the + formatting by dropping and dragging to all the cells in the Total % errors for all the type of errors listed

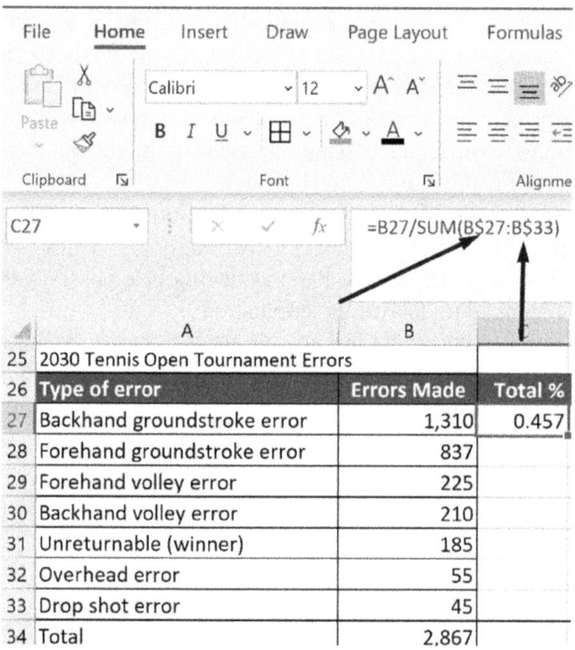

To format the Total % column into a percentage with two decimal places make sure to click on the decimal arrow twice and to select the number drop down arrow.

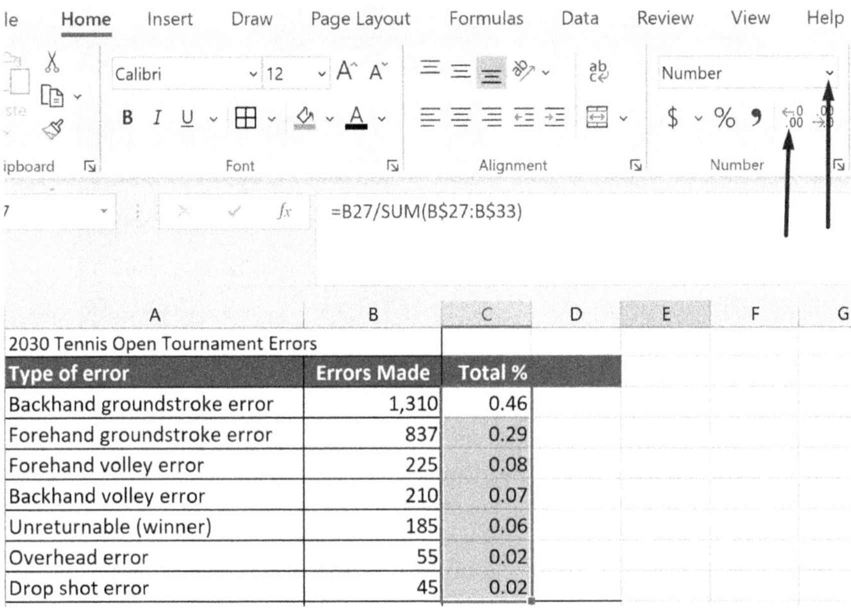

- Scroll down and select the **Percentage** option

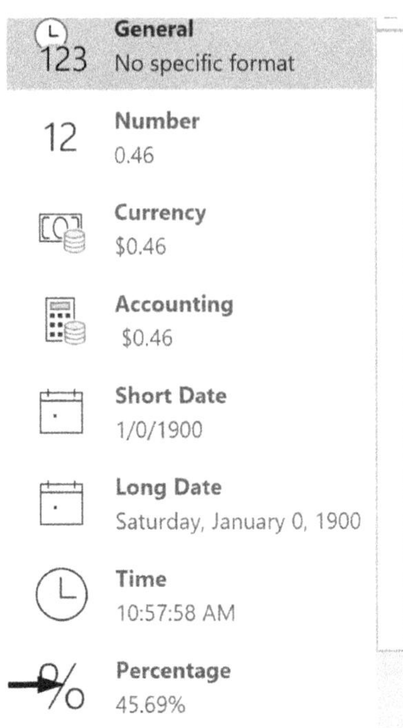

2030 Tennis Open Tournament Errors		
Type of error	**Errors Made**	**Total %**
Backhand groundstroke error	1,310	45.69%
Forehand groundstroke error	837	29.19%
Forehand volley error	225	7.85%
Backhand volley error	210	7.32%
Unreturnable (winner)	185	6.45%
Overhead error	55	1.92%
Drop shot error	45	1.57%

- Add the **Cumulative %** column
- To calculate the backhand groundstroke errors **Cumulative %** you must enter the formula with the correct designated column and row and in this case the formula is = **SUM(C$27:C28)**. This is the total backhand groundstroke errors percentage of 45.69% + 29.19%. The dollar sign allows to keep the constant of that cell number
- To format Cumulative % cells you must move your mouse cursor to the lower right-hand corner of the Cumulative % for Backhand groundstroke error until the + sign appears

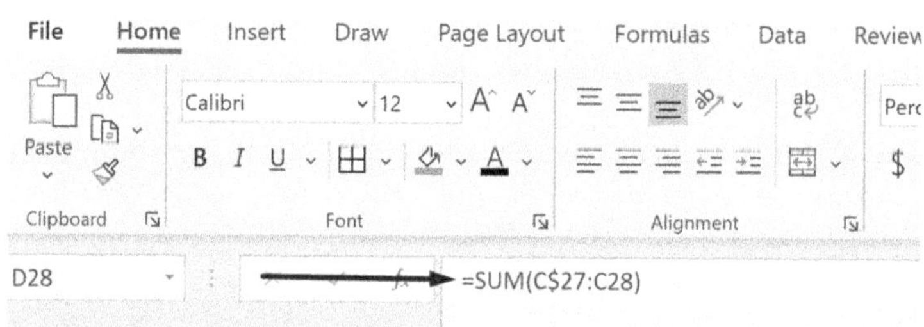

	A	B	C	D
25	2030 Tennis Open Tournament Errors			
26	Type of error	Errors Made	Total %	Cummulative %
27	Backhand groundstroke error	1,310	45.69%	45.7%
28	Forehand groundstroke error	837	29.19%	74.9%
29	Forehand volley error	225	7.85%	
30	Backhand volley error	210	7.32%	
31	Unreturnable (winner)	185	6.45%	
32	Overhead error	55	1.92%	
33	Drop shot error	45	1.57%	
34	Total	2,867	100.00%	

- Make sure to apply the + formatting by dropping and dragging to all the cells in the Cumulative % errors for all the type of errors listed

2030 Tennis Open Tournament Errors			
Type of error	Errors Made	Total %	Cummulative %
Backhand groundstroke error	1,310	45.69%	45.7%
Forehand groundstroke error	837	29.19%	74.9%
Forehand volley error	225	7.85%	82.7%
Backhand volley error	210	7.32%	90.1%
Unreturnable (winner)	185	6.45%	96.5%
Overhead error	55	1.92%	98.4%
Drop shot error	45	1.57%	100.0%
Total	2,867	100.00%	

- To create the Pareto Chart, you must now highlight all your cells in your table as shown

26	Type of error	Errors Made	Total %	Cummulative %
27	Backhand groundstroke error	1,310	45.69%	45.7%
28	Forehand groundstroke error	837	29.19%	74.9%
29	Forehand volley error	225	7.85%	82.7%
30	Backhand volley error	210	7.32%	90.1%
31	Unreturnable (winner)	185	6.45%	96.5%
32	Overhead error	55	1.92%	98.4%
33	Drop shot error	45	1.57%	100.0%

- Click **Insert Histogram** chart type with the cumulative line that mentions Pareto

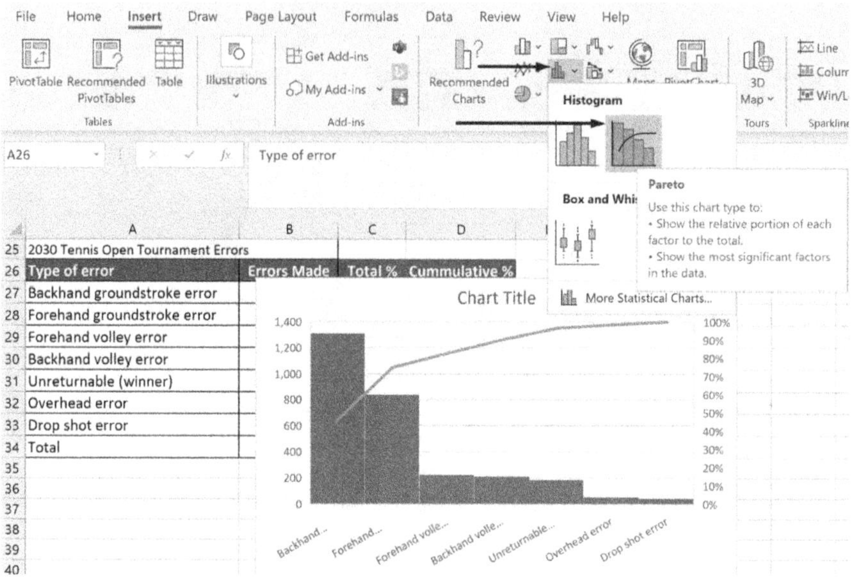

- Click on **Chart Design** with the **Quick Layout** option for **Layout 3**
- This layout option allows for you to label your Chart Title, Horizontal Axis, etc.

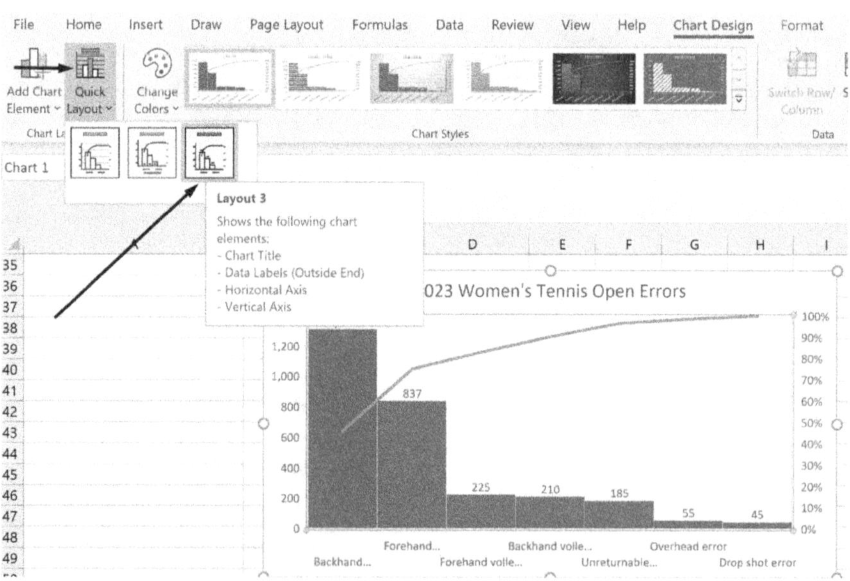

You have completed your Pareto chart. The Pareto chart below shows the most common to least common type of error made or the root cause to most common errors made in the 2030 Women's Tennis Open.

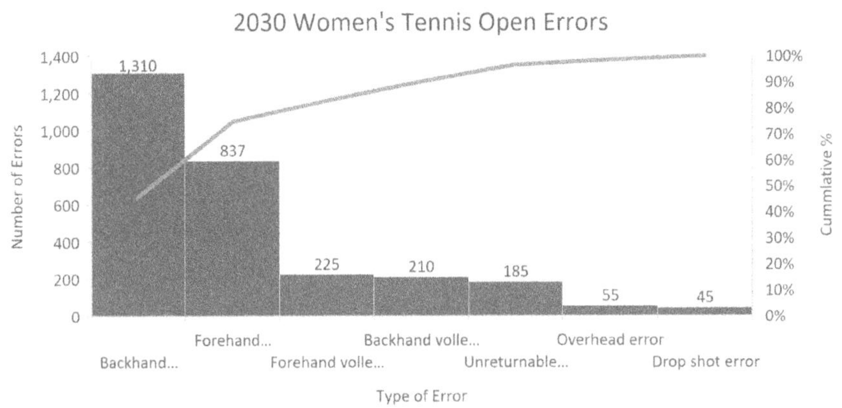

Pictographs

Early forms of pictographs were known as pictograms or ideograms. These ideograms (pictures or symbols that represent ideas) or pictograms now more popularly known as pictographs use images to represent numbers on a chart. It is believed that in 5000 BC, in Sumeria on clay tablets and about 4000 BC in Egypt on papyrus the first ideograms appeared.[4]

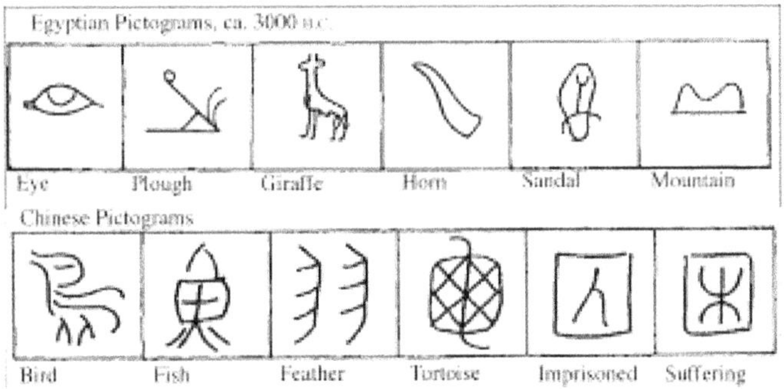

[4] https://en.wikipedia.org/wiki/Pictogram

In Microsoft Excel, pictographs have become a popular way to visually express mathematical charts using pictures to present mathematical ideas. Pictographs have become more widespread because of the phrase, "A picture is worth a thousand words." Pictographs can be used in an international presentation setting because of the ability to communicate pictures to an audience effectively, even if the languages and cultures are completely different. Business presentations are using pictographs as a data visualization technique to present quantitative business information with charts and graphs.

A sales manager is interested in yearly tennis racket sales from 2030 to 2034. The sales manager must present the number of users for a tennis racket product from the 1.0 version to the 3.0 version of the racket design. The upcoming sales meeting requires a presentation to demonstrate the number of users adopting the

Tennis Racket Version by Number of Users

Year	Version 1.0	Version 2.0	Version 3.0
2030	1500	2000	4000
2031	3800	2000	4200
2032	3000	1700	5300
2033	2700	1400	5900
2034	2500	1200	6200

new versions of the racket design from 2030 to 2034. Use the following table to create a pictograph using images of the tennis racket product:

- To create a tennis racket pictograph, you must highlight all your cells in your table as shown

Year	Version 1.0	Version 2.0	Version 3.0
2030	1,500	2,000	4,000
2031	3,800	2,000	4,200
2032	3,000	1,700	5,300
2033	2,700	1,400	5,900
2034	2,500	1,200	6,200

- Click **Insert Charts** with the chart type **2-D Column Clustered Column** chart

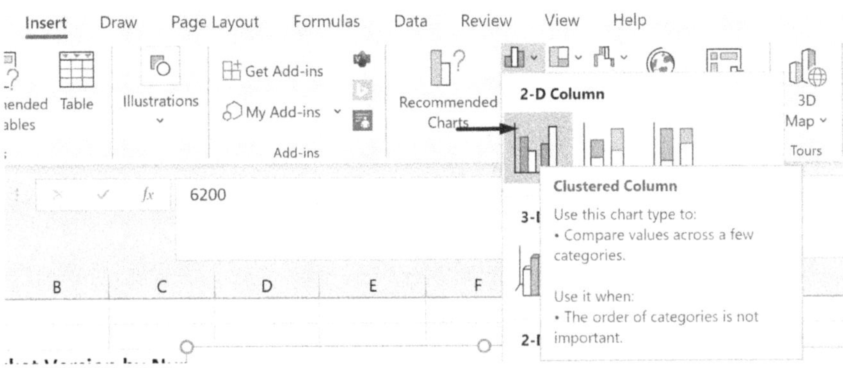

The following 2-D clustered column chart should appear. If not, then make sure to scroll down and **Click More Column Charts** and select the appropriate column clustered chart that includes the year of each racket version from 2030 to 2034 as color coded below.

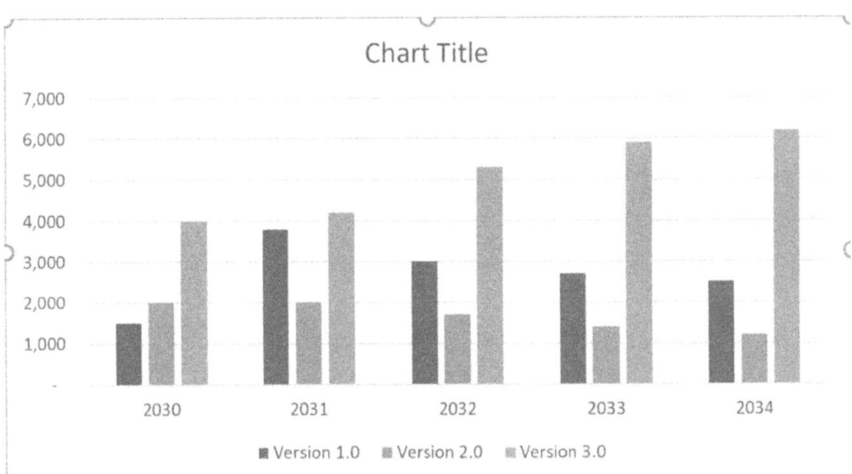

- **Click** on the Version 1.0 column bar to select all the blue column bars
- **Right Click to Format Data Series** and make sure the paint bucket is selected
- **Click Picture or texture fill**

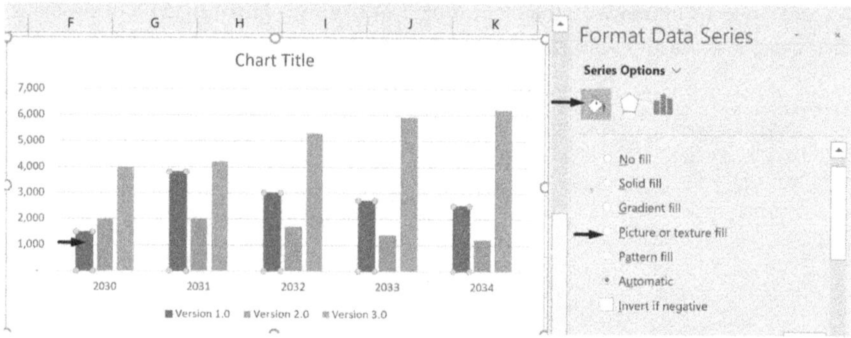

- **Click** on **Insert Picture Source**
- Select the picture source **From a File, Stock Images, Online Files**, etc.
- Select **Stack** picture source to fill the bar graph with an image of the Version 1 tennis racket

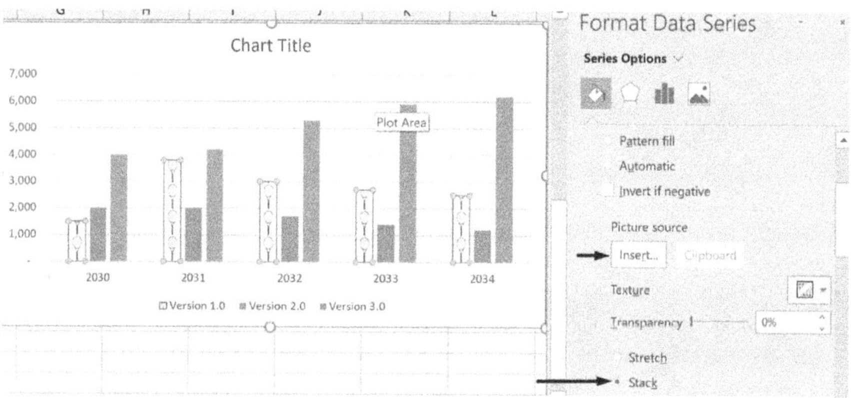

Complete the following steps for Version 2 and Version 3 of the tennis racket images.

- **Click** on **Insert Picture Source**
- Select the picture source **From a File, Stock Images, Online Files**, etc.

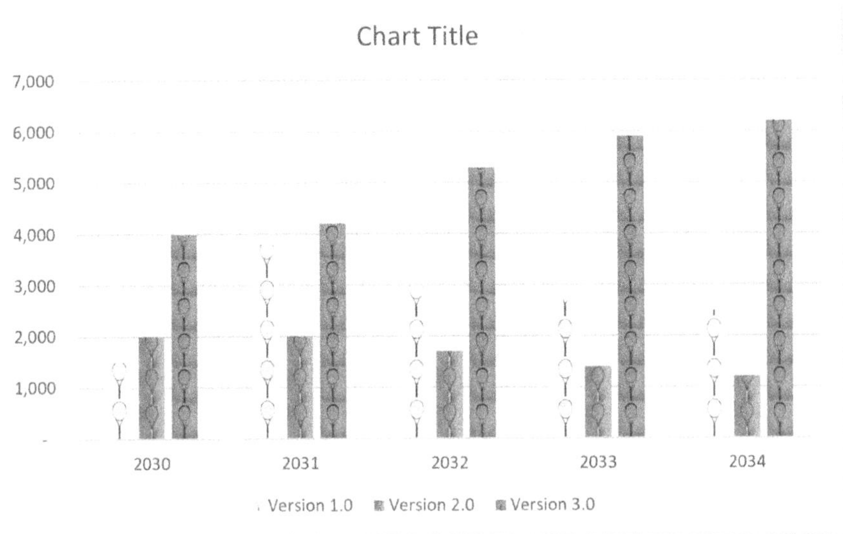

- **Click** on the entire chart
- **Click** on the "+" to include the following **Chart Elements**

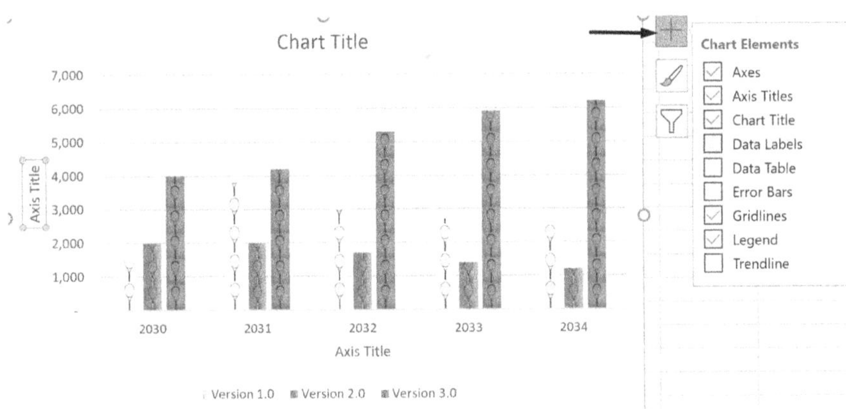

- Add the **Chart Title—**"Tennis Racket Version Users"
- Add the **Axis Titles—**"Number of Users" and "Year"

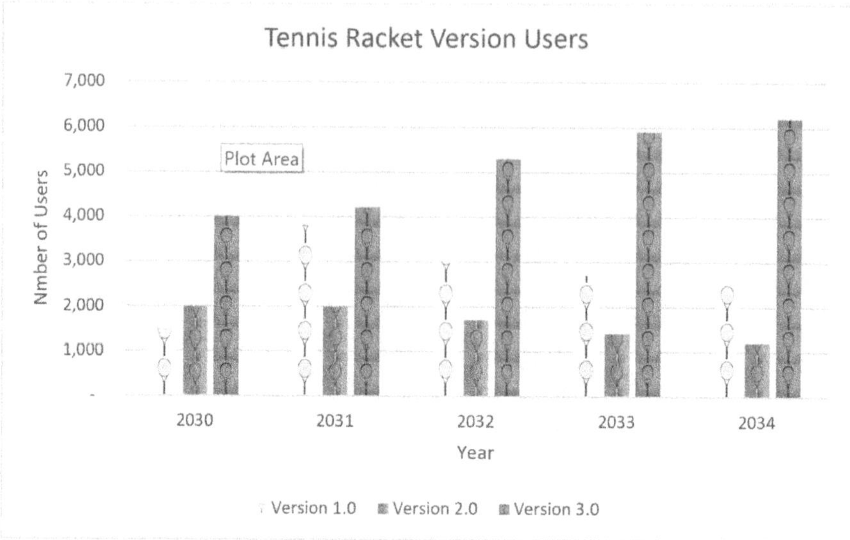

The pictograph is complete. The sales manager can show the fluctuation of tennis racket users from Version 1.0 to Version 3.0 with stacked pictures of the three racket product designs. The pictograph shows how users are adopting Version 3.0 to be the most popular racket design from 2030 to 2034.

People Graph

Microsoft Excel has a simple and quick way to insert a people graph or pictograph. Let us depict the number of units sold on four tennis racket versions in 2034. The following table will be used to create a quick pictograph.

- **Insert Add-ins** and **Select** the people graph icon

Version	**Units sold in 2034**
Version 1	2500
Version 2	1200
Version 3	6200
Version 4	5800
Grand total	15,700

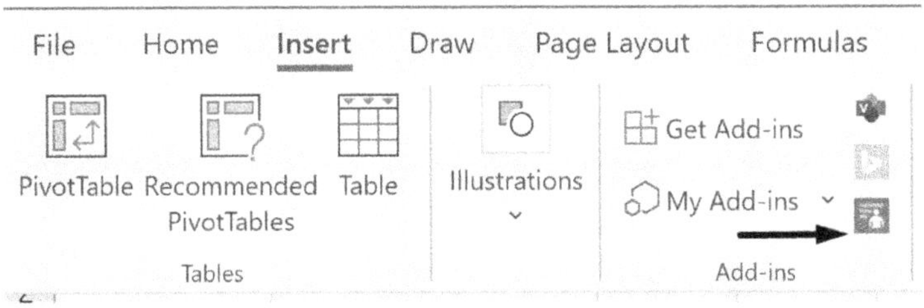

By default, the following people graph will show up. You will use this graph to create your people graph with your units sold spreadsheet.

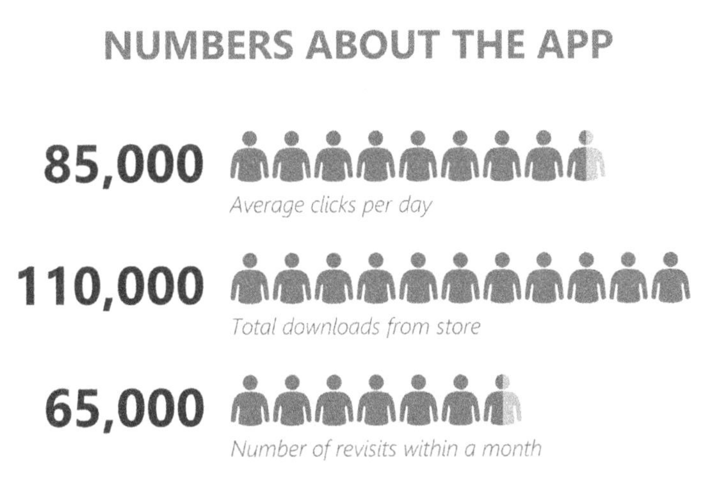

- **Scroll** around on the people graph above until the spreadsheet icon appears below
- **Click** on the spreadsheet icon

The following dialog will appear.

- First, **Type** in the Title, **"Units Sold in 2034"**
- Second, **Click** on the "Select your data" dialog box

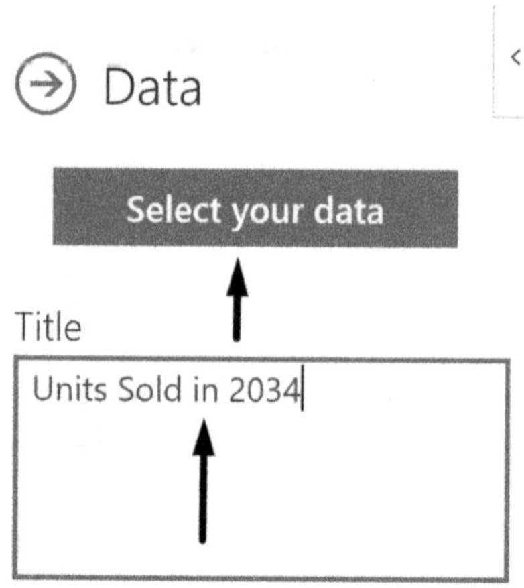

- **Highlight** your data in your units sold spreadsheet

Version	**Units Sold in 2034**
Version 1	2,500
Version 2	1,200
Version 3	6,200
Version 4	5,800
Grand Total	15,700

Upon highlighting your spreadsheet above, you will notice the sample data will notify you that, "You've selected six rows and two columns."

- **Click** on the green **Create** box.

The following People Graph will appear, but feel free to make any format changes necessary by clicking on the format settings icon next to the spreadsheet icon.

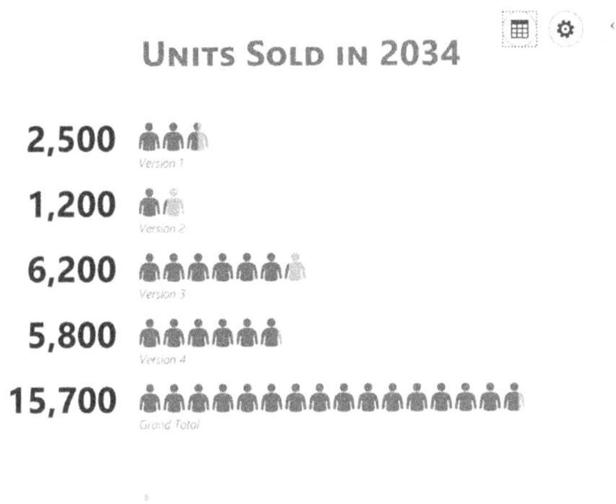

Sparklines

Sparklines are a great way to show data trends using columns, lines, or win/loss chart inside a single or multiple embedded cell/s on an Excel spreadsheet. Business managers can quickly produce a simple chart inside a single cell to visualize trends from a selected source of data cells. Our tennis racket sales manager wants to know the sales trends on four separate racket designs. The new 4.0 Version of the racket has been very popular on social media since its release in January. The sales manager wants to see the sales trends across all four versions of the racket. Use the following table to create sparklines for each racket version:

Racket Version	Jan	Feb	Mar	Apr	May	Jun	Total
Version 1.0	$354,426	$529,923	$245,761	$376,266	$246,345	$146,373	$1,899,094
Version 2.0	$451,535	$779,997	$684,199	$704,804	$405,847	$245,752	$3,272,134
Version 3.0	$458,367	$862,118	$887,255	$879,881	$308,545	$786,555	$4,182,721
Version 4.0	$155,999	$226,015	$346,431	$525,400	$787,823	$2,688,460	$4,730,128
Total	$1,420,327	$2,398,053	$2,163,646	$2,486,351	$1,748,560	$3,867,140	$14,084,077

- To insert sparklines for all racket versions sold from January to June, you must highlight all your cells in your table as shown

Racket Version	Jan	Feb	Mar	Apr	May	Jun	Total
Version 1.0	$354,426	$529,923	$245,761	$376,266	$246,345	$146,373	$1,899,094
Version 2.0	$451,535	$779,997	$684,199	$704,804	$405,847	$245,752	$3,272,134
Version 3.0	$458,367	$862,118	$887,255	$879,881	$308,545	$786,555	$4,182,721
Version 4.0	$155,999	$226,015	$346,431	$525,400	$787,823	$2,688,460	$4,730,128
Total	$1,420,327	$2,398,053	$2,163,646	$2,486,351	$1,748,560	$3,867,140	$14,084,077

- Click **Insert** scroll over to Sparklines and select **Line** for the type of sparkline

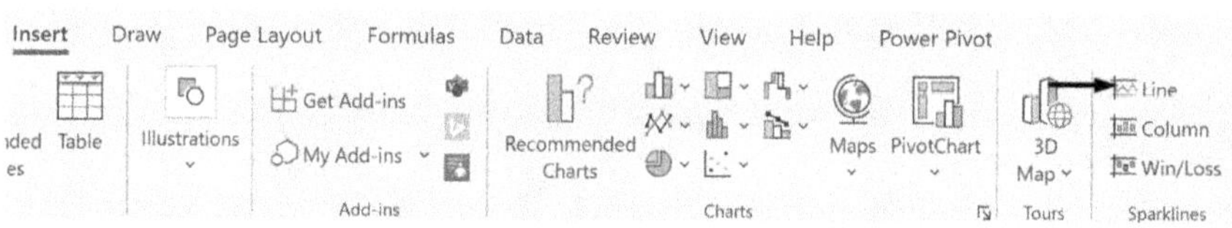

The following dialog box will appear:

- The **Data Range** should be the highlighted range you selected above
- The **Location Range** is where you want the sparklines to be embedded in range of cells
- Click **Ok**

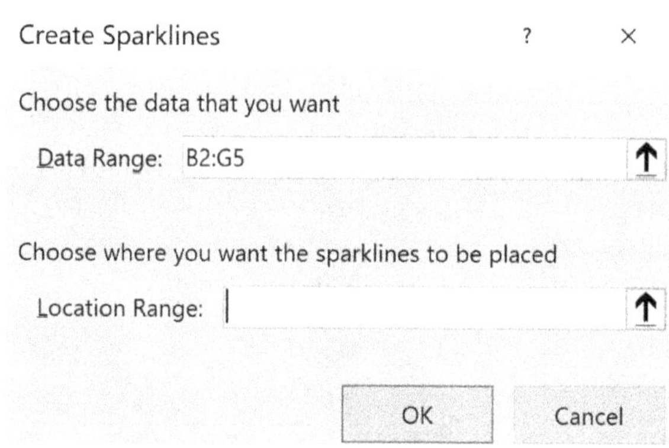

The sparklines now appear on the selected cells.

- **Click** on the sparklines created
- **Click** on show sparkline **Markers**

Racket Version	Jan	Feb	Mar	Apr	May	Jun	Total	Jan. to Jun. Sales Trend
Version 1.0	$354,426	$529,923	$245,761	$376,266	$246,345	$146,373	$1,899,094	
Version 2.0	$451,535	$779,997	$684,199	$704,804	$405,847	$245,752	$3,272,134	
Version 3.0	$458,367	$862,118	$887,255	$879,881	$308,545	$786,555	$4,182,721	
Version 4.0	$155,999	$226,015	$346,431	$525,400	$787,823	$2,688,460	$4,730,128	
Total	$1,420,327	$2,398,053	$2,163,646	$2,486,351	$1,748,560	$3,867,140	$14,084,077	

The sparkline markers show six months sales trend from January to June.

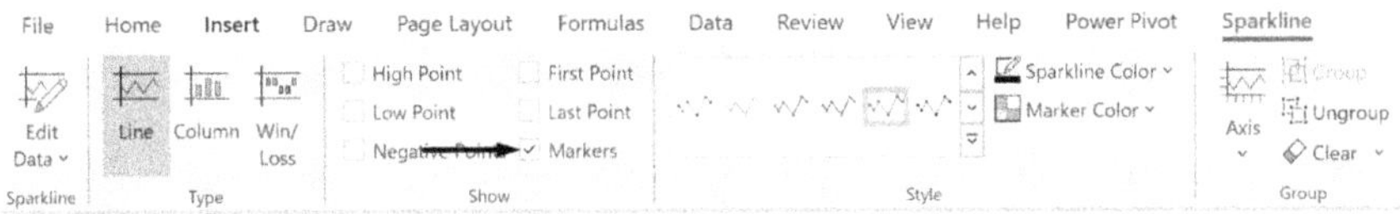

Racket Version	Jan	Feb	Mar	Apr	May	Jun	Total	Jan. to Jun. Sales Trend
Version 1.0	$354,426	$529,923	$245,761	$376,266	$246,345	$146,373	$1,899,094	
Version 2.0	$451,535	$779,997	$684,199	$704,804	$405,847	$245,752	$3,272,134	
Version 3.0	$458,367	$862,118	$887,255	$879,881	$308,545	$786,555	$4,182,721	
Version 4.0	$155,999	$226,015	$346,431	$525,400	$787,823	$2,688,460	$4,730,128	
Total	$1,420,327	$2,398,053	$2,163,646	$2,486,351	$1,748,560	$3,867,140	$14,084,077	

If you wanted to show the sparkline as a column chart:

- **Click** on the sparklines created
- **Click** on the **Column** type of sparkline

You have changed your sparkline from a line type to a column type of chart. See the trend in sales in a column chart type from January to June sales below:

Racket Version	Jan	Feb	Mar	Apr	May	Jun	Total	Jan. to Jun. Sales Trend
Version 1.0	$354,426	$529,923	$245,761	$376,266	$246,345	$146,373	$1,899,094	
Version 2.0	$451,535	$779,997	$684,199	$704,804	$405,847	$245,752	$3,272,134	
Version 3.0	$458,367	$862,118	$887,255	$879,881	$308,545	$786,555	$4,182,721	
Version 4.0	$155,999	$226,015	$346,431	$525,400	$787,823	$2,688,460	$4,730,128	
Total	$1,420,327	$2,398,053	$2,163,646	$2,486,351	$1,748,560	$3,867,140	$14,084,077	

Power BI Desktop

We conclude our data visualization chapter by using Power BI Desktop to enhance your data visualization charts in Excel. Power BI Desktop is a free download that allows users to visually explore data and reporting features while using Excel's graphs and charts. The use of Excel and Power BI Desktop together creates an interactive reporting/dashboard experience to present analytical results.

As a beginner to Power BI Desktop, let us start with a simple experience to create reports that build into an interactive dashboard. Although reports and dashboards are similar, they are often confused on how they work together to present visual results. Typically, a dashboard will use several Excel visual charts (no more than 5 to 9) on a dashboard page to display results.[5] The following image can help distinguish the difference between reports and dashboards when using Power BI Desktop.

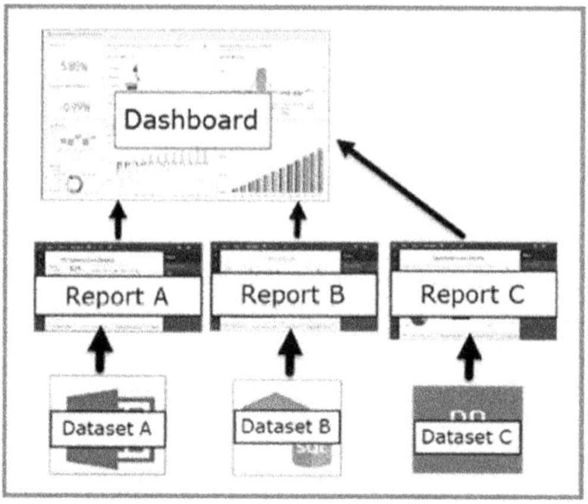

To get started, you need to download the free Power BI Desktop download. Start by going to the Microsoft Store and type in Power BI Desktop in the search bar. If you have trouble searching or installing then, visit www.powerbi.com to install product directly from the Power BI website.

- **Click** on **Download** button

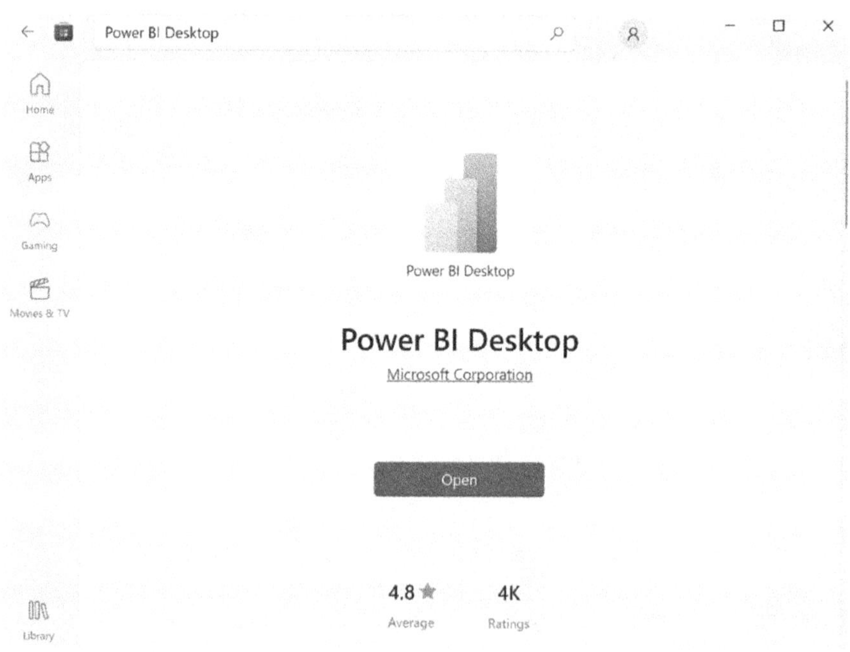

[5] https://www.sisense.com/blog/4-design-principles-creating-better-dashboards/

- **Click** on **Install**

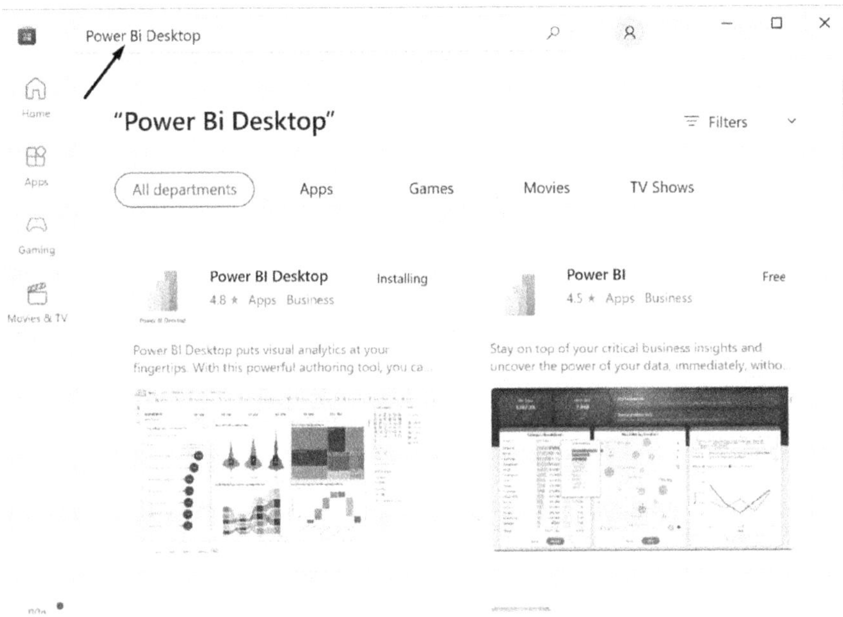

Once you have successfully downloaded the Power BI Desktop App, you should see the following Power BI Desktop icon and the following Power BI Desktop App opened:

- **Click** on the Excel icon below to import your data
- **Select** the **Power BI** Excel data file to import

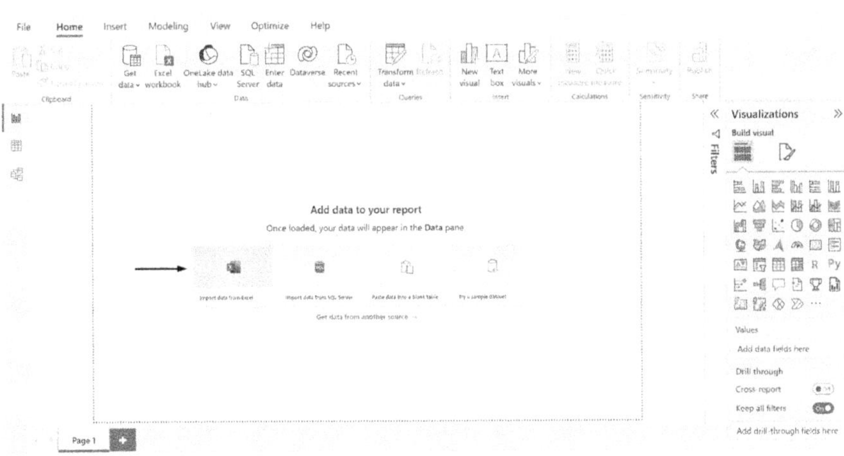

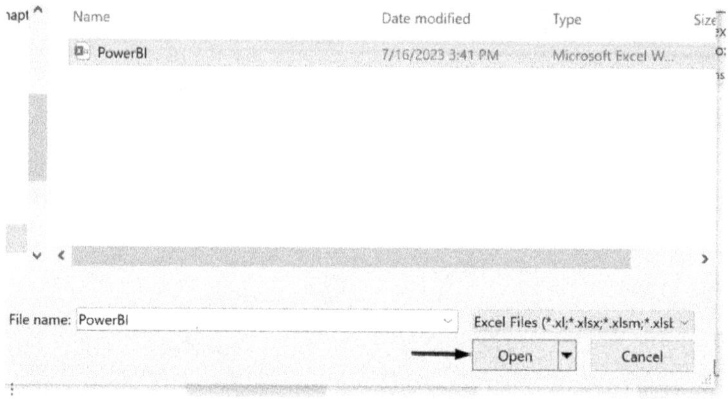

The **Power BI** data file is an expanded sales Excel file with 13 variables (from **Segment** to **Year**) on the tennis racket sales from 2033 to 2034. Upon loading the file, the Navigator button shows up and Power BI automatically creates two files underneath the **Power BI** Excel file (**financials** and **Sheet1**).

Once you select the financials file from the **Navigator** screen, you will notice a preview of the **Power BI** Excel file will show up on the right-side. If you decided that you wanted to make any final changes or recoding of any data, you can select **Transform Data**.

- **Select** the **financials** file and click on **Load**

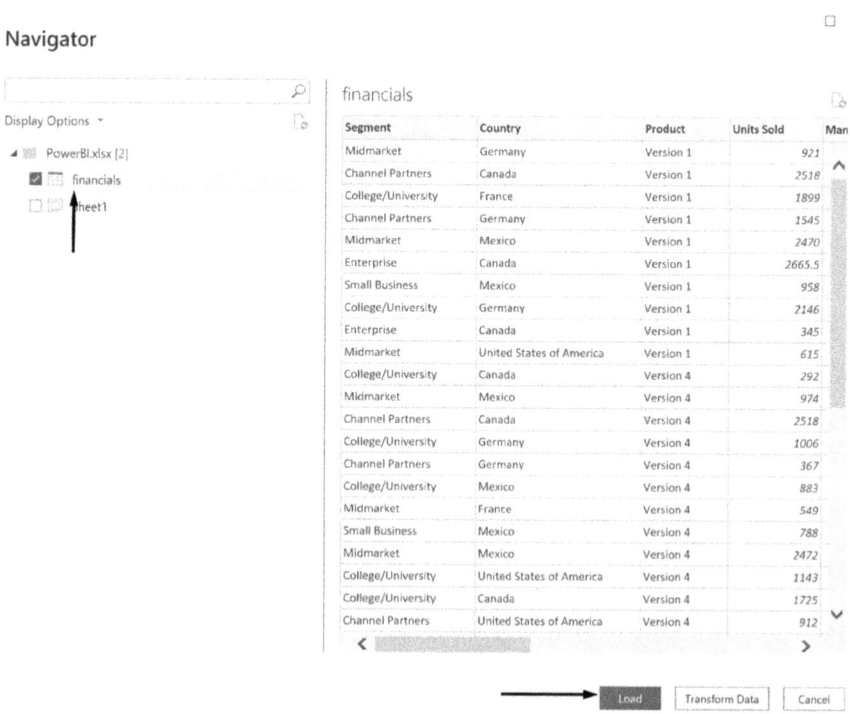

You have loaded one data file to build several reporting charts. However, if you wanted to load multiple data files you would repeat the above steps to create relationship databasing across multiple data files with more variables for analysis. To keep our example simple, we will just analyze the **financials** file to build the report.

- **Click** on the down arrow on the **financials Data**

You should see 13 variables of data appear below the financials table icon. You can use these variables to run any analysis to create visual charts on a dashboard.

The Senior Vice President of Product Sales wants to know the latest sales figures from Version 1 to Version 4 of the tennis rackets. You have been asked to conduct some sales analysis that can be shared as a one-page dashboard to answer the following questions:

1. What are the number of units sold from year to year for each version?
2. What are the sum of profits from each version?
3. What are the sum of sale by month for each version?
4. What are the sum of units sold by channel segment from month to month?
5. What are the sum of profits from the countries selling each version?

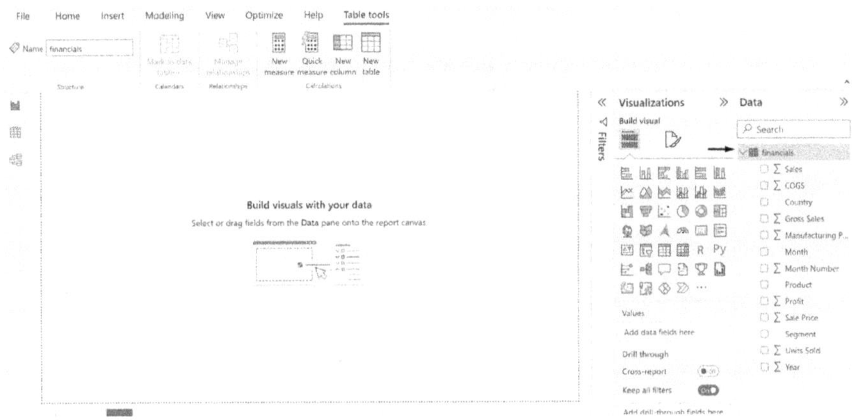

Add Report Title

- **Click** on **Text box Insert** and type, **"Sales Results for Version 1 to Version 4"**
- **Click** on font type or size to adjust appropriate text box title color, size, etc.

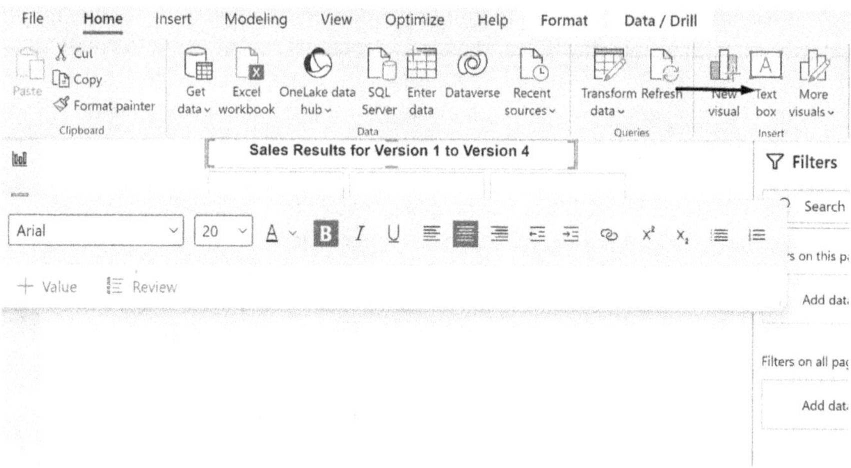

Add Report Visualizations

- From the **Visualizations** pane **Click** on the slicer filter
- From the **Data** fields pane **Select** the **Product** check box

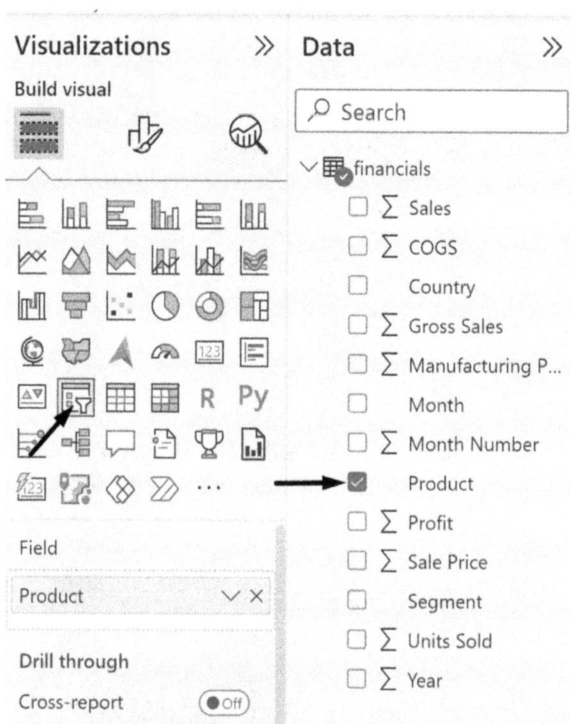

To format the product slicer with a dropdown menu

- **Click** on the **Format** visual icon
- **Click** on the **Options Style** and select **Dropdown**

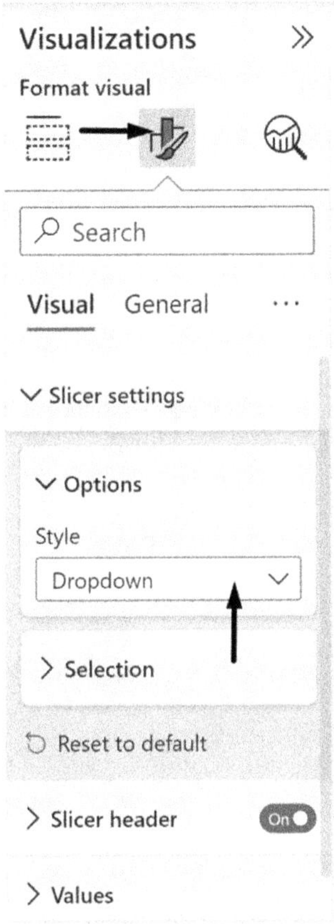

The following product slicer will appear in your report view:

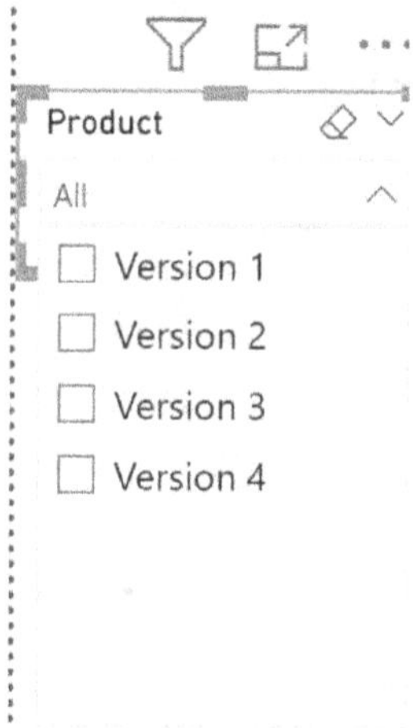

- From the **Visualizations** pane click on the slicer filter
- From the **Data** fields panel select the **Month** and the **∑ Year** check box

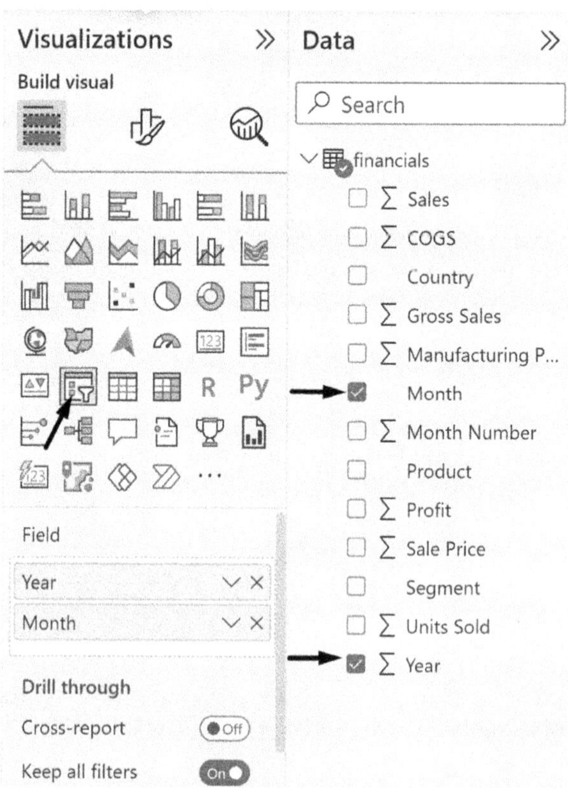

To format the Year and Month slicer with a dropdown menu

- **Click** on the **Format** Visualization icon
- **Click** on the Options Style and select **Dropdown** just as you did above for the product slicer

The following year and month slicer will appear in your report view:

- From the **Visualizations** pane **Click** on the card (new)
- From the **Data** fields pane **Select Product, ∑ Units Sold,** and **∑ Year** check box

The following Card will appear in your report view, but feel free to format the font and stretch out the box size as needed.

- From the **Visualizations** pane click on the **Clustered Column Chart**
- From the **Data** fields pane select **Product** and **∑ Profit** check box

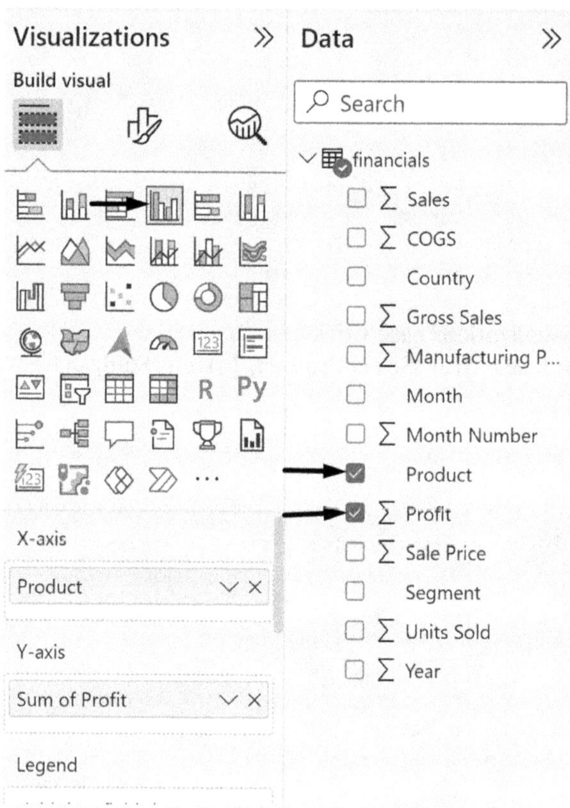

The following Clustered Column Chart will appear in your report view, but feel free to stretch out the box size as needed.

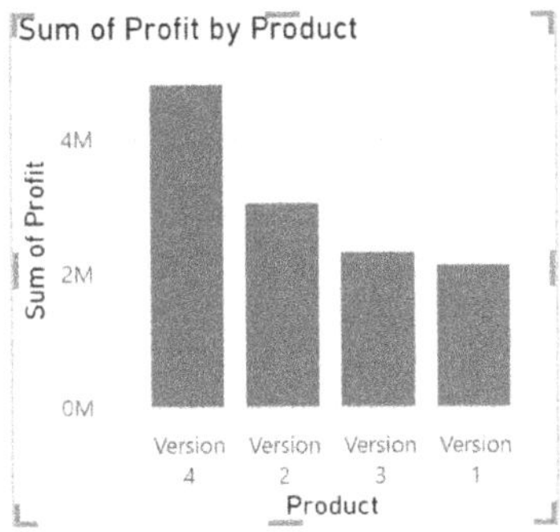

- From the **Visualizations** pane click on the **Line Chart**
- From the **Data** fields pane ∑ **Sales, Month,** and **Product** check box

The following Line Chart will appear in your report view, but feel free to stretch out the box size as needed.

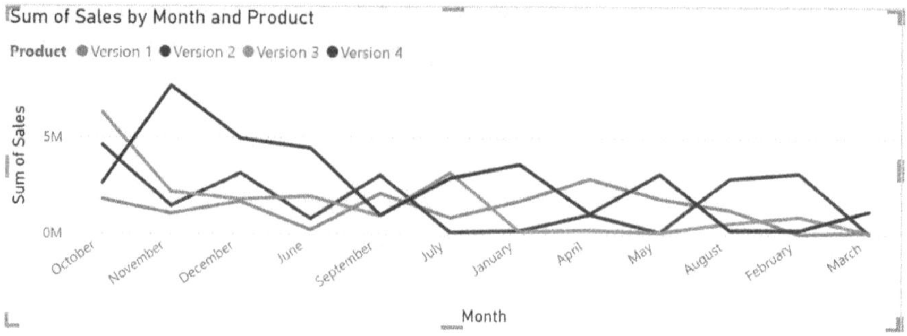

- From the **Visualizations** pane click on the **Clustered Column Chart**
- From the **Data** fields pane **Month**, **Segment**, and **∑ Units Sold**

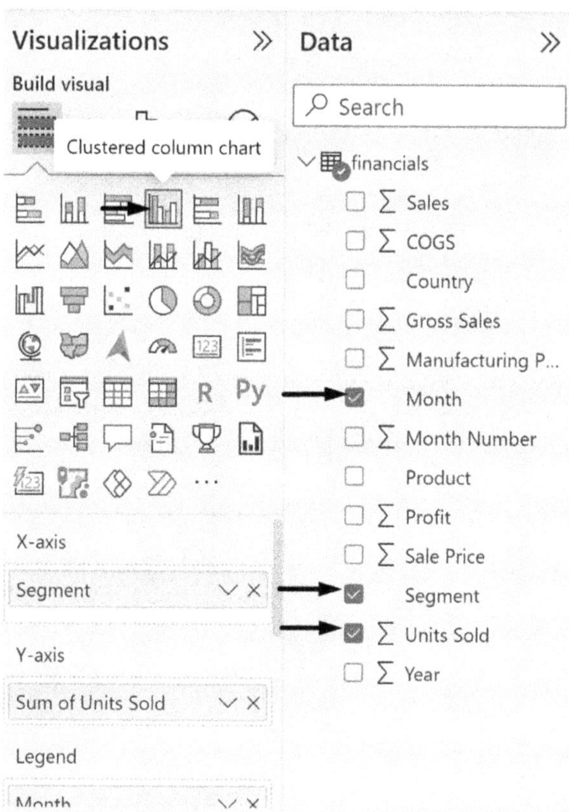

The following clustered column chart will appear in your report view, but feel free to stretch out the box size as needed.

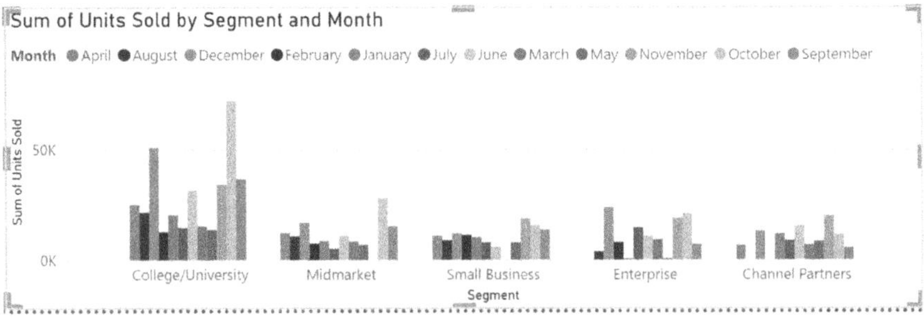

- From the **Visualizations** pane click on the **Map**
- From the **Data** fields pane **Country**, **Product**, and **∑ Profit**

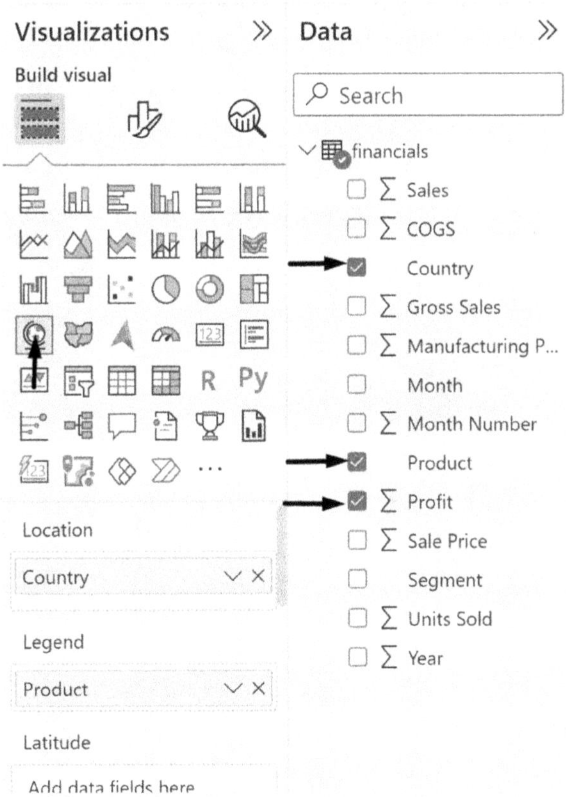

The following map will appear in your report view, but feel free to stretch out the box size as needed.

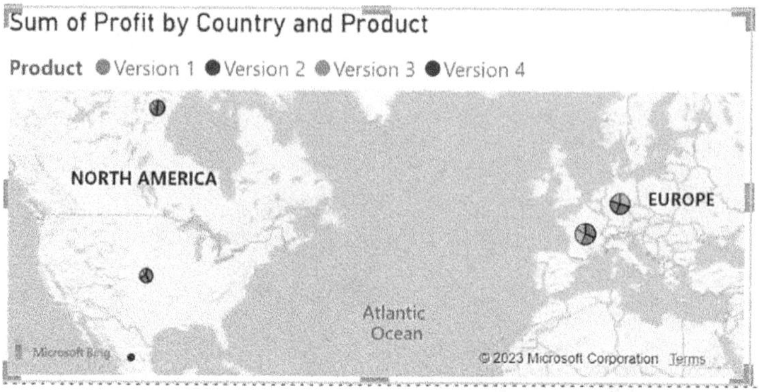

Make sure to save your report. Before publishing your report, you may want to format the report title with an eye catching background or introduce new fonts and colors for each visualization chart you created.

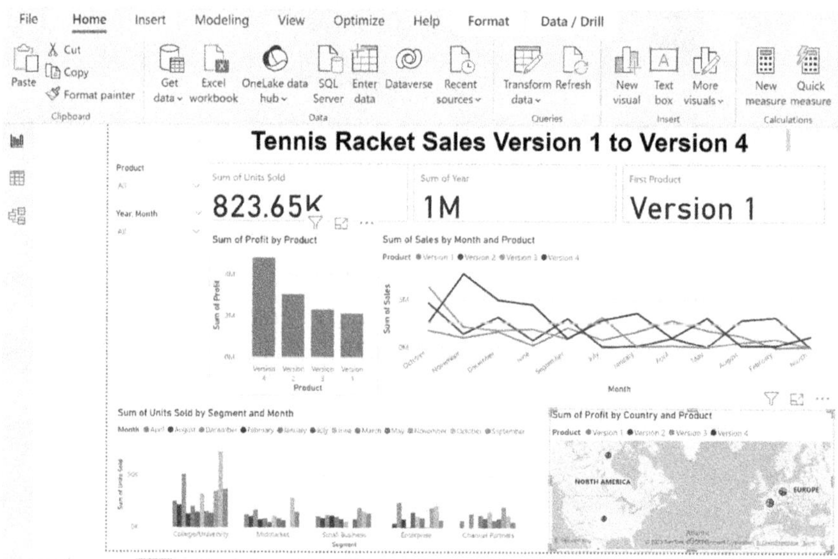

We just created a background transparency for the report title, "Tennis Racket Sales Version 1 to Version 4."

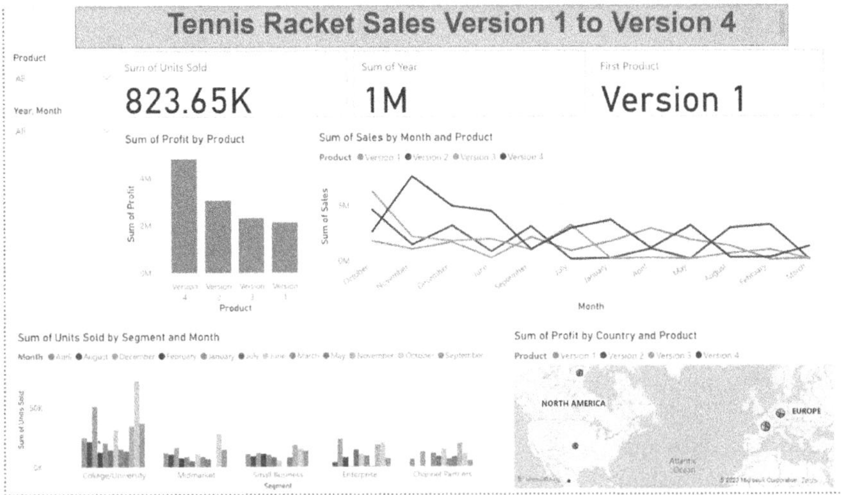

- From the **File** menu, select **Save as** and give your file a name
- From the **File** menu, **Select Publish** to share your Power BI service with your Power BI account colleagues to view your 1-page dashboard from your workspace

Final Thoughts and Activities

Practice Problems

1. **Heat Map Creation**

ToyLand Inc. is a global retailer specializing in toy products such as Motorcycles, Planes, Ships, and Trains. The company wants to analyze its sales performance across different product lines and quarters to identify patterns and make informed business decisions.

Task
Create a heat map to visualize the quarterly sales performance of ToyLand Inc.'s product lines. Each cell in the heat map should represent the total sales amount for a specific product line and quarter combination. Apply color-coding to the cells based on the sales amount to highlight areas of high and low sales

Details
- Rows: Product lines
- Columns: Quarters (Q1, Q2, Q3, Q4) Color-coding
- Use a gradient scale to represent sales amounts, with darker (Green) colors indicating higher sales and lighter colors (Red) indicating lower sales

Utilize the provided **ToyLand** file dataset.

2. **Sparkline Creation**

ToyLand Inc. is interested in tracking the quarterly sales trends for its best-selling product line, Motorcycles, to monitor its popularity over time.

Task
Create a people chart to visualize the distribution of ToyLand Inc.'s customers by country. Each country should be represented by a figure or icon, with the size proportional to the number of customers from that country.

Details
- Each country should be represented by a figure or icon (e.g., human silhouette, flag icon).
- The size of each figure or icon should reflect the number of customers from that country.
- Arrange the figures or icons on the chart to effectively demonstrate the distribution of customers across countries.

Utilize the **ToyLand** file containing customer information, including country of residence.

3. **People Chart Creation**

ToyLand Inc. is interested in visualizing the distribution of its customers across different countries to gain insights into its global customer base.

Task
Create a people chart to visualize the distribution of ToyLand Inc.'s customers by country. Each country should be represented by a figure or icon, with the size proportional to the number of customers from that country.

Details
- Each country should be represented by a figure or icon (e.g., human silhouette, flag icon).
- The size of each figure or icon should reflect the number of customers from that country.
- Arrange the figures or icons on the chart to effectively demonstrate the distribution of customers across countries.

Utilize the **ToyLand** file containing customer information, including country of residence.

Problem Solutions

1. **Problem type: Heat Map Creation**

Step 1: Prepare the Data
Group the Data: Group the data by quarter and product line. This involves organizing the dataset so that you have distinct combinations of quarter and product line, along with the corresponding total sales amount.

Calculate Total Sales: Calculate the total sales amount for each combination of quarter and product line. You can do this by using Excel functions like SUMIFS or PivotTables.

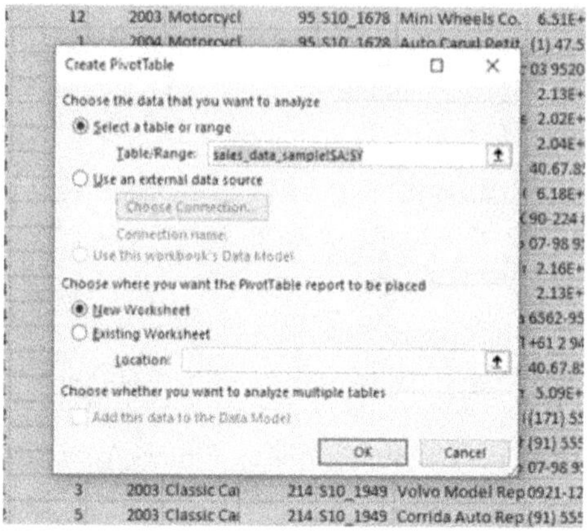

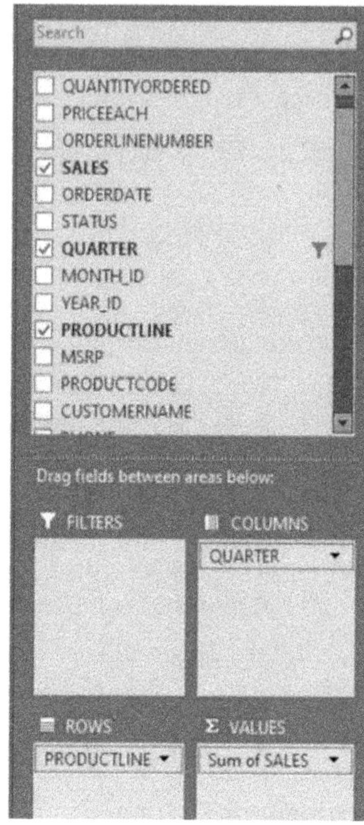

Sum of SALES	Column Labels				
Row Labels	1	2	3	4	Grand Total
Classic Cars	880278.9	766618.73	745268.46	1527449.57	3919615.66
Motorcycles	264385.54	277822.27	213194.7	410985.83	1166388.34
Planes	234356.51	247054.75	143052.95	350539.36	975003.57
Ships	203877.3	143114.28	120498.22	246947.33	714437.13
Trains	58115.8	35284.87	39450.16	93392.64	226243.47
Trucks and Buses	208619.04	240838.42	199511.01	478821.37	1127789.84
Vintage Cars	501184.64	337386.98	297935.31	766643.91	1903150.84
Grand Total	2350817.73	2048120.3	1758910.81	3874780.01	10032628.85

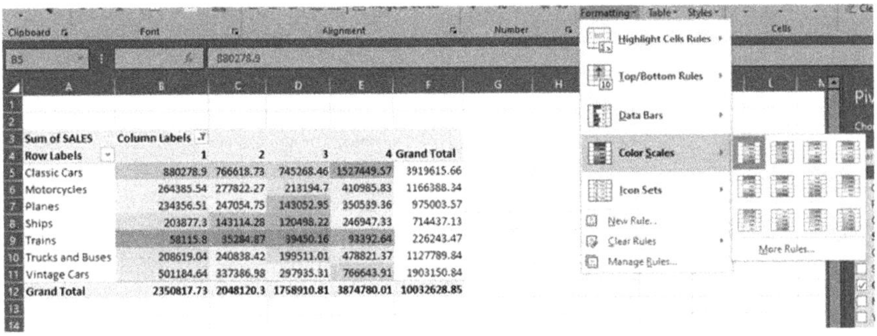

Step 2: Create the Heat Map

Select Data Range: Select the range of cells containing the total sales data for each quarter and product line. Make sure to include row and column headers if applicable. Conditional Formatting: Go to the "Home" tab in Excel.

Apply Conditional Formatting: Click on "Conditional Formatting" in the toolbar. Choose "Color Scales" from the dropdown menu.

Choose Color Scale: Select a suitable color scale, such as green to yellow to red. This color scale will represent the range of sales amounts, with darker colors indicating higher sales.

Apply Color Scale: Excel will apply the selected color scale to the cells in the selected range. Each cell will be color-coded based on the total sales amount it represents.

2. **Problem type: Sparkline Creation**

Step 1: Prepare the Data

Calculate the total quantity ordered for each quarter.

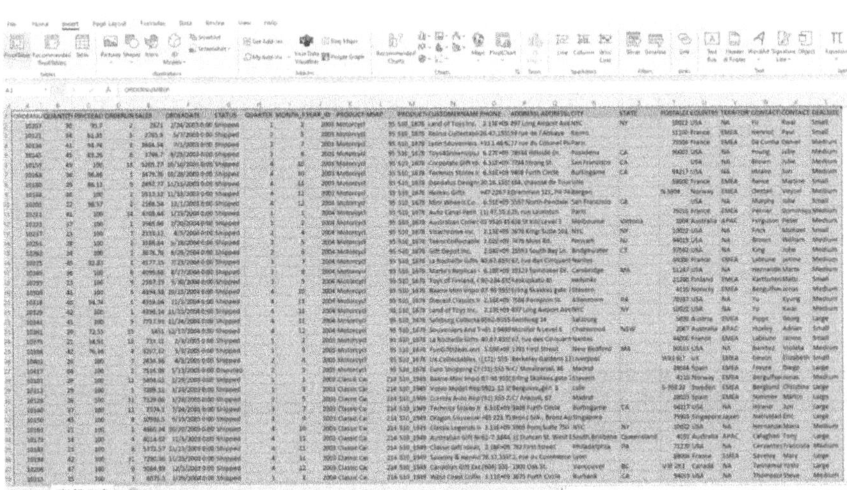

Select the PivotTable Fields as follows

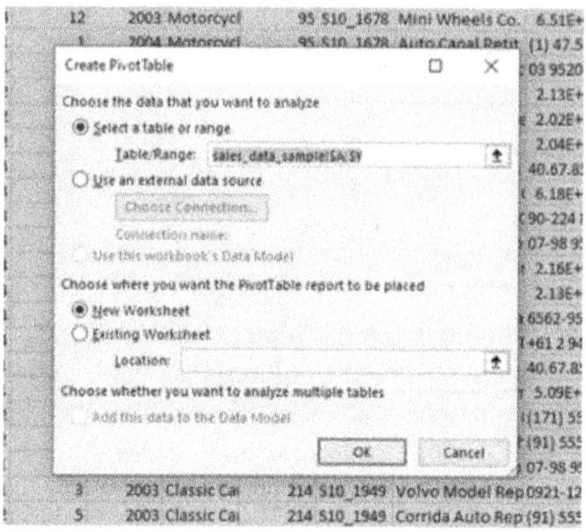

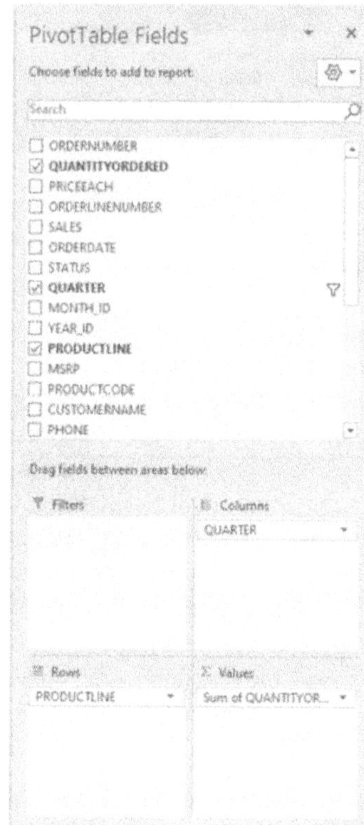

Sum of QUANTITYORDERED	Column Labels				
Row Labels	1	2	3	4	Grand Total
Classic Cars	7784	6447	6190	13571	33992
Motorcycles	2671	2721	2214	4057	11663
Planes	2625	2773	1549	3780	10727
Ships	2163	1620	1416	2928	8127
Trains	693	439	531	1049	2712
Trucks and Buses	2068	2286	1949	4474	10777
Vintage Cars	5481	4030	3269	8289	21069
Grand Total	**23485**	**20316**	**17118**	**38148**	**99067**

Step 2: Insert Sparklines

Select Cells: Select the cells where you want to insert the sparklines.

Sum of QUANTITYORDERED	Column Labels					
Row Labels	1	2	3	4	Grand Total	Sparkline
Classic Cars	7784	6447	6190	13571	33992	
Motorcycles	2671	2721	2214	4057	11663	
Planes	2625	2773	1549	3780	10727	
Ships	2163	1620	1416	2928	8127	
Trains	693	439	531	1049	2712	
Trucks and Buses	2068	2286	1949	4474	10777	
Vintage Cars	5481	4030	3269	8289	21069	
Grand Total	**23485**	**20316**	**17118**	**38148**	**99067**	

Go to Insert Tab: Go to the "Insert" tab in Excel.
Click on Sparklines: Click on "Sparklines" in the toolbar.
Choose Line Sparkline Type: From the Sparklines options, choose "Line" as the sparkline type.
Select Data Range: In the "Data Range" box, select the range containing the quarterly quantity ordered data for the chosen product line. This should exclude any headers and only include the numeric data.
Click OK: Click "OK" to insert the sparklines

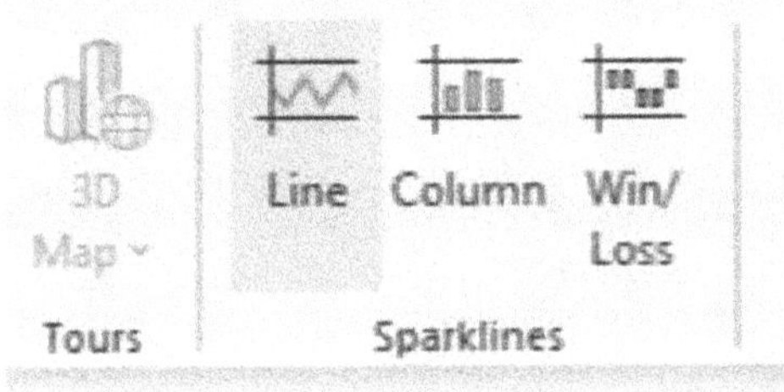

Select Location Range as the new column for sparkline
Select Makers if we want to include them

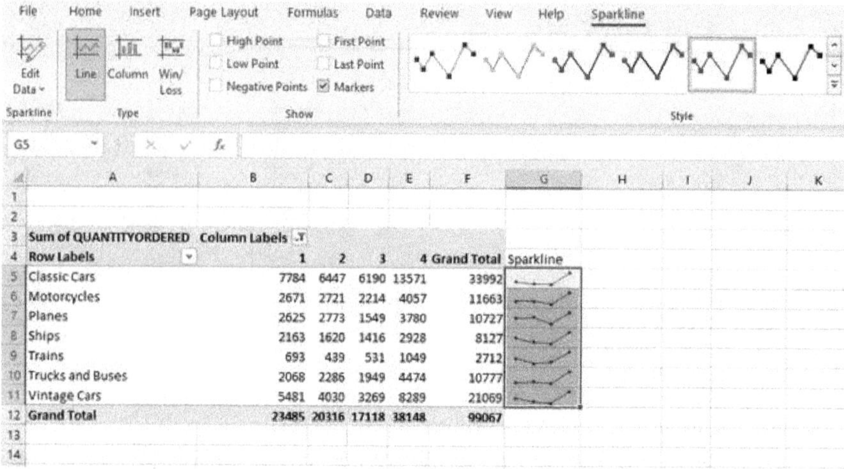

3. **Problem type: People Chart Creation**

Step 1: Prepare the Data

Select the Entire Sheet Click on Remove Duplicates to get unique customers, Select only Customername in the list to get unique customers

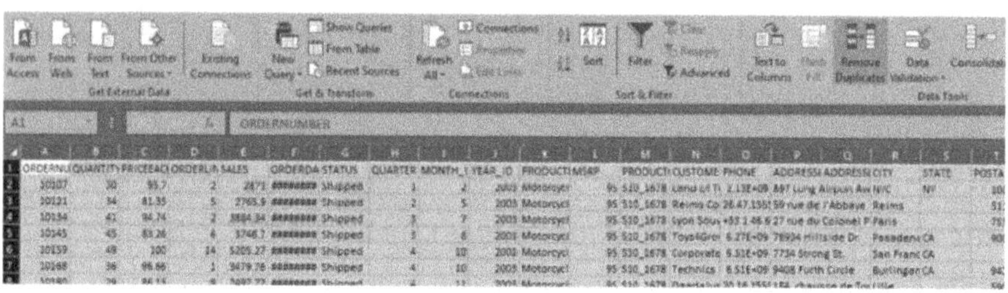

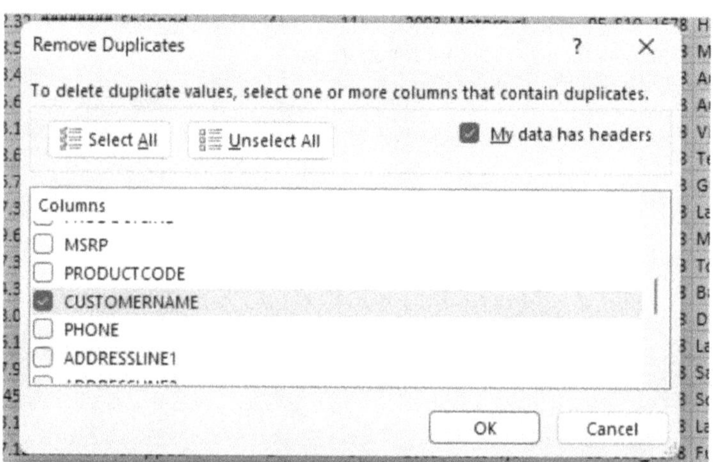

Select the New Data Create a pivot table for the New Data and select the PivotTable Fields as follows

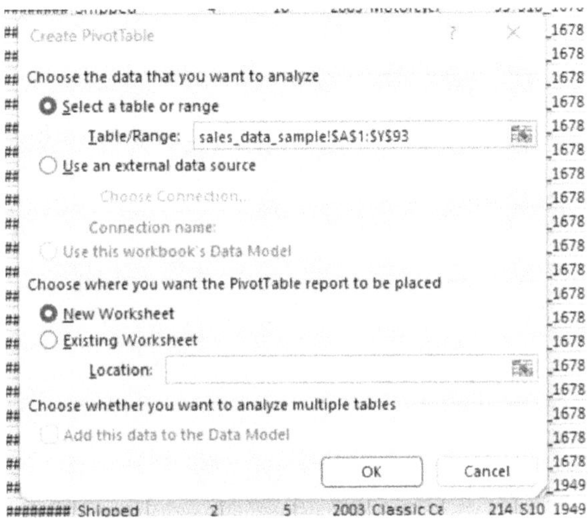

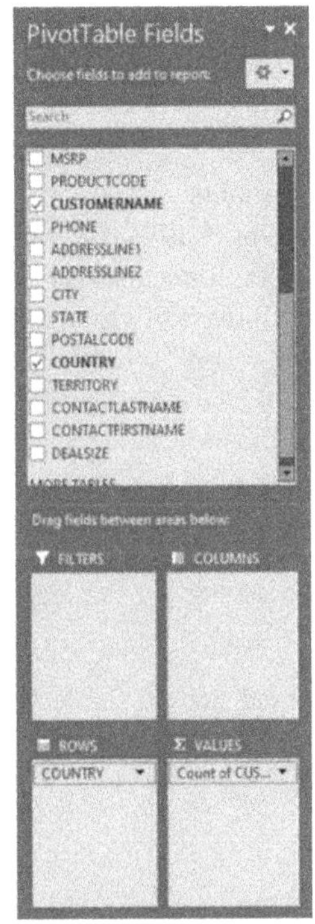

Once the Number is filled for each country, we make the Peoples Graph

Step 2: Create the People Chart

Select the range of cells containing the country names and the corresponding number of customers.

Go to the "Insert" tab in Excel.

Click on "People Graph" in the toolbar.

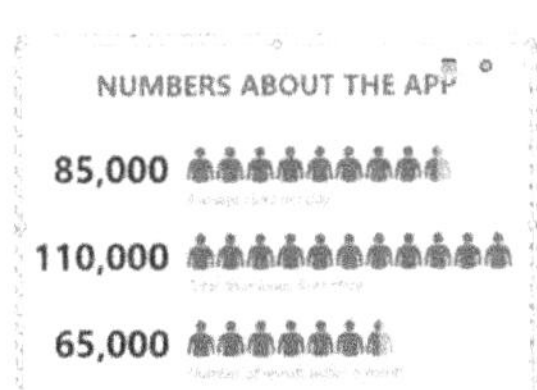

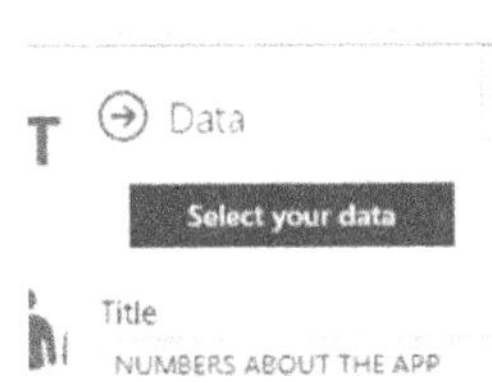

Step 3: Configure the People Chart

In the "People Graph" dialog box that appears:

Choose the layout option for the people chart. For example, you might select "Icons with Data Labels" to show both the icons and the data labels.

Choose an icon style for the chart. You can select from various options provided by Excel.

Select the data range for the country names and the corresponding number of customers.

Optionally, you can customize the chart further by adjusting options such as colors, fonts, and layout settings.

Click "OK" to insert the people chart.

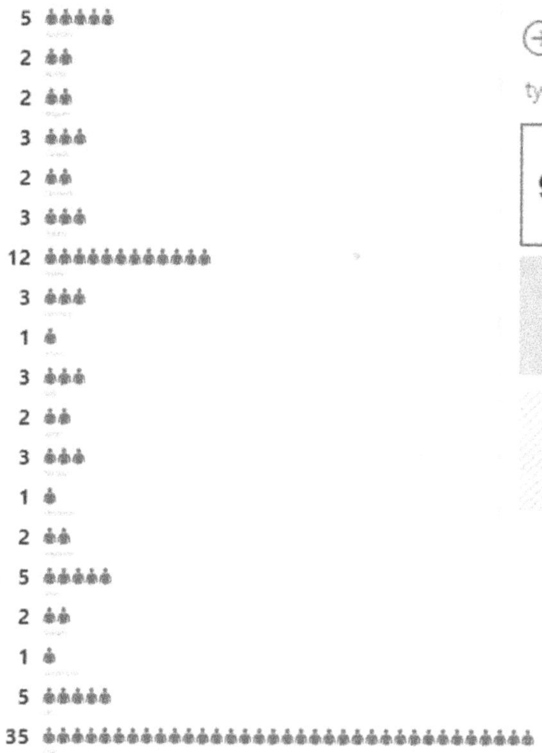

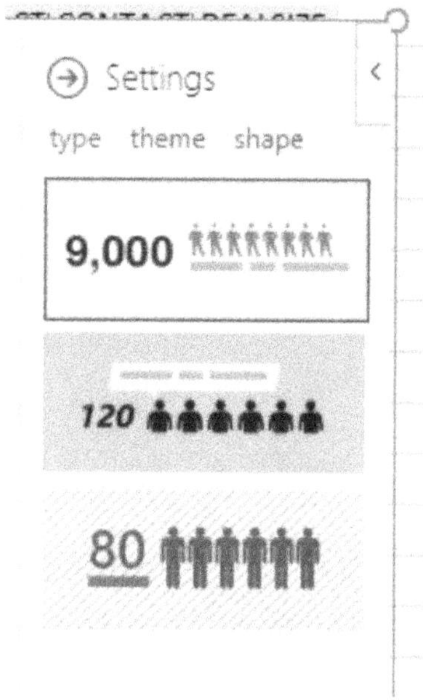

After inserting the people chart, you can further customize it by:

- Adjusting the size of the icons based on the number of customers. To do this, click on the chart to select it, then use the handles to resize the chart as desired. Changing the color of the icons or data labels, adjusting fonts, or modifying other visual aspects of the chart to suit your preferences.

Discussion Board

1. What do you think can be a common problem or pitfall with data visualization charts? Discuss with examples to support your opinion.
2. What are three pros that would support your argument that data visualization provides effective communication of data?

Group Activity

1. Create six data visualization charts discussed above using sales data. Have each group member create one or two charts to best communicate the data. Compile the data and create one data visualization report. What were the findings?
2. Research how the *Journal of Data Science, Statistics, and Visualisation* effectively present data visualization research findings. Choose a paper and discuss why the findings were important.

Parting Thought

Make sure to read the find print when seeing any data visualization chart…

Chapter 17
Qualitative Analysis

Linda Herkenhoff and John Fogli

Key Concepts

Qualitative analysis, Mixed methods, Codes, Themes, Deductive coding, Inductive coding, Intercoder reliability, Document analysis, Ethnography, Saturation, and Segmentation.

Discussion

In the world of business statistics qualitative data plays a key role in providing a comprehensive understanding of complex business phenomena. Qualitative analysis allows us to explore the underlying context, motivations, and sentiments that shape the data while quantitative data provides us with measurable quantities. For example, when we need to understand people's opinions, perspectives, or experiences we usually turn to qualitative data. In this chapter we will discuss qualitative analysis and explore its applications in business statistics which will allows us to uncover valuable insights that go beyond numbers.

Qualitative analysis involves the systematic examination of nonnumerical data to understand the meanings and interpretations behind the information. It differs from

Supplementary Information The online version contains supplementary material available at https://doi.org/10.1007/978-3-031-46371-6_17.

L. Herkenhoff (✉)
Saint Mary's College, Moraga, CA, USA

J. Fogli
Diablo Valley College, Pleasant Hill, CA, USA

L. Herkenhoff, J. Fogli (eds.), *Applied Statistics for Business and Management using Microsoft Excel*, https://doi.org/10.1007/978-3-031-46371-6_17

quantitative analysis by focusing on the underlying context, perceptions, and behaviors associated with the data rather than numerical measurements. By employing qualitative analysis in business statistics we gain a deeper understanding of the human/subjective aspects that impact decision making processes and business outcomes.

Qualitative analysis can also inform quantitative research by helping analysts refine research questions/hypotheses, conduct exploratory qualitative studies, and design research instruments that capture the richness of the phenomena being studied. Qualitative analysis allows analysts to explore the context, gather insights, and refine their measurement instruments before embarking on large-scale quantitative studies. Analysts can improve the relevance/accuracy of their measurements thus leading to more robust statistical analyses by integrating qualitative findings into quantitative studies.

Qualitative analysis has numerous practical applications in the field of business statistics. Qualitative analysis helps uncover consumer preferences, motivations, and emerging market trends in market research and consumer behavior studies. Analysts can better explore the subjective experiences and perceptions of consumers thus providing valuable insights into their decision-making processes by conducting in depth interviews or focus groups.

Qualitative analysis is also invaluable in assessing organizational culture and employee engagement. Analysts can explore workplace dynamics, identify factors that influence employee satisfaction, and gain a deeper understanding of organizational culture through methods such as employee interviews, observation, or analysis of internal documents. These insights can guide human resource management, organizational development, and leadership strategies within businesses.

Furthermore, qualitative analysis supports competitive analysis and strategic planning by evaluating competitors' strengths/weaknesses. Analysts can uncover valuable information about their strategies, capabilities, and market positions by conducting interviews or analyzing competitors' public statements/documents. This knowledge enables businesses to make informed decisions, identify strategic opportunities, and effectively respond to competitive threats. The integration of qualitative analysis with traditional quantitative methods will foster more robust/holistic decision making processes allowing organizations to propel towards success in an ever evolving marketplace.

To fully appreciate qualitative analysis it is crucial to differentiate it from quantitative analysis. Quantitative data is characterized by its numerical nature which allows for mathematical calculations and statistical tests. On the other hand qualitative data consists of nonnumerical information such as text, images, audio recordings, and observations. Qualitative analysis aims to uncover meaning, explore patterns, and identify themes within this rich array of data. Compared to quantitative data, which captures structured information, qualitative data is unstructured and has more depth. Qualitative analysis can be used to complement quantitative findings by providing insights into the "why" behind the numbers. While quantitative analysis can reveal correlations, trends, and statistical significance it often falls short in explaining the underlying reasons and motivations. Qualitative analysis bridges this gap by exploring the subjective experiences, perceptions, and behaviors of individuals or groups.

Qualitative analysis can be conducted through various methods such as interviews, surveys, focus groups, observations, document analysis, and ethnography. Each method offers unique advantages in capturing different aspects of the phenomenon under study. For example, interviews allow analysts to directly interact with participants and gain knowledge about their perspectives while observations provide an opportunity to study behavior in natural settings. By employing multiple methods analysts can gather rich and diverse qualitative data to expand their analysis.

The five steps in qualitative data analysis typically include:

1. Gathering and collecting data
2. Organizing data
3. Coding and identifying themes
4. Data analysis and visualization
5. Reporting on the insights derived from your analysis

Gathering and Collecting Data

The amount of data needed for qualitative analysis can vary depending on various factors including the research question, the complexity of the phenomenon under investigation, the richness of the data, and the research design. Unlike quantitative research, where sample size calculations are often based on statistical power, qualitative research focuses on achieving depth, richness, and understanding rather than generalizability.

The method of saturation serves as a key determinant for sample size to ensure there is enough qualitative data to answer your research questions and achieve data sufficiency. Saturation refers to the point at which new data collection or analysis no longer yields significant additional insights or information to enrich the understanding of the phenomenon under investigation. It signifies reaching a point of theoretical redundancy where the data collected is deemed comprehensive enough to capture the range and diversity of perspectives/experiences related to the research topic. Saturation is typically achieved when new data no longer reveals novel aspects of the subject (such as themes, patterns, or insights) and when existing motifs are consistently reinforced across different participants or contexts. This method emphasizes the importance of data quality/richness over quantity as the goal is to attain in depth insights rather than simply accumulating a large volume of data. Saturation acts as a critical checkpoint in qualitative research indicating that analysts have gathered adequate information to address their research objectives and reach a level of theoretical saturation where further data collection is unlikely to generate substantially new findings.

Saturation can be achieved with a smaller sample size compared to quantitative research. However, there is no fixed rule or specific sample size that guarantees saturation as it depends on the complexity/variability of the research topic. In some cases saturation may be reached with as few as 10–20 participants while in other cases larger sample sizes may be necessary.

Qualitative analysts typically employ purposeful sampling which involves selecting participants who can provide rich and diverse perspectives related to the research question. Analysts continuously assess the quality/depth of the data during the analysis process and may continue data collection until saturation is achieved. Ultimately the goal in qualitative research is to achieve data sufficiency and depth of understanding rather than focusing on a specific numerical sample size.

As mentioned before the most common data collection tools include interviews, focus groups, observations, document analysis, and ethnography.

Interviews: Interviews involve gathering information through direct conversation between an interviewer and a participant or a group of participants. There are different types of interviews which include structured, semi-structured, and unstructured interviews. In structured interviews a predetermined set of questions is asked. Where as semi-structured and unstructured interviews allow for more flexibility and follow-up questions. Interviews can be conducted face-to-face, over the phone, or through video conferencing. This method allows analysts to obtain in depth and personalized responses leading to a deeper exploration of participants' perspectives and ultimately a more rich qualitative data set.

Focus Groups: Focus groups involve bringing together a small group of individuals (typically 6–12 participants) to discuss a specific topic or research question. The participants share their thoughts, opinions, and experiences in a group setting facilitated by a moderator. Focus groups encourage interaction and group dynamics which can lead to the emergence of new ideas/insights. Analysts can explore participants' attitudes, beliefs, and perceptions as well as observe how different individuals respond to and interact with each other. Focus groups are useful for generating qualitative data and exploring diverse perspectives within a specific context.

Observations: Observations involve systematically watching/recording behaviors, events, or interactions in a natural setting. Analysts can engage in either structured or unstructured observations. In structured observations specific variables and behaviors are predefined/measured while unstructured observations allow for more flexibility in exploring various aspects of the observed context. Observations can be done overtly (participants are aware of being observed) or covertly (participants are unaware). This method provides an opportunity to collect firsthand data and view nonverbal cues. Furthermore, it can allow analysts to understand social dynamics, environmental factors, and contextual influences.

Document Analysis: Document analysis involves examining existing documents, texts, or records to extract relevant information. Analysts review written materials such as books, articles, reports, official documents, letters, emails, social media posts, or historical archives. This method allows analysts to gain insights into past events, policies, trends, or public opinions without direct involvement of participants. It assists in the understanding of historical contexts, policy frameworks, organizational practices, or public discourses.

Ethnography: Ethnography is a research method primarily used in anthropology and sociology. It involves immersive long-term fieldwork in a specific cultural or social group to understand their perspectives, behaviors, and cultural practices. Ethnographers participate in the daily lives of the participants allowing them to observe their activities, engage in conversations, and document their experiences. This method aims to provide a holistic understanding of the culture or community being studied by offering insights into social norms, values, and beliefs. Ethnographic research often includes the use of participant observation, interviews, and document analysis to gather qualitative data. This collected data in turn generates rich descriptions/interpretations of the studied group or community.

Organizing Your Qualitative Data

Excel

Excel can be used to organize qualitative data in a structured and systematic manner. Preparing qualitative data for analysis requires careful transcription and cleaning to ensure data integrity/reliability. Analysts must meticulously transfer data from its original form, such as interview transcripts or field notes, into a format suitable for analysis. This process involves removing any identifying information and ensuring the accuracy of the transcriptions. The classic method of structuring qualitative data is to input all the data you've gathered into a spreadsheet.

1. Create a new Excel worksheet: Open Excel and create a new worksheet specifically dedicated to organizing your qualitative data.
2. Define column headers: Identify the different categories or themes that you want to use in your data organization. These can be topics, concepts, or any other relevant criteria. Assign each category as a column header in the first row of your Excel sheet. To begin coding in Excel analysts typically allocate one column for the raw qualitative data such as interview transcripts or textual responses. Each row represents a unit of analysis such as a participant or a specific segment of data. You can enter direct quotes, summaries, or any other relevant information in the respective cells. Analysts can format the data column to wrap text and adjust the row height to accommodate longer responses.

	A	B	C
1		**Qualitative Data from Interviews**	
2			
3		pleased with product	
4		love it	
5		great performance	
6		well built	
7		terrible product	
8		bad customer service	
9		unreliable	
10		does not work right	
11		don't care	
12		sometimes like and don't like	
13		no interest to me it was a gift	
14		no opinion	
15			

If all data is easily accessible in one place and analyzed in a consistent manner you will have an easier time summarizing and making decisions based on this data.

Coding and Themes

Coding becomes the important next step in the qualitative analysis process once the data has been loaded into an Excel spreadsheet. Coding involves the identification and labeling of meaningful patterns/themes within the data. The phrase "to code" means to identify key words/phrases and assign them to a category of meaning. "I really hate the customer service of this software company" would be coded as "poor customer service." Analysts read through the qualitative data line by line and assign descriptive codes to capture the essence of each segment. This reassignment of descriptive codes is referred to as **deductive coding**. These codes act as labels that help organize and categorize the data. Alternatively, **inductive coding** allows new codes to emerge from the data itself thus enabling analysts to explore unanticipated themes/patterns. The coding framework serves as a guide thus ensuring that all relevant aspects of the data are systematically examined/analyzed. Establishing **intercoder reliability** is crucial to ensure the consistency of coding across multiple analysts. This involves having multiple coders independently code a subset of the data and calculating the agreement between their coding decisions. By achieving a high level of intercoder reliability, Analysts can enhance the credibility/trustworthiness of their qualitative analysis by achieving a high level of intercoder reliability.

Excel offers several features that make it suitable for coding qualitative data. While it may not be specifically designed for qualitative analysis its flexibility, familiarity, and widespread availability make it a viable option for analysts who prefer to work within the Excel environment. Coding qualitative data in Excel involves adapting the software's functions and features to create a systematic coding framework. Excel provides various formatting options to enhance the readability and organization of your data. You can adjust column widths, apply text formatting (e.g., bold, italics), and use different cell fill colors to visually distinguish different categories/themes. Analysts often allocate additional columns for each coding category. For example, when analyzing interview data one can code categories that represent key topics, themes, or concepts present in the data. Each row in the coding category columns is then populated with codes that correspond to the relevant groups. You can assign codes to segments of data by manually typing them into the respective coding category columns. Alternatively, you can use Excel's **search and replace** functionality to expedite the coding process. This allows for quick and efficient coding by automatically replacing specific text strings with designated codes.

Excel allows you to sort and filter data based on specific criteria. You can sort the data in ascending or descending order based on a particular column. Filtering enables you to display only specific rows that meet certain conditions thus making it easier to analyze/extract relevant information. Excel's filtering and sorting capabilities enable analysts to perform initial exploratory analysis of the coded data. Filtering allows analysts to focus on specific codes or combinations of codes thus making it easier to identify patterns/themes within the data. Sorting the data based on codes allows for a structured view thus facilitating further analysis/interpretation.

Filter your Excel data if you only want to display records that meet certain criteria.

- Click any single cell inside a dataset

A	B	C	D
Participan	**Emotio**	**Service**	**Produc**
3	angry	bad	dissatified
2	happy	good	satisfied
1	sad	great	satisfied
4	happy	great	satisifed

- In the **Data** tab look for the **Sort & Filter** group
- Click **Filter**

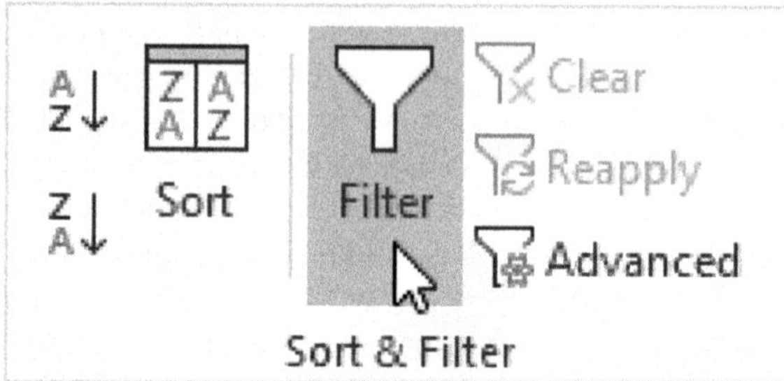

Select whatever criteria you want to include. Click **OK**. You can also select **all** as an option.

If you had only selected **satisfied** the following would be displayed

A	B	C	D
Participan	Emotio	Service	Produc
2	happy	good	satisfied
1	sad	great	satisfied
4	happy	great	satisifed

You can click on each drop down arrow to select your filter criteria on each column
To remove the filter go to the **Data** tab and look in the **Sort & Filter** group

- Click **Filter Clear**

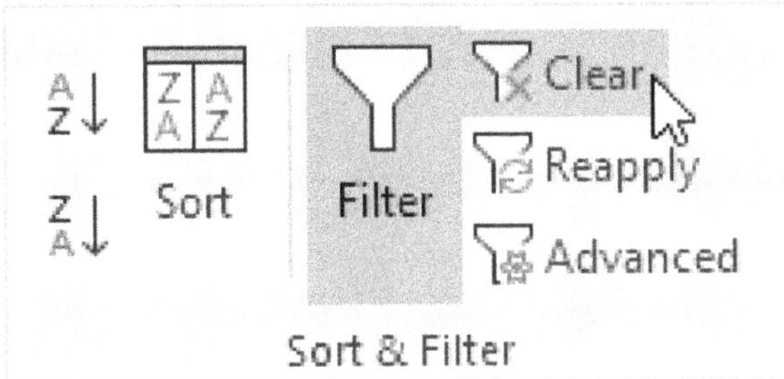

There's a quicker way to filter Excel data. Select a cell

A	B	C	D
Participant	**Emotion**	**Service**	**Product**
1	sad	great	satisfied
2	happy	good	satisfied
3	angry	bad	dissatified
4	happy	great	satisifed

Right click, and then click

- **Filter**

- **Filter by Selected Cell's Value**

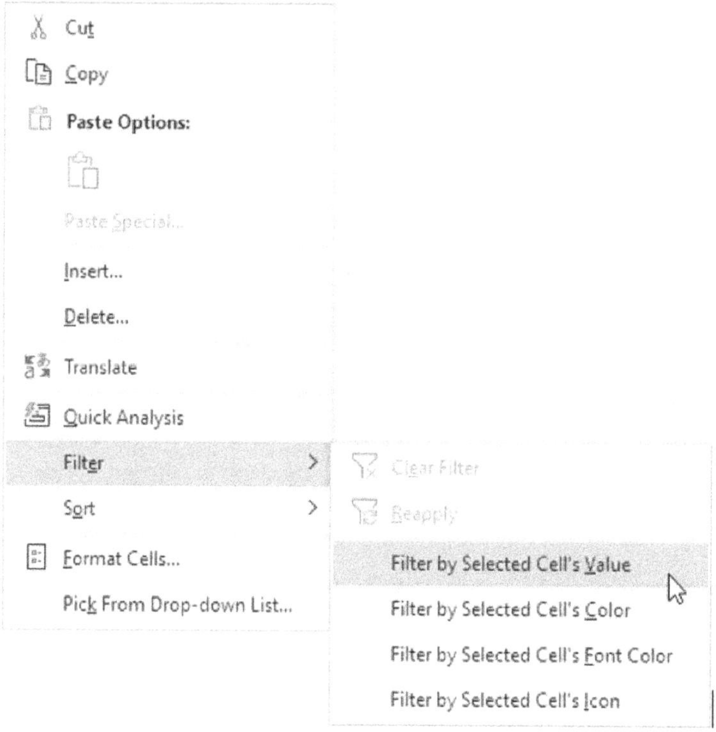

Select another cell in another column to further filter this dataset

You can also filter by specifying conditions. This will allow you to create custom filters that narrow down the data in the exact way that you want. You do this by building a filter. If you've ever queried data in a database this will look familiar to you

- Point to either **Number Filters** or **Text Filters** in the list. A menu appears that allows you to filter on various conditions
- Choose a condition and then select or enter criteria. Click the **And** button to combine criteria (two or more criteria that must both be met) and the **Or** button to require only one of multiple conditions to be met
- Click **OK** to apply the filter and get the results you expect

If you have a large amount of qualitative data or multiple sources you can create multiple worksheets/tabs within the Excel file. Each worksheet can be dedicated to a specific research participant, interview, or source. This helps keep your data organized and easily accessible.

Using Excel for qualitative coding offers several advantages. Excel offers a straightforward and customizable interface for coding. Analysts have the flexibility to create their coding categories and labels thus allowing for a tailored coding framework that aligns with their research objectives. The ability to adjust column widths, wrap text, and use various formatting options ensures that the data remains organized/readable.

Despite its advantages Excel has certain limitations when it comes to qualitative coding. One limitation is the lack of advance features specifically designed for qualitative analysis. Excel does not provide automated coding assistance (such as text search or code suggestion tools) unlike dedicated qualitative analysis software. Analysts must manually enter codes and rely on their own expertise/knowledge of the data. Another limitation is the potential for limited data management capabilities. Excel may not be the optimal choice for large-scale qualitative studies with extensive datasets. Excel's performance may be affected as the data volume increases and it may become more challenging to manage/navigate through the data efficiently. Furthermore, Excel may not be well-suited for complex coding hierarchies or visual representations of themes.

Once the coding is complete, analysts move on to the process of **theme** development. Themes are higher level patterns or concepts that emerge from the coded data. Analysts analyze the coded segments (looking for similarities and connections) and group them into coherent themes. These themes provide a structured framework for interpreting and understanding the data (Fig. 17.1).

Analysts refine themes through an iterative process after identifying them. This involves reviewing the themes in relation to the original data while considering their internal coherence and external relevance. Analysts may modify, combine, or split themes based on their analysis, ensuring that the themes accurately capture the complexity and richness of the data. You can now see the common themes in your feedback data based on the number of times a particular code occurs.

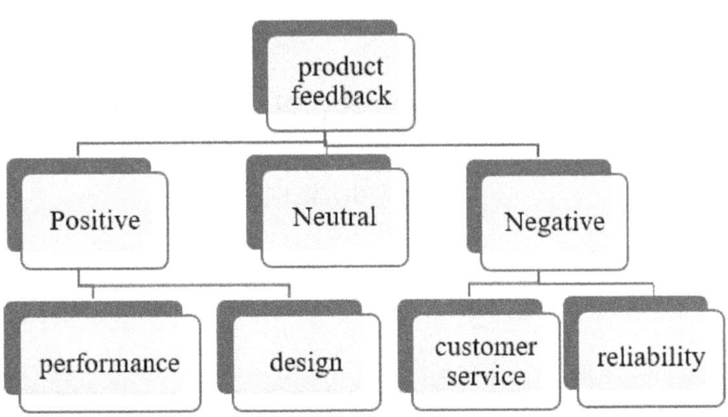

Fig. 17.1 Coding and themes

Data Analysis and Visualization

The task of uncovering insights involves reviewing the codes that emerge from the data and drawing meaningful correlations from them. It is also about making sure each insight is distinct and has enough data to support it. If you find that your codes are too broad to be able to extract meaningful insights you will need to create sub-codes to your primary codes. This process is sometimes referred to as meta coding. If you take an inductive coding approach you can create sub-codes as you are reading through your feedback data. You need to carefully read your qualitative data to create quality sub-codes. By calculating the frequency of these sub-codes you can get insights into which ones are the most common. Although time-consuming, this exercise will improve the quality of your analysis (Fig. 17.2).

Excel allows you to perform basic data analysis on your qualitative data. You can calculate summary statistics such as counts, averages, or percentages using functions like **Count**, **Average**, or **Sum**. These calculations can provide insights into the distribution/prevalence of different themes or categories within your data. You can also use the **Concatenate** function to combine data from different cells or the **Countif** function to count the occurrence of specific themes/categories.

Visual representations can help in identifying trends or making comparisons across different categories. Excel offers various chart types that can visually represent your qualitative data (for more details on these charts, see Chaps. 2 and 16). While qualitative data may not have the same numerical properties as quantitative data there are several visualization techniques that can be employed to enhance the analysis/interpretation of qualitative data in Excel. By leveraging visual representations such as word clouds, bar charts, pie charts, heat maps, network diagrams, concept maps, timeline visualizations, and scatterplots analysts can gain valuable insights, identify patterns, and communicate their findings effectively. The choice of visualization technique depends on the nature of the qualitative data, research objectives, and the specific insights analysts seek to extract from the data.

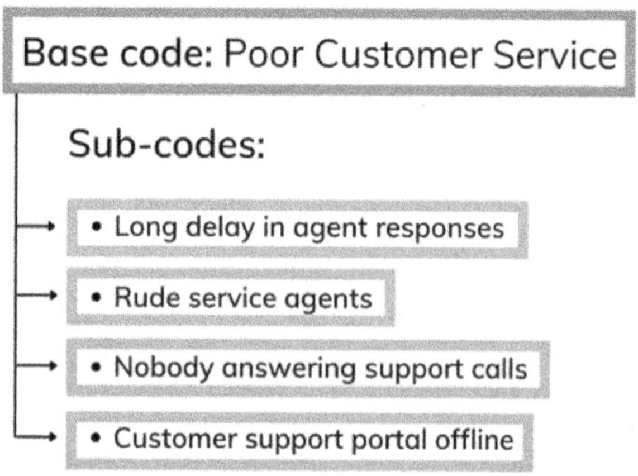

Fig. 17.2 Sub-codes

- Bar Charts, Pie Charts, Pareto Charts: Excel's charting capabilities include bar charts (Fig. 17.3), pie charts (Fig. 17.4), and Pareto charts (Fig. 17.5) which can be employed to visualize qualitative data. Bar charts can be utilized to represent the frequency or distribution of different categories/themes within qualitative data. Each category or code is represented by a bar while the height or length of the bar indicates its frequency or occurrence. Bar charts are useful for comparing the frequency of different categories/themes and identifying patterns/variations in the data. For a review of how to create these charts see Chaps. 2 and 16.
  A special type of bar graph is a **Pareto chart** which is just a bar chart with the bars sorted with the highest frequencies on the left (Fig. 17.5). The advantage of Pareto charts is that you can visually see the most popular answer to the least popular. This is especially useful in business applications where you want to know what services your customers like the most, what processes result in more injuries, which issues employees find more important, and other type of questions like these.
- Word Clouds: While Excel does not have a built-in word cloud feature, you can create word clouds using additional add-ins or external tools that generate word clouds from Excel data such as WordItOut.com or WordArt.com. By extracting the text data from Excel and applying word cloud generators, you can create visual representations where the size/color of each word reflects its frequency/significance within the data. Word clouds are useful for quickly identifying key terms, themes, or topics present in qualitative data (Fig. 17.6).
- Network Diagrams: Network diagrams are valuable for visualizing connections or relationships between different entities in qualitative data. In network dia-

Fig. 17.3 Bar graph

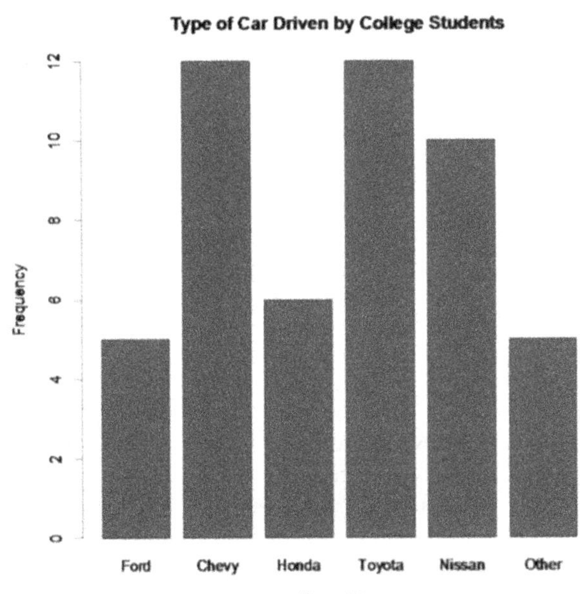

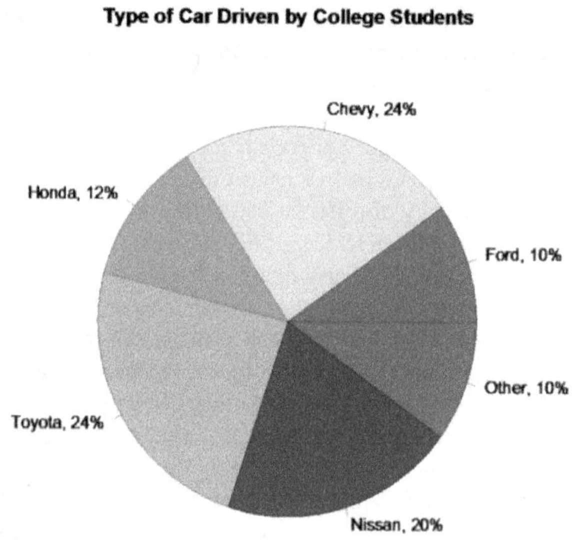

Fig. 17.4 Pie chart

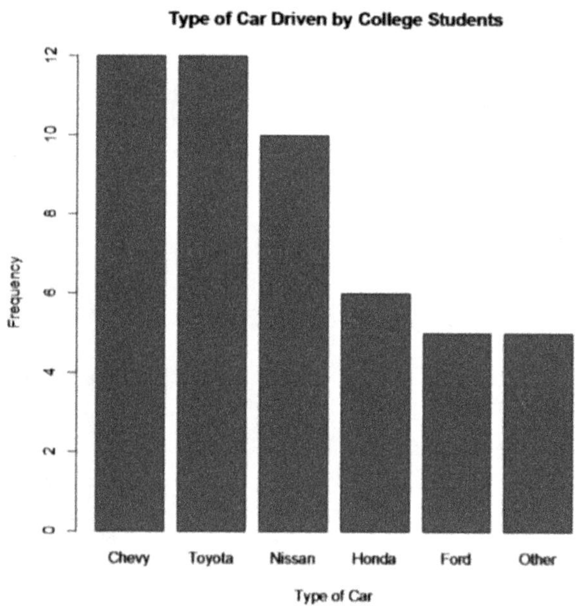

Fig. 17.5 Pareto chart

grams, nodes represent entities (e.g., people, concepts, themes) and edges/lines indicate the connections or relationships between them. Network diagrams are particularly useful when analyzing qualitative data with complex interdependencies or when exploring social networks, communication patterns, and conceptual frameworks. You can import the Excel data into Word and create your own network model or import the Excel data into a more sophisticated software such as

Fig. 17.6 Word cloud

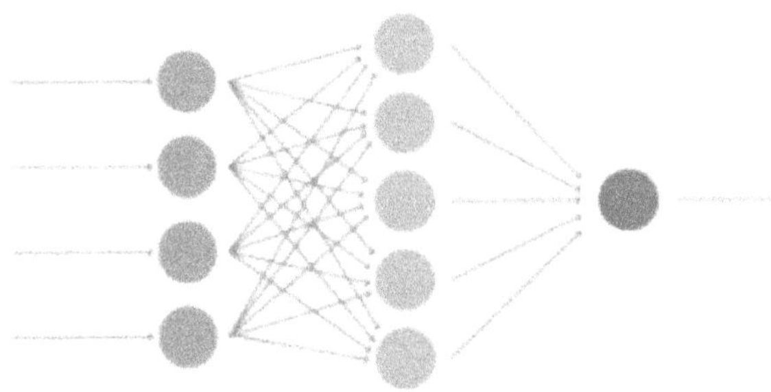

Fig. 17.7 Network diagram

MAXQDA. Concept maps help analysts understand the structure, hierarchy, and interrelationships between different themes/concepts in their data (Fig. 17.7).

- Scatter Plots: Scatter plots can be adapted to represent qualitative data in Excel. Analysts can assign categorical variables to the X/Y axes and plot data points accordingly. Scatter plots can be useful for exploring relationships, clusters, or patterns within qualitative data. For example, analysts can plot the frequency of different codes/themes on the X and Y axes to identify any correlations/associations between them.
- Timeline Visualizations: Timeline visualizations are effective for representing the chronological order or temporal aspects of qualitative data. They can be used to visualize the progression of events, changes over time, or the development of themes/concepts. Timeline visualizations help analysts identify patterns, trends, or critical moments within their qualitative data.
- Sparklines: Sparklines are small data-rich visualizations that can be embedded within a cell in Excel. Analysts can use sparklines to display trends/patterns in qualitative data over time or across categories. Sparklines can be particularly useful when analyzing temporal aspects such as tracking the frequency, occurrence of codes, and themes across different time periods.
- Conditional Formatting: Excel's conditional formatting feature allows analysts to apply formatting rules to cells based on specific criteria/values. Analysts can use conditional formatting to visually highlight specific codes or themes within their qualitative data. For example, they can apply different colors or font styles

to cells containing specific codes thus making them stand out for easier identification and analysis.

- Data Bars and Icon Sets: Excel's data bars and icon sets are conditional formatting options that analysts can use to represent qualitative data visually. Data bars provide horizontal bars within cells with the length of the bar indicating the value/frequency of a code/theme. Icon sets allow analysts to assign different icons/symbols to different categories/levels of a variable. These visual representations facilitate quick comparisons and assessments of qualitative data.
- Heat maps: A heat map is a visual representation of data where values are depicted using different colors to indicate the intensity/density of certain characteristics/variables. Heat maps are typically used with quantitative data, but they can also be applied to qualitative data to provide a visual overview of patterns/trends. Heat maps can be employed to illustrate the co-occurrence or association between codes/themes in qualitative data analysis. Each cell in the heat map represents the frequency/strength of the relationship between two variables and the color intensity reflects the magnitude of the relationship. Heat maps can help summarize/highlight key themes or frequencies within the data. The example data was collected from 126 participants who were asked to describe usability issues on an airline website of their choice. The airlines were defined as "amazing", "better", or "crummy" from a prior survey. Input the data and code it into different categories/themes. This involves identifying recurring ideas, concepts, or topics that emerge from the data. Each theme represents a specific aspect of the qualitative data that you want to visualize using the heat map (Fig. 17.8).
- Assign a frequency or intensity value to each category based on how frequently/prominently it appears in the data. This could be the number of occurrences, the percentage of responses, or any other relevant measure that captures the importance of the category. You can use the **frequency** function in Excel.

			Themes (usability issues encountered)				
UZ ID	**Site**	**User Comment**	**Technical issues**	**Could not enter details in to the booking panel**	**Unable 'sort' the flights by departure time**	**Difficulty locating the search function**	**Could not enter details in to the forms**
UZ ID1	Amazing airlines	page froze	1				
UZ ID2	Amazing airlines	site froze, then could not enter details on first page.	1	1			
UZ ID3	Amazing airlines	Couldn't find search bar				1	
UZ ID4	Better airlines	where was search???				1	
UZ ID5	Better airlines	page went grey and stopped working	1				
UZ ID6	Better airlines	site crashed	1				
UZ ID7	Crummy airlines	sruggled to select flight details in panel		1			
UZ ID8	Crummy airlines	Could not enter personal details in to the passenger info page					1
UZ ID9	Crummy airlines	Could not sort flight search results by time			1		

Fig. 17.8 Input data for heat map

- 21% experienced technical issues
- 15% struggled to enter details into the booking panel
- 10% struggled to locate the search functionality
- 9% could not work out how to proceed to the next page
- 8% were unable to sort the flights by departure time
- 8% struggled to understand the content on pages
- 6% were unable to enter their personal details into the forms
- 2% were unable to navigate back to previous steps

Assign colors to the categories/themes based on the assigned frequency/intensity values. Usually a color gradient or scale is used with lighter/cooler colors indicating lower values and darker/warmer colors representing higher values. This color scheme helps visually represent the distribution/intensity of the qualitative data across the identified categories.

Plot the heat map typically in a tabular format or grid-like structure. The rows of the table represent the categories/themes and the columns represent the different data sources, time periods, or any other relevant dimensions. Each cell in the table is filled with the appropriate color based on the assigned frequency/intensity value (Fig. 17.9).

Analyze the heat map to gain insights into the qualitative data. Identify the categories/themes that appear more frequently or with higher intensity as they represent the areas of greater importance/focus within the data. Look for patterns, trends, and relationships between different categories that can provide valuable information or support decision-making processes. Heat maps can be particularly useful when working with large qualitative datasets or when you want to summarize/visualize the overall distribution and emphasis of certain themes. However, it's important to note that heat maps should be used as a complementary tool alongside more in depth qualitative analysis methods (such as thematic coding or narrative interpretation) to ensure a comprehensive understanding of the data.

	Amazing airlines	Better airlines	Crummy airlines	Disastrous airlines
Task Success	90%	72%	80%	45%
Ease of use	6.1	5.2	5.8	4.5
Time on task	190	220	200	188
Ease of Navigation	6.3	6.0	5.5	4.5
Search Functionality	6.1	5.8	5.8	2.0
Design and Imagery	6.5	5.6	6.0	4.8

Fig. 17.9 Heat map

- You can also use Excel's **Pivot Table** feature to summarize/visualize the data. See Chap. 2 for a review on using this tool. Pivot tables allow for aggregating, grouping, and summarizing the coded data based on themes or other variables. This provides a structured/organized view of the data thus enabling analysts to identify trends, patterns, and relationships between different themes. Pivot tables allow analysts to examine the frequency/distribution of codes, identify patterns, and compare different categories. Analysts can dynamically rearrange/explore their data thus gaining insights and summary statistics by dragging/dropping fields.

While manual analysis is possible in Excel, when one uses dedicated qualitative software options like NVivo, Atlas.ti, MAXQDA, Dedoose, QDA Miner, or MAXMaps one can greatly enhance the efficiency/effectiveness of the analysis process. Analysts can choose the software that best fits their specific needs and research objectives thus allowing them a more systematic/comprehensive exploration of their qualitative data. You can easily import your Excel data into these software offerings.

Reporting

At this point the codes are fully developed and the focus is on communicating the narrative to your audience. A coherent outline of the qualitative research, the findings, and the insights is vital for stakeholders to discuss/debate before they can devise a meaningful course of action.

When reporting qualitative data in a business report it is important to present the information in a clear/concise manner that allows the reader to understand the findings and draw meaningful insights.

Begin the report with a brief introduction that sets the context for the qualitative data. Provide a background on the research objectives, methodology, and the purpose of the qualitative data in the report. Suggestions for effective qualitative reports include:

1. Organize the data: Group similar themes/categories together to make it easier for the reader to understand the patterns or trends in the qualitative data. This can be done through careful analysis/coding of the data.
2. Summarize findings: Provide a summary of the key findings from the qualitative data analysis. Highlight the main themes/patterns that emerged and present them in a logical/coherent manner. Use descriptive language to convey the essence of the findings.
3. Use quotes or excerpts: Include specific quotes/excerpts from the qualitative data to support and illustrate the findings. These quotes should be relevant, representative, and provide insights into the thoughts, opinions, or experiences of the participants.
4. Provide context: Offer background information about the participants or the research setting to help the reader understand the qualitative data better. This

could include demographic details, participant characteristics, or relevant contextual information.

5. Use visuals sparingly: While the focus of reporting qualitative data is primarily on text there may be instances where visual aids such as charts, diagrams, and tables can enhance understanding. Use visuals sparingly and only when they add value to the interpretation of the qualitative findings.
6. Interpretation and implications: Provide an interpretation of the qualitative findings and discuss their implications for the business. Explain how the findings align with or challenge existing theories/concepts. Furthermore, explain how the findings offer insights into potential opportunities, challenges, or recommendations for future actions.
7. Provide transparency: Ensure transparency in reporting by being clear about the limitations/biases associated with qualitative data. Acknowledge any potential biases in data collection/analysis and discuss how they may have influenced the findings.
8. Maintain confidentiality and anonymity: If the qualitative data includes sensitive or confidential information emphasize the need for maintaining confidentiality/anonymity. Use pseudonyms or participant codes to protect the identities of the participants.
9. Conclusion: Summarize the main points of the qualitative findings and their implications in the conclusion section. Emphasize the importance of the qualitative data in informing decision making and future strategies.

Remember that the exact structure/format of reporting qualitative data may vary depending on the specific requirements of the business report and the intended audience.

Pitfalls: Qualitative Data

Working with qualitative data requires careful attention/consideration to ensure accurate and meaningful analysis. However, there are several pitfalls that analysts should be aware of to avoid potential pitfalls and biases. Here are some key pitfalls in working with qualitative data:

1. Bias and Subjectivity: Analysts bring their own biases, perspectives, and interpretations to the data analysis process. It is crucial to be aware of personal biases and strive for objectivity during coding, analysis, and interpretation of qualitative data. Techniques such as member checking and peer debriefing can help mitigate the impact of bias.
2. Insufficient Sampling: Qualitative data analysis often involves a smaller sample size compared to quantitative research. It is important to select a representative sample that adequately captures the diversity of experiences and perspectives related to the research question. Failure to do so can lead to limited generalizability or validity of findings.

3. Lack of Transparency: Maintaining transparency throughout the research process is essential. It is crucial to document/report the steps taken during data collection, analysis, and interpretation. Providing clear/detailed descriptions of coding decisions, themes, and interpretations enhances the credibility/reproducibility of the research.
4. Inadequate Data Organization: Poor data organization can hinder the analysis process. It is important to develop a systematic approach for organizing/managing qualitative data including transcripts, field notes, and documents. Failure to effectively organize data can result in difficulties in retrieving and analyzing relevant information.
5. Overlooking Negative Cases: Analysts may unintentionally focus on confirming their initial hypotheses or preconceived notions leading to a bias towards positive cases that support their expectations. It is important to actively seek/analyze negative/contradictory cases to ensure a comprehensive understanding of the data and avoid confirmation bias.
6. Overreliance on Coding Software: While coding software can facilitate qualitative data analysis, it is essential to maintain a critical and active role in the analysis process. Overreliance on software without scrutiny/interpretation of the data can lead to the risk of overlooking important nuances, connections, or patterns.
7. Inadequate Data Interpretation: Qualitative data analysis involves thorough and complete interpretation of the data. It is important to approach data interpretation with caution and rigor. Analysts should provide rich contextual explanations, engage in reflexivity, and consider alternative interpretations to ensure the validity/reliability of findings. Inadequate data interpretation can lead to misleading conclusions.
8. Lack of Reflexivity: Reflexivity refers to reflecting on the analyst's influence and biases in the research process. Failing to critically examine one's own positionality and how it may shape the research can lead to the overlooking of alternative perspectives or the misinterpretation of data.
9. Inadequate Participant Voice: Qualitative research aims to capture the perspectives and experiences of participants. Failing to adequately represent participant voices or not involving them in the analysis process can undermine the authenticity/credibility of the findings. Analysts should strive to include participant perspectives through techniques such as member checking, involving participants in data analysis, and having participants involved with data interpretation.
10. Neglecting Ethical Considerations: Ethical considerations are crucial in qualitative research. Analysts must ensure participant confidentiality, informed consent, and adhere to ethical guidelines throughout the research process. Neglecting ethical considerations can compromise the trust and integrity of the research.

By being aware of these pitfalls and taking proactive measures to address them you can enhance the quality/validity of your qualitative data analysis/interpretation.

Final Thoughts and Activities

Practice Problems

1. An MBA student is tasked with analyzing the market entry strategy for a multinational company planning to expand into a new region. The student needs to consider various qualitative factors such as cultural differences, regulatory environment, competitive landscape, and consumer behavior to determine the most suitable market entry approach.
2. You are tasked with evaluating the brand perception of a company in the wake of a public relations crisis. The student needs to assess qualitative data such as customer sentiment, media coverage, and social media discussions to understand how the crisis has impacted the brand's reputation and customer loyalty.
3. Why are these focus group questions not useful? Explain in detail.
   - "Do you like our product?"
   - "What do you think about our company?"
   - "Would you recommend our services to your friends?"

Problem Solutions

1. The student should conduct in-depth qualitative research to understand the cultural nuances, legal frameworks, and market dynamics of the target region. This may involve conducting interviews with local experts, analyzing government policies, and studying competitor strategies. Based on this analysis the student can recommend an appropriate market entry strategy whether it be through joint ventures, acquisitions, partnerships or organic growth. The aforementioned organic growth can be achieved by aligning with the company's objectives and minimizing risks associated with market entry.
2. You can employ qualitative research methods such as focus groups, interviews, and sentiment analysis of online conversations to gauge public perception of the brand. The student can identify key themes, sentiment trends, and areas of concern among stakeholders by analyzing qualitative data. Based on these insights, the student can develop strategies to rebuild trust, address negative perceptions, and enhance the brand's image through effective communication and targeted marketing campaigns.
3. Lack of Specificity:

   (a) "Do you like our product?" is overly broad and does not provide specific criteria for respondents to evaluate. It does not prompt participants to provide detailed feedback or insights into what aspects of the product they like or dislike. Therefore, the question may yield superficial responses that are not actionable for improving the product.

(b) Subjectivity and Vagueness: "What do you think about our company?" is vague and subjective. It does not specify which aspects of the company's performance/offerings the respondents should focus on. Participants may provide unfocused or generic opinions that are difficult to interpret and act upon without clear guidance.

(c) Leading Question: "Would you recommend our services to your friends?" is a leading question that assumes a positive response. It may bias participants towards giving favorable feedback even if they have reservations or criticisms about the services. This can result in skewed data that does not accurately reflect the participants' true opinions/experiences.

Discussion Boards

1. Qualitative data cannot be quality checked and therefore is not that useful. Do you agree or disagree? Provide examples to support your position.
2. Working with human subjects involves a degree of uncertainty especially in measuring attitudes. How can you collect this data and minimize uncertainty in it?
3. Ethical protocols involved with how data can be collected involving human subjects are critical. But not all organizations and countries follow these protocols. Discuss with examples.

Group Activity

1. Create a word cloud about government communication with pandemics. Have each group member interview ten people about the top three characteristics of effective communication on this subject. Compile the data and create one word cloud. What were the findings?
2. Research how the *Qualitative Report* effectively shares qualitative research findings. Choose a paper and discuss why the findings were important.
3. Research the type of data and methods used to statistically analyze customer complaints about difficulty in using cellphone technology for taking photos.

Parting Thought

Surveys show that surveys never lie…

Appendix I: Z-Table/Standard Normal Distribution Table

Standard normal distribution: Table values represent AREA to the LEFT of the *Z* score

Z	.00	.01	.02	.03	.04	.05	.06	.07	.08	.09
−3.9	.00005	.00005	.00004	.00004	.00004	.00004	.00004	.00004	.00003	00003
−3.8	00007	.00007	.00007	.00006	.00006	.00006	.00006	.00005	.00005	00005
−3.7	.00011	.00010	.00010	.00010	.00009	.00009	.00008	.00008	.00008	00008
−3.6	.00016	.00015	.00015	.00014	.00014	.00013	.00013	.00012	.00012	.00011
−3.5	.00023	.00022	.00022	.00021	.00020	.00019	.00019	.00018	.00017	.00017
−3.4	.00034	.00032	.00031	.00030	.00029	.00028	.00027	.00026	.00025	00024
−3.3	.00048	.00047	.00045	.00043	.00042	.00040	.00039	.00038	.00036	.00035
−3.2	.00069	.00066	.00064	.00062	.00060	.00058	.00056	.00054	.00052	.00050
−3.1	.00097	.00094	.00090	.00087	.00084	.00082	.00079	.00076	.00074	.00071
−3.0	00135	.00131	.00126	00122	.00118	.00114	00111	.00107	.00104	00100
−2.9	.00187	.00181	.00175	.00169	.00164	.00159	.00154	.00149	.00144	.00139
−2.8	.00256	.00248	.00240	.00233	.00226	.00219	.00212	.00205	.00199	.00193
−2.7	.00347	.00336	00326	.00317	.00307	.00298	.00289	.00280	.00272	.00264
−2.6	.00466	.00453	.00440	.00427	.00415	.00402	.00391	.00379	.00368	.00357
−2.5	00621	.00604	.00587	.00570	.00554	.00539	.00523	.00508	.00494	.00480
−2.4	.00820	.00798	00776	.00755	.00734	.00714	.00695	.00676	.00657	.00639
−2.3	.01072	.01044	01017	.00990	.00964	.00939	.00914	.00889	.00866	00842
−2.2	.01390	.01355	.01321	.01287	.01255	.01222	.01191	.01160	.01130	.01101
−2.1	.01786	.01743	.01700	.01659	.01618	.01578	.01539	.01500	.01463	.01426
−2.0	02275	.02222	.02169	.02118	.02068	.02018	.01970	.01923	.01876	01831
−1.9	.02872	.02807	.02743	.02680	.02619	.02559	.02500	.02442	.02385	02330
−1.8	.03593	.03515	.03438	.03362	.03288	.03216	.03144	.03074	.03005	.02938

(continued)

L. Herkenhoff, J. Fogli (eds.), *Applied Statistics for Business and Management using Microsoft Excel*, https://doi.org/10.1007/978-3-031-46371-6

−1.7	.04457	.04363	.04272	.04182	.04093	.04006	.03920	.03836	.03754	.03673
−1.6	.05480	.05370	.05262	.05155	.05050	.04947	.04846	.04746	.04648	.04551
−1.5	06681	.06552	.06426	.06301	.06178	.06057	.05938	.05821	.05705	.05592
−1.4	.08076	.07927	.07780	.07636	.07493	.07353	.07215	.07078	.06944	.06811
−1.3	.09680	.09510	.09342	.09176	.09012	.08851	.08691	.08534	.08379	.08226
−1.2	11507	11314	.11123	.10935	.10749	.10565	10383	.10204	.10027	09853
−1.1	13567	.13350	.13136	.12924	.12714	.12507	12302	.12100	.11900	11702
−1.0	15866	.15625	15386	.15151	.14917	.14686	14457	.14231	.14007	.13786
−0.9	18406	.18141	17879	17619	.17361	.17106	16853	.16602	.16354	16109
−0.8	.21186	.20897	20611	.20327	.20045	.19766	19489	.19215	.18943	18673
−0.7	24196	.23885	23576	.23270	22965	.22663	.22363	.22065	21770	.21476
−0.6	.27425	.27093	.26763	.26435	.26109	.25785	.25463	.25143	.24825	.24510
−0.5	30854	.30503	.30153	.29806	.29460	29116	.28774	.28434	.28096	27760
−0.4	34458	34090	.33724	.33360	32997	32636	32276	31918	.31561	.31207
−0.3	38209	.37828	.37448	.37070	.36693	.36317	.35942	.35569	.35197	.34827
−0.2	.42074	.41683	41294	.40905	.40517	.40129	.39743	.39358	38974	.38591
−0.1	46017	.45620	.45224	.44828	.44433	.44038	.43644	.43251	.42858	42465
−0.0	50000	.49601	.49202	.48803	.48405	.48006	.47608	.47210	.46812	.46414
0.0	.50000	.50399	.50798	.51197	.51595	.51994	.52392	.52790	.53188	.53586
0.1	53983	.54380	.54776	.55172	.55567	.55962	.56356	.56749	.57142	.57535
0.2	57926	.58317	.58706	.59095	.59483	.59871	.60257	.60642	61026	.61409
0.3	61791	.62172	.62552	62930	.63307	.63683	.64058	.64431	64803	.65173
0.4	65542	.65910	.66276	.66640	.67003	.67364	67724	.68082	.68439	.68793
0.5	69146	.69497	.69847	.70194	70540	70884	.71226	.71566	.71904	.72240
0.6	72575	.72907	.73237	.73565	73891	74215	.74537	.74857	.75175	.75490
0.7	75804	.76115	.76424	.76730	.77035	.77337	.77637	.77935	.78230	.78524
0.8	78814	.79103	.79389	.79673	.79955	80234	80511	.80785	.81057	.81327
0.9	.81594	.81859	.82121	82381	82639	.82894	83147	.83398	.83646	.83891
1.0	84134	84375	.84614	.84849	85083	85314	.85543	.85769	.85993	.86214
1.1	86433	.86650	.86864	87076	.87286	.87493	87698	.87900	88100	.88298
1.2	88493	.88686	.88877	89065	.89251	89435	89617	.89796	89973	.90147
1.3	90320	.90490	.90658	.90824	.90988	91149	.91309	.91466	91621	.91774
1.4	91924	.92073	.92220	92364	.92507	92647	92785	.92922	93056	.93189
1.5	93319	.93448	.93574	.93699	.93822	.93943	.94062	.94179	94295	.94408
1.6	94520	.94630	.94738	94845	94950	95053	.95154	.95254	95352	.95449
1.7	95543	.95637	.95728	.95818	95907	.95994	.96080	.96164	.96246	.96327
1.8	96407	.96485	.96562	.96638	.96712	.96784	.96856	.96926	.96995	.97062
1.9	97128	.97193	.97257	97320	97381	97441	.97500	.97558	97615	.97670
2.0	97725	.97778	.97831	97882	.97932	.97982	.98030	.98077	98124	.98169
2.1	98214	.98257	98300	.98341	.98382	.98422	.98461	.98500	.98537	.98574
2.2	.98610	.98645	.98679	.98713	.98745	98778	.98809	.98840	.98870	.98899
2.3	98928	.98956	.98983	99010	.99036	99061	.99086	.99111	.99134	.99158
2.4	99180	.99202	99224	99245	99266	99286	99305	.99324	.99343	.99361

2.5	.99379	.99396	.99413	.99430	.99446	.99461	.99477	.99492	.99506	.99520
2.6	99534	.99547	.99560	99573	.99585	99598	.99609	.99621	.99632	.99643
2.7	99653	.99664	.99674	.99683	.99693	.99702	.99711	.99720	.99728	.99736
2.8	99744	.99752	.99760	.99767	.99774	.99781	.99788	.99795	.99801	.99807
2.9	99813	.99819	.99825	99831	.99836	.99841	99846	99851	99856	.99861
3.0	.99865	.99869	.99874	.99878	.99882	.99886	.99889	.99893	.99896	.99900
3.1	99903	.99906	.99910	99913	.99916	.99918	.99921	.99924	.99926	.99929
3.2	.99931	.99934	.99936	99938	.99940	.99942	99944	.99946	99948	.99950
3.3	.99952	.99953	.99955	99957	.99958	.99960	99961	.99962	99964	.99965
3.4	99966	.99968	.99969	99970	.99971	.99972	99973	.99974	99975	.99976
3.5	.99977	.99978	.99978	.99979	.99980	.99981	.99981	.99982	.99983	.99983
3.6	99984	.99985	.99985	99986	.99986	99987	.99987	.99988	99988	.99989
3.7	99989	.99990	.99990	99990	99991	.99991	99992	.99992	99992	.99992
3.8	.99993	.99993	.99993	.99994	.99994	.99994	99994	.99995	99995	.99995
3.9	.99995	.99995	.99996	.99996	.99996	.99996	.99996	.99996	.99997	.99997

Appendix II: Critical Values of *F* Table

L. Herkenhoff, J. Fogli (eds.), *Applied Statistics for Business and Management using Microsoft Excel*, https://doi.org/10.1007/978-3-031-46371-6

Table entries are critical values for F^* with probably p in right tail of the distribution

			Degrees of freedom in numerator (df1)										
		p	1	2	3	4	5	6	7	8	12	24	1000
Degrees of freedom in numerator (df2)	**1**	0.100	39.86	49.50	53.59	55.83	57.24	58.20	58.91	59.44	60.71	62.00	63.30
		0.050	161.4	199.5	215.7	224.6	230.2	234.0	236.8	238.9	243.9	249.1	254.2
		0.025	647.8	799.5	864.2	899.6	921.8	937.1	948.2	956.6	976.7	997.3	1017.8
		0.010	4052	4999	5404	5624	5764	5859	5928	5981	6107	6234	6363
		0.001	405312	499725	540257	562668	576496	586033	593185	597954	610352	623703	636101
	2	0.100	8.53	9.00	9.16	9.24	9.29	9.33	9.35	9.37	9.41	9.45	9.49
		0.050	18.51	19.00	19.16	19.25	19.30	19.33	19.35	19.37	19.41	19.45	19.49
		0.025	38.51	39.00	39.17	39.25	39.30	39.33	39.36	39.37	39.41	39.46	39.50
		0.010	98.50	99.00	99.16	99.25	99.30	99.33	99.36	99.38	99.42	99.46	99.50
		0.001	998.38	998.84	999.31	999.31	999.31	999.31	999.31	999.31	999.31	999.31	999.31
	3	0.100	5.54	5.46	5.39	5.34	5.31	5.28	5.27	5.25	5.22	5.18	5.13
		0.050	10.13	9.55	9.28	9.12	9.01	8.94	8.89	8.85	8.74	8.64	8.53
		0.025	17.44	16.04	15.44	15.10	14.88	14.73	14.62	14.54	14.34	14.12	13.91
		0.010	34.12	30.82	29.46	28.71	28.24	27.91	27.67	27.49	27.05	26.60	26.14
		0.001	167.06	148.49	141.10	137.08	134.58	132.83	131.61	130.62	128.32	125.93	123.52
	4	0.100	4.54	4.32	4.19	4.11	4.05	4.01	3.98	3.95	3.90	3.83	3.76
		0.050	7.71	6.94	6.59	6.39	6.26	6.16	6.09	6.04	5.91	5.77	5.63
		0.025	12.22	10.65	9.98	9.60	9.36	9.20	9.07	8.98	8.75	8.51	8.26
		0.010	21.20	18.00	16.69	15.98	15.52	15.21	14.98	14.80	14.37	13.93	13.47
		0.001	74.13	61.25	56.17	53.43	51.72	50.52	49.65	49.00	47.41	45.77	44.09

		Degrees of freedom in numerator (df1)										
	p	1	2	3	4	5	6	7	8	12	24	1000
5	0.100	4.06	3.78	3.62	3.52	3.45	3.40	3.37	3.34	3.27	3.19	3.11
	0.050	6.61	5.79	5.41	5.19	5.05	4.95	4.88	4.82	4.68	4.53	4.37
	0.025	10.01	8.43	7.76	7.39	7.15	6.98	6.85	6.76	6.52	6.28	6.02
	0.010	16.26	13.27	12.06	11.39	10.97	10.67	10.46	10.29	9.89	9.47	9.03
	0.001	47.18	37.12	33.20	31.08	29.75	28.83	28.17	27.65	26.42	25.13	23.82
6	0.100	3.78	3.46	3.29	3.18	3.11	3.05	3.01	2.98	2.90	2.82	2.72
	0.050	5.99	5.14	4.76	4.53	4.39	4.28	4.21	4.15	4.00	3.84	3.67
	0.025	8.81	7.26	6.60	6.23	5.99	5.82	5.70	5.60	5.37	5.12	4.86
	0.010	13.75	10.92	9.78	9.15	8.75	8.47	8.26	8.10	7.72	7.31	6.89
	0.001	35.51	27.00	23.71	21.92	20.80	20.03	19.46	19.03	17.99	16.90	15.77
7	0.100	3.59	3.26	3.07	2.96	2.88	2.83	2.78	2.75	2.67	2.58	2.47
	0.050	5.59	4.74	4.35	4.12	3.97	3.87	3.79	3.73	3.57	3.41	3.23
	0.025	8.07	6.54	5.89	5.52	5.29	5.12	4.99	4.90	4.67	4.41	4.15
	0.010	12.25	9.55	8.45	7.85	7.46	7.19	6.99	6.84	6.47	6.07	5.66
	0.001	29.25	21.69	18.77	17.20	16.21	15.52	15.02	14.63	13.71	12.73	11.72
8	0.100	3.46	3.11	2.92	2.81	2.73	2.67	2.62	2.59	2.50	2.40	2.30
	0.050	5.32	4.46	4.07	3.84	3.69	3.58	3.50	3.44	3.28	3.12	2.93
	0.025	7.57	6.06	5.42	5.05	4.82	4.65	4.53	4.43	4.20	3.95	3.68
	0.010	11.26	8.65	7.59	7.01	6.63	6.37	6.18	6.03	5.67	5.28	4.87
	0.001	25.41	18.49	15.83	14.39	13.48	12.86	12.40	12.05	11.19	10.30	9.36

(continued)

		Degrees of freedom in numerator (df1)										
	p	1	2	3	4	5	6	7	8	12	24	1000
9	0.100	3.36	3.01	2.81	2.69	2.61	2.55	2.51	2.47	2.38	2.28	2.16
	0.050	5.12	4.26	3.86	3.63	3.48	3.37	3.29	3.23	3.07	2.90	2.71
	0.025	7.21	5.71	5.08	4.72	4.48	4.32	4.20	4.10	3.87	3.61	3.34
	0.010	10.56	8.02	6.99	6.42	6.06	5.80	5.61	5.47	5.11	4.73	4.32
	0.001	22.86	16.39	13.90	12.56	11.71	11.13	10.70	10.37	9.57	8.72	7.84
10	0.100	3.29	2.92	2.73	2.61	2.52	2.46	2.41	2.38	2.28	2.18	2.06
	0.050	4.96	4.10	3.71	3.48	3.33	3.22	3.14	3.07	2.91	2.74	2.54
	0.025	6.94	5.46	4.83	4.47	4.24	4.07	3.95	3.85	3.62	3.37	3.09
	0.010	10.04	7.56	6.55	5.99	5.64	5.39	5.20	5.06	4.71	4.33	3.92
	0.001	21.04	14.90	12.55	11.28	10.48	9.93	9.52	9.20	8.45	7.64	6.78
12	0.100	3.18	2.81	2.61	2.48	2.39	2.33	2.28	2.24	2.15	2.04	1.91
	0.050	4.75	3.89	3.49	3.26	3.11	3.00	2.91	2.85	2.69	2.51	2.30
	0.025	6.55	5.10	4.47	4.12	3.89	3.73	3.61	3.51	3.28	3.02	2.73
	0.010	9.33	6.93	5.95	5.41	5.06	4.82	4.64	4.50	4.16	3.78	3.37
	0.001	18.64	12.97	10.80	9.63	8.89	8.38	8.00	7.71	7.00	6.25	5.44
14	0.100	3.10	2.73	2.52	2.39	2.31	2.24	2.19	2.15	2.05	1.94	1.80
	0.050	4.60	3.74	3.34	3.11	2.96	2.85	2.76	2.70	2.53	2.35	2.14
	0.025	6.30	4.86	4.24	3.89	3.66	3.50	3.38	3.29	3.05	2.79	2.50
	0.010	8.86	6.51	5.56	5.04	4.69	4.46	4.28	4.14	3.80	3.43	3.02
	0.001	17.14	11.78	9.73	8.62	7.92	7.44	7.08	6.80	6.13	5.41	4.62

		Degrees of freedom in numerator (df1)										
	p	1	2	3	4	5	6	7	8	12	24	1000
16	0.100	3.05	2.67	2.46	2.33	2.24	2.18	2.13	2.09	1.99	1.87	1.72
	0.050	4.49	3.63	3.24	3.01	2.85	2.74	2.66	2.59	2.42	2.24	2.02
	0.025	6.12	4.69	4.08	3.73	3.50	3.34	3.22	3.12	2.89	2.63	2.32
	0.010	8.53	6.23	5.29	4.77	4.44	4.20	4.03	3.89	3.55	3.18	2.76
	0.001	16.12	10.97	9.01	7.94	7.27	6.80	6.46	6.20	5.55	4.85	4.08
18	0.100	3.01	2.62	2.42	2.29	2.20	2.13	2.08	2.04	1.93	1.81	1.66
	0.050	4.41	3.55	3.16	2.93	2.77	2.66	2.58	2.51	2.34	2.15	1.92
	0.025	5.98	4.56	3.95	3.61	3.38	3.22	3.10	3.01	2.77	2.50	2.20
	0.010	8.29	6.01	5.09	4.58	4.25	4.01	3.84	3.71	3.37	3.00	2.58
	0.001	15.38	10.39	8.49	7.46	6.81	6.35	6.02	5.76	5.13	4.45	3.69
20	0.100	2.97	2.59	2.38	2.25	2.16	2.09	2.04	2.00	1.89	1.77	1.61
	0.050	4.35	3.49	3.10	2.87	2.71	2.60	2.51	2.45	2.28	2.08	1.85
	0.025	5.87	4.46	3.86	3.51	3.29	3.13	3.01	2.91	2.68	2.41	2.09
	0.010	8.10	5.85	4.94	4.43	4.10	3.87	3.70	3.56	3.23	2.86	2.43
	0.001	14.82	9.95	8.10	7.10	6.46	6.02	5.69	5.44	4.82	4.15	3.40
30	0.100	2.88	2.49	2.28	2.14	2.05	1.98	1.93	1.88	1.77	1.64	1.46
	0.050	4.17	3.32	2.92	2.69	2.53	2.42	2.33	2.27	2.09	1.89	1.63
	0.025	5.57	4.18	3.59	3.25	3.03	2.87	2.75	2.65	2.41	2.14	1.80
	0.010	7.56	5.39	4.51	4.02	3.70	3.47	3.30	3.17	2.84	2.47	2.02
	0.001	13.29	8.77	7.05	6.12	5.53	5.12	4.82	4.58	4.00	3.36	2.61

(continued)

		Degrees of freedom in numerator (df1)										
	p	1	2	3	4	5	6	7	8	12	24	1000
50	0.100	2.81	2.41	2.20	2.06	1.97	1.90	1.84	1.80	1.68	1.54	1.33
	0.050	4.03	3.18	2.79	2.56	2.40	2.29	2.20	2.13	1.95	1.74	1.45
	0.025	5.34	3.97	3.39	3.05	2.83	2.67	2.55	2.46	2.22	1.93	1.56
	0.010	7.17	5.06	4.20	3.72	3.41	3.19	3.02	2.89	2.56	2.18	1.70
	0.001	12.22	7.96	6.34	5.46	4.90	4.51	4.22	4.00	3.44	2.82	2.05
100	0.100	2.76	2.36	2.14	2.00	1.91	1.83	1.78	1.73	1.61	1.46	1.22
	0.050	3.94	3.09	2.70	2.46	2.31	2.19	2.10	2.03	1.85	1.63	1.30
	0.025	5.18	3.83	3.25	2.92	2.70	2.54	2.42	2.32	2.08	1.78	1.36
	0.010	6.90	4.82	3.98	3.51	3.21	2.99	2.82	2.69	2.37	1.98	1.45
	0.001	11.50	7.41	5.86	5.02	4.48	4.11	3.83	3.61	3.07	2.46	1.64
1000	0.100	2.71	2.31	2.09	1.95	1.85	1.78	1.72	1.68	1.55	1.39	1.08
	0.050	3.85	3.00	2.61	2.38	2.22	2.11	2.02	1.95	1.76	1.53	1.11
	0.025	5.04	3.70	3.13	2.80	2.58	2.42	2.30	2.20	1.96	1.65	1.13
	0.010	6.66	4.63	3.80	3.34	3.04	2.82	2.66	2.53	2.20	1.81	1.16
	0.001	10.89	6.96	5.46	4.65	4.14	3.78	3.51	3.30	2.77	2.16	1.22

Appendix III: Critical Values of *t* Table

t Table

Cum. prob	$t_{.50}$	$t_{.75}$	$t_{.80}$	$t_{.85}$	$t_{.90}$	$t_{.95}$	$t_{.975}$	$t_{.99}$	$t_{.995}$	$t_{.999}$	$t_{.9995}$
One-tail	**0.50**	**0.25**	**0.20**	**0.15**	**0.10**	**0.05**	**0.025**	**0.01**	**0.005**	**0.001**	**0.0005**
Two-tails	**1.00**	**0.50**	**0.40**	**0.30**	**0.20**	**0.10**	**0.05**	**0.02**	**0.01**	**0.002**	**0.001**
df											
1	0.000	1.000	1.376	1.963	3.078	6.314	12.71	31.82	63.66	318.31	636.62
2	0.000	0.816	1.061	1.386	1.886	2.920	4.303	6.965	9.925	22.327	31.599
3	0.000	0.765	0.978	1.250	1.638	2.353	3.182	4.541	5.841	10.215	12.924
4	0.000	0.741	0.941	1.190	1.533	2.132	2.776	3.747	4.604	7.173	8.610
5	0.000	0.727	0.920	1.156	1.476	2.015	2.571	3.365	4.032	5.893	6.869
6	0.000	0.718	0.906	1.134	1.440	1.943	2.447	3.143	3.707	5.208	5.959
7	0.000	0.711	0.896	1.119	1.415	1.895	2.365	2.998	3.499	4.785	5.408
8	0.000	0.706	0.889	1.108	1.397	1.860	2.306	2.896	3.355	4.501	5.041
9	0.000	0.703	0.883	1.100	1.383	1.833	2.262	2.821	3.250	4.297	4.781
10	0.000	0.700	0.879	1.093	1.372	1.812	2.228	2.764	3.169	4.144	4.587
11	0.000	0.697	0.876	1.088	1.363	1.796	2.201	2.718	3.106	4.025	4.437
12	0.000	0.695	0.873	1.083	1.356	1.782	2.179	2.681	3.055	3.930	4.318
13	0.000	0.694	0.870	1.079	1.350	1.771	2.160	2.650	3.012	3.852	4.221
14	0.000	0.692	0.868	1.076	1.345	1.761	2.145	2.624	2.977	3.787	4.140
15	0.000	0.691	0.866	1.074	1.341	1.753	2.131	2.602	2.947	3.733	4.073
18	0.000	0.690	0.865	1.071	1.337	1.746	2.120	2.583	2.921	3.686	4.015
17	0.000	0.689	0.863	1.069	1.333	1.740	2.110	2.567	2.898	3.646	3.965
18	0.000	0.688	0.862	1.087	1.330	1.734	2.101	2.552	2.878	3.610	3.922
19	0.000	0.688	0.861	1.086	1.328	1.729	2.093	2.539	2.861	3.579	3.883

(continued)

L. Herkenhoff, J. Fogli (eds.), *Applied Statistics for Business and Management using Microsoft Excel*, https://doi.org/10.1007/978-3-031-46371-6

20	0.000	0.687	0.860	1.084	1.325	1.725	2.086	2.528	2.845	3.552	3.850
21	0.000	0.686	0.859	1.063	1.323	1.721	2.080	2.518	2.831	3.527	3.819
22	0.000	0.686	0.858	1.061	1.321	1.717	2.074	2.508	2.819	3.505	3.792
23	0.000	0.685	0.858	1.060	1.319	1.714	2.069	2.500	2.807	3.485	3.768
24	0.000	0.685	0.857	1.059	1.318	1.711	2.064	2.492	2.797	3.467	3.745
25	0.000	0.684	0.856	1.058	1.316	1.708	2.060	2.485	2.787	3.450	3.725
26	0.000	0.684	0.856	1.058	1.315	1.706	2.056	2.479	2.779	3.435	3.707
27	0.000	0.684	0.855	1.057	1.314	1.703	2.052	2.473	2.771	3.421	3.690
28	0.000	0.683	0.855	1.056	1.313	1.701	2.048	2.467	2.763	3.408	3.674
29	0.000	0.683	0.854	1.055	1.311	1.699	2.045	2.462	2.756	3.396	3.659
30	0.000	0.683	0.854	1.055	1.310	1.697	2.042	2.457	2.750	3.385	3.646
40	0.000	0.681	0.851	1.050	1.303	1.684	2.021	2.423	2.704	3.307	3.551
60	0.000	0.679	0.848	1.045	1.296	1.671	2.000	2.390	2.660	3.232	3.460
80	0.000	0.678	0.846	1.043	1.292	1.684	1.990	2.374	2.639	3.195	3.416
100	0.000	0.677	0.845	1.042	1.290	1.680	1.984	2.364	2.626	3.174	3.390
1000	0.000	0.675	0.842	1.037	1.282	1.646	1.962	2.330	2.581	3.098	3.300
Z	0.000	0.674	0.842	1.038	1.282	1.645	1.960	2.326	2.576	3.090	3.291
	0%	50%	60%	70%	80%	90%	95%	98%	99%	99.8%	99.9%
	Confidence level										

Appendix IV: Critical Values of Chi-Square Table

Table C: Chi-Square distributions

Cum. probability	0.025	0.80	0.90	0.95	0.975	0.99	0.995	0.999	0.9995
Right tail	**0.975**	**0.2**	**0.1**	**0.05**	**0.025**	**0.01**	**0.005**	**0.001**	**0.0005**
df									
1	0.00098	1.64	2.71	3.84	5.02	6.63	7.88	10.83	12.12
2	0.051	3.22	4.61	5.99	7.38	9.21	10.60	13.82	15.20
3	0.216	4.64	6.25	7.81	9.35	11.34	12.84	16.27	17.73
4	0.48	5.99	7.78	9.49	11.14	13.28	14.86	18.47	20.00
5	0.83	7.29	9.24	11.07	12.83	15.09	16.75	20.51	22.11
6	1.24	8.56	10.84	12.59	14.45	16.81	18.55	22.46	24.10
7	1.69	9.80	12.02	14.07	16.01	18.48	20.28	24.32	26.02
8	2.18	11.03	13.36	15.51	17.53	20.09	21.95	26.12	27.87
9	2.70	12.24	14.68	16.92	19.02	21.67	23.59	27.88	29.67
10	3.25	13.44	15.89	18.31	20.48	23.21	25.19	29.59	31.42
11	3.82	14.63	17.28	19.68	21.92	24.73	26.76	31.26	33.14
12	4.40	15.81	18.55	21.03	23.34	26.22	28.30	32.91	34.82
13	5.01	16.98	19.81	22.36	24.74	27.69	29.82	34.53	36.48
14	5.63	18.15	21.06	23.68	26.12	29.14	31.32	36.12	38.11
15	6.26	19.31	22.31	25.00	27.49	30.58	32.80	37.70	39.72
16	6.91	20.47	23.54	26.30	28.85	32.00	34.27	39.25	41.31
17	7.56	21.61	24.77	27.59	30.19	33.41	35.72	40.79	42.88
18	8.23	22.76	25.99	28.87	31.53	34.81	37.16	42.31	44.43
19	8.91	23.90	27.20	30.14	32.85	36.19	38.58	43.82	45.97
20	9.59	25.04	28.41	31.41	34.17	37.57	40.00	45.31	47.50
21	10.28	26.17	29.62	32.67	35.48	38.93	41.40	46.80	49.01
22	10.98	27.30	30.81	33.92	36.78	40.29	42.80	48.27	50.51

(continued)

L. Herkenhoff, J. Fogli (eds.), *Applied Statistics for Business and Management using Microsoft Excel*, https://doi.org/10.1007/978-3-031-46371-6

23	11.69	28.43	32.01	35.17	38.08	41.64	44.18	49.73	52.00
24	12.40	29.55	33.20	36.42	39.36	42.88	45.56	51.18	53.48
25	13.12	30.68	34.38	37.65	40.65	44.31	46.93	52.62	54.95
30	16.79	36.25	40.26	43.77	46.98	50.89	53.67	59.70	62.16
40	24.43	47.27	51.81	55.76	59.34	63.09	66.77	73.40	76.10
50	32.36	58.16	63.17	67.50	71.42	76.15	79.49	86.66	89.56
60	40.48	88.97	74.40	79.08	83.30	88.38	91.95	99.61	102.7
80	57.15	80.41	98.58	101.9	106.6	112.3	116.3	124.8	128.3
100	74.22	111.7	118.5	124.3	129.6	135.8	140.2	149.4	153.2

Appendix V: Sample Insurance Phone Survey

Infinity Auto Insurance Phone Script

Interviewee Name_____________Location/City_______

Interviewer Name_____________Date______ /______/______

<u>STEP 1. INTRODUCTION</u>

Hello, am I speaking with_______? Hello, my name is_______and I am calling on behalf of an independent survey research company. We are a marketing research firm conducting random interviews with auto insurance consumers who, like you, are living in location_______. Are you able to take a moment (less than 5 minutes) to answer a short survey about your opinions of the car insurance industry? We are offering a $5.00 gas card if you complete today's interview. **IF NOT, ASK WHEN WOULD BE A GOOD TIME TO CALL BACK**. I promise we are not trying to sell you anything but are only interested in your opinions about auto insurance companies.

Your participation in this interview is completely voluntary and your answers and comments will be kept strictly confidential. We are very interested in your feedback so please be as honest as possible in your responses.

<u>STEP 2. BACKGROUND</u>

1. Do you own a car, truck, SUV, or van? **Yes or No? (IF NO to this question Skip to STEP 7, and conclude the survey)**
2. Do you currently own car insurance? Please be honest in your response for our research. **Yes or No?**

L. Herkenhoff, J. Fogli (eds.), *Applied Statistics for Business and Management using Microsoft Excel*, https://doi.org/10.1007/978-3-031-46371-6

3. If No to the question above ask: On a 5-point scale 1- Not at All Likely, 3- Somewhat Likely, and 5- Extremely Likely, How likely are you to buy car insurance during the next 12 months? **Not at all Likely, Somewhat Likely, or Extremely Likely – skip to Step 3**

   A. If somewhat likely to extremely likely, ask "Why are you [somewhat likely, likely, or extremely likely] to buy car insurance during the next 12 months? **(type response)**

4. If Yes to question 2, ask: "What company are you insured with?" **write name**
5. "On a 5-point scale 1- Not at All Likely, 3- Somewhat Likely, and 5- Extremely Likely, How likely are you to switch car insurance companies during the next 12 months?" **Not at all Likely, Somewhat Likely, or Extremely Likely**

   A. If somewhat likely to extremely likely, ask "Why are you [somewhat likely, likely, or extremely likely] to switch car insurance during the next 12 months? **(type response)**

STEP 3. COMPETITORS

1. What companies come to mind when you think of car insurance (like insurance for cars, trucks, SUVs, or vans)? **(List names or type "none").**
2. Please tell me the names of any car insurance companies whose advertisements you have heard or seen in the past year. **(List names or type "none").**
3. I am now going to read a list of insurance companies to you. Please say "yes" if you have heard of them or "no" if you have not.

   A. Have you heard of Geico? **Yes or No?** If Yes, ask: "Have you seen or heard any advertisements for Geico in the past year? **Yes or No?** If Yes, ask "What do you remember about the advertising?" **(Type in their response, if nothing is remembered type "none")**
   B. What about State Farm? Have you heard of them? **Yes or No?** If Yes, ask: "Have you seen or heard any advertisements for State Farm in the past year? **Yes or No?** If Yes, ask "What do you remember about the advertising?" **(Type in their response, if nothing is remembered type "none")**
   C. Allstate? **Yes or No?** If Yes, ask: "Have you seen or heard any advertisements for Allstate in the past year? **Yes or No?** If Yes, ask "What do you remember about the advertising?" **(Type in their response, if nothing is remembered type "none")**
   D. Progressive? **Yes or No?** If Yes, ask: "Have you seen or heard any advertisements for Progressive in the past year? **Yes or No?** If Yes, ask "What do you remember about the advertising?" **(Type in their response, if nothing is remembered type "none")**
   E. Infinity? **Yes or No?** If Yes, ask: "Have you seen or heard any advertisements for Infinity in the past year? **Yes or No?** If Yes, ask "What do you remember about the advertising?" **(Type in their response, if nothing is remembered type "none")**

4. How favorable would you rate each car insurance company based on their reputation and how well they treat their customers?ÆOn a 5 point scale 1 being Very Unfavorable and 5 being Very Favorable , how favorable is: A. Geico **check one: Very Unfavorable □ Unfavorable □ Somewhat Favorable □ Favorable □ Very Favorable □**
   A. State Farm **check one: : Very Unfavorable □ Unfavorable □ Somewhat Favorable □ Favorable □ Very Favorable □**
   B. Allstate **check one: : Very Unfavorable □ Unfavorable □ Somewhat Favorable □ Favorable □ Very Favorable □**
   C. Progressive **check one: : Very Unfavorable □ Unfavorable □ Somewhat Favorable □ Favorable □ Very Favorable □**
   D. Infinity **check one: : Very Unfavorable □ Unfavorable □ Somewhat Favorable □ Favorable □ Very Favorable □**

STEP 4. IMPRESSIONS OF INFINITY

1. Have you ever had a claim where an Infinity customer was at fault? **Yes or No? IF NO to this question and Step 3, question #3 C., PLEASE PROCEED TO LANGUAGE PREFERENCE.)**
2. "Please tell me your impression of Infinity Insurance in each of the following areas: On a 5 point scale 1 being Very Unfavorable and 5 being Very Favorable , how favorable is
   check one: very unfavorable □ unfavorable □ somewhat favorable □ favorable □ very favorable

a.	**Trust**	***Infinity has been in business for over 60 years***
b.	**Dependability**	***Infinity is always there when you need them***
c.	**Reputation**	***Infinity is a strong and stable company***
d.	**Community Involvement**	***Infinity is actively involved in local communities***
e.	**Coverage**	***Infinity offers many coverage options***
f.	**Cost**	***Infinity offers price discounts for good students, good drivers and multiple family discounts***
g.	**Personal**	***Infinity gives you customized service with your needs in mind***
h.	**Value**	***Infinity offers quality coverage at fair prices***
i.	**Convenience**	***Infinity has hundreds of local agents and claims offices***
j.	**Accessible**	***Infinity is open 24/7 and offers both English and Spanish language representatives***
k.	**Service**	***Infinity has quick claims response***

3. "Please tell me your overall impression of Infinity Insurance:
   check one:
   very unfavorable □ unfavorable □ somewhat favorable □ favorable □ very favorable

STEP 5. ADVERTISING PREFERENCES

1. What **language** do you primarily speak in your home? **(English Only, Mostly English, Both English and Spanish Equally, Spanish Mostly, Spanish Only?)**

2. Do you prefer **English-language advertising** or **Spanish-language advertising**? **(English Only, Mostly English, Both English and Spanish Equally, Spanish Mostly, Spanish Only?)**
3. What type of advertising do you most prefer? **(Newspaper/Print, Radio, Television,** or **Internet)**

STEP 6. DEMOGRAPHIC INFORMATION

1. What is your marital status? **Married, Single?**
2. How many children under the age of 16are living with you at home **(Type in the number of children, if none type "0").**
3. What is your current employment status? **(Full-time, Part-Time, Unemployed, Student, Retired)**
4. What is your highest level of education completed? **(High school or GED, Some College, College Degree, Some Graduate Course work, Graduate Degree)**

STEP 7. CLOSE INTERVIEW

Bibliography

Freedman D, Pisani R, Purves R (1998) Statistics, 3rd edn. W. W. Norton & Company, New York
Nelson D, Quick J (2006) Organizational behavior, 5th edn. Thomson/South-Western, Mason

L. Herkenhoff, J. Fogli (eds.), *Applied Statistics for Business and Management using Microsoft Excel*, https://doi.org/10.1007/978-3-031-46371-6

Index

L. Herkenhoff, J. Fogli (eds.), *Applied Statistics for Business and Management using Microsoft Excel*, https://doi.org/10.1007/978-3-031-46371-6

GPSR Compliance
The European Union's (EU) General Product Safety Regulation (GPSR) is a set of rules that requires consumer products to be safe and our obligations to ensure this.

If you have any concerns about our products, you can contact us on

ProductSafety@springernature.com

In case Publisher is established outside the EU, the EU authorized representative is:

Springer Nature Customer Service Center GmbH
Europaplatz 3
69115 Heidelberg, Germany

www.ingramcontent.com/pod-product-compliance
Lightning Source LLC
Chambersburg PA
CBHW051923050226
39253CB00003B/50
* 9 7 8 3 0 3 1 4 6 3 7 0 9 *